Paleoalgology

Contemporary Research and Applications

Edited by
D. F. Toomey and M. H. Nitecki

With 157 Figures and 40 Plates

Springer-Verlag
Berlin Heidelberg New York Tokyo

Dr. DONALD F. TOOMEY
Conoco Incorporated
Ponca City, OK 74603, U.S.A.

Dr. MATTHEW H. NITECKI
Field Museum of Natural History
Chicago, IL 60605, U.S.A.

ISBN 3-540-15312-8 Springer-Verlag Berlin Heidelberg New York Tokyo
ISBN 0-387-15312-8 Springer-Verlag New York Heidelberg Berlin Tokyo

Library of Congress Cataloging in Publication Data. Main entry under title: Paleoalgology: contemporary research and applications. Proceedings of the Third International Symposium on Fossil Algae (ISFA), held at Colorado School of Mines in Golden, Colorado, Aug. 15–17, 1983. Bibliography: p. Includes index. 1. Algae, Fossil – Congresses. 2. Paleobotany – United States – Congresses. 3. Paleobotany – Congresses. I. Toomey, Donald F. II. Nitecki, Matthew H. III. International Symposium on Fossil Algae (3rd: 1983: Colorado School of Mines). QE955.P35. 1985. 561'.93.85-7941.

Offsetprinting and bookbinding: Graphischer Betrieb Konrad Triltsch, Würzburg.
2132/3130-543210

Preface

Prior to the 3rd International Symposium on Fossil Algae, held in Golden, Colorado in August of 1983, a 3-day fieldtrip concentrating on Recent algal distribution in southern Florida was led by Robert N. Ginsburg and his associates from Miami. A post-symosium fieldtrip to the West Texas–southern New Mexico region was led by Don Toomey and Jack Babcock. This fieldtrip examined Precambrian and Paleozoic algal carbonates in a region characterized by superb outcrop exposures, and included the famous Permian Reef complex. The Toomey-Babcock fieldtrip resulted in a unique fieldguide published by the Colorado School of Mines (1983), and issued to all participants of the symposium.

The symposium itself was held from August 15 through Augsut 17 on the campus of the Colorado School of Mines where Professor J. Harlan Johnson, to whose memory this volume is dedicated, spent so much of his career studying fossil algae. The symposium sessions were held for 3 days, and during that time 40 papers were presented and discussed. A 1-day fieldtrip, led by John L. Wray, to examine Late Paleozoic phylloid algal carbonates exposed in the nearby Rocky Mountains, concluded the formal meetings.

The aim of the symposium was to discuss the recent developments contributing to a better understanding of fossil algae. The organizing committee consisted of Robert N. Ginsburg, Matthew H. Nitecki, Donald F. Toomey and John L. Wray. However, the actual mechanics of running the meeting, and the initiation of the symposium was in the hands of Wray, the President of the International Fossil Algae Association. The editors take particular pleasure, surely shared with all those present in Golden, in thanking Jack Wray for the success of the symposium, and for this publication resulting from ISFA.

The present volume represents the Proceedings of the 3rd ISFA, and includes the papers read at the symposium, and those manuscripts that were submitted to the editors at the time of the meeting. The arrangement of the book differs from the arrangement of the ISFA. While it is possible to arrange the work of paleoalgologists according to the taxonimic expertise of the authors, it appears to be more profitable to organize it according to the philosophy and methodology of the research practice. This arrangement is followed here, and is discussed in detail in the essay written by the editors.

We acknowledge the kindness of those of our colleagues who reviewed the individual contributions, and those who helped with various aspects of preparing this volume. We are particularly grateful to Zbigniew Jastrzebski of the Field Museum, who prepared many figures, and to Ms. Elaine Zeiger, who typed many manuscripts many times over. To all those our most sincere thanks.

Donald F. Toomey
Matthew H. Nitecki

Reference

Toomey DF, Babcock JA (1983) Precambrian and Paleozoic algal carbonates, West Texas–southern New Mexico: Fieldguide to selected localities of Late Proterozoic, Ordovician, Pennsylvanian, and Permian ages, including the Permian Reef complex: Colorado School of Mines Prof Contrib No 11, 345 p

Contents

Diversity and Evolution

Contributors

You will find the addresses at the beginning of the respective contributions

Introduction

J. Harlan Johnson, preeminent American researcher on fossil algae.
Faculty of Geology, Colorado School of Mines, 1920–1957

1 J. Harlan Johnson (1892-1974), Father of American Paleoalgology

J.L. WRAY[1]

It is fitting that the Third International Symposium on Fossil Algae in 1983 convened in Golden at the Colorado School of Mines, where J. Harlan Johnson carried out research on fossil algae and algal limestones for nearly forty years begining in the early 1930's. It is also appropriate that this volume be dedicated to his memory.

Professor Johnson discovered, systematically described and interpreted fossil algae of all groups and ages from around the world. He did much to raise the status of fossil algae above that of paleontological curiosities and emphasized the fundamental importance of calcareous algae in the development of sedimentary carbonate facies. Johnson's contributions to the science were indeed outstanding, and he was known worldwide as an eminent authority on paleoalgology.

Johnson was born on March 21, 1892, in Philadelphia, Pennsylvania, where he attended public schools between 1898 and 1910. Later he enrolled in the South Dakota School of Mines, but his higher education was interrupted by World War I. As a noncombatant in the Army Medical Corps in France he was wounded and gassed. He recuperated in France and in 1919 earned a certificate in geology and language at the University of Montpellier. Returning to the South Dakota School of Mines he received an Engineer of Mines degree in 1920. Later that same year he accepted an appointment to the geology faculty at the Colorado School of Mines in Golden, where he was to begin an unusually productive career in teaching and research that lasted for nearly half a century.

Johnson received his Master of Science degree from Colorado School of Mines in 1923, and in 1925 he became Associate Professor of Geology. He earned his Ph.D. in geology and biology from the University of Colorado in 1936. Johnson became a full Professor of Geology at Mines in 1947 and retired from the faculty in 1957.

Harlan Johnson's career began with a wide range of geological interests, in addition to archeology and natural history in general. To illustrate, in the 1920's he published work on mineral deposits, Pleistocene man, evolution, invertebrate paleontology, and the geology of the Golden, Colorado area; compiled works on geophysical principles; and reviewed and commented on one of Wegener's articles on continental drift. Regarding his early years at Mines, he said, "I had no specialty then. The School expected every faculty member to be able to teach all the courses in his department. Mine was geology."

Shortly after this period, field studies of Paleozoic and Mesozoic strata in Colorado focused his attention on sedimentary rocks and paleontology, and ultimately led

---

1 Marathon Oil Research Center, P.O. Box 269, Littleton, CO 80160, U.S.A.

Paleoalgology: Contemporary Research and Applications
(eds. D.F. Toomey and M.H. Nitecki)

to his interest on fossil algae which was to dominate his research for the remainder of his life. During the summer of 1930, while engaged in field mapping of Late Paleozoic sediments in central Colorado for the U.S. Geological Survey, Johnson noted, "I ran into considerable deposits of algal limestone. At that time I was unable to find any specialist who could assist me in their study or to advise me as to their meaning with regards to conditions of deposition and other geologic aspects. Consequently, I became interested in their study." This then, was the beginning of Johnson's lifetime of research on fossil algae and algal limestones.

Field studies and basic geology continued to occupy much of Johnson's research time during the 1930's and early 1940's, primarily Paleozoic stratigraphy of Colorado, especially in the Mosquito and Sawatch Ranges and the Sangre de Cristo Mountains. He collaborated with others in studies of the geology of South Park, Colorado, which was published as Geological Society of America Memoir 33. He devoted particular attention to Tertiary sediments of South Park, especially Oligocene stromatolitic algal limestones, and these rocks served as the subject for his Ph.D. dissertation. Although his earlier studies of algae were confined mainly to formations in Colorado, soon he began to expand his research into other regions, including the Permian "lime-secreting" algae of West Texas and New Mexico.

Starting in 1947 and continuing into the mid-1950's, Johnson was engaged by the Military Geology Branch of the U.S. Geological Survey to study modern and fossil calcareous algae of Pacific reefs. This program, in which K.O. Emery, J.I. Tracey, and H.S. Ladd, among many others, were involved, was a part of a comprehensive USGS examination of modern reefs. The investigations began at the site of the Bikini Atoll 1946 nuclear test, and subsequently covered intensive studies of nearby islands, including Saipan, Eniwetok, and Guam. This research enhanced his knowledge of the petrography of organic limestones and Cenozoic algae, and served as a base to expand his studies of algae into other regions of the Pacific.

Consulting for the St. Joseph Lead Company in Missouri in 1953, Johnson demonstrated a preferred relationship of mineralization and algal carbonates, specifically the localization of ore deposits in "roll" structures, which he interpreted as stromatolitic algal bioherms. He often cited this as the beginning of his applied use of calcareous algae in exploration geology. Soon he was to be in great demand to help the petroleum industry with the interpretation of sedimentary carbonate rocks in oil exploration.

Many petroleum companies in the 1950's substantially increased the emphasis on understanding carbonate rocks in their research programs. The discoveries and development of enormous reserves of oil in limestones and dolomites in the Middle East and Libya, Devonian of Canada, and late Paleozoics of West Texas and the Four Corners region, contributed to this new interest. Johnson was well prepared because he had pioneered in the modern approach to describing organic limestones, and, in fact, had already written his now classic *Introduction to the Study of Organic Limestones*. In addition he was the only fossil algae specialist in North America, and calcareous algae were becoming appreciated as very significant elements in carbonate sedimentology. Consequently, Johnson's expertise was much sought after, and he was retained as a research constultant for several major oil companies in the 1950's and 1960's.

Following his retirement from the faculty at Mines in 1957 at age 65, Johnson worked diligently on fossil algae for a further 12 fully productive years. (Over half of his articles on algae were published during this period!) Now freed of teaching responsibilities, he was able to devote full time to his research and consulting. Consulting work did not diminish his basic research on algae; on the contrary, it gave him a vast reservoir of new materials and problems to investigate. In 1958 he began a long-term project to compile the existing knowledge on fossil algae by geological periods, managing to complete the Paleozoic and most of the Mesozoic. He reviewed in detail certain families and genera of fossil calcareous algae, and, in addition, he compiled three bibliographies on fossil algae and algal limestones.

Lecturing and consulting projects carried him to nearly every continent, and his crossings of the Atlantic and Pacific are too numerous to recall. He spoke to scores of far-flung societies and industry groups, and held visiting lectureships at the universities of Tokyo, Paris (Sorbonne), and London, in addition to many in the United States.

Johnson's scientific publications number approximately 175, of which several are book-length contributions. His bibliography is proof of the extent and depth of his research, and many of his articles on fossil algae will remain the most definitive, if not the only, statements on these aspects of paleontology for generations.

In addition to his scientific contributions, Johnson was conscious of a larger responsibility to the scientific community and thus devoted much time to the affairs of geological societies in the United States and abroad. He was a fellow of The Geological Society of America, and a member of The American Association of Petroleum Geologists, The American Institute of Mining, Metallurgical and Petroleum Engineers, and the Paleontological Society. He was President of the Society of Economic Paleontologists and Mineralogists in 1943 and was elected to honorary membership in 1969. An early member of the Rocky Mountain Association of Geologists, he served as Vice President in 1929, President in 1930 and 1931, and later was elected an honorary member. He was Vice President of the Société Géologique de France in 1934.

J. Harlan Johnson died at his home in Crowley, Texas, on October 31, 1974 at the age of 82. He was a highly regarded geologist/paleontologist and respected teacher. Because of his pioneering insight and dedicated hard work on fossil algae he laid a firm foundation for future research. His published works on calcareous algae and organic limestones remain fundamental references for scientists concerned with broader aspects of interpreting Earth history.

Selected Bibliography of J. Harlan Johnson

Johnson JH (1925) The geology of the Golden area, Colorado. Colo Sch Mines Q 20:no 3, 25

Johnson JH (1929) Origin of the Sangre de Cristo conglomerates, Colorado. Am Assoc Pet Geol Bull 13:177–178

Johnson JH (1933) A Permian algal reef in South Park, Colorado. Am Assoc Pet Geol Bull 17: 863–865

Johnson JH (1934) Paleozoic formations of the Mosquito Range, Colorado. US Geol Surv Prof Pap 185-B:15–43

Johnson JH (1935) Stratigraphy of northeastern and east-central parts of South Park, Colorado. Am Assoc Pet Geol Bull 19:1339–1358
Johnson JH (1937) Algae and algal limestones from the Oligocene of South Park, Colorado. Geol Soc Am Bull 48:1227–1236
Johnson JH (1940) Lime-secreting algae and algal limestones from the Pennsylvanian of central Colorado. Geol Soc Am Bull 51:571–596
Johnson JH, Dorr ME (1942) The Permian algal genus *Mizzia*. J Paleontol 16:63–77
Johnson JH (1942) Permian lime-secreting algae from the Guadalupe Mountains, New Mexico. Geol Soc Am Bull 53:195–226
Johnson JH (1943) Geologic importance of calcareous algae, with annotated bibliography. Colo Sch Mines Q 38: no 1, 102
Johnson JH (1944) Paleozoic stratigraphy of the Sawatch Range, Colorado. Geol Soc Am Bull 55:303–378
Johnson JH (1944) Fossil algae, In: Shimer HW, Shrock RR (eds) Index fossils of North America. Wiley, New York, pp 711–719
Johnson JH (1945) A résumé of the Paleozoic stratigraphy of Colorado. Colo Sch Mines 40: no 3, 109
Johnson JH (1945) Calcareous algae as useful microfossils. J Paleontol 19:350–354
Johnson JH (1945) Calcareous algae of the upper Leadville Limestone near Glenwood Springs, Colorado. Geol Soc Am Bull 56:829–848
Johnson JH (1946) Mississippian algal limestones from the vicinity of St. Louis, Missouri. J Paleontol 20:166–171
Johnson JH (1946) Late Paleozoic algae of North America. Am Midl Nat 36: no 2, 264–274
Johnson JH (1946) Lime-secreting algae from the Pennsylvanian and Permian of Kansas. Geol Soc Am Bull 57:1087–1120
Johnson JH, Ferris BJ (1948) Eocene algae from Florida. J Paleontol 22:762–766
Johnson JH (1949) An introduction to the study of organic limestones. Colo Sch Mines Q 44: no 4, 139
Johnson JH, Stark JT et al. (1949) Geology and origin of South Park, Colorado. Geol Soc Am Mem 33:188
Johnson JH, Ferris BJ (1949) Tertiary coralline algae from the Dutch East Indies. J Paleontol 23:193–198
Johnson JH, Ferris BJ (1950) Tertiary and Pleistocene coralline algae from Lau, Fiji. Bernice P Bishop Mus Bull 201:27
Johnson JH (1950) A Permian algal-foraminiferal consortium from West Texas. J Paleontol 24: 61–62
Johnson JH (1951) Permian calcareous algae from the Apache Mountains, Texas. J Paleontol 25:21–30
Johnson JH (1951) Fossil algae, In: Smith GM (ed) Manual of phycology. Chronica Botanica Co, pp 191–202
Johnson JH (1952) Ordovician rock-building algae. Colo Sch Mines Q 47: no 2, 29–56
Johnson JH (1953) *Microcodium* Glück est-il un organisme fossil? CR Acad Sci Paris 237: no 1, 84–86
Johnson JH, Stewart WA (1953) Eocene coralline algae from the Meganos Formation, California. J Paleontol 27:130–136
Johnson JH (1954) Geology of Bikini and nearby atolls – Fossil calcareous algae from Bikini Atoll. US Geol Surv Prof Pap 260-M:537–545
Johnson JH (1954) An introduction to the study of rock-building algae and algal limestones. Colo Sch Mines Q 49: no 2, 117
Johnson JH (1955) Early Tertiary coralline algae from Trinidad, British West Indies. Eclogae Geol Helv 48:69–78
Johnson JH (1956) *Archaeolithophyllum*, a new genus of Paleozoic coralline algae. J Paleontol 30: 53–55
Johnson JH (1956) Ancestry of the coralline algae. J Paleontol 30:563–567

Johnson JH, Konishi K (1956) Studies of Mississippian algae – Part 1: A review of Mississippian algae; part 2: Mississippian algae from the western Canada basin and Montana. Colo Sch Mines Q 51: no 4, 1–107
Johnson JH (1957) Geology of Saipan, Mariana Islands – Calcareous algae. US Geol Surv Prof Pap 280-E:209–246
Johnson JH (1957) Bibliography of fossil algae – 1942–1955. Colo Sch Mines Q 52: no 2, 92
Johnson JH (1957) Geology of Saipan, Mariana Islands – Petrography of limestones. US Geol Surv Prof Pap 280-C:177–187
Johnson JH (1957) Calcareous algae – Annotated bibliography. In: Ladd HS (ed) Paleoecology. Geol Soc Am Mem 67:699–701
Johnson JH, Konishi K (1958) A review of Devonian algae. Colo Sch Mines Q 53: no 2, part 1, 1–84
Johnson JH, Konishi K (1959) A review of Silurian (Gotlandian) algae. Colo Sch Mines Q 54: no 1, 1–82
Johnson JH, Konishi K (1959) Some Silurian calcareous algae from northern California and Japan; Studies of Silurian (Gotlandian) algae. Colo Sch Mines Q 54: no 1, 131–162
Johnson JH (1960) Paleozoic Solenoporaceae and related red algae. Colo Sch Mines Q 55: no 3, 77
Johnson JH (1961) Bikini and nearby atolls – Fossil calcareous algae from Eniwetok, Funafuti and Kita-Daito-Jima. US Geol Surv Prof Pap 260-Z:907–950
Johnson JH (1961) Jurassic algae from the subsurface of the Gulf Coast. J Paleontol 35:147–151
Johnson JH (1961) Studies of Ordovician algae. Colo Sch Mines Q 56: no 2, 1–101
Johnson JH (1961) Limestone-building algae and algal limestones. Colo Sch Mines, 297 p
Johnson JH (1962) Calcareous algae from Sarawak – Geology and mineral resources of the Suai-Baram area, north Sarawak. Br Borneo Geol Surv Mem 13:151–168
Johnson JH (1962) The algal genus *Lithothamnium* and its fossil representatives. Colo Sch Mines Q 57: no 1, 1–111
Johnson JH (1963) The algal genus *Archaeolithothamnium* and its fossil representatives. J Paleontol 37:175–211
Johnson JH (1963) Pennsylvanian and Permian algae. Colo Sch Mines Q 58: no 3, 211
Johnson JH (1964) Lower Devonian algae and encrusting Foraminifera from New South Wales. J Paleontol 38:98–108
Johnson JH (1964) Paleocene calcareous red algae from northern Iraq. Micropaleontology 10: 207–216
Johnson JH (1964) The Jurassic algae. Colo Sch Mines Q 59: no 2, 129
Johnson JH (1964) Miocene coralline algae from northern Iraq. Micropaleontology 10:477–485
Johnson JH (1964) Eocene algae from Ishigaki-shima, Ryukyu-retto. US Geol Surv Prof Pap 399-C:13
Johnson JH (1964) Fossil and recent calcareous algae from Guam. US Geol Surv Prof Pap 403-G: 40
Johnson JH (1965) The genus *Lithophyllum* and its fossil representatives. Colo Sch Mines Q 60: no 2, 1–65
Johnson JH (1965) Coralline algae from the Cretaceous and early Tertiary of Greece. J Paleontol 39:802–814
Johnson JH (1965) Tertiary red algae from Borneo. Br Mus (Nat Hist) Bull Geol 2: no 5, 257–280
Johnson JH, Kaska HV (1965) Fossil algae from Guatemala. Colo Sch Mines Prof Contrib no 1: 152
Johnson JH, Adey WH (1966) Studies of *Lithophyllum* and related algal genera. Colo Sch Mines Q 60: no 2, 105
Johnson JH (1966) A review of the Cambrian algae. Colo Sch Mines Q 61: no 1, 162
Johnson JH (1967) Bibliography of fossil algae, algal limestones, and the geological work of algae, 1956–1965. Colo Sch Mines Q 62: no 4, 148
Johnson JH (1968) Lower Cretaceous algae from Texas. Colo Sch Mines Prof Contrib no 4:71
Johnson JH (1968) Lower Cretaceous algae from the Blake Escarpment, Atlantic Ocean and from Israel. Colo Sch Mines Prof Contrib no 5:46

Johnson JH, Toomey DF (1968) *Ungdarella americana*, a new red alga from the Pennsylvanian of southeastern New Mexico. J Paleontol 42:556–560
Johnson JH (1969) A review of the Lower Cretaceous algae. Colo Sch Mines Prof Contrib no 6: 180
Johnson JH (1969) Lower Permian algal stromatolites from Kansas and Oklahoma – part 2: Algal stromatolites from the Wellington Formation of Kansas and Oklahoma. Kansas Univ Paleontol Contribu Pap 43:9–19
Johnson JH, Achauer CW (1969) Algal stromatolites in the James Reef Complex (Lower Cretaceous), Fairway Field, Texas. J Sediment Petrol 39:1466–1472

2 Paleoalgology

M.H. NITECKI[1] and D.F. TOOMEY[2]

The papers presented in this volume do not lend themselves to a central theme, but, instead, reflect the overall interests and research direction of the various participants of the symposium. This volume does not pretend to consider the entire field of paleoalgology, but is a sampling representing the state-of-the-art at this particular point in time.

As the introductory contribution to this volume, Wray's "*J. Harlan Johnson (1892–1974), Father of American Paleoalgology*," is both a tribute and a synopsis of the career of Professor Johnson, the foremost American worker on fossil algae. Johnson was a highly regarded scientist and an inspiration to many workers interested in algae. He did much to encourage younger researchers with a kindness and respect that was generously extended to all. Johnson's pioneering study of fossil algae indeed created a firm foundation for later generations to build upon. In this tribute Wray has also included a selected bibliography of Professor Johnson's published works spanning the interval from 1925–1969, and documenting the breadth and scope of his achievements – a model for all of us to emulate.

Three papers comprise the section on *Stromatolites and Precambrian Algae.* Hofmann (*Precambrian Carbonaceous Megafossils*) is interested in the origin of eukaryotic algae, and in the history of megascopic algae. He believes that the development of organisms of megascopic size, and the related acquisition of the nucleus in cells, are landmark evolutionary events. Based on the presently known paleontological record of megafossils of probable algal affinities, the origin of eukaryotic algae occurred by Middle Proterozoic time before the development of Ediacaran metazoans. Hofmann describes the fossils as being generally preserved as small compressions on bedding planes. They include circular, elliptical, ovate, and angulate films, and rectilinear, curvilinear, spiraliform, and twisted ribbons. Among his fossils are some that Hofmann interprets as megascopic colonies of prokaryotes. Others, such as the distinct and cosmopolitan *Chuaria* and *Tawuia*, he regards as probable algal megafossils.

Horodyski (*Stromatolites of the Middle Proterozoic Belt Supergroup, Glacier National Park, Montana*) is concerned with stromatolites common in ca. 1.1–1.4 billion-year-old strata of the Belt Supergroup, Montana. Stromatolites are very common within five of the six formations exposed in Glacier National Park. They include mound-shaped stromatolites, branched columnar stromatolites referable to the group

1 Field Museum of Natural History, Chicago, IL 60605, U.S.A.
2 Conoco Inc., Ponca City, OK 60603, U.S.A.

Paleoalgology: Contemporary Research and Applications
(eds. D.F. Toomey and M.H. Nitecki)

Baicalia, and conically laminated stromatolites referable to the group *Conophyton.* Since these stromatolites have diverse morphologies, and are associated with different lithologies deposited in a variety of depositional settings, they provided a good opportunity for Horodyski to evaluate the relationship between stromatolite morphology and paleoenvironmental conditions. This relationship, like other interpretations of the paleoenvironments, is a major concern of most paleoalgologists.

Wright and Wright (*A Stromatolite Built by a Phormidium-Like Alga from the Lower Carboniferous of South Wales*) ask questions concerning the evolutionary trends of stromatolites: Are the similarities of Recent to ancient stromatolites the result of conservation or convergence? Were motile microalgae the builders of the stromatolites? If so, how can we tell? Wright and Wright describe stromatolites from the Lower Carboniferous of south Wales, which were built by a community of microscopic, filamentous, phototropic, and probably motile cyanobacteria related to the living oscillatoriacean *Phormidium.* There are great similarities between these Carboniferous and the Recent stromatolites. The cyanobacterial mats grew on oolite bars, influenced the fabric and grain size distribution of the sediments, and stabilized and preserved bed forms.

The following section of the volume is made up of six papers relating to *Systematics and Morphology*, and notes that paleoalgology, with the exception of Precambrian microorganisms and stromatolites, concentrates mostly on calcareous algae, which appear near the Precambrian-Cambrian boundary, from whence they become increasingly abundant and varied. Their classification, however, is not well established. They have been classified by the differences in their morphological characters and by affinities among them. These subjective evaluations are based on somewhat dubious similarities with extant algae and can be a cause for future taxonomic problems. It is this subjective element of classification of Cambrian algae that Riding and Voronova (*Morphological Groups and Series in Early Cambrian Calcareous Algae*) are attempting to resolve. They propose a morphological classification of Cambrian algae based on six morphological characters: external morphology, internal organization, wall-structure, wall thickness, wall layers, and wall surfaces. The authors, utilizing these six characters, place the Cambrian algae into six morphological series: spherical, botryoidal, dendritic, tubiform, tuberous, and cups. They note that similarities may be due to mutual affinities or to morphological convergence, and the presence of a particular series should not, be taken as sol, indication of mutual affinity. They also caution that taxa juxtapositioned within a series should not necessarily be regarded as prime candidates for synonymy.

During Ordovician time certain dasycladaceous genera became common and cosmopolitan. These are well known fossils and much have been written on them. Now Johnson and Sheehan (*Late Ordovician Dasyclad Algae of the Eastern Great Basin*) extend our knowledge of these to the late Ordovician of the Great Basin area, where they find the genera *Vermiporella, Rhabdoporella,* and *Cyclocrinites* occurring in rocks laid down in relatively shallow, open, clean marine waters.

There is still much room for descriptive work on algal microfloras, including descriptions of genera and species, which can be considered as the building blocks of the science. In this context Groves and Mamet (*Masloviporidium, a Cosmopolitan*

Middle Carboniferous Red Alga) have erected a new genus and species based on a form originally recognized by Maslov, as the type species of a new genus of Middle Carboniferous algae. This descriptive approach is continued by Mamet and Pinard (*Carboniferous Algae from the Peratrovich Formation, Southeast Alaska*) in a paper in which a rich, cosmopolitan, well-preserved microflora is described from the Carboniferous of southeast Alaska. Thirty-four genera and *incertae sedis* have been recognized, and one new species is described.

In a major contribution, the Romanian paleoalgologist Dragastan presents a significant review of Tethyan porostromatic algae (*Review of Tethyan Borostromatic Algae*). He is particularly concerned with the taxonomic criteria used for identifying these Mesozoic and Cenozoic algae, and accordingly numerous taxonomic changes are included. Results of this much needed review is the erection of seven new genera and 18 new species.

The remaining paper in this section is the presentation by Poignant of identification keys for the specific determinations of Cretaceous *Archaeolithothamnium* (*Key to Specific Determination of Cretaceous Archaeolithothamnium*). These determinations are based upon an assemblage of several features according to a well-chosen hierarchy.

In the following section on *Reefs, Buildups and Other Frameworks*, Magathan (*Devonian Reef-Associated Articulate Red Algae from Western Canada*) has documented the existence in Upper Devonian reefs of articulate calcareous red algae that are homologous with the modern articulate coralline *Amphiroa.* She has profiled the habitat range of these fossil algae across the Frasnian Alexandra Reef complex in Canada. The algae in this reef complex are confined to marginal organic shale facies, where they are associated with the reef-building stromatoporoid *Actinostroma*, encrusting coralline-like alga, crinkly-surfaced stromatolites, and *Renalcis.* Significantly, their occurrences are inversely related to occurrences of dasycladaceans, codiaceans, calcispheres, and *Amphiroa.* Dolomite rhombs replacing some algal fragments suggests high-magnesian calcite such as characterizes modern coralline algae.

Close correspondence in structure, mineralogy, and habitat between this alga and the living *Amphiroa* suggests a phylogenetic linkage, and Magathan recommends that "*Archaeoamphiroa*" should be classified with the articulate coralline subfamily Amphiroideae. Furthermore, because the ontologic history of articulate coralline algae indicates an encrusting precursor, documentation of these algae supports evidence previously cited for the existence of crustose coralline algae in the Devonian.

The role of Permian algae in building carbonate structures is very well known. Three papers in this volume deal with the importance of Permian algae in the formation of carbonate mounds. Cys (*Lower Permian Phylloid Algal Mounds, Southern Tatum Basin, Southeastern New Mexico*) studies the Lower Permian (Wolfcamp-Assel) rocks of the subsurface Tatum Basin of southeastern New Mexico and finds that they are characterized by shoaling-upward carbonate cycles of which phylloid algal mound are an integral component. These mounds are biohermal hydrocarbon reservoirs in the area and are paleogeographically situated on a shallow carbonate shelf. Cys points out that the codiacean *Eugonophyllum* is the primary biotic constituent and mound builder. *Tubiphytes* and foraminifers are common, whereas gastropods, bryozoans, and crinoids are less common constituents. These mounds are situated on underlying paleostructural features that provided optimum water depth parameters for the abundant growth of phylloid algae.

Pol (*Sedimentation of an Upper Pennsylvanian (Virgilian) Phylloid Algal Mound Complex, Hueco Mountains, El Paso County, West Texas*) is concerned with sedimentation of phylloid algal mounds, which are common in Late Paleozoic carbonates of West Texas and southern New Mexico. The published literature, according to Pol, suggests a close relationship between carbonate sedimentation and progradational clastic sequences, as well as the strong influence of local tectonics, an idea noted by a number of earlier workers. Study by various workers indicate that distinct community succession and characteristic sedimentological processes are common to algal mound deposition. These tend to produce similar vertical and lateral facies relationships. However, minor differences in environmental conditions (i.e., climate, water circulation, local tectonics, etc.) ensure that no two algal mound complexes are identical.

Wahlman (*Middle Wolfcampian-aged Archaeolithoporella-Tubiphytes-Sponge Boundstones from the Subsurface of West Texas*) as a petroleum geologist, is interested in Lower Permian (Wolfcamp) boundstones from the subsurface of West Texas, which have a basic organic framework of calcareous sponges, phylloid algae, and branching bryozoans bound by the problematical forms *Archaeolithoporella* and *Tubiphytes*, radial fibrous cement, laminar red algae, and encrusting bryozoans. To him, this boundstone community appears to be ancestral to that of the younger Guadalupe-age Capitan reef of West Texas and southeastern New Mexico.

Tertiary coralline algal limestones can be easily compared with Recent analogs. In the next contribution Bosence (*The "Coralligène" of the Mediterranean – a Recent Analog for Tertiary Coralline Algal Limestones*) looks at the coralline algal buildups "Coralligène" that occur in the intertidal zone off the rocky Albére coastline of southern France. The deep-water examples form areas of small reefs which are surrounded by basement rocks, locally derived carbonate gravels, rhodoliths, and terrigenous muds and sands. Bosence shows that reef frameworks are constructed by foliaceous crusts of the coralline algae: *Mesophyllum, Lithothamnium, Pseudolithophyllum, Neogoniolithon,* and *Tenarea.* Lightly calcified *Peyssonnelia* and *Halimeda* also occur as epiphytes. Bioerosion by sponges, bivalves, polychaetes, and algae is common, and the cavities are usually infilled with sediment. Bosence argues that the Recent reefs have many features in common with Miocene coralline limestones from Tethyan areas and that they are the best known Recent analogs for these limestones.

Corals and coralline algae are the dominant reef-building organisms in the Recent as well as in the Tertiary. Taberner and Bosence (*Ecological Succession from Corals to Coralline Algae in Eocene Patch Reefs, Northern Spain*) examine how middle Eocene corals are repeatedly overgrown by algae – a situation identical to that in the Recent intertidal zones where coralline algal ridges form. The authors show that middle Eocene sedimentation on the southeastern margin of the Catalan Basin in northeast Spain is alluvial and deltaic. Coralgal reefs develop in an arcuate band on preexisting delta lobes. The patch reefs are built by corals and coralline algae. To them, the coralline algae overgrow the corals with densely superposed crusts, some of which are constructed laterally around coral, and others are eroded and redeposited. Submarine cements are present in the form of calcite micrite and spar botryoids. The association of corallines within the crusts is typical of deep tropical waters, which contrasts with the shallow marine corals and surrounding fan delta and beach sediments. Taberner and Bosence believe that the coralline crusts represent a shallow, shaded flora growing over and between corals in relatively turbid waters.

These examples demonstrate that paleoalgology, like all other geological sciences, is restrained by the principle of uniformitarianism, which requires the comparison of fossils with their living counterparts. This is why in the succeeding contribution by Minnery, Rezak and Bright (*Depth Zonation and Growth Form of Crustose Coralline Algae: Flower Garden Banks, Northwestern Gulf of Mexico*) the authors analyze depth zonations and the growth form of living crustose coralline algae. They study the Flower Garden Banks in the northwestern Gulf of Mexico, which are bathymetric highs of salt domes. Although the crests of the Flower Garden Banks are capped by a coral reef of hermatypic corals, crustose coralline algae are the dominant sediment contributors and framework builders below 50 m. Minnery et al., establish depth zonation spanning 20–100 m on the banks and slopes by using eleven genera of corallines, one squamariacean, and an encrusting foraminifer. They correlate the growth form and abundance of corallines with depth, currents, available substrate, and grazing. The deep-water algal reefs are a living coralline veneer over drowned algal reefs that began in shallower water during the Pleistocene.

The following section of the volume comprises three papers related to *Calcification, Microstructure, and Growth.* Since almost all that we know about past life is derived from the study of skeletal remains, nothing would seem more fundamental to our interpretation of the path of evolution than our understanding of the processes of calcification. The process of calcification, the nature of the skeleton, and the ultra-fine structures are actively being investigated in many parts of the world. Growth, an equally important biological process, is also studied by paleoalgologists.

The mineralogy of skeletons, a subject of extensive research, may perhaps represent a new subdivision of paleoalgology. As in other fields, the studies of microscopic and ultra-thin sections may provide new answers to old questions. For example, the article by Buchbinder and Halley (*Occurrence and Preservation of Eocene Squamariacean and Coralline Rhodoliths; Eua, Tonga*) asks how the original mineralogy influences the preservation of anatomy. They study the squamariacean and coralline rhodoliths in middle Eocene platform limestones in Eua, Tonga. Buchbinder and Halley point out that although both types of rhodolith are now altered to low magnesian calcite, the degree of preservation of the original microstructure is higher in the original aragonitic *Polystrata* than in the original magnesian calcite coralline algae. Although the magnesian-calcite skeletal material is better preserved than skeletal aragonite, Eua squamariaceans suggest to them that preservation detail of recrystallized skeletal material should not be solely attributed to original mineralogy.

That calcareous algae are major producers of carbonate sediments has been known for a long time. However, the quantitative measurements of that productivity have not been readily available. In order to understand the rate of algal productivity in the past, it is necessary to study the rates of this productivity in Recent seas. To measure the rate of accumulation of algal remains, their rate of growth must first be determined. Hudson (*Growth Rate and Carbonate Production in Halimeda opuntia: Marquesas Keys, Florida*) has measured growth rates and carbonate production in the codiacean *Halimeda opuntia*, which is a major contributor to marine sediments. His studies were conducted at the Marquesas Keys, 29 km west of Key West, Flordia. Hudson determined growth rates of entire colonies of *Halimeda.* He calculated the increase in weight of *Halimeda* by weighing over 10,000 plates. Plate formation was most rapid in summer months and slowest in late fall. One colony in three months increased its original weight by 85%, a most unusual increase in growth!

Dasyclad algae, while not very common today, have a long geological record, with evidence of having been cosmopolitan and abundant, and of exhibiting great taxonomic diversity. Génot (*Calcification in Fossil Neomereae*) is interested in the calcification of one group of Dasycladales, the Neomerae. His study is based on the observation of 18 Cenozoic species of Neomereae and the comparison of two living species. For each of these Génot is able to define accurately the variability in the degree of calcification of the skeletal elements, and to study the effects of this variability on the external morphology of the fossils. His ultrastructure analysis of walls of numerous specimens provides new information on the aspect and orientation of the aragonite crystals within the calcified skeleton.

In the succeeding chapter of this volume, *Algae and Sediments*, there are six contributions which emphasize that algae are producers of carbonate rocks by directly precipitating calcium carbonate from sea water through the photosynthetic activities, by contributing their skeletal material to the sedimentary basins, and by being the bottom of the food-chain for all calcareous marine invertebrates. Yet, few geologists, other than paleoalgologists, recognize the importance of algae in the carbonate budget.

Interpretations of the influence of biota upon the environment and vice versa swing like a pendulum. It is only relatively recently that we have begun to look at the enormous influences that the biota may have upon the formation of the sedimentary mantle. Thus Tsien's (*Origin of Stromatactis – a Replacement of Colonial Microbial Accretions*) suggestion that stromatactis is algally produced may be better accepted today than it would have been in the past. Tsien states that most specialists regard stromatactis as a cavity system, while others see it as a recrystallized calcite mudstone. His work in Belgium suggests that stromatactis represents organic colonies, probably a replacement by algae and cyanobacteria of colonial microbial concretions. The next chapter, also by Tsien (*Algal-Bacterial Origin of Micrites in Mud Mounds*), is less controversal than the previous article, and may be standing upon somewhat firmer grounds when discussing the algal-bacterial origin of micrite found in Devonian mud mounds.

Toomey (*Paleodepositional Setting of Rhodoliths from the Upper Pennsylvanian (Virgil) Salem School Limestone of Northcentral Texas*) notes that algal rhodoliths found in the Late Pennsylvanian Salem School Limestone of northcentral Texas, are algally coated grains composed of concentrically laminar thalli of the ancestral coralline alga *Archaeolithophyllum lamellosum.* These rhodoliths are associated with a diverse biota and occur within a shallow-water depositional setting, whose rock lithology can be described as a burrowed skeletal wackestone. Based on the biota, especially the occurrence of distinctive fusulinid foraminifers, the Salem School Limestone is shown to be the time-stratigraphic equivalent of the widespread Leavenworth Limestone of the midcontinental United States.

Algae are very well represented in Permian rocks, and Cys (*Taxonomic Survey of Lower Permian Algae from the Southern Tatum Basin, Southeastern New Mexico*) in his taxonomic survey finds that they are important biotic constituents in the Lower Permian subsurface of southeastern New Mexico. He emphasizes that the codiacean phylloid alga *Eugonophyllum* frequently forms organic buildups that are hydrocarbon reservoirs.

Toomey (*Dasyclad Algae Within Permian (Leonard) Cyclic Shelf Carbonates ("Abo"), Northern Midland Basin, West Texas*) describes the occurrence of dasyclad algae in the subsurface Permian ("Abo") of the northern Midland Basin, West Texas, and notes that these occurrences are restricted to thin, cyclic shelf carbonates, and generally occur in those sediments thought to have been laid down within the low intertidal regime. The accurate placing of dasyclad-rich intervals within a paleodepositional sequence is made less difficult when both overlying and underlying sediments can be characterized by distinctive rock types and biota.

In the final article of this subdivision, Wright (*Algal Marsh Deposits from the Upper Jurassic of Portugal*) is interested in the fresh water algal deposits in the fossil record. Nonmarine algal marsh deposits form today in Florida, the Bahamas, and other regions. Wright describes comparable series of nonmarine limestone from the Upper Jurassic of Portugal. His comparisons are based on the study of the biota, petrography, and sedimentology. He also demonstrates a climatic shift toward dry seasons.

The concluding section of the volume deals with *Diversity and Evolution*, and is represented by two contributions. Flügel's (*Diversity and Environments of Permian and Triassic Dasycladacean Algae*) contribution represents an attempt to examine the pattern of diversity of Phanerozoic dasyclads. Most attempts to deal with the pattern of diversity of fossil dasyclads have been made by comparing their diversity with that of living dasyclads. But such comparisons, just like paleoenvironmental comparisons, are beset with difficulties. Flügel is well aware of these problems, and also cautions against uncritical comparisons of Recent with fossil dasyclads. He sees the differences in the diversity and distribution of fossil dasyclads as caused by the differences of availabilities of various substrates, rather than by strict ecological adaptations of fossil algae. It is in this light that Flügel explains the Permian maximum diversity of dasyclads as against the early Triassic low diversity.

In the final article of the volume, Steneck (*Adaptations of Crustose Coralline Algae to Herbivory: Patterns in Space and Time*) examines the coexistence and coadaptation of crustose coralline algae with grazing herbivores. He observes that crustose coralline algae are among the most abundant organisms, whether plant or animal, occupying the hard substrate in the photic zone. In fact, since coralline are more diverse and abundant today than in any other geological period, Steneck calls the Holocene an *Age of Crustose Corallines*. The Holocene also appears to be a time of maximum intensity of marine herbivory. Steneck correlates temporally the radiation of crustose corallines with coralline-grazing herbivores. His ecological and paleontological research suggests that there are several anatomical and morphological adaptations to algal herbivory, identified by finding convergently evolved characters of similar function under different physical conditions. These specific adaptations are crust thickness, branching morphology, conceptacle shape, and protection of the growth region. Today, geographic patterns of these characters correspond with geographic patterns in the distribution and abundance of herbivore groups. Coralline taxa with these adaptations appear to have progressively increased throughout the Phanerozoic, corresponding to the increased pressure of herbivory.

Many, but not all paleoalgological fields are represented in the present volume, but what actually constitutes a paleoalgological field is still a matter of some uncertainty. Surely, the term "algae," where has little taxonomic validity, and for purists, paleoalgology as a science should hardly exist. Yet it is here and well, and the term "algae" is not only useful, but its usage is widespread. In paleontology, in general, "algae" include almost all fossils that are neither plants nor animals, and those problematical organisms that cannot be assigned to either plant or animal kingdom. Important "algal" groups omitted here are mostly skeletonized microalgae, for example, diatoms and dinoflagellates – excluded simply because most paleontologists working on these taxa call themselves micropaleontologists, and simply have not associated themselves with paleoalgologists.

We, the paleoalgologists, can be well satisfied with most parts of our science, and in some areas we can show considerable accomplishments. In the sphere of Precambrian paleoalgology we have done very well, and have established quite a firm reputation. We have addressed a wide spectrum of questions from the origin of life to the history and evolution of life. We have also explained the nature of stromatolites and demonstrated the richness of Precambrian microalgal biotas.

It may seem redundant to stress the importance of systematics since all paleontologists make that the core of their fields. But some paleontologists nevertheless imagine that systematics has been bypassed and is outdated. The volume of systematic work from the Soviet Union, through France and England to North America, is an obvious indication of the everyday concern of paleoalgology with taxonomy and morphology. The appearance of the first eukaryotes is today a controversial issue extending far outside paleoalgology. There are difficulties to be resolved, mainly those of interpretation of the internal morphology of microscopic biota – certainly the future labors of algal taxonomists, who also will have to resolve the unknown effects of physical and chemical alteration during the process of preservation.

Equally important are the problems regarding the ancestors of the first animals, surely hidden somewhere among the unknown groups of "algae." The theory of evolution seems to require a long time span for the appearance of less complex forms that must have preceded the now known fossils, yet the only known organisms are algae.

The biomass of the Precambrian and Phanerozoic algae has never been calculated. Although the problems of measurements of such mass are difficult, their solutions will be of some value in explaining the evolution of the atmosphere, and the content of oxygen and carbon dioxide in it.

Geological time is conveniently divided into the Precambrian without fossil animals, but with a good algal record, and the Phanerozoic with a good animal record, but with poor, by comparison, algal record. There are many boundaries still not well defined but with a good and continuous geological record, and with fossil algae available for correlation. These boundaries should be investigated and we suspect that they may need redefinitions and adjustments.

The fossil record of algae strongly indicates an entirely new concept of durability and permanence of the constantly changing life on earth.

Stromatolites and Precambrian Algae

3 Precambrian Carbonaceous Megafossils

H.J. HOFMANN[1]

Introduction

Carbonaceous compressions have been known for more than a century from rocks now considered pre-Cambrian. They include remains with regular rounded outlines, as well as irregular angulate fragments, and ribbon-like films (Fig. 1), all usually preserved on bedding planes in fine-grained clastic rocks, and most showing some wrinkling due to compaction. Only the youngest Proterozoic ones have primarily longitudinal and punctate markings with biologic function. The fossils range in age from the Early Proterozoic (2 Ga) into the Phanerozoic, and their geographic distribution includes all continents except South America and Africa (Figs. 2 and 3). The most ancient known megascopic compressions are millimetric films from black shales in Aphebian rocks of North America (localities 12–14 in Fig. 2).

Descriptions of carbonaceous fossils are widely scattered through the literature, often in obscure publications. Some have received Linnéan designations, making them rather more accessible: they appear in indices and compilations, and have found their way into text books. Some others, without formal names, have become buried in the literature, and are only occasionally rediscovered as paleontologists interested in the Precambrian come across them by chance and recognize them as belonging to named groups erected much later. Many uncertainties continue to exist concerning the biological affinities of these fossils, though almost all are generally considered to be remains of photosynthesizing organisms. The purpose of this paper is to present an inventory of the known Precambrian megascopic compressions and to briefly discuss the significance of the 0.7–1.1-Ga-old *Chuaria-Tawuia* assemblage.

Historical Background

The first of the references thus far encountered in the literature is to the ribbon-like films named *Laminarites antiquissimus* by Eichwald in 1854, now referred to *Vendotaenia antiqua* (Gnilovskaya 1971), from the Vendian of the East European Platform. Other early references are to millimetric carbonaceous discs from the Vindhyan Kurnool Supergroup of India (King 1872, p 68–69), the Chuar Group of the Grand Can-

1 Department of Geology, University of Montreal, Montreal, Quebec, Canada, H3C 3J7

Paleoalgology: Contemporary Research and Applications
(eds. D.F. Toomey and M.H. Nitecki)

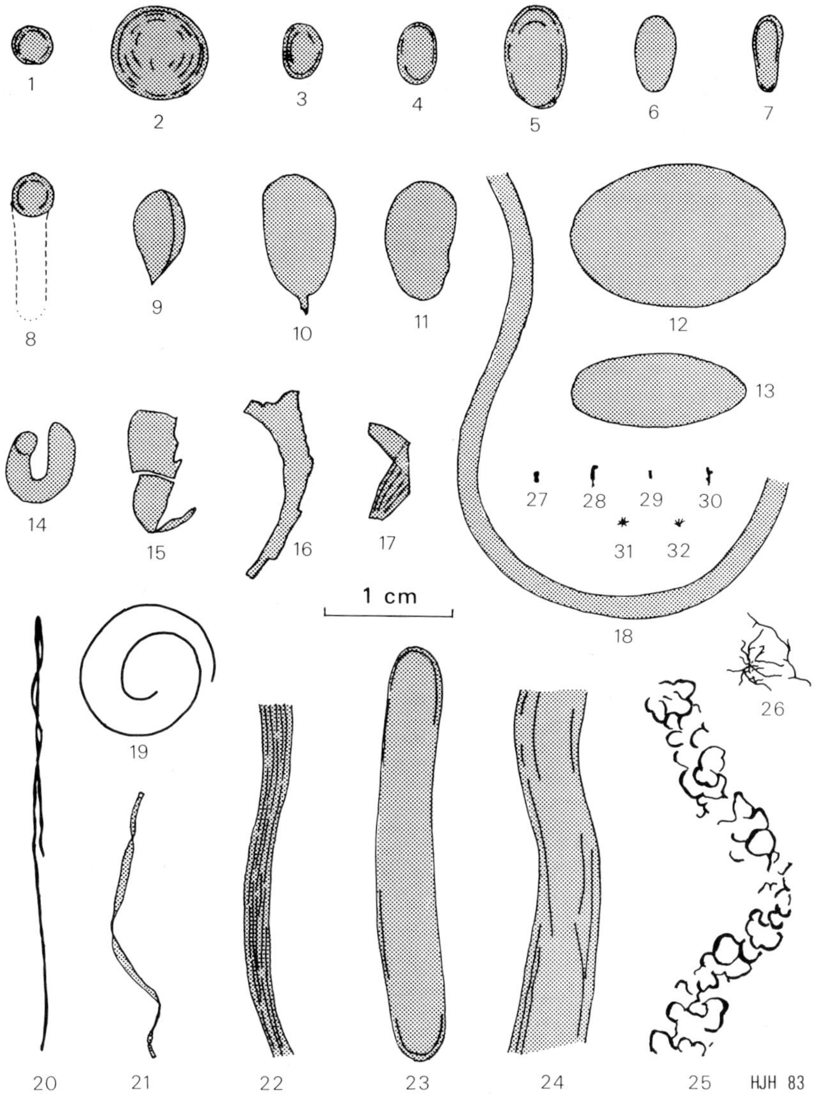

Fig. 1. Precambrian megascopic compressions: *1 Chuaria* Walcott 1899; *2 Beltanelloides* Sokolov 1965, 1973; *3 Ovidiscina* Zheng 1980; *4 Ellipsophysa* Zheng 1980; *5 Nephroformia* Zheng 1980; *6 Shouhsienia* Xing 1979 from Du 1982; *7 Pumilibaxa* Zheng 1980; *8 Fermoria* with "filament" Sahni 1977; *9 Krishnania* Sahni and Shrivastava 1954: *10 Longfengshania* Du 1982; *11 Morania*? Walcott 1919 in Fenton and Fenton 1937; *12* Unnamed graphitic compression, Stinchcomb et al. 1965; *13* Unnamed graphitic compression, Tyler et al. 1957; *14 Tawuia* Hofmann 1979 in Hofmann and Aitken 1979, recurved specimen; *15 Beltina* Walcott 1899; *16 Radicula*? Zheng 1980; *17 Lanceoforma* Walter et al. 1976; *18 Helminthoidichnites*? (Walcott 1899) Walter et al. 1976; *19 Grypania* (Walcott 1899) Walter et al. 1976; *20 Proterotainia* (Walcott 1899) Walter et al. 1976; *21 Tyrasotaenia* Gnilovskaya 1971; *22 Vendotaenia* (Eichwald 1854) Gnilovskaya 1971, *23 Tawuia* Hofmann in Hofmann and Aitken 1979, slightly curvilinear specimen; *24 Mezenia* Sokolov 1976; *25 Orbisiana* Sokolov 1976; *26 Eoholynia* Gnilovskaya 1975; *27 Lakhandinia* Timofeev and Hermann 1979; *28 Caudina* Gnilovskaya 1979; *29 Aataenia* Gnilovskaya 1976; *30 Majaphyton* Timofeev and Hermann in Timofeev et al. 1976; *31 Sarmenta* Gnilovskaya 1979; *32 Ulophyton* Timofeev and Hermann 1979

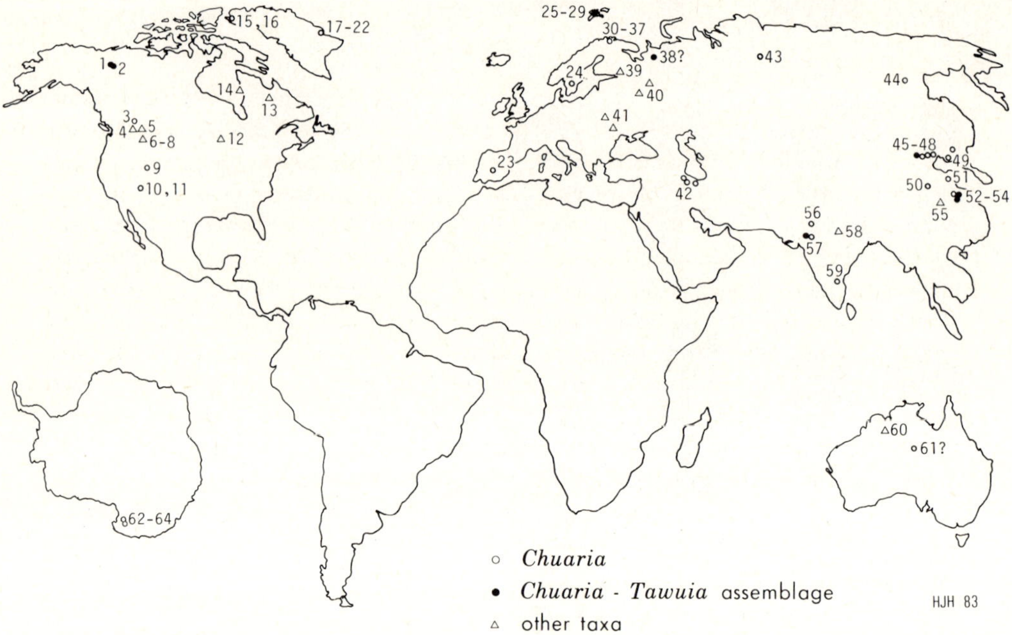

Fig. 2. Occurrences of Precambrian carbonaceous megafossils

yon in Arizona (Powell 1876, p 79), and the late Riphean Visingsö Formation of southern Sweden (Nathorst 1879, p 216, Wiman 1894). It was not until the end of the nineteenth century that Walcott (1899) provided a more comprehensive view of Mid- and Late-Proterozoic carbonaceous remains, including an additional occurrence from the Belt Supergroup of Montana. He erected new formal taxa to accommodate them: *Chuaria circularis* for the discs, and *Beltina danai* for angulate fragments (which he thought were eurypterids). He also named 3 species of ribbon-like films, which he attributed to the Phanerozoic trace fossil genus *Helminthoidichnites* Fitch: *H.? neihartensis* (=*Proterotainia neihartensis* and *P. montana* Walter et al. 1976), *H.? spiralis* (=*Grypania spiralis* Walter et al. 1976) and *H. meeki* (=*H.? meeki* Walter et al. 1976, = *Tawuia? meeki* Hofmann and Aitken 1979, p 160).

During the next 6 decades, descriptions of a few more occurrences were added, particularly from the Belt and Vindhyan Supergroups. Also, the older, Aphebian, remains became known (Low 1903, Tyler et al. 1957, Stinchcomb et al. 1965). However, it is only in the last 15 years that systematic studies have really begun. More than two dozen form genera and over 40 form species have entered the literature, some evidently synonyms of others, or *nomina nuda* (Fig. 1); of these, only about 17 genera and less than 20 species are probably valid.

Morphologic Categories

The simplicity of structure, the absence of characteristic microstructure of the remains, the widely scattered nature of reports on individual occurrences, and the difficulty in obtaining type specimens for examination have all been impediments to a better understanding of these problematic structures. Several schemes for suprageneric classification have been proposed. For example, *Chuaria* and its synonyms have been considered as belonging to such diverse groups as brachiopods, gastropods, foraminifera, plants, prasinophytes, acritarchs, and even as inorganic structures (for review see Ford and Breed 1973, Duan 1982). A number of family-level taxa have been erected: Fermoriidae Sahni 1936, Chuariidae Wenz 1938, Megasphaeromorphida Timofeev 1970, Chuariamorphida Sokolov 1974 (1976, p 137), Chuariaceae Wenz *nom. trans., emend.* Duan 1982.

Others, such as the various filamentous remains from the Vendian of the East European Platform, have been assigned to the single metaphyte Group Vendotaenides Gnilovskaya (1971), despite great divergence in size and shape between some genera; these include large, smooth, unbranched ribbons as well as patterned structures with sporangia and smaller branched forms.

Still others, irregular and nondescript films and fragments such as *Morania*? and *Beltina*, have been thought to be eurypterids or algae, or they have remained unclassified.

The compressions are the orphans of paleontology; many are classed as "unrecognized and unrecognizable 'genera'" in the Treatise (Häntzschel 1975). Although the best-known genus, *Vendotaenia*, is most probably a phaeophyte or rhodophyte, much work remains to be done before it will be possible confidently to assign most of the taxa to major groups. I therefore recognize here, solely on the basis of morphology, the *informal* categories of genus-level taxa of Precambrian megascopic and mesoscopic carbonaceous remains set out below in Table 1, that is, those visible to the unaided eye (> 0.2 mm); the categories are named after the dominant genus. It should be obvious that such a morphologic categorization is likely to cut across phylogenetic lines, but it is of some use in their description until their affinities become better known.

Chuarid remains include *Chuaria* and *Beltanelloides*. Despite the arguments by Duan (1982, pp 65–66), I here follow Ford and Breed (1973) in regarding most of the taxa cited in their synonymy as synonyms, except that I exclude *Krishnania,* which I place tentatively in the longfengshanid remains, and *Beltanelloides*, which I consider a separate, larger, taxon. *Ovidiscina* appears to be deformed *Chuaria.*

Among the tawuid remains I place the nominate genus *Tawuia* and several taxa I consider to be short specimens of *Tawuia.* Also included is *Fermoria*, sensu Sahni (1977, Pl. 2 Fig. 4; Pl. 3 Fig. 1; see also Häntzschel 1975, Fig. 108-4), who illustrated specimens I would assign to *Tawuia.* Further taxa are the small structures named *Lakhandinia*, which seem to be miniature versions of *Tawuia*, and fragments of large *Tawuia*-like structures called *Mezenia. Helminthoidichnites*?, possibly a new species of *Tawuia*, may also belong in this category or be a vendotaenid.

TAXA (listing includes various synonyms)	OCCURRENCES
Chuaria circularis Walcott 1899	1,9,10,11,15,16, 19-25,28-37,42,45, 48,49,52,53,57
Chuaria wimani Brotzen 1941	24
Chuaria annularis Zheng 1980	52
Chuaria circularis? Vidal 1979	17,18,62-64
Chuaria sp.	2,3,43,44,46,47, 50,51,54
cf. Chuaria Hofmann et Schopf 1983	60
Neobolus minima Chapman 1932	57
Obolella jonesi Chapman 1933	57
Fermoria minima Chapman 1933	57
Fermoria granulosa Chapman 1933	57
Fermoria tripartita Chapman 1933	57
Fermoria capsella Chapman 1935	57
Protobolella jonesi Chapman 1935	57
Vindhyanella jonesi Sahni 1936	57
Kildinella magna Timofeev 1969	24
Kildinella jacutica Timofeev 1969	43
Trachysphaeridium vetterni Timofeev 1969	24,43
Trachysphaeridium lachandinum Timof.1969	44
Trachysphaeridium maicum Timofeev 1969	44
Ovidiscina bagongshanica Zheng 1980	52
Beltanelloides sorichevae Sokolov 1965,1973	23?,38?,61?
Beltanelliformis brunsae Menner 1968,1974	23?,38?,61?
Tawuia dalensis Hofmann 1979	1,2,26,52,53
Tawuia striatia Zheng 1980	52
Tawuia sinensis Duan 1982	52,53
Tawuia sp. (see Sahni 1977, pls. 2,3)	57
Mezenia kossovoyi Sokolov 1976	38
Ellipsophysa axicula Zheng 1980	52
Ellipsophysa proceriaxis Zheng 1980	52
Shouhsienia shouhsienensis Xing 1979 ex Du 1982	45
Shouhsienia longa Xing 1979 ex Du 1982	45
Nephroformia liulaobeiensis Zheng 1980	52
Pumilibaxa huaiheiana Zheng 1980	52
Lakhandinia prolata Timof.et Hermann 1979	44
Krishnania acuminata Sahni et Shrivastava 1954	57
Longfengshania stipitata Du 1982	45
Morania? antiqua Fenton et Fenton 1937	1,5,47
Morania spp. Zheng 1980	52
Unnamed round compressions	6,12,13,14? 56, 58,59
Beltina danai Walcott 1899	1,4,5,8,47
Beltina sp. Aitken 1981	2
cf. Beltina Hofmann et Schopf 1983	60
Radicula? Zheng 1980	52
Unnamed films Horodyski 1980	7
Helminthoidichnites? meeki (Walcott 1899) Walter et al. 1976	8
Proterotainia neihartensis (Walcott 1899) Walter et al. 1976	8
Proterotainia montana Walter et al. 1976	8
Lanceoforma striata Walter et al. 1976	8
Grypania spiralis (Walcott 1899) Walter et al. 1976	8
Laminarites antiquissimus Eichwald 1854	39
Vendotaenia antiqua Gnilovskaya 1971	39,41
Vendotaenia sp. Xing et Liu 1978	55
Vendotaenia? sp. Du 1982	45
Tyrasotaenia podolica Gnilovskaya 1971	39,41
Tyrasotaenia sp., cf. T. podolica	47,55
Aataenia reticularis Gnilovskaya 1975	39,40
Eoholynia mosquensis Gnilovskaya 1975	39,40
Caudina cauda Gnilovskaya 1979	39
Sarmenta capitata Gnilovskaya 1979	39
Majaphyton antiquaum Timofeev et Hermann 1976	44
Ulophyton rifeicum Timof. et Hermann 1979	44
Orbisiana simplex Sokolov 1976	40

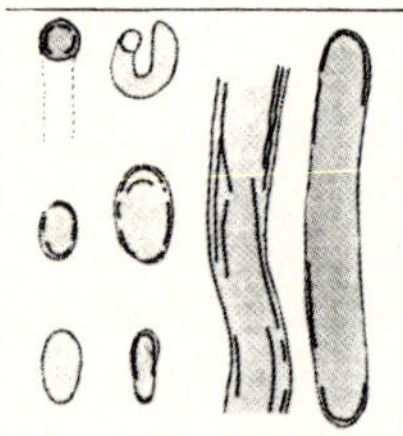
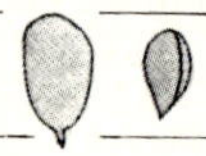
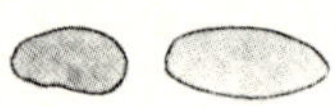
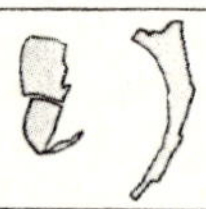
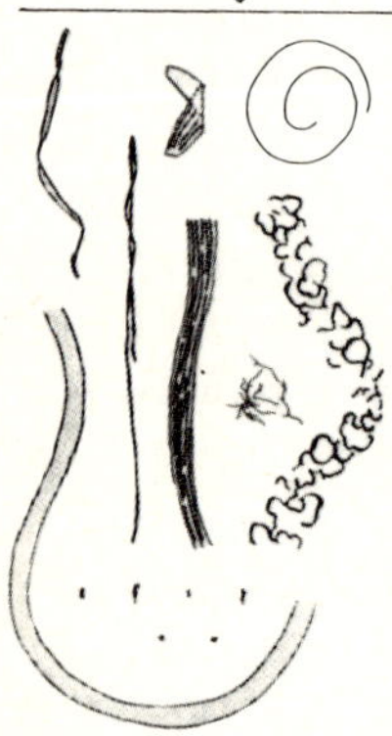
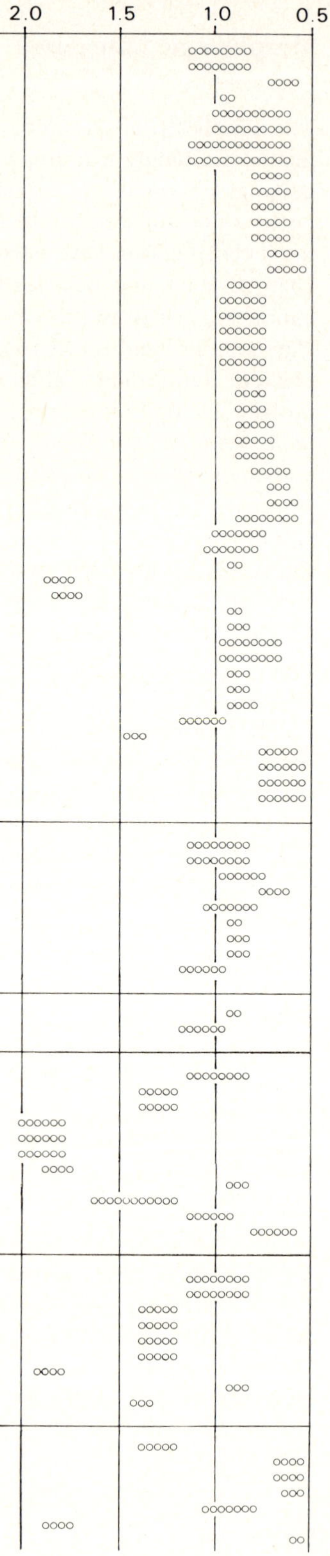

0.5

1 Little Dal Gp., basinal sequence
2 Little Dal Gp., rusty shale
3 Windermere Gp., Hector Fm.
9 Uinta Mountain Gp., Red Pine Sh.
10 Chuar Gp., Galeros Fm.
11 Chuar Gp., Kwagunt Fm.
15 Thule Gp., Wolstenholme Fm.
16 Thule Gp., Dundas Fm.
17 Eleonore Bay Gp., Calc-Argillaceous Ser.
18 Eleonore Bay Gp., U. Argillaceous-Arenaceous Ser.
19 Eleonore Bay Gp., Quartzite Ser.
20 Eleonore Bay Gp., Multicoloured Ser.
21 Eleonore Bay Gp., Limestone-Dolomite Ser.
22 Lower Tillites
23 Pusa Sh.
24 Visingsö Fm.
25 Franklinsundet Gp., Westmanbukta Fm.
26 Franklinsundet Gp., Kapp Lord Fm.
27 Celsiusberget Gp., Flora Fm.
28 Roaldtoppen Gp., Hunnberg Fm.
29 Roaldtoppen Gp., Ryssö Fm.
30 Vadsø Gp., Klubbnes Fm.
31 Vadsø Gp., Andersby Fm.
32 Vadsø Gp., Golneselv Fm.
33 Vadsø Gp., Ekkerøy Fm.
34 Tanafjord Gp., Dakkovarre Fm.
35 Tanafjord Gp., Grasdal Fm.
36 Vestertana Gp., Lower Tillite
37 Vestertana Gp., Nyborg Fm.
38 Redkino Ser.
42 Soltanieh Fm., Chapoghlu Sh.
43 Miroedikha Fm.
44 Lakhanda Fm.
45 Qingbaikou Syst., Changlongshan Fm.
46 Changcheng Syst., Chuanlingkou Fm.
47 Changcheng Syst., Tuanzhanzi Fm.
48 Qingbaikou Syst., Jingeryü Fm.
49 Xihe Gp., Nanfen Fm.
50 Paoyü Fm.
51 Sinian
52 Bagongshan Gp., Liulaobei Fm.
53 Sidingshan Gp., Jiuliqiao Fm.
54 Huainan Gp., Xingxing Fm.
57 Semri Gp., Suket Sh.
60 Bungle Bungle Dol.
61 Central Mount Stuart Beds
62 Robertson Bay Gp.
63 Bowers Supgp., Molar Fm.
64 Clasts in Houliston Glacier Beds

---

1 Little Dal Gp., basinal sequence
2 Little Dal Gp., rusty shale
26 Franklinsundet Gp., Kapp Lord Fm.
38 Redkino Ser.
44 Lakhanda Fm.
45 Qingbaikou Syst., Changlongshan Fm.
52 Bagongshan Gp., Liulaobei Fm.
53 Sidingshan Gp., Jiuliqiao Fm.
57 Semri Gp., Suket Sh.

---

45 Qingbaikou Syst., Changlongshan Fm.
57 Semri Gp., Suket Sh.

---

1 Little Dal Gp., basinal sequence
5 Belt Supgp., Altyn Fm.
6 Belt Supgp., Chamberlain Sh.
12 Animikie Gp., Michigamme Fm.
13 Kaniapiskau Supgp., Attikamagen Fm.
14 Nastapoka Gp.
47 Changcheng Syst., Tuanshanzi Fm.
52 Bagongshan Gp., Liulaobei Fm.
56 Delhi Syst., Alwar Qte.
58 Semri Gp., Rohtas Ls.
59 Kurnool Supgp., Auk Sh.

---

1 Little Dal Gp., basinal sequence
2 Little Dal Gp., rusty shale
4 Purcell Supgp., Aldridge Fm.
5 Purcell Supgp., Altyn Fm.
7 Belt Supgp., Newland Ls.
8 Belt Supgp., Greyson Sh.
47 Changcheng Syst., Tuanshanzi Fm.
52 Bagongshan Gp., Liulaobei Fm.
60 Bungle Bungle Dol.

---

8 Belt Supgp., Greyson Sh.
39 Valday Ser.
40 Valday Ser., various fms.
41 Kanilovka Fm.
44 Lakhanda Fm.
47 Changcheng Syst., Tuanshanzi Fm.
55 Tengying Fm.

HJH 83

Fig. 3. Stratigraphic distribution of occurrences

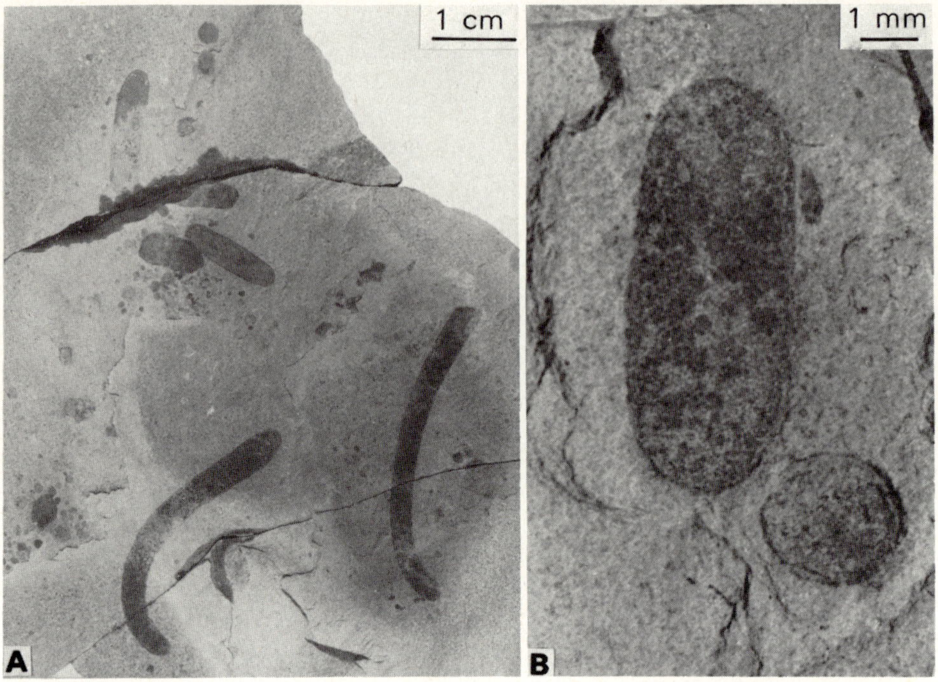

Fig. 4A,B. *Chuaria-Tawuia* assemblage, deemed useful for broad correlation of 0.7–1.1 Ga sequences. Little Dal Group, Mackenzie Mountains, Northwest Territories, Canada (Locality 1; = section 76AC-2 in Hofmann and Aitken 1979, Figs. 4 and 6; 64°48'N 128°52.5W). **A** Note wide range of shape and size of *Tawuia* specimens associated with small and large *Chuaria* specimens. Geol. Surv. Canada, hypotype 66185 (large curved specimen at *lower left*). **B** Juxtaposed individuals of *Tawuia dalensis* and *Chuaria circularis.* Note small circular markings in *Tawuia* specimen (spores, or immature *Chuaria* ?). Geol. Surv. Canada, hypotype 66186 (*Tawuia*)

Longfengshanid remains comprise *Longfengshania*, questionably *Krishnania* (which has also been considered a deformed chuarid) and the very small structures called *Sarmenta* and *Caudina*. They all appear to be epibenthonic or epiplanktonic.

Moranid remains are *Morania*? and unnamed elliptical compression that lack annular wrinkles and that appear to be flattened colonies of coccoid and filamentous microbes.

Beltinid remains are represented by *Beltina*, originally interpreted as eurypterids, but now thought to be most likely fragments of algal mats or larger algal taxa similar to the modern *Monostroma* and *Porphyridium. Radicula*? is a probable synonym for *Beltina.*

Vendotaenid remains include two groups of taxa, one from the Helikian (1.3 Ga) and the other from Vendian deposits. All are now generally viewed as megascopic algae (Phaeophyta or Rhodophyta). The Vendian forms are reported to have well preserved reproductive structures (Gnilovskaya 1971, 1975, 1979) not known from the Helikian ones (Walter et al. 1976).

Table 1. Morphologic categorization of Precambrian carbonaceous remains (*Incertae sedis* – most probably photosynthesizers)

A. Chuarid remains: spheroids or compressed spheroids (discs), normally with concentric wrinkles (Chuariaceae Duan 1982, *partim*)

Chuaria Walcott 1899
=*Neobolus* Chapman 1932
=*Obolella* Chapman 1933
=*Fermoria* Chapman 1935
=*Protobolella* Chapman 1935
=*Vindhyanella* Sahni 1936 (*nom. superfl.*)
=*Kildinella* Timofeev 1969, *partim*
=*Trachysphaeridium* Timofeev 1969, *partim*
=*Ovidiscina* Zheng 1980 (cf. deformed *Chuaria*)

? *Beltanelloides* Sokolov 1965, 1973, 1976
?=*Beltanelliformis* Menner 1968 from Menner in Keller et al. 1974

B. Tawuid remains: tomaculate (sausage-shaped) or compressed tomaculate forms, straight or bent, with rounded extremities; generally with marginal wrinkles (Chuariaceae Duan 1982, *partim*)

Tawuia Hofmann 1979 (in Hofmann and Aitken 1979)
=*Fermoria* with "filament", Sahni 1977, p. 293, 298
=*Ellipsophysa* Zheng 1980 (cf. short *Tawuia*)
=*Pumilibaxa* Zheng 1980 (cf. short *Tawuia*)
=*Nephroformia* Zheng 1980 (cf. short *Tawuia*)
=*Shouhsienia* Xing 1978 from Du 1982 (cf. short *Tawuia*)

? *Mezenia* Sokolov 1976

? *Helminthoidichnites*? (Walcott 1899) Walter et al. 1976

? *Lakhandinia* Timofeev and Hermann 1979

C. Longfengshanid remains: round to oblong structures with stipe or appendage

Lonfengshania Du 1982

? *Krishnania* Sahni and Shrivastava 1954

? *Sarmenta* Gnilovskaya 1979

? *Caudina* Gnilovskaya 1979

D. Moranid remains: relatively large, elliptical to irregular round films without wrinkles

Morania? Walcott 1919 (in Fenton and Fenton 1937)

E. Beltinid remains: irregular, angulate forms, fragments

Beltina Walcott 1899
=*Radicula*? Zheng 1980

F. Vendotaenid remains: slender filamentous structures, generally unbranched, twisted or untwisted, smooth or with primary striate or speckled pattern (e.g., sporangia); (Vendotaenides Gnilovskaya 1971, *partim*)

Vendotaenia Gnilovskaya 1971
=*Laminarites* Sternberg 1833 (in Eichwald 1854, *partim*)

Tyrasotaenia Gnilovskaya 1971

Aataenia Gnilovskaya 1971

Table 1 (cont.)

Proterotainia (Walcott 1899) Walter et al. 1976
Lanceoforma Walter et al. 1976
Grypania (Walcott 1899) Walter et al. 1976
? *Helminthoidichnites*? (Walcott 1899) Walter et al. 1976

G. **Eoholynid remains**: noticeably branched filamentous structures, smooth or with primary structures; (Vendotaenides Gnilovskaya 1971, *partim*)

Eoholynia Gnilovskaya 1976
? *Orbisiana* Sokolov 1976
? *Majaphyton* Timofeev and Hermann, in Timofeev et al. 1976
? *Ulophyton* Timofeev and Hermann 1979

The eoholynid remains of branching filaments are represented by *Eoholynia*, which is sporangiate, and *Ulophyton, Majaphyton* and *Orbisiana*. They appear to have been attched to some substrate. Gnilovskaya (1979) includes *Eoholynia* in her Group Vendotaenides and considers it to be a phaeophyte or rhodophyte and a possible descendant of the upper Riphean *Ulophyton. Orbisiana* is very poorly known.

Discussion

Three important observations need to be made: one deals with the *Chuaria-Tawuia* relationship, a second with the *Beltanelloides-Mezenia* relationship, and the third with biostratigraphic aspects.

Based on a study of the size distributions of specimens of co-occurring *Tawuia* and *Chuaria* in certain occurrences, it is realized that *Tawuia* growth is distinctly allometric (Hofmann and Aitken 1979, p. 162; Knoll 1982, p. 275), and that the shape of *Tawuia* grades into *Chuaria* through intermediate forms (named *Ellipsophysa, Pumilibaxa, Nephroformia,* and *Shouhsienia*). Moreover, some small disc-like structures are sometimes associated with the carbon film of *Tawuia* (e.g., Fig. 4B). This has led to the speculation that the megascopic forms called *Chuaria* and *Tawuia* may represent an alternation of generations of eucaryotic organisms (Hofmann 1981, p. 42). This biologic affinity is recognized also by Duan (1982), who includes both genera in his biological concept of the Family Chuariaceae. While I support such a treatment, I regard both *Chuaria* and *Tawuia* as compressed spheroidal and sausage-shaped bodies, based on three-dimensionally preserved specimens in the type assemblage of *Tawuia dalensis* (Hofmann 1985), and not as primary discs and ribbons as implied by Duan (1982, p. 65). This matter is complicated somewhat further by the existence of the coeval *Lakhandinia*, which looks like a miniature edition of *Tawuia.* If small individuals of *Chuaria* grow into large ones and then develop into *Tawuia*, it is difficult to see why there should be coeval large *Chuaria* and small *Lakhandinia*, unless the latter represents a dwarfed occurrence.

The second interesting point is the nature of the relationship of *Beltanelloides* with fragements identified as *Mezenia*. Both genera (still apparently *nomina nuda*) are from the Vendian Redkino Series of the East European Platform, and both are considered to be macrophytoplankton by Sokolov (1976, p. 138). Given the possible relationship between *Chuaria* and *Tawuia*, I would propose considering a similar affinity for structures called *Beltanelloides* and *Mezenia*, and regarding these to be the morphologically equivalent Vendian descendants of the Helikian taxa. The larger size of both Vendian forms is in keeping with the general increased in size of Precambrian organisms with time.

The third point concerns the biostratigraphic aspect of carbonaceous megafossils. The data available (Fig. 3) show that the oldest films are late Early Proterozoic elliptical compressions, probably flattened microbial colonies. Filamentous, discoid, and angulate forms appear slightly later. The first morphologically diversified biota of large unornamented filamentous taxa is found in 1.3 Ga Beltian rocks of North America (Walter et al. 1976). This is followed by the 1.1–0.7 Ga rocks characterized by the *Chuaria-Tawuia* assemblage, and by the introduction of forms with stipes that are most likely epibionts. Finally, the morphologically more complex vendotaenid filaments characterize the Vendian-Early Cambrian interval, in which 4 successive floras have been recognized (Gnilovskaya 1979, p. 616).

Summary

The development of organisms of megascopic size and the related acquisition of the nucleus in cells are landmark evolutionary events. Based on the presently known paleontologic record of megafossils of probable algal affinities, these events occurred by Mid-Proterozoic time, well before the development of Ediacaran metazoans. While the first accounts of Precambrian carbonaceous fossils date back to the middle of the 19th century, relatively few new finds were made during the next hundred years. Only in the past 20 years, and particularly in the last 6, have such reports become more frequent. The fossils are generally preserved as millimetric compressions on bedding planes. They include circular, elliptical, ovate, and angulate films, and rectilinear, curvilinear, spiraliform, and twisted ribbons. Most of them are unornamented, except for variably oriented folds and markings that are best regarded as compactional artifacts; some Late Proterozoic types exhibit branching and striate or granular structures with presumed biologic function. More than two dozen genera and over 40 species have now been formally erected, but these include synonyms. Among the megascopic compressions are some that can be interpreted as megascopic colonies of procaryotes. Others, such as the distinct and cosmopolitan *Chuaria* and *Tawuia*, are regarded as probable algal megafossils with some value (as a biota) for broad intercontinental correlation of Middle to Late Proterozoic sequences.

Acknowledgment. Financial support from the Natural Sciences and Engineering Research Council of Canada (Grant no. A7484) is gratefully acknowledged.

References

Locality

Aitken JD (1981) Stratigraphy and sedimentology of the Upper Proterozoic Little Dal Group, Mackenzie Mountains, Northwest Territories. In: Campbell FHA (ed) Proterozoic basins of Canada. Geol Surv Can, Pap 81–10, pp 47–71 — 2

Allan JA (1913) Rocky Mountains (Bankhead to Golden). Geol Surv Can Guide Book 8, Transcontinental Excursion Cl, part 2, pp 167–201 — 3

Assereto R (1963) The Paleozoic formations in Central Elburz, Iran. Riv Ital Paleontol 69:503–543 — 42

Brasier MD, Perejon A, de San José MA (1979) Discovery of an important fossiliferous Precambrian-Cambrian sequence in Spain. Estud Geol 35:379–383 — 23

Brotzen F (1941) Några bidrag til visingsöformationens stratigrafi och tektonik. Geol Fören Stockholm Förh 63: no 3, 245–261 — 24

Chapman F (1932) In: Paleontol Dep Rep. Rec Geol Surv India 1931 66:28–29 — 57

Chapman F (1933) In: Paleontol Dep Rep. Rec Geol Surv India 1932 67:20–21 — 57

Chapman F (1935) Primitive fossils, possible atrematous and neotrematous brachiopods, from the Vindhyans of India. Rec Geol Surf India 1935–1936 69:109–120 — 57

Chen Jinbiao, Zhang Huimin, Zhu Shixing, Zhao Zhen, Wang Zhengang (1980) Research on Sinian Suberathem of Jixian, Tianjin. In: Research on Precambrian geology. Sinian suberathem in China. Tianjin Sci Technol Press, Tianjin, China, pp 56–114 (in Chinese with English abstract) — 48

Choubert G, Faure-Muret A (1980) The Precambrian in North Peri-Atlantic and South Mediterranean mobile zones; general results. Earth-Sci Rev 16:85–219 — 23

Copper RA, Jago JB, MacKinnon, DI, Shergold JH, Vidal G (1982). Late Precambrian and Cambrian fossils from Northern Victoria Land and their stratigraphic implications. In: Craddock C (ed) Antarctic geoscience. Univ Wisconsin Press, pp 629–633 — 62–64

Daly RA (1912) Geology of the North American Cordillera at the forty-ninth parallel. Geol Surv Can Mem 38: pt 1, 528 — 5

Drysdale CW (1917) Investigations in British Columbia. Geol Surv Can Sum Rep 1916:44–63 — 4

Du Rulin (1982) The discovery of the fossils such as *Chuaria* in the Qingbaikou System in northwestern Hebei and their significance. Geol Rev (Beijing) 28:no 1, 1–7 (in Chinese with English abstract) — 45

Du Rulin, Li Peiju (1980) Sinian Suberathem in the western Yenshan Ranges. In: Research of Precambrian geology, Sinian suberathem in China. Tianjin Sci Technol Press, Tianjin, China, pp 341–357 (in Chinese with English abstract) — 45

Duan Cheng-hua (1982) Late Precambrian algal megafossils *Chuaria* and *Tawuia* in some areas of eastern China. Alcheringa 6:57–68 — 49, 52, 53

Dutt GN, Shrivastava RN (1975) Fossil flora in the Alwar Quartzite, Ferozpur Jhirka, Gurgaon District, Haryana. Geol Surv India Misc Publ no 23: pt 1, 149–156 — 56

Eichwald E (1854) The paleontology of Russia. The ancient period. 1. Flora of the graywacke, calcareous rocks and cuprous-shale formation of Russia. St. Petersburg, 245 p (not seen; *fide* Gnilovskaya 1971) — 39

Eisenack A (1966) Über *Chuaria wimani* Brotzen. Neues Jahrb Geol Paläontol Stuttgart, pp 52–56 — 24

Fenton CL, Fenton MA (1931) Algae and algal beds in the Belt series of Glacier National Park. J Geol 39:670–686 — 5

Fenton CL, Fenton MA (1937) Belt series of the north: stratigraphy, sedimentation, paleontology. Bull Geol Soc Am 50:1873–1970 — 5

Ford TD, Breed WJ (1972) The problematical Precambrian fossil *Chuaria*. Int Geol Congr 24th Sess Proc Sect 1:11–18 — 10, 11

Ford TD, Breed WJ (1973) The problematical Precambrian fossil *Chuaria*. Palaeontology 16: pt 3, 535–550 10, 11, 42, 61
Ghare MA, Badve RM (1979) *Chuaria circularis* Walcott, a new taxonomic position for *Protobolella* Chapman and *Fermoria* Chapman. Biovigyanam J Life Sci 5: no 1, 91–92 57
Gnilovskaya MB (1971) Drevneyshiye vodniye rasteniya venda Russkoy platformy pozdniy dokembriy). [The oldest aquatic plants of the Vendian of the Russian Platform (Late Precambrian)]. Paleontol Zh 1971: no 3, 101–107 (English transl, pp 372–378 39, 41
Gnilovskaya MB (1975) Noviye danniye or prirode vendotaenid (New data on the nature of vendotaenids). Dokl Akad Nauk SSSR 221:953–955 (Am Geol Inst transl. pp 258–261) 40
Gnilovskaya MB (1976) Drevneishie Metaphyta (The oldest Metaphyta). Geol Congr 25th Sess Rep Sov Geol Paleontol Mar Geol. Nauka, Moscow, pp 10–14 40
Gnilovskaya MB (1979) The Vendian Metaphyta. Bull Cent Rech Explor-Prod Elf-Aquitaine 3:611–618 39
Gussow WC (1973) *Chuaria* sp. cf. *C. circularis* Walcott from the Precambrian Hector Formation, Banff National Park, Alberta, Canada. J Paleontol 47:1108–1112 3
Häntzschel W (1975) Trace fossils and Problematica. In: Teichert C (ed) Treatise on Invert Paleontol, part W (Miscellanea) Suppl 1. Geol Soc Am and Univ Kansas, 269 p 57
Hofmann HJ (1971) Precambrian fossils, pseudofossils, and problematica in Canada. Geol Surv Can Bull 189:146 3–5, 13
Hofmann HJ (1977) The problematic fossil *Chuaria* from the Late Precambrian Uinta Mountain Group, Utah. Precambrian Res 4:1–11 9
Hofmann HJ (1981) Precambrian fossils in Canada – the 1970s in retrospect. Geol Surv Can Pap 81-10:419–443 1
Hofmann HJ (1985) The mid-Proterozoic Little Dal macrobiota, Mackenzie Mountains, north-west Canada. Palaeontol 28 (part 2):331–353 1
Hofmann HJ, Aitken JD (1979) Precambrian biota from the Little Dal Group, Mackenzie Mountains, northwestern Canada. Can J Earth Sci 16:150–166 1
Hofmann HJ, Chen Jinbiao (1981) Carbonaceous megafossils from the Precambrian (1800 Ma) near Jixian, northern China. Can J Earth Sci 18:443–447 46, 47
Hofmann HJ, Schopf JW (1983) Early Proterozoic microfossils. Chap 14. In: Schopf JW (ed) Earth's earliest biosphere: its origin and evolution. Princeton Univ Press 60
Horodyski RJ (1980) Middle Proterozoic shale-facies microbiota from the lower Belt Supergroup, Little Belt Mountains, Montana. J Paleontol 54:649–663 6,7
Jones HJ (1909) In: Gen Rep. Rec Geol Surv India 38:66 57
Keller BM, Menner VV, Stepanov VA, Chumakov NM (1974) Novye nakhodki metazoa v vendomii Russky platformy. Izv Akad Nauk SSSR Serv Geol no 12:132
King W (1872) On the Kadapah and Karnul Formations in the Madras Presidency. Mem Geol Surv India 8:313 59
Knoll AH (1982) Microfossil-based biostratiography of the Precambrian Hecla Hoek sequence, Nordaustlandet, Svalbard. Geol Mag 119:269–279 25–29
Low AP (1903) Report on the geology and physical character of the Nastapoka Islands, Hudson Bay. Geol Surv Can Annu Rep pt. DD 13:31 14
Misra RC (1957) *Fermoria*, the enigma of Indian palaeontology. J Paleontol Soc India 2:54–57 57
Misra RC, Bhatnagar GS (1950) On carbonaceous discs and 'algal dust' from the Vindhyans Pre-Cambrian. Curr Sci 19:88–89 58
Misra RC, Dube SN (1952) A new collection and a restudy of the organic remains from the Suket Shales (Vindhyans), Rampura, Madhya Barat. Sci Cult 18:46–48 57
Nathorst AG (1879) En egendomlig strukturvarietet af lerhaltig kalksten fran Grennatrakten. Geol Fören Stockholm Förh 4: no 8, 216 24
Powell JW (1876) Report on the geology of the eastern portion of the Uinta Mountains. US Geol Geog Surv, 218 11

Rajurkar ST (1963) Discoidal impressions akin to Fermoria, from the Owk shales of Kurnool District Andra Pradesh. Indian Miner 17: no 3, 306–307 59

Sahni MR (1936) *Fermoria minima*: a revised classification of the organic remains from the Vindhyans of India. Rec Geol Surv India 69:458–468 57

Sahni MR (1977) Vindhyan palaeobiology, stratigraphy and depositional environments: a critical review. J Paleontol Soc India 20:289–304 57

Sahni MR, Shrivastava RN (1954) New organic remains from the Vindhyan System and the probable systematic position of *Fermoria* Chapman. Curr Science 23: 39–41 57

Sokolov BS (1965) Drevneyshie otlozheniya rannego kembriya; sabelliditidy. Tezisy Dokl Vses Simp Paleontol Dokem 25–30 Oct 1965, pp 78–91

Sokolov BS (1973) Vendian of northern Eurasia. Am Assoc Petrol Geol Mem 19: 204–218 38

Sokolov BS (1976) Organicheskiy mir zemli na puti k fanerozoiskoy differentziatzii (The organic world on the path of Phanerozoic differentiation). Vestn Akad Nauk SSSR 1976: no 1, 126–143 38

Stinchcomb BL, Levin HL, Echols DJ (1965) Precambrian graphitic compressions of possible biologic orgin from Canada. Science 148:75–76 12, 13

Stöcklin J, Ruttner A, Nabavi M (1964) New data on the Lower Paleozoic and Precambrian of North Iran. Geol Surv Iran Rep 1:29 42

Timofeev BV (1969) Sferomorfidy proterozoya. Nauka, Leningrad, 146 p 24, 43, 44

Timofeev BV (1970) Sphaeromorphida géants dans le Précambrien avancé. Rev Palaeobot Palaeobot Palynol 10:157–160

Timofeev BV, German [Hermann] TN (1979) Dokembriyskaya mikrobiota Lakhandinskoy svity (Precambrian microbiota of the Lakhanda Formation). In: Sokolov S (ed) Paleontologiya dokembriya i rannego kembriya (Paleontology of the Precambrian and Lower Cambrian). Nauka, Leningrad, pp 137–147 44

Timofeev BV, German [Hermann] TN, Mikhailova NS (1976) Mikrofitofossilii dokembriya, kembriya i ordovika. Nauka, Leningrad, 107 p 44

Tyler SA, Barghoorn ES, Barret LP (1957) Anthracitic coal from Precambrian Upper Huronian black shale of the Iron River district, northern Michigan. Bull Geol Soc Am 68:1293–1304 12

Vidal G (1974) Late Precambrian microfossils from the basal sandstone unit of the Visingsö beds, South Sweden. Geol Palaeontol 8:1–14 24

Vidal G (1976a) Late Precambrian microfossils from the Visingsö Beds in southern Sweden. Fossils Strata no 9:57 24

Vidal G (1976b) Late Precambrian acritarchs from the Eleonore Bay Group and Tillite Group in East Greenland. Geol Surv Greenl Rep 78:19 19–21

Vidal G (1979) Acritarchs from the Upper Proterozoic and Lower Cambrian of East Greenland. Geol Surv Greenl Bull 134:40 17–22

Vidal G (1981) Micropalaeontology and biostratigraphy of the Upper Proterozoic and Lower Cambrian sequence in East Finmark, northern Norway. Nor Geol Unders no 362: (Bull 59) 53 30–37

Vidal G, Dawes PR (1980) Acritarchs from the Proterozoic Thule Group, Northwest Greenland. Geol Surv Greenl Rep 100:24–29 15, 16

Wade M (1969) Medusae from uppermost Precambrian or Cambrian sandstones, central Australia. Palaeontology 12: pt 3, 351–365 61

Walcott CD (1899) Pre-Cambrian fossiliferous formations. Bull Geol Soc Am 10: 199–244 8, 11

Walter MR, Oehler JH, Oehler DZ (1976) Megascopic algae 1300 million years old from the Belt Supergroup, Montana: a reinterpretation of Walcott's *Helminthoidichnites*. J Paleontol 50:872–881 8

Wang Yuelun, Lu Zongbin, Xing Yusheng, Gao Zhenjia, Lin Weixing, Ma Guogan, Zhang Luyi, Lu Songnian (1980) Subdivision and correlation of the Upper Precambrian in China. In: Research on Precambrian geology, Sinian suberathem in 48, 50, 52

China. Tianjin Sci Technol Press, Tianjin, China, pp 1–30 (in Chinese with English abstract)
Wenz W (1938) Gastropoda. In: Schindewolf (ed) Handbuch der Paläozoologie, vol VI, part 1. Borntraeger, Berlin, 240 pp 24
Willis B (1902) Stratigraphy and structure, Lewis and Livingston Ranges, Montana. Bull Geol Soc Am 13:305–352 5
Wiman C (1894) Paleontologische Notizen, 1 und 2. Bull Geol Inst Upsala 2:109–117 24
Xing Yusheng, Liu Kueichi (1978) Sinian microfossils. In: Sanshia Stratigr Div (ed) Hubei Geological Survey. Stratigraphy and paleontology, Sinian to Permian, East Gorge area, 381 p, 113 pls 55
Yang Qihe, Zhang Youli, Zheng Wenwu, Xu Xuesi (1980) Subdivision and correlation of Sinian Suberathem in Northern Jiangsu and Anhui. In: Research on Precambrian geology, Sinian suberathem in China. Tianjin Sci Technol Press, Tianjin, China, pp 231–265 (in Chinese, with English abstract) 52–54
Zheng Wenwu (1980) A new occurrence of fossil group of *Chuaria* from the Sinian System in North Anhui and its geological meaning. Bull Chin Acad Geol Sci Ser 4 1:no 1, 49–69 (in Chinese with English abstract) 52–54

4 Stromatolites of the Middle Proterozoic Belt Supergroup, Glacier National Park, Montana: a Summary and a Comment on the Relationship Between Their Morphology and Paleoenvironment

R.J. HORODYSKI[1]

Geologic Setting

The Belt Supergroup is a thick sequence of Middle Proterozoic strata occurring in western Montana, northern Idaho, and adjacent parts of Washington state, Alberta, and British Columbia (Ross 1963, Harrison 1972). In Glacier National Park the Belt strata are well exposed, relatively unmetamorphosed, and in many places structurally simple. In the central and northeastern part of the park the exposed Belt section is about 2900 m thick and consists of argillite, sandstone, and sandy and muddy limestone and dolostone. Most of these strata were deposited in shallow subtidal, intertidal, and alluvial settings, with offshore deposits occurring only in portions of the Appekunny Argillite (for a summary of the sedimentary geology of the Belt Supergroup in Glacier National Park see Horodyski 1983).

Stromatolites

Stromatolites are prominent in several formations of the Belt Supergroup in Glacier National Park (Fig. 1), and were the subject of some classical early studies of stromatolites in North America (Walcott 1914, Fenton and Fenton 1937, Rezak 1957). Branched columnar stromatolites referable to the group *Baicalia* and inclined, highly elongate, unbranched columns occur locally in the Altyn Limestone (Horodyski 1976a). These stromatolites formed in a shallow subtidal setting and were produced by a combination of sediment stabilization and in situ carbonate precipitation.

Stromatolites are absent from the overlying Appekunny Argillite and are rare in the Grinnell Argillite, where two horizons occur in the lower part of the formation. These stromatolites are mound-shaped dolomitic structures, which originated largely by in situ carbonate precipitation in a very shallow marine or lacustrine setting.

The Siyeh Limestone contains the greatest number and variety of stromatolites in the Belt Supergroup. Mound-shaped, partially silicified, dolomitic stromatolites occur in the lower part of the formation. Mound-shaped stromatolites composed of calcite or dolomite occur throughout the formation; they locally attain heights of 2 m, and they rarely exhibit an internal structure consisting of branched columns (Horodyski

1 Department of Geology, Tulane University, New Orleans, LA 70118, U.S.A.

Paleoalgology: Contemporary Research and Applications
(eds. D.F. Toomey and M.H. Nitecki)

SHEPARD FORMATION	>270m	Green and greenish gray argillite, dolomitic argillite, muddy sandstone, and muddy dolostone. Purcell Lava present locally near the base of the formation	Shallow subtidal to intertidal	Stromatolites are relatively uncommon. Dolomitic, mound-shaped structures 5-40 cm high
SNOWSLIP FORMATION	385 m	Green and red argillite, sandy argillite, and muddy sandstone, commonly somewhat dolomitic. Subordinate muddy dolostone	Largely intertidal. Parts shallow subtidal	Stromatolites are locally common. Calcitic, mound-shaped structures 5-50 cm high. Formed by in situ carbonate precipitation. Some may be largely nonbiogenic
SIYEH LIMESTONE [Helena Dolostone]	780m	Muddy and sandy dolostone, muddy and sandy dolomitic limestone, muddy and sandy limestone, dolomitic and calcitic argillite, and dolomitic and calcitic sandstone. Rare mudstone	Subtidal to intertidal. Occasional elevated salinity indicated by halite casts 250-370m above the base of the formation	Stromatolites are common throughout the formation. Partially silicified dolomitic stromatolites occur 110-180m above the base of the formation. Calcitic and calcitic-dolomitic mound-shaped stromatolites 5-200 cm high are common in the middle and upper part of the formation. Prominent, 24-32m thick unit composed of Baicalia and Conophyton occurs 200m below the top of the formation
GRINNELL ARGILLITE [Spokane & Empire Formations]	605 m	Red argillite, sandy argillite, siltite, and muddy sandstone. White quartzose sandstone present and locally abundant. Subordinate green argillite and sandy argillite	Largely alluvial plain with deposition by sheetflooding or flooding of extremely shallow channels. Portions at margin of lacustrine or marine shoreline	Two discontinuous stromatolite horizons, 67 and 84 m above the base of the formation. Dolomitic, mound-shaped stromatolites 5-20 cm high
APPEKUNNY ARGILLITE [Greyson Shale]	~700m	Green argillite, sandy argillite, siltite, and muddy fine-grained sandstone. Quartzose sandstone locally forms prominent units. Subordinate red-colored strata and rare black pyritic mudstone	Largely offshore and below wave base. Portions shallow subtidal to intertidal. Possibly offshore of and marginal to an alluvial plain	No stromatolites are known from this unit
ALTYN LIMESTONE	>150m	Sandy dolarenite and silty and clayey dololutite.	Shallow subtidal to intertidal	Stromatolites are locally common. Dolomitic, Baicalia-like columnar stromatolites and highly elongate, inclined, unbranched columnar stromatolites

Fig. 1. Summary of the stratigraphy, lithology, depositional setting, and stromatolites of the Middle Proterozoic Belt Supergroup in the central and northeastern part of Glacier National Park, Montana. Note that neither the base nor the top of the Belt Supergroup is present, and that the Altyn Limestone and Shepard Formation are represented by partial sections. Stratigraphic units are after Ross (1959) and Childers (1963), with correlatives to the south (after Harrison 1972) given in parentheses. Observations were made at the following localities: Shepard Formation, Reynolds Mtn.; Snowslip Formation, Piegan Mtn.; Siyeh Limestone, Mt. Grinnell and Piegan Mtn.; Grinnell Argillite, Mt. Henkel; Appekunny Argillite, Altyn Peak and Appekunny Mtn.; Altyn Limestone, Appekunny Falls

1976b). The most interesting of the stromatolites in the Belt Supergroup are those found in a 24–32-m-thick unit located about 200 m below the top of the Siyeh Limestone. This unit is densely crowded with branched columnar stromatolites generally referable to the group *Baicalia*, unbranched columnar stromatolites, conically laminated stromatolites referable to the group *Conophyton*, and mound-shaped stromatolites. It can be subdivided into six distinct units (Fig. 2): Unit A – the lower *Baicalia* Unit, consisting of divergently branched columnar stromatolites (*Baicalia*) at the base

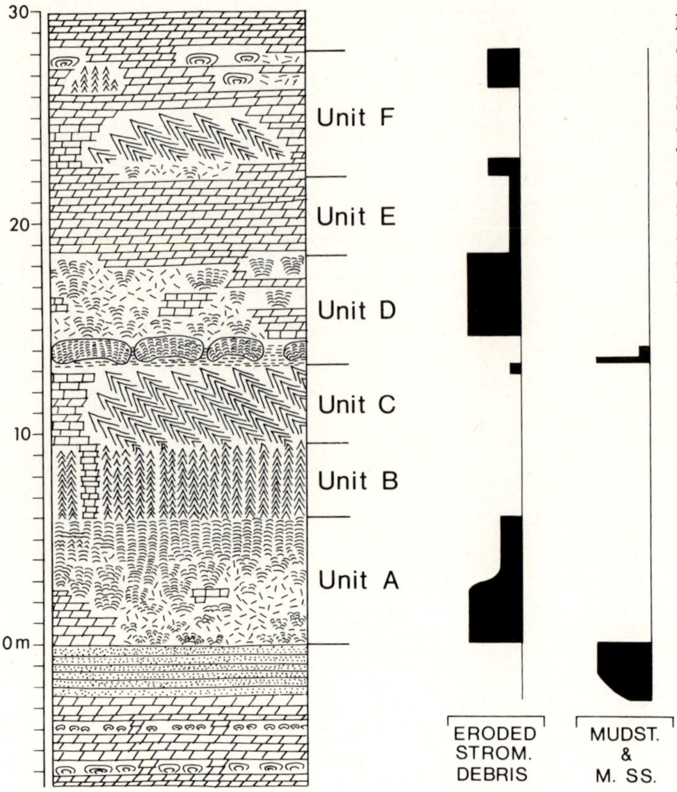

Fig. 2. Diagrammatic section of the *Baicalia-Conophyton* stromatolite cycles, South Swiftcurrent Glacier, northeastern Glacier National Park. The relative abundance and distribution of eroded stromatolite debris and mudstone and muddy sandstone is indicated by the width of the columns on the right

and unbrached or parallel branched columnar stromatolites at the top; Unit B – the Lower Small-Diameter *Conophyton* Unit, consisting of erect, 3–15 cm diameter *Conophyton*; Unit C – the Lower Large-Diameter *Conophyton* Unit, consisting of inclined, 10–60 cm diameter *Conophyton*; Unit D – the Middle *Baicalia* Unit, consisting of branched columnar stromatolites of the group *Baicalia*; Unit E – the Middle Sedimentary Unit, consisting of sandy and muddy dolomitic limestone; and Unit F – the Upper Mixed Stromatolite Unit, consisting of inclined, 10–50 cm diameter *Conophyton*, as well as branched columnar stromatolites of the group *Baicalia* and mound-shaped stromatolites. The *Conophyton* in Units B, C, and F typically form closely spaced bioherms ranging from 3 to 20 m across. Poorly developed *Conophyton*-like lamination is locally present in the upper 2 m of Unit A, and these laminae can be traced laterally into gently convex lamination.

The nature of the non-stromatolitic sediment varies with position in the *Baicalia-Conophyton* stromatolite cycles. Eroded stromatolite debris beds commonly are associated with the more divergently branched stromatolites of the lower part of Unit A and the middle and upper part of Unit D. They are less commonly associated with the unbranched and parallel branched columnar stromatolites in the upper part of Unit A, and they are essentially absent from the *Conophyton* horizons (Fig. 2). The ratio of thinly bedded micritic limestone to sandy dolomitic limestone also appears to be related to stromatolite type, with micritic limestone being relatively more abundant within *Conophyton*-bearing units, where it occurs between bioherms.

Units A through C appear to represent a single vertical stratification sequence, which terminates abruptly at the top of Unit C. Overlying units at the Swiftcurrent Glacier locality of Fig. 2 are less well organized; however, *Conophyton* appears at the top of Unit D in the southeastern part of the park, forming a similar sequence. These sequences appear to be generally shallowing upward.

Calcitic and dolomitic mound-shaped stromatolites are relatively common in the Snowslip Formation (Horodyski 1975) and less common in the Shepard Formation. These stromatolites were produced largely by in situ carbonate precipitation, and some may be largely nonbiogenic.

Actual Abundance of Stromatolite. Stromatolites have long been considered to be prominent features of several units of the Belt Supergroup. For example, in their description of the Piegan Group, Fenton and Fenton (1937, p. 1890) state, "The group is characterized by great development of calcareous [sic] algae." This section in Glacier National Park is one of the most stromatolite-rich sections of the Belt Supergroup, with the Siyeh Limestone, Snowslip Formation, and Altyn Limestone being the most stromatolite-rich units. However, when the actual volumetric abundance of stromatolites is determined, it is found that well-defined stromatolites constitute but a small proportion of these "stromatolite-rich" units (Fig. 3). Furthermore, horizontal stromatolitic lamination in uncommon in the Siyeh Limestone and is essentially absent from the Snowslip Formation. The paucity of stromatolites in these peritidal units may largely be due to high sedimentation rates of the strata that are preserved. In parts of the Siyeh Limestone, for example, much of the strate occur as stacked regressive sequences, with each sequence probably having been deposited quite rapidly.

	thickness of unit	number of stromatolite horizons	cumulative thickness of stromatolites	approximate percent stromatolites(a)
Snowslip Formation	385 m	21	3.1 m	0.2
upper part of Siyeh Limestone	180 m	10	10.2 m	3
<u>Baicalia-Conophyton</u> stromatolite cycles	28 m	not applicable	21 m (b)	75
lower part of Siyeh Limestone	572 m	30	9.4 m	0.6

Fig. 3. Table showing the number of stromatolite horizons and cumulative thickness of these stromatolites for measured sections of the Snowslip Formation and Siyeh Limestone in the central part of Glacier National Park; (a) estimation of the actual volume of the rock unit occupied by stromatolites takes into consideration the horizontal separation between individual stromatolites; (b) includes eroded stromatolite debris beds

Environmental Influence on Stromatolite Morphology. Studies of modern cyanophycean mats have shown that mat morphology is influenced by both the biologic makeup of the mat and the physical environment (Hoffman 1976a), and studies of ancient strata have shown that stromatolite morphology can be a useful paleoenvironmental indicator (Hoffman 1976b). To further refine the use of stromatolites as paleoenvironmental indicators, it is necessary to conduct additional studies on both recent cyanophycean mats and ancient stromatolites. The Belt Supergroup in Glacier National Park is an excellent unit for such a study because of the diversity of stromatolites, the variety of lithologies that contain stromatolites, the variety of depositional environments in which stromatolites developed, and the excellent exposure. In a previous study the branching pattern of columnar stromatolites in Glacier National Park was found to be influenced by the accumulation of detritus on growth surfaces, the spacing of columns, the growth of profile of bioherms, and the variability of environmental conditions (Horodyski 1977). Subsequent studies have further elucidated the effect of depositional energy on stromatolite morphology. In the *Baicalia-Conophyton* stromatolite cycles the abundance of eroded stromatolite debris beds (Fig. 2) and the erosive truncation of stromatolitic laminae may provide an indication of the relative energy under which these stromatolites originated. If this is so, then these features would indicate that the divergently branched columnar stromatolites at the base of Unit A originated under high energy conditions than the parallel branched and unbranched stromatolites at the top of Unit A, and that *Conophyton* originated under the lowest energy conditions. Study of the stromatolites in the other units indicate that a high rate of sedimentation of detrital material relative to the rate of microbial growth inhibits the development of columns and favors the development of mound-shaped structures. In addition, highly elongate stromatolites may not necessarily indicate high energy conditions, but only high directional currents.

Summary

Stromatolites are common in the Middle Proterozoic Belt Supergroup in Glacier National Park, Montana, occurring in 5 of the 6 formations that are exposed in the central part of the park. They include mound-shaped stromatolites, branched columnar stromatolites referable to the group *Baicalia*, and conically laminated stromatolites referable to the group *Conophyton*. Because these stromatolites have diverse morphologies, are associated with different lithologies, and occur in strata deposited in a variety of settings, they provide a good opportunity to evaluate the relationship between stromatolite morphology and paleoenvironmental conditions.

Acknowledgment. To the Donors of the Petroleum Research Fund, administered by the American Chemical Society, for support of this research.

References

Childers MO (1963) Structure and stratigraphy of the southwest Marias Pass area, Flathead County, Montana. Geol Soc Am Bull 74:141–164

Fenton CL, Fenton MA (1937) Belt series of the north: stratigraphy, sedimentation, paleontology. Geol Soc Am Bull 48:1873–1969

Harrison JE (1972) Precambrian Belt Basin of northwest United States: its geometry, sedimentation, and copper occurrences. Geol Soc Am Bull 83:1215–1240

Hoffman P (1976a) Stromatolite morphology in Shark Bay, Western Australia. In: Walter MR (ed) Stromatolites. Elsevier, Amsterdam, pp 261–271

Hoffman P (1976b) Environmental diversity of Middle Precambrian stromatolites. In: Walter MR (ed) Stromatolites. Elsevier, Amsterdam, pp 599–611

Horodyski RJ (1975) Stromatolites of the lower Missoula Group (Middle Proterozoic), Belt Supergroup, Glacier National Park, Montana. Precambrian Res 2:215–254

Horodyski RJ (1976a) Stromatolites from the Middle Proterozoic Altyn Limestone, Glacier National Park, Montana. In: Walter MR (ed) Stromatolites. Elsevier, Amsterdam, pp 585–597

Horodyski RJ (1976b) Stromatolites of the upper Siyeh Limestone (Middle Proterozoic), Belt Supergroup, Glacier National Park, Montana. Precambrian Res 3:517–536

Horodyski RJ (1977) Environmental influences on columnar stromatolite branching patterns: examples from the Middle Proterozoic Belt Supergroup, Glacier National Park, Montana. J Paleontol 51:661–671

Horodyski RJ (1983) Sedimentary geology and stromatolites of the Middle Proterozoic Belt Supergroup, Glacier National Park, Montana. Precambrian Res 20:391–425

Rezak R (1957) Stromatolites of the Belt Series in Glacier National Park and vicinity, Montana. US Geol Surv Prof Pap 294D:127–154

Ross CP (1959) Geology of Glacier National Park and the Flathead region, northwestern Montana. US Geol Surv Prof Pap 292:125

Ross CP (1963) The Belt Series in Montana. US Geol Surv Prof Pap 346:122

Walcott CD (1914) Cambrian geology and paleontology, III. No 2 – Precambrian Algonkian algal flora. Smithson Misc Collect 64:77–156

5 A Stromatolite Built by a Phormidium-Like Alga from the Lower Carboniferous of South Wales

V.P. WRIGHT[1] and J.M. WRIGHT[2]

Introduction

Much is now known about the compositions of modern stromatolite-building communities (see review by Golubic 1976). Such stromatolites, although from widely separated areas, are built by a limited number of micro-organisms. These often possess special adaptations which allow them to survive under the rather special conditions in which they live. Many forms inhabiting intertidal settings for example produce gel envelopes which enables them to resist desiccation, while other forms exhibit motility which helps prevent burial by sediment (Golubic and Campbell 1979).

Many ancient stromatolite-building micro-organisms, identified from silicified material, are very similar to extant forms. This led Golubic and Campbell (1979) to speculate on whether such similarities were the result of conservatism or convergence. It seems likely that ancient stromatolites were built by micro-organisms that possessed specialized morphological and behavioural properties similar to Recent stromatolite-building forms, and that such stromatolites were also probably built by a limited range of micro-organisms. Since ancient stromatolites with silicified micro-biotas are rare, evidence for these conclusions must be sought by careful morphogenetic analyses of unsilicified forms. For example, Wright and Mayall (1981) and Mayall and Wright (1981) were able to show that some Upper Triassic stromatolites from south-west England, even though lacking any identifiable organic remains, were built by micro-organisms showing strong motility and clumping which resulted in fabrics and features identical to those seen in living *Lyngbya* mats. This type of approach is followed here, using stromatolites from the Lower Carboniferous (Mississippian) of South Wales. These stromatolites possess fabrics which suggest that they were built by microorganisms exhibiting strong phototropism with a marked diurnal growth pattern, strikingly like the extant examples of *Phormidium hendersoni*.

Geological Setting

The stromatolites occur in the Penllwyn Oolite Member, part of the Llanelly Formation of probable Arundian age (Lower Carboniferous) in South Wales. The Llanelly

1 Department of Geology, University of Bristol, University Walk, Bristol BS8 1TR, Great Britain
2 9, Cae Mawr Gardens, Porth, Rhondda, Mid Glamorgan, Great Britain

Paleoalgology: Contemporary Research and Applications
(eds. D.F. Toomey and M.H. Nitecki)

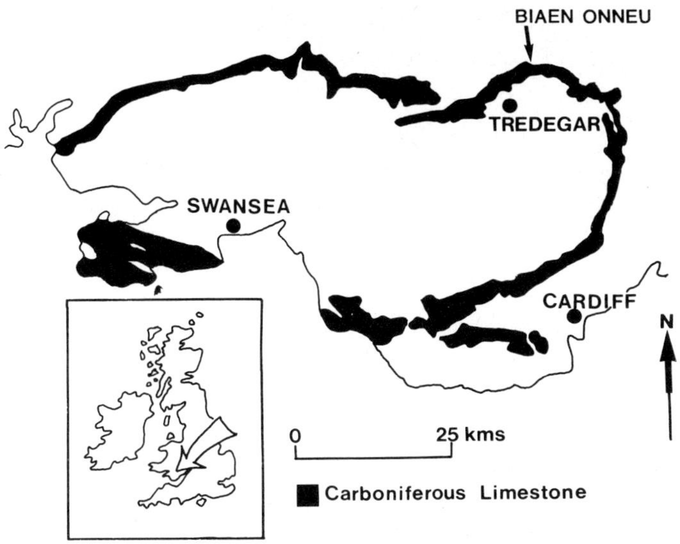

Fig. 1. Location of Blaen Onneu in South Wales, U.K.

Formation is a sequence of peritidal limestones, alluvial sandstones, and clays with calcretes. Various aspects have been described by Wright (1981a,b, 1982). One of the limestone units, the Penllwyn Oolite Member, consists of two sub-units with a lower thin (under 30 cm) bioclastic grainstone containing large pisoids (Wright 1981a), overlain by an oolitic sub-unit up to 2.5 m thick. Over most of the outcrop this oolitic sub-unit consists of bioturbated peloidal and oolitic grainstones to wackestones, with grapestone, botryoidal and flake aggregates. It is interpreted (Wright 1981b) as a lagoonal deposit comparable to many backshoal oolites from the Bahamas Banks (Ball 1967, Bathurst 1975, Hine 1977, Harris 1979) or the Persian Gulf (Loreau and Purser 1973). The occurrence of these various grain aggregates suggests that the sediment was at times probably bound by algal mats (see Winland and Matthews 1974). More convincing evidence of algal mats is to be seen as thin crypt-algal laminites that occur in the middle of the sub-unit (Wright 1981b). These horizons contain gypsum pseudomorphs or may be replaced by an irregular solution surface. However, at Blaen Onneu (Fig. 1; Wright 1981b) the oolitic sub-unit consists of 2.5 m of quartz-rich, cross-bedded oolitic grainstone with a high percentage of superficial ooids. The cross-bedding is planar and bipolar, and low angle cross-lamination and reactivation surfaces also occur. The top 30 cm of the unit contains ripples coated by stromatolitic laminae (Fig. 2). This sub-unit probably represents a small oolite bar, influenced by tidal currents, waves, and swash processes.

Morphology of the Stromatolites

The stromatolites at Blean Onneu are mainly LLH structures (in sense of Logan et al. 1964), which coated ripples (Fig. 3); but rarer, small isolated SH-type columnar growths also occur (Fig. 4). The former grew over symmetrical or slightly asymmetri-

Fig. 2. Outcrop photograph of interference ripple marks that are covered by thin stromatolitic laminae

Fig. 3. Polished slab showing stromatolitic laminae coating ripple marks

Fig. 4. Thin section photomicrograph of small columnar growth; *arrows* on the *stromatolite* indicate orientation of micrite threads, and *arrows* in *spar-filled fenestra* indicate attached vermiform gastropods. Compare the fenestrae with that figured by Monty (1976, Fig. 19). Field of view is 18 mm wide

cal interference wave ripples (Fig. 2), which have average heights of 10 mm and wave lengths of 70 mm. The stromatolitic coatings reach thicknesses of 20–25 mm and, judging from bedding surfaces, entended over many square metres. Some of these LLH structures are domed over sparry calcite filled fenestrae. The LLH forms are separated vertically by oolitic grainstone. The SH stromatolites form individual columns up to 20 mm high and up to 10 mm in diameter with a moderate to highly convex form in the sense of Hofmann (1969).

The laminations are prominent in both types and are smooth in form, from flat to convex over the ripples and concave between. The lamination in the algal layers is brought out because of the alternation of light and dark laminae. The former have a more open fabric with micritic threads (see below), while the latter have a denser sediment-rich fabric.

Microstructure

The stromatolitic laminae contain very thin micritic threads, 7–29 μm wide (Fig. 5). The preservation of these threads is variable, they can occur as well-defined straight or sinuous threads (Fig. 5a–c), several hundred microns long, or as less distinct lines of minute micrite aggregates (Fig. 5c,d), which are degraded threads.

These two types of thread occur in three growth patterns (Fig. 6).

1. Erect, vertically oriented and parallel threads, reaching lengths of several hundred microns (Fig. 5a,b). These commonly occur forming the central part of SH struc-

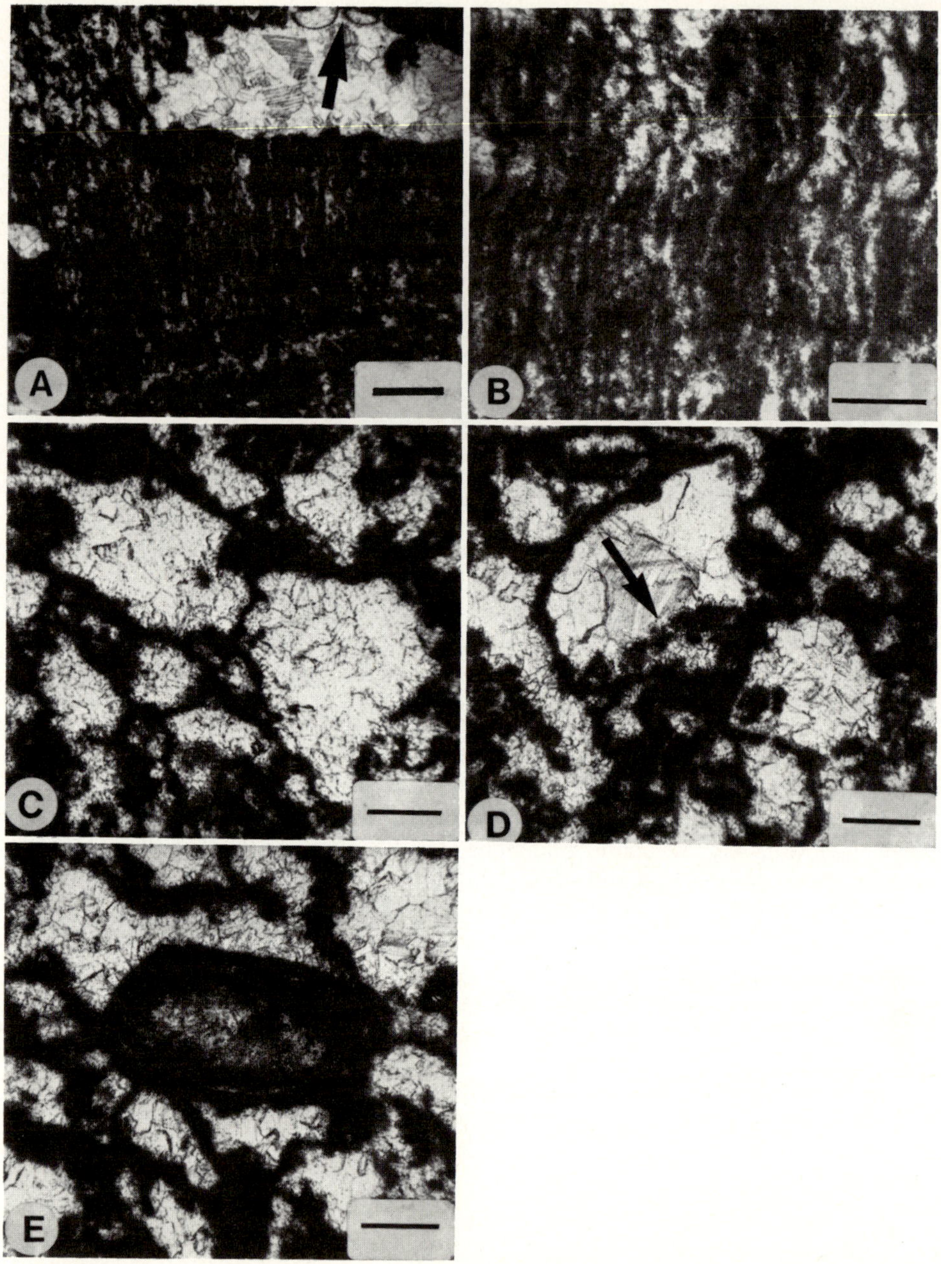

Fig. 5A–E. Thin section photomicrograph of micrite thread patterns. **A** Parallel, erect threads; note dome shaped fenestra at top with attached vermiform gastropod (*arrowed*) = 500 μm. **B** Parallel erect threads; **C** reticulate pattern of micrite threads; **D** micrite threads in a reticulate pattern with some of the threads having degraded to micrite blebs (*arrowed*); **E** superficial ooids trapped by micrite threads. Scale bars in **B–E** = 200 μm

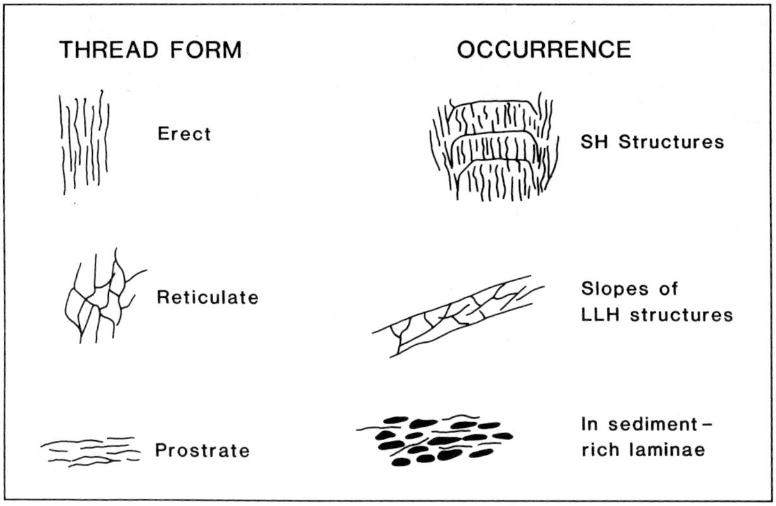

Fig. 6. Schematic diagram of micrite thread forms and their occurrence

tures. In the margins of such columns the threads are curved. They rarely contain sediment grains such as ooids or peloids.

2. Anastomosing threads, either individual threads or bundles of threads, which form a reticulate pattern (Fig. 5c,d). Sediment grains, such as superficial ooids (Fig. 5e) occur "trapped" within them, but are not common.

   The density and orientation of these threads vary from one part of the laminae to another; for example, on the sides of the LLH structures the threads display a crude imbricated pattern (Fig. 6) with the threads oriented up the slope of the structure (Fig. 7a,b). At the tops of these ripple-coated LLH structures the threads are more vertically oriented and form small intertwined tufts.

   This type of thread pattern occurs in laminae 230–800 μm in thickness. The open fabric of the thread pattern gives rise to the lighter laminae seen in hand specimen.

3. The third type comprises the dark laminae seen in hand specimen, which are 116–800 μm in thickness (average 320 μm). They consist of threads lying parallel or sub-parallel to the lamination (Fig. 6). These laminae contain large numbers of sediment grains that are intertwined by the threads. These grains, especially the more elongate pellets, commonly display a preferred orientation (Fig. 8) with their long axes parallel to the lamination.

The lighter and darker laminae – those with an open reticulate thread fabric type A) and those with a prostrate, sediment grain-rich fabric (type B) – alternate in a regular way (Fig. 9). The ratio of thicknesses between these two is on average 2:1 or 3:1 (A:B) (Fig. 9). Where type B laminae lack sediment grains this ratio can be as high as 7:1, but where type B laminae incorporate a lot of sediment grains this ratio may be less than 1:1 (Fig. 10). A comparison of grain sizes between

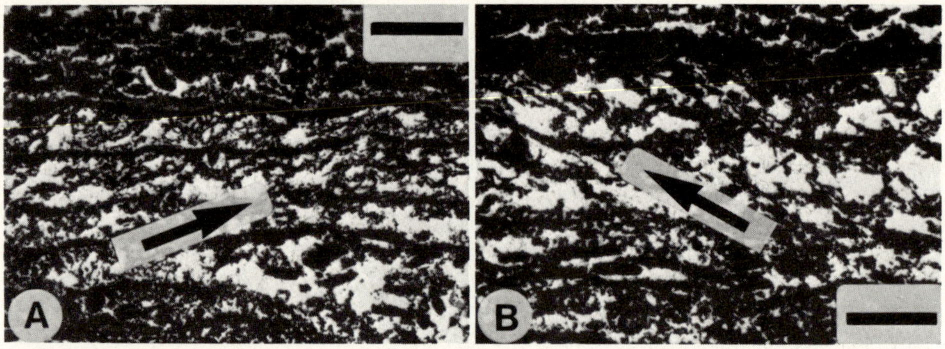

Fig. 7A,B. Thin section photomicrograph of laminae from the sides of low amplitude stromatolites domes coating ripples. The two types of laminae can be seen with the darker laminae of prostrate threads separating lighter, thicker laminae with a reticulate thread pattern (see Fig. 6). *Arrows* indicate the preferred direction of thread orientation. Scale bar in **A** = 1.5 mm; **B** = 1 mm; compare **A** with Monty (1976, Fig. 2B)

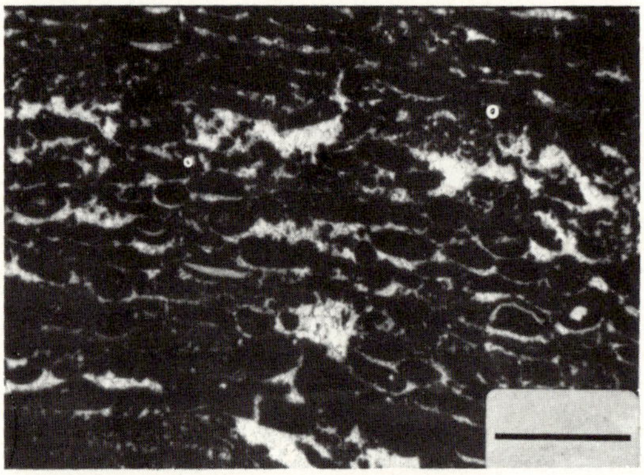

◁ **Fig. 8.** Thin section photomicrograph of elongate peloids (fecal pellets?) with a preferred orientation (compare with Monty 1976, Fig. 15). Scale bar = 2 mm

Fig. 9. Thin section photomicrograph of lamination in a stromatolite dome showing the alternation of the two types of laminae. The wider, lighter bands contain threads with a reticulate pattern that often show a preferred orientation (*arrowed*). The thin dark bands have prostrate threads. Several sediment rich laminae occur (*S*). Field of view is 17 mm wide
▽

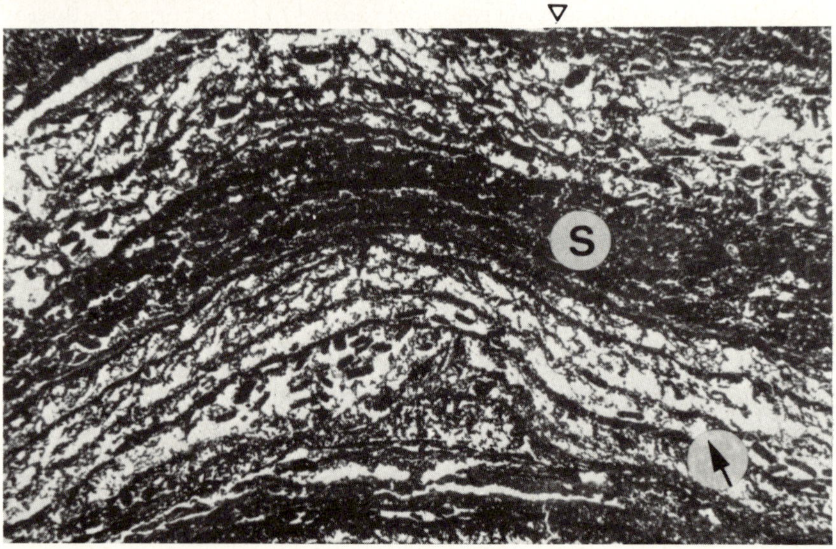

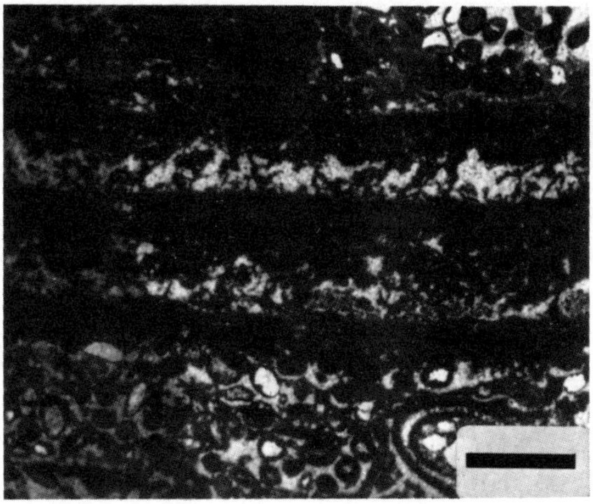

Fig. 10. Thin section photomicrograph of thick, sediment-rich laminae separated by lighter laminae with a more open thread pattern. Scale bar = 1 mm

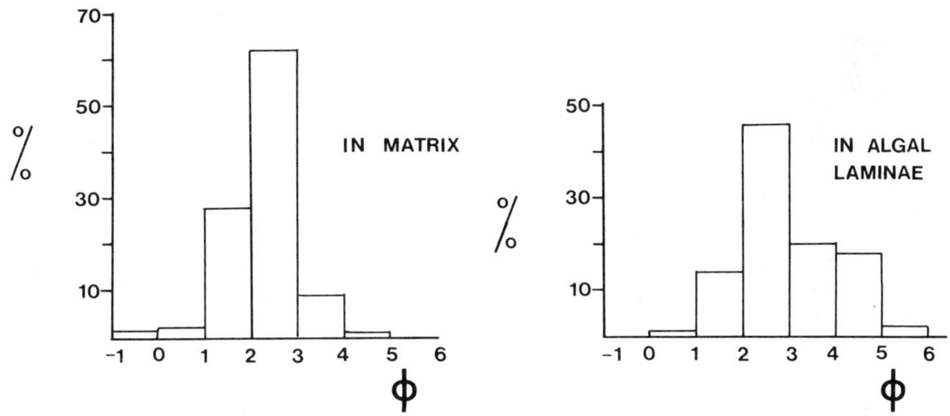

Fig. 11. Histograms showing the grain size distribution between sediment grains in the matrix, between the stromatolites, and that within the stromatolitic laminae. Percentage frequency is plotted against Phi scale. Even though the measurements are from thin sections, they clearly show a shift to the finer grain sizes in the stromatolitic laminae

the grains within the stromatolitic laminae and those outside (comprising the ripples and inter-ripple fill) is shown in Fig. 11. The grain sizes, though measured from thin sections, clearly show a shift to the finer grain sizes in the stromatolitic laminae.

Fenestrae

Fenestrae, which are filled by drusy spar, are common and consist of three types:

1. Fine, laminoid fenestrae, 200–300 μm high and up to 6 mm long (Fig. 12a), which extend parallel to the lamination and can be traced laterally into thread-bearing laminae.

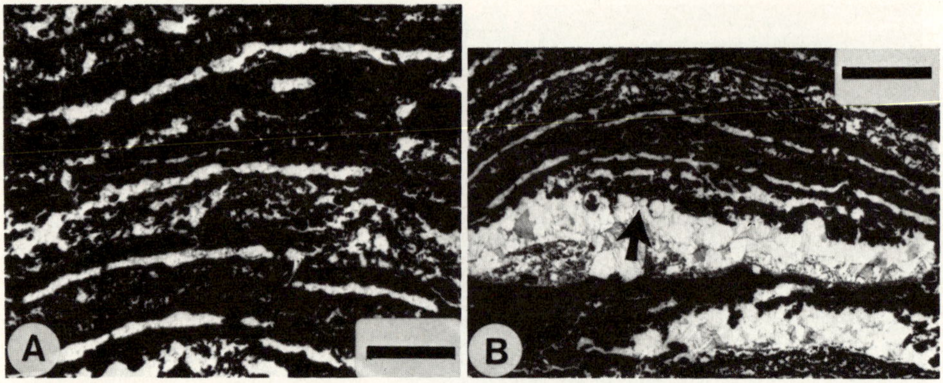

Fig. 12A,B. Thin section photomicrograph of fenestrae. **A** Laminoid fenestrae; scale bar = 1 mm. **B** Dome-shaped fenestra overlain by laminoid fenestrae. Note attached vermiform gastropod (*arrowed*); scale bar = 3 mm

2. Larger dome shaped fenestrae, 2–3 mm high and up to 15 mm long (Figs. 5a and 12b), with arched roofs and flat floors. They appear to have caused the arching of the laminae above and may be both encrusted by vermiform gastropods (Fig. 5a) and have partial geopetal fills.
3. Large, irregular fenestrae (Fig. 4) that may reach 30 mm in diameter. They have highly irregular scalloped walls and may contain "floating" fragments of thread-bearing laminae within spar. The spar fill is usually of a void-filling drusy spar, but more irregular crystal mosaics lacking typical neomorphic features such as curved crystal boundaries also occur. The irregular edges of these larger fenestrae are encrusted by vermiform gastropods (Fig. 4), and some fenestrae contain partial geopetal fills. There is no evidence of evaporite dissolution.

Interpretation

The micrite threads in these stromatolites are not typical of ancient stromatolites. In many such stromatolites any relic algal structures occur as micrite-walled tubes, such as is seen in the form genera *Ortonella* and *Girvanella.* Such micrite-walled tubes are also common in Recent calcified algal mats (Monty 1976) and are formed by the precipitation of micritic carbonate in or on the mucuous sheaths around algal trichomes (see Riding 1977) and Fig. 13. If these micrite threads are algal in origin, then they have not been formed by such a mechanism. The precipitation of micritic carbonate can also occur within the algal filaments in one of two ways (Fig. 13). First, carbonate could have been precipitated within the vacated sheaths after the trichomes have moved or decayed, and second, but rarely in the Recent, the actual trichome itself becomes calcified (Monty 1980, personal communication). The thickness and lengths of the micrite threads described here argues against their being actual calcified trichomes; they probably represent the sites of empty sheaths. This cementation must have

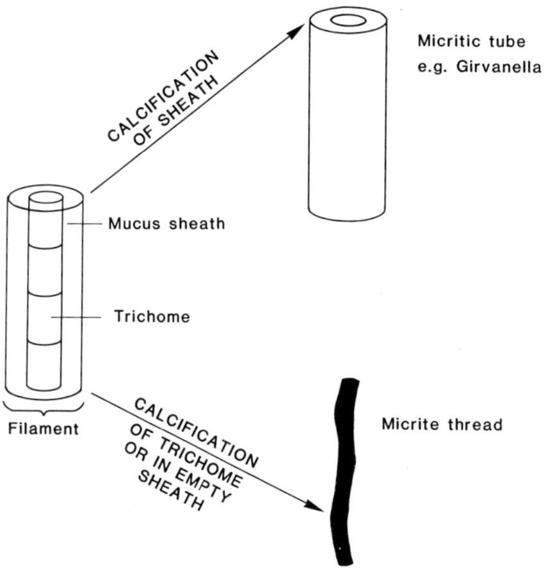

Fig. 13. Schematic diagram of calcification types and products in cyanobacteria

been rapid, before the empty sheaths collapsed. Adolphe and Rolfes (1973) have also described similar micritic thread structures from fossil stromatolites.

The thread patterns seen in this Lower Carboniferous stromatolite are similar to those seen in stromatolites built by the extant cyanobacterium *Phormidium.* The most common thread pattern seen is that of anastomosing threads, often forming a reticulate pattern in which the individual threads or thread bundles diverge and fuse together but have a dominant vertical or subvertical orientation (Monty 1976, p. 196, 1979, p. 169, Golubic and Focke 1978, and Walter et al. 1976). The layers of anastomosing filaments are called hyaline layers (Monty 1979) and typically contain dispersed sediment grains trapped during the growth of the mats, as indeed do these Carboniferous stromatolites (Fig. 5e).

In *Phormidium* mats the hyaline layers form during the day when the trichomes are active. The orientation of these trichomes relative to the growth laminae varies and may be perpendicular at the tops of domes and inclined upwards (imbricated) on the sides (see Golubic 1973, pp. 451–452). These orientations reflect the phototropic response of this cyanobacterium. Similar phototropic orientations occur in this Carboniferous stromatolite with vertically oriented threads in the SH and on the top of LLH structures, and with the upward imbricated threads on the sides of the LLH domes (Fig. 6).

The type B laminae in the Carboniferous stromatolites, with the prostrate threads and a high concentration of detrital particles, also have an analog in Recent *Phormidium* mats. The prominent lamination seen in modern *Phormidium* mats is caused by thin, dark layers consisting of dense prostrate trichomes incorporating many sediment grains. These layers are known as corpuscular layers (Monty 1976, pp. 196–197,

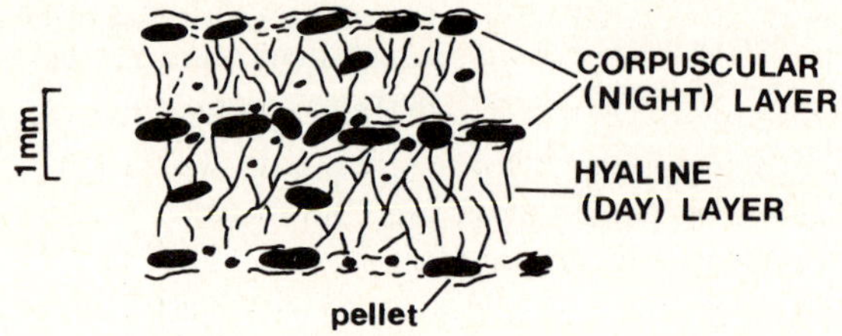

Fig. 14. Interpretation of the stromatolitic laminae, showing the different filament orientations and sediment distributions between day and night

1979, Golubic and Focke 1978), and are formed at night when the trichomes, no longer phototropically oriented, become prostrate and glide over the surface of the stromatolite. The higher amounts of particulate matter seen in these layers is believed to result from the fact that sedimentation is probably constant throughout the day and night in the subtidal settings where most *Phormidium* mats occur, but the sediment particles become less dispersed in the thinner, night layers, which are often only a tenth of the thickness of the hyaline day layer (Golubic and Focke 1978, p. 756).

The couplets of subvertically and horizontally oriented threads in the stromatolites described here are therefore interpreted as representing diurnal laminations, with each couplet representing a daily growth cycle (Fig. 14). Recent stromatolites built by *Phormidium* are well known for showing such a diurnal or noctihemeral lamination (see earlier references). The occurrence of only one type of thread structure in the Penllwyn Oolite stromatolites suggests that either only one type of alga was originally present or that only one type calcified. It is interesting to note that many stromatolites built by *Phormidium* are monospecific (Golubic and Focke 1978, Monty 1979).

Fenestrae

The fine laminoid fenestrae which occur, and are laterally contiguous with, the micrite thread-bearing laminae were most likely the sites where algal filaments were not calcified or where early decay destroyed the organic matter before calcification.

The second type of fenestrae, which appear to have caused the doming of the laminations above, are most likely to have formed from gas bubble blisters. These form by gas generation during photosynthesis or bacterial decomposition and are common in Recent algal mats (Davies 1970, Schartz et al. 1975, Monty 1976). The absence of any micrite threads within the fenestrae probably precludes their being formed by internal growth processes (see Monty 1976, p. 218).

The third type of fenestrae, the large irregular forms, are similar to the large "oxidation" fenestrae described in stromatolites by Monty (1976, p. 223) and are here interpreted as areas where decay occurred before micrite precipitation, as in the first type of fenestrae described, but on a larger scale. The scalloped edges of these larger fenestrae suggests that some later dissolution may have occurred, perhaps associated with subaerial exposure, for these stromatolites are overlain by flood-plain deposits (Wright 1981a).

The absence of evaporite pseudomorphs at Blaen Onneu and the occurrence of vermiform gastropods in predominantly subtidal settings elsewhere in the Llanelly Formation (Wright and Wright 1981) suggest that these stromatolites grew in a shallow subtidal setting or one with a relatively low degree of exposure. Where similar stromatolites occur at other localities they are poorly developed and contain evaporite pseudomorphs. Living *Phormidium* stromatolites occur in a wide variety of settings from turbulent reef flats to protected intertidal areas (Golubic and Focke 1978, Monty 1979).

Discussion

It is tempting to try to estimate the longevity of the individual stromatolites using the diurnal couplets, e.g., did they exist for 3 days, 4 days, or a weak or more? Some LLH stromatolites contain three or four couplets, which suggests growth for 3 or 4 days before burial by the oolitic sediments; it is difficult to be more precise. There have been many attempts to recognize periodicity in ancient stromatolites, whether daily (Pannella et al. 1968), or annual (Hudson 1970), but their reliability is suspect. Studies of modern algal mats has shown that lamination does not always occur in daily increments (see Hofmann 1973, p. 356–357), and in some stromatolites several laminae can form per day e.g., in *Phormidium* mats (Monty 1979); and similarly, the reverse case can occur where mats can stop growing for weeks at a time (Golubic 1973). Therefore, even though the biolamination (in the sense of Golubic and Focke 1978) seen in the Carboniferous stromatolites is probably diurnal, it need not necessarily represent growth on consecutive days. Subtidal algal mats are more reliable for recording diurnal lamination then intertidal mats (Park 1976, p. 380), but there is some uncertainty over the exact setting of these Carboniferous stromatolites in this respect.

The variability in the thicknesses of the individual laminae described in this ancient stromatolite is also common to modern *Phormidium* mats (Golubic and Focke 1978, p. 754). We resist the temptation to speculate that the lengths of the days and nights in the Lower Carboniferous varied from day to day. These differences probably resulted from variations in growth rate, sediment supply, and tidal regime as happens in modern mats (Monty 1979, p. 166).

The recognition of periodicity in ancient stromatolites becomes more difficult when one takes into account the influence of diagenetic overprinting such as compaction, oxidation, and recrystallization (Park 1976, 1977). The excellent preservation of the microstructure in the Carboniferous stromatolites must have resulted from early lithification, presumably the calcification of the empty filament sheaths, as discussed earlier. Evidence of this early calcification is seen in the attachment of vermiform gastropods on the laminae, which elsewhere in the Llanelly Formation encrust a variety of hard substances such as lithoclasts, orthocones, and *Ortonella* bushes. The presumably delicate micrite threads are now encased in sparry calcite, which must have formed as a void filling cement in the spaces between the filaments. In Recent *Phormidium* mats, the spaces between the trichomes are empty of mucilage and are filled with sea water (Golubic and Focke 1978).

The micro-organisms which built these stromatolites influenced the sedimentary fabric of the rock in several ways. The differences in grain-size distribution between the algal laminae and the inter-laminae sediment (Fig. 11) is evidence of some grain sorting. It is interesting to note that *Phormidium* mats also exhibit similar grain-size sorting, e.g., Gebelein (1969, p. 57) and Frost (1974, p. 535) describe the concentration of the finer grain sizes in modern stromatolites built by *Schizothrix* as compared to the surrounding sediment. In both theses cases, the alga involved has since been re-identified as *Phormidium hendersonii* Howe (Golubic and Focke 1978).

Secondly, the stromatolitic mats that acted as sediment stabilizers for the preservation potential of the ripples would have been very low, and they would have been removed by the next phase of bed movement if not coated and "fossilized" by algal mats. The sediment stabilizing ability of algal mats is well known (Bathurst 1975, p. 122, Gebelein 1969, Scoffin 1970) and indeed Neumann et al. (1970), in their study of various types of subtidal algal mats, found that some types of mat were able to withstand current velocities up to five times that required to erode unbound sediment.

Thirdly the algal mats influenced sediment packing, for in the corpuscular laminae (type B) the pellets have their long axes aligned parallel to the lamination. This is known as laminoid boundstone fabric and is caused by the surface properties of algal mats (see Mont 1976, p. 217).

We have noted only one record of a similar ancient stromatolite from the Viséan of the Namur Basin of Belgium, described by Hance and Hennebert (1980). The authors describe stromatolites growing over small ripples, and their plate 1 shows the very clear alignment of pellets in the laminae as well as a lamination which crudely resembles that in the Penllwyn Oolite.

Summary

The stromatolites described from the Penllwyn Oolite Member were built by a monospecific community of microscopic, filamentous, phototropic, and probably motile organisms, almost certainly a form of blue-green alga (cyanophyte, cyanobacterium) related to the extant oscillatoriacean *Phormidium.* This conclusion is reached mainly on the similarities between the filament patterns and filament orientations of these Carboniferous stromatolites and of those in Recent *Phormidium* mats.

The former presence of the micro-organisms is indicated by thin micrite threads which represent precipitation of carbonate within the empty cyanobacterial mucus sheaths. This precipitation occurred at an early stage before the oxidation and compaction of the sheaths and provided a hard substrate for the attachment of encrusting gastropods.

These ancient stromatolite-building micro-organisms influenced the resulting carbonate rock as regards sediment fabric, and grain size distribution, and stabilized and preserved bed forms such as interference ripples.

Recent stromatolites are built by a limited number of often specialized micro-organisms that commonly leave distinctive microstructural traces that may be reflected in ancient stromatolites by differences in lamination form or sediment particle distribution. Such subtle features may enable specific mat types to be recognized in ancient stromatolites.

Acknowledgments. This work was carried out while VPW was a post-graduate student at University College, Cardiff. We would like to thank John Taylor and Helen Boxall for preparing the diagrams and Donna Evans and Beryl West for typing the manuscript. The senior author gratefully acknowledges the award of a Royal Society travel grant which enabled him to present this paper at the Third International Symposium on Fossil Algae.

References

Adolphe JP, Rolfes G (1973) Les concretionnements calcaires de la Levriere (Eure). Assoc Fr Etude Quet Bull 10: no 35, 79–87

Ball MM (1967) Carbonate sand bodies of Florida and the Bahamas. J Sediment Petrol 37:556–591

Bathurst RGC (1975) Carbonate sediments and their diagenesis, 2nd ed. New York, Elsevier, 658 p

Davies GR (1970) Algal laminated sediments, Gladstone embayment, Shark Bay, Western Australia. Am Assoc Petrol Geol Mem 13:169–205

Frost JG (1974) Subtidal algal stromatolites from the Florida backreef environment. J Sediment Petrol 44:532–537

Gebelein CD (1969) Distribution, morphology, and accretion rate of recent subtidal algal stromatolites, Bermuda. J Sediment Petrol 39:49–69

Golubic S (1973) The relationship between blue-green algae and carbonate deposits. In: Carr NG, Whitton BA (eds) The biology of blue green algae. Bot Monogr 9:434–447

Golubic S (1976) Organisms that build stromatolites. In: Walter MR (ed) Stromatolites. Amsterdam, Elsevier, pp 113–126

Golubic S, Campbell SE (1979) Analogous microbial forms in Recent subaerial habitats and in Precambrian cherts: *Gloeuthece coerulea* Geitler and Eosynechococcus moorei Hofman. Precambrian Res 8:201–217

Golubic S, Focke JW (1978) *Phormidium hendersoni* Howe: identity and significance of a modern stromatolite building micro-organism. J Sediment Petrol 48:751–764

Hance L, Hennebert M (1980) On some lower and middle Viséan carbonate deposits of the Namur Basin, Belgium. Meded Rijks Geol Dienst 32: no 9, 66–68

Harris PM (1979) Facies anatomy and diagenesis of a Bahamian ooid shoal. Sedimenta, vol IV. University of Miami, 163 p

Hine AC (1977) Lily Bank, Bahamas: history of an active oolite sand shoal. J Sediment Petrol 47:1554–1581

Hofmann HJ (1969) Attributes of stromatolites. Geol Surv Can Pap 69–39, 58 p

Hofmann HJ (1973) Stromatolites: characteristics and utility. Earth Sci Rev 9:339–373

Hudson JD (1970) Algal limestones with pseudomorphs after gypsum from the Middle Jurassic of Scotland. Lethaia 3:11–40

Logan BW, Rezak R, Ginsburg RN (1964) Classification and environmental significance of algal stromatolites. J Geol 72:68–83

Loreau JP, Purser BH (1973) Distribution and ultrastructure of holocene ooids in the Persian Gulf. In: Purser BH (ed) The Persian Gulf. Springer, Berlin Heidelberg New York, pp 279–328

Mayall M, Wright VP (1981) Algal tuft structures from the Upper Triassic of Southwest England. Palaeontology 24:655–660

Monty CLV (1976) The origin and development of cryptalgal fabrics. In: Walter MR (ed) Stromatolites. Amsterdam, Elsevier, pp 193–249

Monty CLV (1979) Monospecific stromatolites from the Great Barrier Reef tract and their palaeontological significance. Ann Soc Geol Belg 101:163–171

Neumann AC, Gebelein CD, Scoffin TP (1970) The composition, structure and erodability of subtidal mats, Abaco, Bahamas. J Sediment Petrol 40:274–297

Pannella G, MacClintock C, Thompson MN (1968) Paleontological evidence of variations in length of the synodic month since the Late Cambrian. Science 162:792–796

Park RK (1976) A note on the significance of lamination in stromatolites. Sedimentology 23: 379–393

Park RK (1977) The preservation potential of some stromatolites. Sedimentology 24:485–506

Riding R (1977) Calcified *Plectonema* (blue-green algae), a Recent example of *Girvanella* from Aldabra Atoll. Palaeontology 20:33–46

Schwartz HU, Einsele G, Herm D (1975) Quartz-sandy, grazing-contoured stromatolites from coastal embayments of Mauritania, West Africa. Sedimentology 22:539–560

Scoffin TP (1970) The trapping and binding of subtidal crabonate sediments by marine vegetation in Bimini Lagoon. J Sediment Petrol 40:249–273

Walter MR, Bauld J, Brock TD (1976) Microbiology and morphogenesis of columnar stromatolites (*Conophyton, Vacerrilla*) from hot springs in Yellowstone National Park. In: Walter MR (ed) Stromatolites. Amsterdam, Elsevier, pp 273–310

Winland HD, Matthews RK (1974) Origin and significance of grapestone, Bahama Islands. J Sediment Petrol 44:921–927

Wright VP (1981a) Algal aragonite – encrusted pisoids from a Lower Carboniferous schizohaline lagoon. J Sediment Petrol 51:479–489

Wright VP (1981b) The stratigraphy and Sedimentology of the Llanelly Formation between Penderyn and Blorenge, South Wales: Unpubl Ph D thesis, Univ Wales, 408 p

Wright VP (1982) The recognition and interpretation of paleokarsts: two examples from the Lower Carboniferous of South Wales. J Sediment Petrol 52:83–94

Wright VP, Mayall M (1981) Organism-sediment interactions in stromatolites: an example from the Upper Triassic of Southwest England. In: Monty CLV (ed) Phanerozoic stromatolites. Springer, Berlin Heidelberg New York, pp 74–84

Wright VP, Wright EVG (1981) The palaeoecology of some algal-gastropod bioherms in the Lower Carboniferous of South Wales. Neues Jahrb Geol Palaontol Monatsh 10:546–558

Systematics and Morphology

6 Morphological Groups and Series in Cambrian Calcareous Algae

R. RIDING[1] and L. VORONOVA[2]

Introduction

Calcareous algae first appear as a well-defined array of fossils near the Precambrian-Cambrian boundary, and immediately assume importance as both paleontological and sedimentological constituents of shallow marine carbonate biotas and deposits. They constitute a wide range of morphological types with very different sizes, from microscopic tubes and spheres to cup-shaped skeletons tens of centimetres in size. Classification has been based on both their comparative morphology and also on evaluations of their affinity based on similarities to Recent algae. This latter aspect of research, while in itself a useful and necessary part of the study of these fossils, has also introduced a subjective element that has tended to destabilize classifications. Voronova (1976, Table 1) compares several earlier classificatory schemes which show the strong influence of interpretations of affinity. For example, Vologdin (1962) places *Renalcis, Epiphyton, Chabakovia, Razumovskaya*, and *Bija* in the Rhodophyta, whereas Luchinina (1975) places all the same genera in the Cyanophyta. At other times these genera have been split in various ways between both divisions.

Although progress is being made in establishing the affinities of these and other Early Cambrian algae by searching for Recent analogues, it remains true in many cases, even though a number of these fossils range into the Mesozoic and Cenozoic, that their systematic positions relative to modern algae are doubtful and cannot yet be used as a reliable guide in constructing a working taxonomic scheme. Consequently, the current situation is that workers in this field are faced by several available classifications, none of which is generally accepted as superior to any other, and which offer a confusing variety of groupings and interpretations.

One method of examining this problem, and therefore of assessing the relative merits and features of current classifications, is to analyse the morphological bases for groupings of these fossils. In order to do this, it is necessary to overlook information from Recent analogues concerning the affinity of particular genera, and to establish groupings purely in terms of the comparative morphology of the fossils themselves. An important requirement is the use of as many morphological characters as possible, rather than reliance upon a few. This last aim is constrained by the fact that many of these fossils, and certainly the commonest ones, are often rather simple ob-

1 Department of Geology, University College, Cardiff CF1 1XL, Great Britain

2 Palaeontological Institute, USSR Academy of Sciences, ul. Profsoyuznaya 113, Moscow 117321, USSR

Paleoalgology: Contemporary Research and Applications
(eds. D.F. Toomey and M.H. Nitecki)

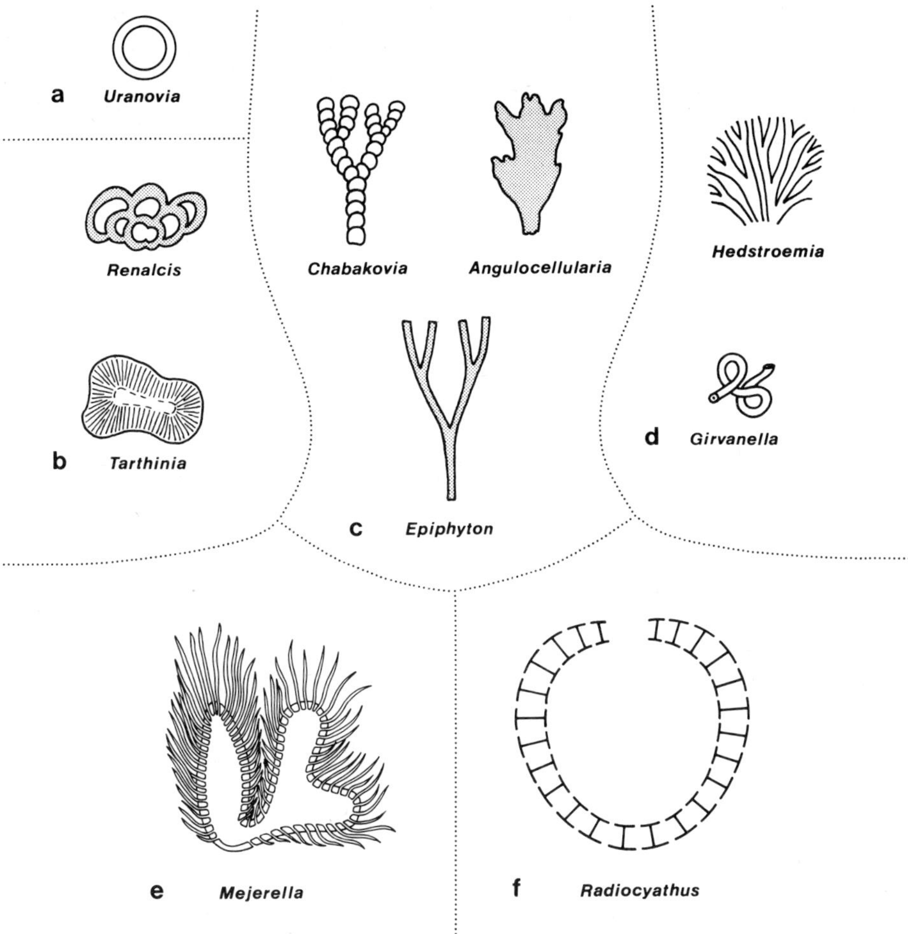

Fig. 2a–f. Six major morphological groups among Early Cambrian calcareous algae, exemplified by typical genera. **a** Spherical, **b** botryoidal, **c** dendritic, **d** tubiform, **e** tuberous, **f** cups

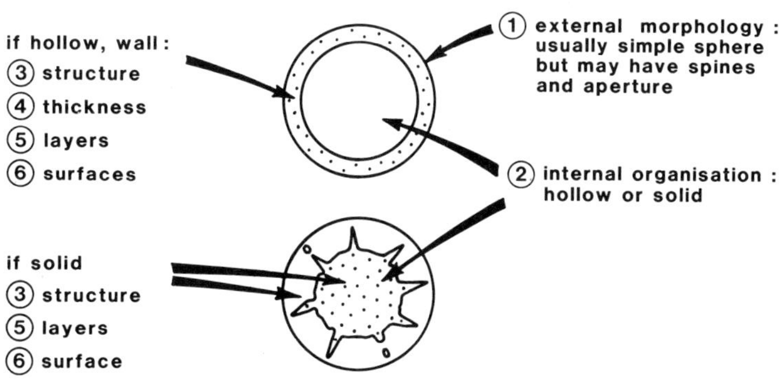

Fig. 3. Morphological features of spherical forms (calcispheres and *Foninia* groups)

1. Spherical

a) Hollow. Globular, hollow, relatively thin-walled calcispheres, 100–400 μm in diameter; wall apparently micritic, e.g., *Archaesphaera* Suleimanov, *Calcisphaera* Williamson, *Vicinesphaera* Antropov (Reitlinger 1959).

b) Solid. Globular bodies with micritic, peloidal, radial-fibrous, or stellate interiors, e.g., *Foninia* Korde, *Marenita* Korde, *Globuloella* Korde, *Marenita* Korde (Korde 1973); most of these appear to be ooids, or microfabrics of ooids and stromatolites.

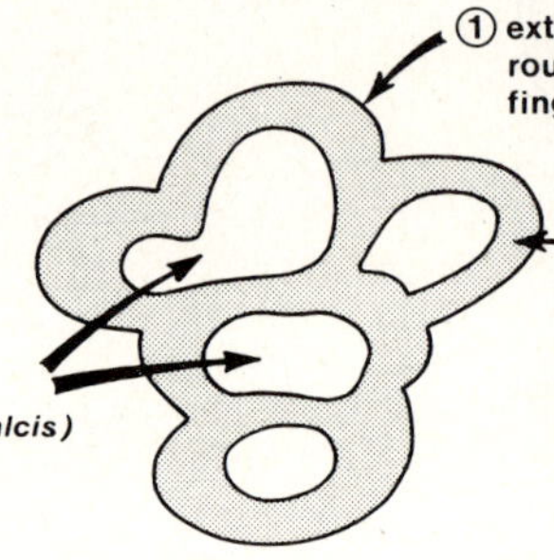

Fig. 4. Morphological features of the botryoidal forms (*Renalcis* and *Tarthinia* groups). Only general differences are indicated. For details see text. Generic names refer to generic groups, not individual genera

2. Botryoidal

a) Micritic Wall (Fig. 10a). Irregularly rounded bubble-like, hollow bodies; wall relatively thick and sometimes discontinuous, occasionally with a dark outer layer; the bubble-like bodies may form chains with poorly organized branches resulting in large internally subdivided masses, e.g., *Renalcis* Vologdin, *Nephelostroma* Dangeard & Doré, *Nubecularites* Maslov, (*Izhella* Antropov, best known from the Devonian).

Nubecularites is doubtfully included here, it may be a catagraph, i.e., an apparently hollow grain of possibly inorganic origin. *Izhella* is characterized by deep, narrow internal clefts in the wall.

The *Renalcis* group is distinguished from the *Chabakovia* group by relatively thick walls and poor development of branching.

b) Fibrous Wall (Fig. 10b). Simple, rounded forms with small protuberances on the outer surface and sometimes with finger-like projections. Wall thick, sometimes discontinuous, consisting of fibres normal to the surface. A dark outer border may be present; the outer surface is sharp, but the inner surface may be vague, e.g.., *Tarthinia* Drosdova which is distinguished from *Renalcis* by thicker, structurally different, walls and relatively simple internal organization.

c) Peloidal Wall (Fig. 10c). Irregular, rounded forms with thick micritic wall containing distinct peloidal or hollow spheres, e.g., *Gemma* Luchinina.

Table 1. General morphological features distinguishing important genera in the botryoidal and dendritic groups

<table>
<tr><th rowspan="2" colspan="2">External form</th><th rowspan="2">Internal organization</th><th rowspan="2">Genus or genus group</th><th colspan="4">Wall</th></tr>
<tr><th>Structure</th><th>Thickness</th><th>Layers</th><th>Surfaces</th></tr>
<tr><td rowspan="3" colspan="2">Botryoidal</td><td>Rounded compartments</td><td>Renalcis</td><td>Micritic</td><td>Thick</td><td>Possibly</td><td>Sharp</td></tr>
<tr><td rowspan="2">Hollow</td><td>Tarthinia</td><td>Fibrous</td><td rowspan="2">Very thick</td><td>Yes</td><td>Inner vague</td></tr>
<tr><td>Gemma</td><td>Peloidal</td><td rowspan="5">No</td><td rowspan="6">Sharp</td></tr>
<tr><td rowspan="5">Dendritic</td><td rowspan="2">Short irregular branches</td><td rowspan="2">Solid</td><td>Angulocellularia</td><td rowspan="5">Micritic</td><td rowspan="2">–</td></tr>
<tr><td>Epiphyton</td></tr>
<tr><td rowspan="3">Long delicate branches</td><td>Tube</td><td>Tubomorphophyton</td><td rowspan="2">Thick</td></tr>
<tr><td>Disc-shaped compartments</td><td>Gordonophyton</td></tr>
<tr><td>Rounded compartments</td><td>Chabakovia</td><td>Thin</td><td>Possibly</td></tr>
</table>

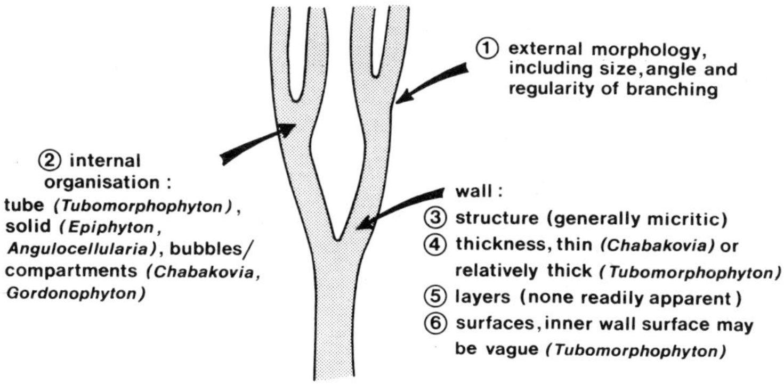

Fig. 5. Morphological features of dendritic forms (*Angulocellularia, Chabakovia*, and *Epiphyton* groups). For details of differences between groups and genera see text

3. Dendritic

a) Formed by Large Bubble-Like Bodies (Fig. 11). Branched dendritic series of bubble-like bodies. Each bubble has a relatively thin, distinct micritic wall which may be layered. The bubbles are round to lunate in form, and the branches may be long or short, resulting in either spindly or squat bushes. The angle of branching is relatively wide. For example, *Chabakovia* Vologdin, *Chomustachia* Korde, *Parachabakovia* Korde, *Taninia* Korde, (*Shugaria* Antropov, best known from the Devonian).

The Chabakovia group is distinguished from the *Epiphyton* group by the large bubble-like bodies forming the branches.

b) Delicate Filaments with Micritic, Tubiform, or Chambered Interiors (Fig. 10d–f). Branched narrow, straight or slightly bent, bushy dendritic filaments, either solid, tubular, or chambered. Branching – dichotomous to irregular, angles – acute to wide; e.g., *Epiphyton* Bornemann, *Cambrina* Korde, *Confervites* Bornemann, *Epiphytonoides* Korde, *Erbina* Korde, *Filaria* Korde, *Gordonophyton* Korde, *Kordephyton* Radugin & Stepanova, *Korilophyton.* Voronova, *Kyzassia* Korde, *Potentillina* Korde, *Sajania* Korde, *Sporinula* Korde, *Tomentula* Korde, *Tubomorphophyton* Korde, *Umbellula* Korde, *Vologdinia* Korde, (*Paraepiphyton* Wray, and *Tharama* Wray, best known from the Devonian).

The *Epiphyton* group is distinguished from the *Angulocellularia* group by the delicate, many-branched filaments, but it is very variable in style of branching and internal organization, expressed in part by the number of genera it contains.

c) Irregular Micritic Bushes (Fig. 10g). Bush-like form, very irregular and variable in shape and size, composed of micrite. Usually elongate, with angular to rounded outline and short branch-like projections that change width abruptly; e.g., *Angulocellularia* Vologdin, (*Frutexites* Maslov).

The type material of *Frutexites*, from the Ordovician of Siberia, occurs in haematitic limestone (Maslov 1960). Silicified organic matter also referred to this genus has been described from the Precambrian Gunflint Formation (Walter and Awramik 1979).

The *Angulocellularia* group is distinguished from the *Tarthinia* group by its solid micritic structure without fibres, spheres, or internal spaces. However, transition exists between *Angulocellularia* and *Tarthinia*, and this, together with the only weakly dendritic form of *Angulocellularia*, suggests that the latter may be better placed in the botryoidal group than in the dendritic group.

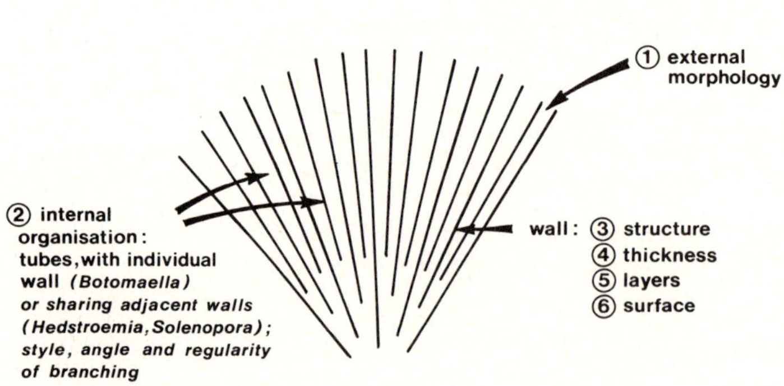

Fig. 6. Morphological features of the tubiform fan subgroup, *Botomaella*, *Hedstroemia*, and *Solenopora* groups). Subdivisions are based mainly on details of internal organization (see Table 2)

4. Tubiform

a) Fans

Delicate Tubes (Fig. 12a,b). Fine, straight to slightly curved, tubular filaments forming fan-shaped masses, branching irregular to dichothomous at acute angles; branches separate, walls thin. Tubes circular in cross-section; diameter relatively constant or increases gradually distally, circular in cross-section; e.g., *Botomaella* Korde, *Kenella* Korde (*Ortonella* Garwood, best known from the Carboniferous). *Kenella*, placed by Korde (1973, p. 216) with *Girvanella*-like genera, is here included in the *Botomaella* group because Korde describes it as branched and because of the apparently radiating organization of the type material (Korde 1973, pl. 45, Fig. 4). The type-species *Fistulella* Korde (*F. decipiens* Korde 1973, p. 217) belongs with *Botomaella*. It is not congeneric with the other species (*F. sanashtykgolica* Korde 1973, p. 218), which therefore becomes the type-species for *Fistulella* Korde and belongs with the *Hedstroemia* group (see below).

The *Botomaella* group is distinguished from tubular forms of the *Epiphyton* group (*Tubomorphophyton*, see Fig. 10e) by thinner walls and usually less frequent, more closely-spaced, branching.

Table 2. General morphological features distinguishing important genera in the tube-group

<table>
<tr><th rowspan="2"></th><th rowspan="2">External form</th><th rowspan="2">and</th><th rowspan="2">Internal organization</th><th rowspan="2">Genus or genus group</th><th colspan="4">Wall</th></tr>
<tr><th>Structure</th><th>Thickness</th><th>Layers</th><th>Surfaces</th></tr>
<tr><td rowspan="7">Tubes</td><td rowspan="3">Branched radiating fans</td><td>Tubes fine separate</td><td>Individual walls</td><td>Botomaella</td><td rowspan="7">Micrite</td><td rowspan="2">Variable</td><td rowspan="6">None</td><td rowspan="6">Sharp</td></tr>
<tr><td rowspan="2">Tubes coarse appressed</td><td>Individual and shared walls multiple branching</td><td>Hedstroemia</td></tr>
<tr><td>Shared walls</td><td>Solenopora</td><td rowspan="5">Thin</td></tr>
<tr><td rowspan="4">Generally unbranched</td><td rowspan="3">Tubes fine</td><td>Bundles, tubes ± parallel</td><td>Batinevia</td></tr>
<tr><td>Irregular masses</td><td>Girvanella</td></tr>
<tr><td>Solitary: Coiled</td><td>Obruchevella</td></tr>
<tr><td>Tubes coarse</td><td>Solitary: Straight</td><td>Proaulopora</td><td>Yes</td><td>+ Flanges on outer surface</td></tr>
</table>

Coarse Tubes (Fig. 12c). Generally coarse, irregularly, dichothomously, or many branched diverging tubes that may be separate but close together, or mutually attached and sharing adjacent walls. Tubes rapidly change thickness; circular to irregular in cross-section; sometimes set in a dark matrix, obscuring the wall. For example, *Bajanophyton* Drosdova, *Hedstroemia* Rothpletz, *Fistulella* Korde (*pars*). *Bajanophyton* is doubtfully included here. The type material figured by Drosdova (1980, pl. 13, Fig. 1) shows coarse tubes dividing into bunches of very fine tubes.

Wall character can be difficult to evaluate in this group because of the presence of dark matrix between filaments, which emphasizes the cement-fill of the tubes but obscures the wall (see Fig. 12c).

The *Hedstroemia* group is principally distinguished from the *Botomaella* group by the variable tube diameter. *Hedstroemia* has coarse branches that expand distally. *Botomaella* has fine branches that widen only gradually. Multiple branching and sharing of adjacent walls are additional differences, clearly shown in *Hedstroemia* itself. This is the first record of *Hedstroemia* from the Cambrian. *Bija* Vologdin is based on a transverse section that shows features resembling both *Hedstroemia* and *Solenopora* (see below).

Juxtaposted Tubes (Fig. 12d,e). Coarse to fine branching tubes forming fan-shaped, radiating masses. Tubes usually mutually attached, sharing walls; sometimes slightly separated; cross-sections round to polygonal and irregular. Walls thin, sometimes uneven; tube diameter expands gradually distally. Septa-like cross partitions may be present, e.g., *Solenopora* Dybowski, *Bija* Vologdin, *Metasolenopora* Yabe, *Parachaetetes* Deninger, *Petrophyton* Yabe, *Pseudochaetetes* Peterhans, *Solenoporella* Rothpletz. This is a relatively large and internally inhomogeneous group. The type material of *Bija*, from the Lower Cambrian of southern Siberia, is difficult to distinguish from *Solenopora* (see also Johnson 1966, p. 15), but the name *Bija* has also been applied to specimens more closely resembling members of the *Hedstroemia* group (e.g., Korde 1973, pl. 61, Fig. 1a).

The *Solenopora* group is distinguished from the *Hedstroemia* group by more constant tube diameter and less conspicuous branching.

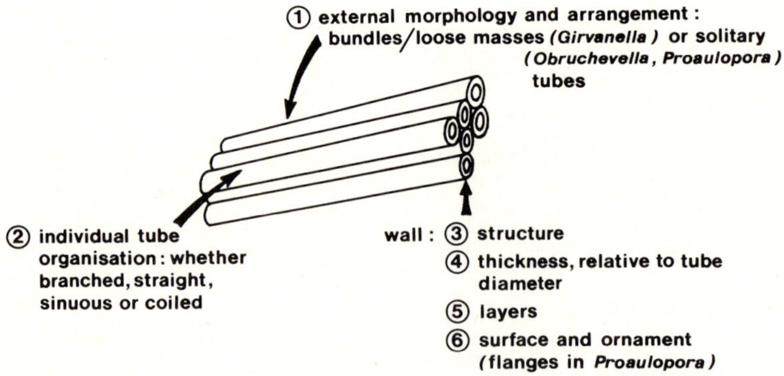

Fig. 7. Morphological features of the tubiform non-fan subgroup (massed and solitary tubes, *Girvanella, Obruchevella,* and *Proaulopora* groups)

b) Tangled, Coiled and Straight Tubes

Bundles and Irregular Masses (Fig. 12f–h). Generally fine tubes of relatively constant diameter, straight to twisted, in bundles or loose irregular masses. Branching rarely, if at all. Wall micritic, relatively thin, e.g., *Girvanella* Nicholson and Etheridge, *Batinevia* Korde, *Botominella* Reitlinger, *Nicholsonia* Korde, *Subtiflora* Maslov (*Cladogirvanella* Ott, best known from the Triassic, but also present in the Middle Palaeozoic).

Although the basic tube-structure of members of this group is very simple, there are clear differences in their mutual arrangement, expressed in several of these genera. *Batinevia* (Fig. 12g), *Botominella* (Fig. 12h) and *Subtifloria* form cable-like bundles of sub-parallel tubes; *Girvanella* (Fig. 12f) is characteristically loosely tangled. Another arrangement, with loosely associated tubes curving vertically upwards from a prostrate felted mass, has been referred to as *Razumovskia* by some workers (e.g., Korde 1973, p. 126, and Pl. 10, Fig. 1a; Drosdova 1980, p. 20 and Pl. 5, Fig. 1) although it is possible that the type-material of *Razumovskia* (Vologdin 1939, p. 216) does not consist of tubes.

The *Girvanella* group is distinguished from the *Botomaella* group by the non-radial, non-fanlike arrangement of the tubes and virtual absence of branching.

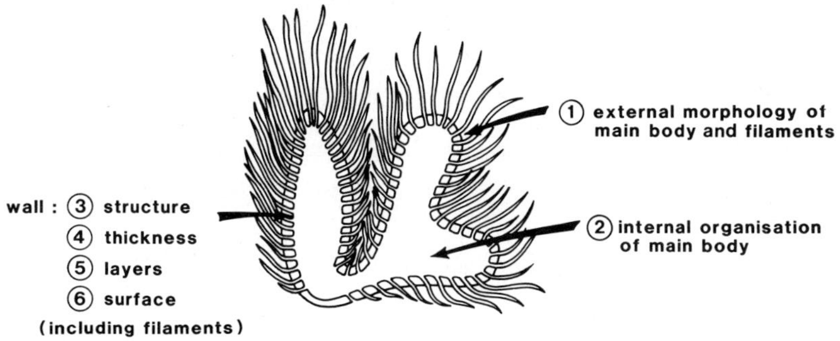

Fig. 8. Morphological features of tuberous forms (Seletonellaceae s.s.)

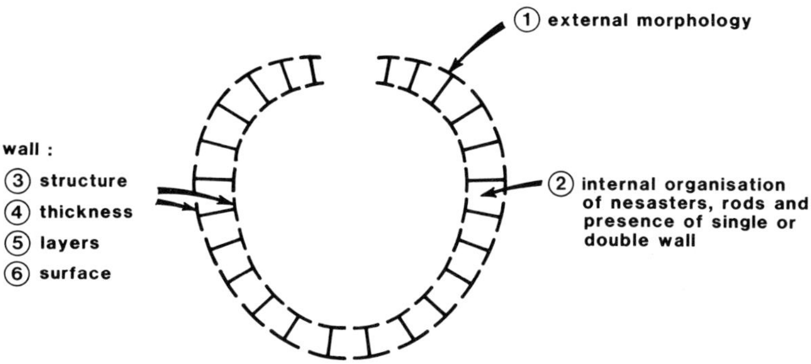

Fig. 9. Morphological features of cup-like forms (Radiocyathids)

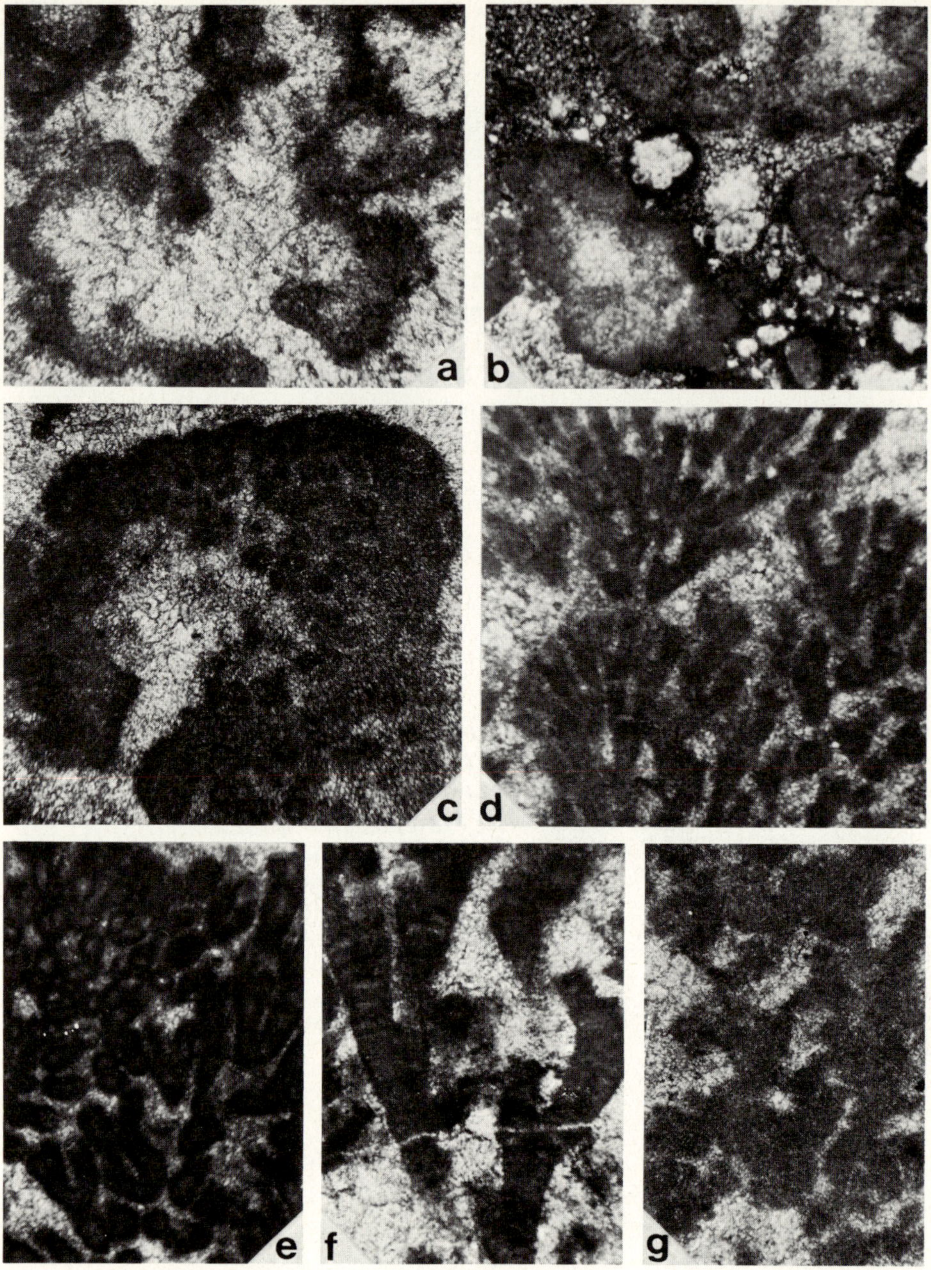

Fig. 10a–g. Botryoidal (**a–c**) and dendritic (**d–g**) Lower Cambrian calcareous algae. **a** *Renalcis polymorphus* (Maslov) Reitlinger. Sample F-42, ?Atdabanian; Salani-Gol, west Mongolia (x 51). **b** *Tarthinia rotunda* Drosdova Sample. M71-6/19, Tommotian; Olenek River, Siberia (x 19). **c** *Gemma inclusa* Luchinina. Sample B79/73, Tommotian; Salani-Gol, west Mongolia (x 51). **d** *Epiphyton* sp. (cf. *Tubomorphophyton* sp.). Sample OM, Atdabanian; Oi-Muraan, Lena River, Siberia (x 46). **e** *Tubomorphophyton* sp. Sample M47/55 I, Atdabanian; Zhurinsky Mys, Lena River, Siberia (x 46). **f** *Gordonophyton durum* (Korde) Korde. Sample 229A 3175-186 (5) (coll'n N.A. Drosdova), Atdabanian; Ser Mountains west Mongolia (x 51). **g** *Angulocellularia anisotoma* Vologdin. Sample B79/30(2), Tsaganalom Formation, Tommotian; Salani-Gol, west Mongolia (x 46)

Coiled (Fig. 13). Fine, apparently solitary, spirally, cylindrically, or irregularly, relatively tightly coiled tubes; e.g., *Obruchevella* Reitlinger, *Cavifera* Reitlinger, *Glomovertella* Reitlinger. These genera were originally described as foraminifers but, whatever their affinity, they can logically be compared on morphological grounds with the *Girvanella* group.

The *Obruchevella* group is distinguished from the *Girvanella* group by the small, compact, tight, usually coiled form.

Straight, Solitary (Fig. 14). Relatively coarse tubes, straight to slight curved, very occasionally branched; sometimes with distinctive whorl-like flanges on the external surface. Wall micritic, relatively thick, especially adjacent to flanges; may be layered. Tubes often broken into short stick-like fragments; e.g., *Proaulopora* Vologdin, *Amganella* Reitlinger, *Palaeonites* Maslov, *Tubophyllum* Krasnopeeva, *Vologdinella* Korde.

The *Proaulopora* group is distinguished from the *Girvanella* and *Obruchevella* groups by generally larger tube diameter and stick-like form.

5. Tuberous

Hollow, elongate, sometimes branched bodies enclosed by thin or thick walls usually pierced by pores and covered by numerous external unbranched filaments; e.g., *Amgaella* Korde, *Mejerella* Korde, *Seletonella* Korde.

These and some other genera comprise the Seletonellaceae, regarded by Korde (1973, p. 239) as dasycladalean although some reservations have been expressed concerning this (Bassoullet et al. 1979, p. 432, footnote).

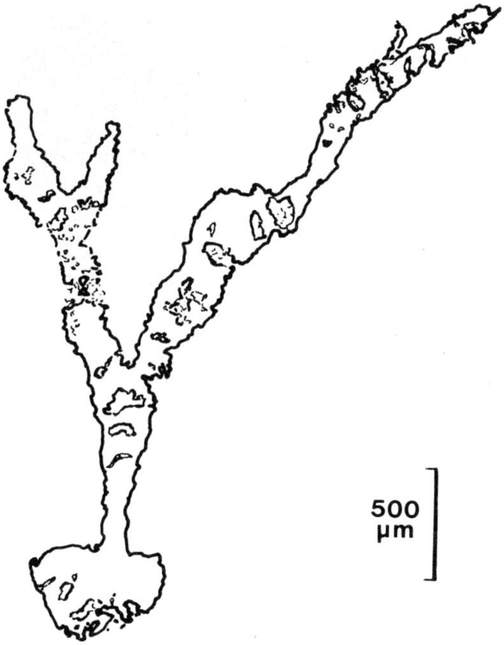

Fig. 11. *Chabakovia nodosa* Korde, Lower Cambrian, Bazaikha River, Eastern Sayan. (After Korde 1973, Pl. XXVI, Fig. 3)

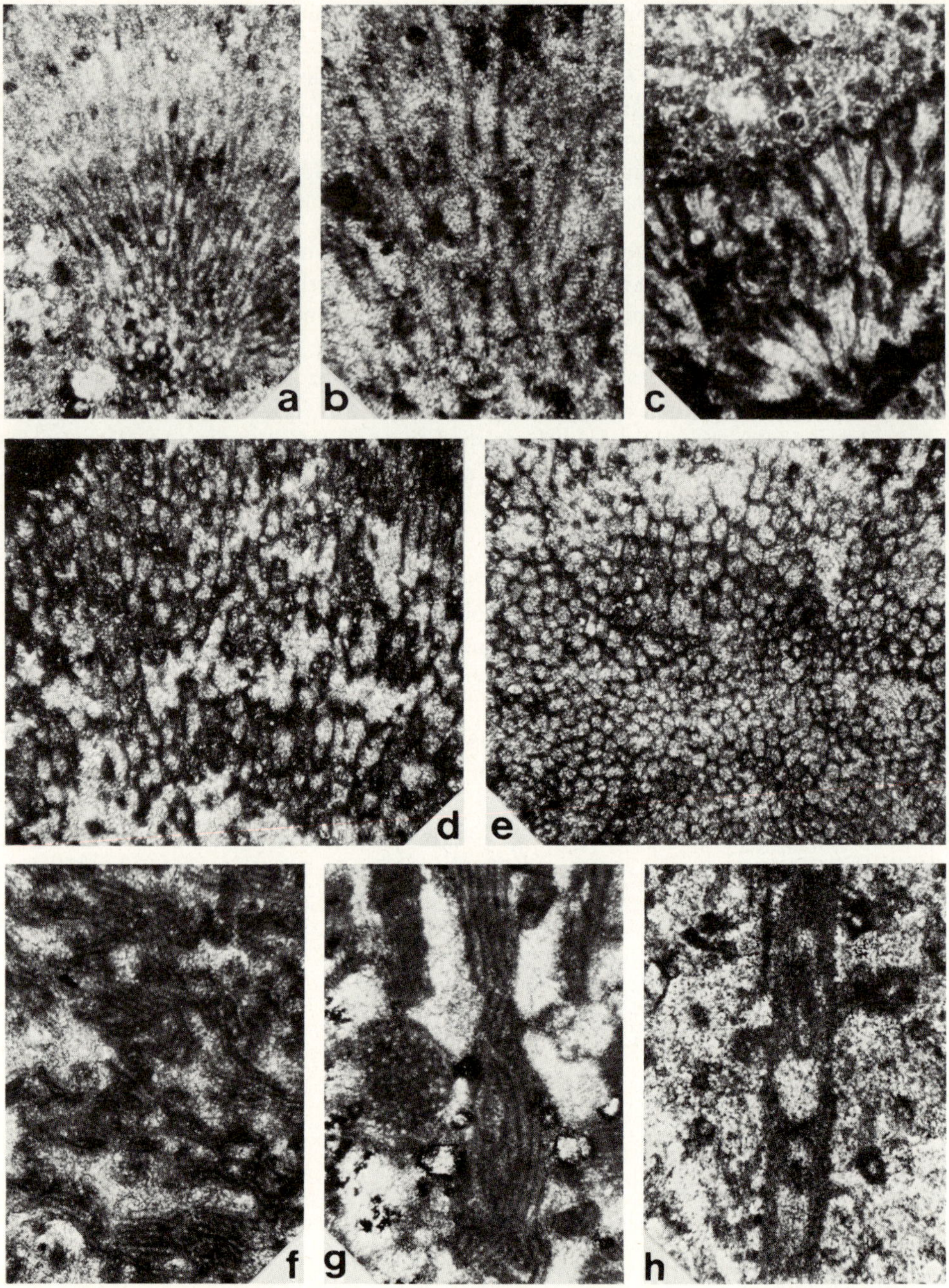

Fig. 12a–h. Lower Cambrian calcareous algae with tubiform organization. **a–e** Specimens with fan-like arrays of tubes; **f–h** specimens with tangled or cable-like organization. **a,b** *Botomaella sibirica* Voronova. Sample M71-4/110, Tommotian; Kessyusin Formation; Olenek River, Siberia (**a** x 20, **b** x 51). Longitudinal section. **c** *Hedstroemia* sp. Sample 3457-161(a) (coll'n N.A. Drosdova) Botomian, 30 km south of Khadkhal, north Mongolia (x 20). Longitudinal section. **d,e** *Solenopora* sp. Sample 402/6 3457–162(b) (coll'n N.A. Drosdova). Botomian, 30 km south of

6. Cups

Globular or conical, double- or single-walled cup-shaped bodies with a circular opening. The walls are formed by a network of stellate plates (nesasters) and joined by rods; e.g., *Gonamispongia* Korshunov, *Radiocyathus* Okulitch, *Uranosphaera* Bedford and Bedford.

These genera comprise the Radiocyathidae Okulitch (Radiocyatales) which shows morphological similarities with receptaculitids and also, to some degree, Archaeocyatha (Nitecki and Debrenne 1979).

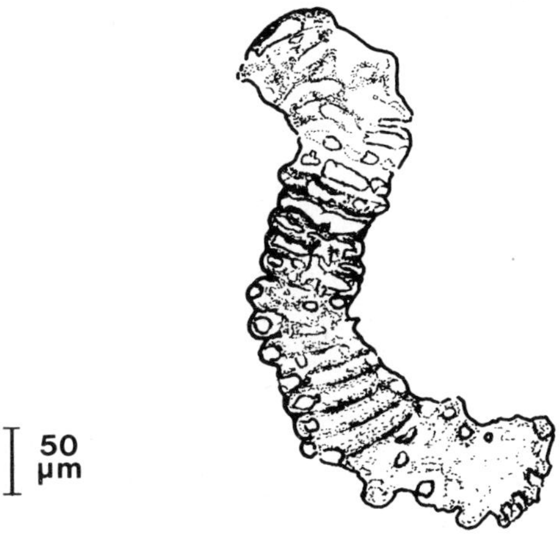

Fig. 13. *Obruchevella sibirica* Reitlinger, Upper Cambrian or Ordovician. (After Reitlinger 1959, Pl. VII, Fig. 5)

Morphological Series

The members of the botryoidal, dendritic, and tubiform groups can be placed in series that express their morphological similarities, and these three groups can themselves be linked together. The result is a keyhole-shaped diagram (Fig. 15) composed by the *Renalcis-Epiphyton* ring (botryoidal and dendritic groups) juxtaposed against the two linear *Botomaella-Solenopora* and *Girvanella-Proaulopora* series of the tubiform group.

Khadkhal, north Mongolia; **d** longitudinal section (x 20). **e** transverse section (x 20). **f** *Girvanella problematica* Nicholson and Etheridge. Sample B70/49(2), Atdabanian; Tjuser Formation, Bulkurskaya anticline, Tjuser stream, lower Lena River, Siberia. Longitudinal sections (x 51). **g** *Botominella* sp. Sample D108-1, Tommotian; Bayangol Formation, Salani-Gol, west Mongolia. Longitudinal section (x 47). **h** *Batinevia ramosa* Korde. Sample B380, Atdabanian; Krasny Porog Formation, Sukharikha River, Siberia. Longitudinal and transverse sections (x 47)

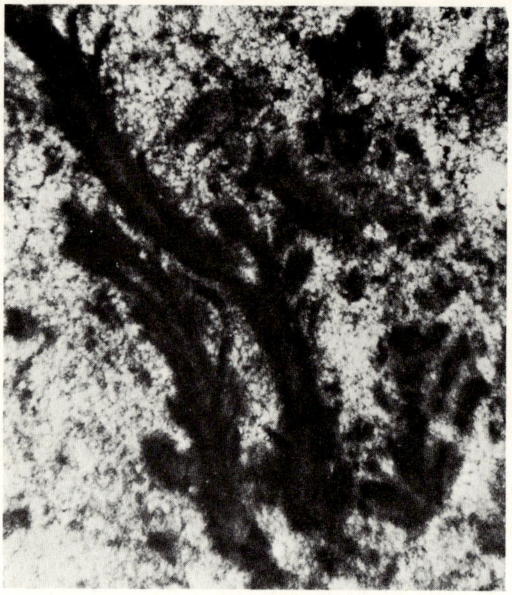

Fig. 14. *Proaulopora rarissima* Vologdin. Sample B524a, Botomian; Shumnin Beds, Sukharikha River, Siberia (x 66)

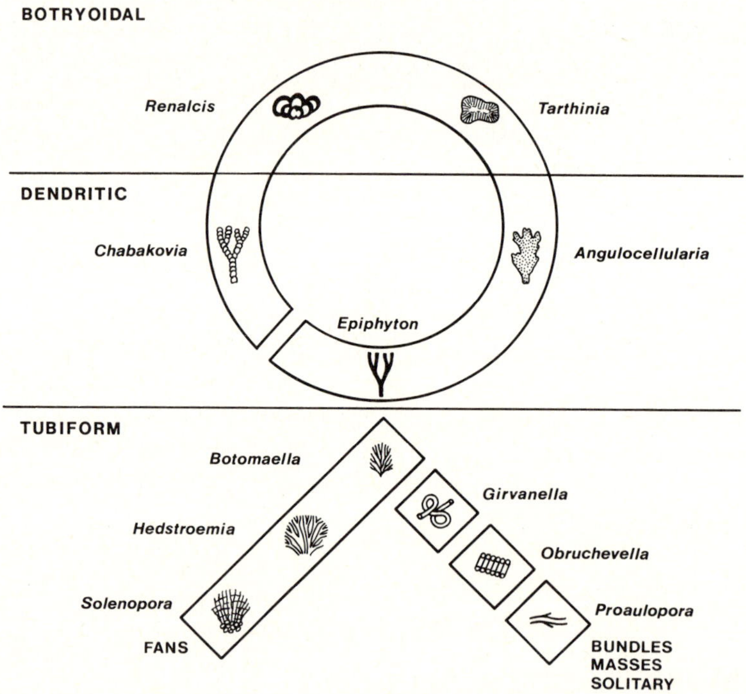

Fig. 15. Morphological classification of common Early Cambrian calcareous algae with botryoidal, dendritic, and tube organization. Each fossil name denotes a generic group and their relative positions on the diagram indicate degree of morphological similarity. Discontinuities are represented by *breaks* in the series

The spherical, tuberous and cup groups are more discrete although they, too, show similarities with other groups which probably also relate to the existence of morphological series: the spherical group relates to the diverse array of calcispheres known in post-Cambrian rocks, the tuberous group has similarities with Ordovician and Silurian dasycladalean fossils (the Seletonellaceae as emended by Bassoullet et al. 1979, p. 432), and the cup group can be placed in morphological series with the receptaculitids and archaeocyaths. However, these connections go far beyond the Early Cambrian and open up more questions than can be dealt with here. In order to limit the field we concentrate on an examination of the most common of Cambrian calcareous algae: those belonging to the botryoidal, dendritic, and tubiform groups.

Renalcis-Epiphyton Ring

The botryoidal and dendritic groups constitute a relatively continuous morphological series forming a ring (Fig. 16): *Renalcis, Tarthinia, Angulocellularia, Epiphyton*, and *Chabakovia*. Transitions between *Renalcis* and *Chabakovia*, through *Shugaria*, and between *Tarthinia* and *Angulocellularia*, through *Renalcis gelatinosus*, which we regard as a *Tarthinia* species (Riding and Voronova 1984), both cross the boundary between botryoidal and dendritic forms. *Renalcis* and *Tarthinia* also show intergradation (Fig. 10a,b). Transition between the *Angulocellularia* and *Epiphyton* groups is less gradational. Possibly it may be exemplified by the Late Precambrian genus *Floribundaphyton* Kolosov. The only clear break in the ring at present appears to occur between the *Epiphyton* and *Chabakovia* groups.

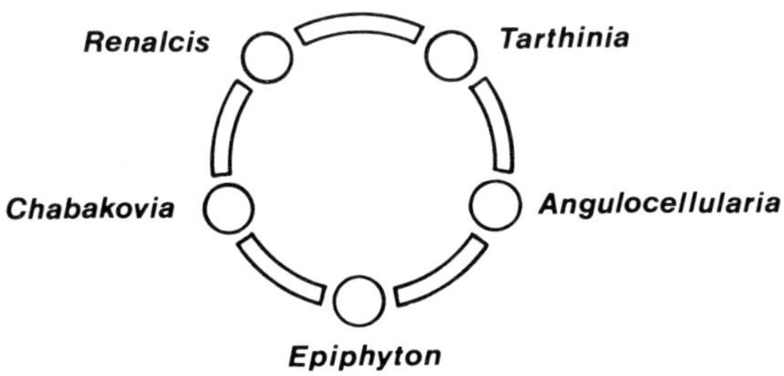

Fig. 16. The *Renalcis-Epiphyton* ring: a nearly continuous morphological series linking botryoidal (*Renalcis* and *Tarthinia* groups) and dendritic (*Chabakovia-Epiphyton-Angulocellularia*) forms

Botomaella – Solenopora Series

The two tubiform sub-groups each form linear series (Fig. 17 which in the case of the *Botomaella-Solenopora* series is relatively continuous. The fan-forms of the *Botomaella* group have branched tubes that are separated and do not share a common wall. In contrast, the *Solenopora* group generally does show common walls between adjacent tubes. The *Hedstroemia* group mixes these characters, sometimes having common walls, sometimes showing separation of adjacent tubes, even though a discrete wall may not be clear. Thus, this series appears to be continuous.

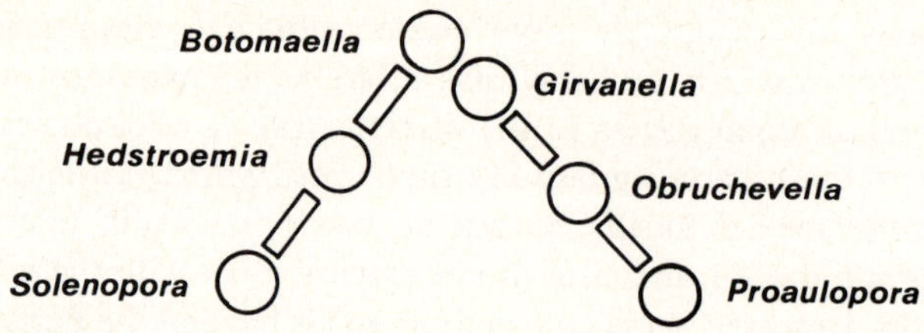

Fig. 17. The *Botomaella-Solenopora* (fan) and *Girvanella-Proaulopora* (non-fan) series: linear morphological trends among fossils having a basic tube-structure. The *Botomaella* and *Girvanella* groups have more features in common than do the *Solenopora* and *Proaulopora* groups and so are juxtaposed

Girvanella – Proaulopora Series

In contrast to the *Botomaella-Solenopora* series the non-fan forms show discontinuities between each generic group. The *Girvanella* group has tubes aggregated into bundles or masses; those of the *Obruchevella* group are solitary and coiled; those of the *Proaulopora* group solitary and straight to slightly curved.

Batinevia (*Girvanella* group) and *Botomaella* both consist of aggregates of fine tubes, and so it is reasonable to juxtapose these two ends of the series. However, the other end-members, *Solenopora* and *Proaulopora*, are quite distinct. Thus, the two linear series together have an L-shaped arrangement (Fig. 17). The two components of Fig. 15 – the botryoidal and dendritic forms in a ring and the tube forms in an L-shape – can be related because there is some morphological similarity between members of the *Epiphyton* and *Botomaella* groups, especially between *Tubomorphophyton* (*Epiphyton* group) and *Botomaella* itself. Both these genera have erect, tubiform filaments, even though *Tubomorphophyton* is dendritic while *Botomaella* is fan-shaped. These relationships between the ring and the L-shape can be shown by placing the angle of the L near to the *Epiphyton* group to give the keyhole-shaped arrangement.

Discussion

a) Transitional Series

Gradation between genera presents problems of taxonomic definition and also suggests either mutual affinity or morphological convergence. The variety of conflicting classifications available for these fossils is a direct result of the presence of transitional series, although this has rarely been acknowledged. Recognition of these series is an important starting point for taxonomic stabilization and will assist, although not necessarily resolve, discussion of systematic position.

The degree of transition shown by some members of the *Renalcis-Epiphyton* ring is extreme, and there can be little doubt that it has contributed to the taxonomic splitting, confusion, and synonymy that is rampant in these genera. *Epiphyton* and

Renalcis between them have more than one hundred described species, despite the fact that morphologically they are relatively simple fossils. The *Botomaella-Solenopora* series is much less complicated, and it is possible to use this morphological analysis to clarify taxonomic problems, particularly those involving *Hedstroemia* and *Solenopora*. The *Girvanella-Proaulopora* series is discontinuous, a contrast with the others, which is all the more interesting in that the constituent genera of the *Girvanella-Proaulopora* series are, if anything, morphologically simpler and therefore might be expected to more readily show intergradation.

Renalcis-Epiphyton Ring. *Renalcis, Tarthinia, Angulocellularia*, and *Epiphyton* show intergradation. *Tarthinia* was long regarded as *Renalcis, R. gelatinosus* now being placed in *Tarthinia* (Riding and Voronova 1984), and *Angulocellularia* was confused with both *Renalcis* and *Epiphyton* (see Riding and Voronova 1982, p. 107 for examples). Thus, the steps in the series are becoming more clear, but the transition remains and is further complicated by the fact that the series appear to change with time. In the Cambrian, intergradation exists between *Renalcis* and *Tarthinia, Tarthinia* and *Angulocellularia*, and to a lesser degree between *Angulocellularia* and *Epiphyton*. But neither *Tarthinia* nor *Angulocellularia* appear to be present in the Devonian, and transition instead exists directly between *Renalcis* and *Epiphyton*. Saltovskaya (1975), following an earlier suggestion by Maslov (1956) that *Renalcis* shows ecophenotypic variation, has interpreted transition between *Renalcis* and *Epiphyton* as being a reflection of mutual affinity of these genera and consequently has placed both in the Epiphytaceae.

One reason why *Renalcis* and *Epiphyton* do not show such a close relationship in Cambrian material is that the species are different from those involved in the Devonian transition. In fact, such is the taxonomic uncertainty that even the genera may actually be different. Devonian specimens of *Renalcis* commonly resemble two forms which are uncommon, and possibly absent, in the Cambrian: the multi-chambered forms are similar to *Shugaria* Antropov, and the thick-walled forms with internal v-shaped clefts are similar to *Izhella* Antropov. Devonian specimens of *Epiphyton* are commonly small forms with delicate, radially divergent branches, which have been named *Paraepiphyton* Wray. Interestingly, this form is very similar to the shrub-like colonies present in Quaternary pisoids reported to be bacterial in origin by Folk and Chafetz (1983).

Consequently, the similarities emphasized by Saltovskaya (1975) are best seen between *Izhella* and *Paraepiphyton* rather than between the Cambrian type-material of *Renalcis* and *Epiphyton*. She concluded that construction of the wall of *Renalcis* is fundamentally filamentous, and indicates such a close relationship with *Epiphyton* that not only should both taxa be placed in the Hormogonophyceae, but that *Renalcis* is a junior synonym of *Epiphyton*. This transition described by Saltovskaya is striking, but it should be noted that it relates two quite different morphological features of these fossils: the *external* branching organization of *Epiphyton* and the *internal* wall structure of *Renalcis*. These are much more difficult to relate in Cambrian forms, in which the clusters of *Epiphyton* branches greatly exceed the size of the *Renalcis* wall. Pratt and James (1982) also recognize a ring-series involving *Chabakovia, Renalcis, Epiphyton, Gordonophyton*, and related forms, and they emphasize the transition between *Renalcis* and *Epiphyton*. However, they take the view that the chambered *Renalcis* and dendritic *Epiphyton* are intergrown and consider the entire transitional series to represent post-mortem calcification of coccoid cyanophytes.

Thus, both Saltovskaya (1975) and Pratt and James (1982) regard morphological similarity between members of the *Renalcis* and *Epiphyton* groups as being an indication of mutual systematic affinity within the Cyanophyta, but they reach signficantly different conclusions concerning their basic morphological structure: Saltovskaya regarding both as being filamentous in organization, Pratt and James interpreting them as coccoid. Resolution of this question is likely to lie in discovery of Recent analogues of these fossils, but it is worth pointing out that an alternative explanation, not considered by these authors, is that transition may reflect morphological convergence rather than systematic affinity. This is particularly to be expected in such simple fossils where the scope for morphological variation is restricted, and it is perhaps more plausible to regard the mutual association of somewhat similar forms as reflecting similar environmental preferences of convergent morphotypes rather than ecophenotypic variation. Ecophenes will occupy separate habitats, whereas Saltovskaya (1975, Pl. 4) shows a variety of forms, all regarded as variants of *Epiphyton buldyricum*, in a single thin section. But in any case, whatever the mutual relationships of these organisms, Saltovskaya's view that *Renalcis* is a synonym of *Epiphyton* seems to us to be unacceptable, because it denies the presence of mutually discrete specimens. Recognition of a transitional series should not in itself be regarded as a reason for synonymizing all the taxa involved, especially where there are still so many uncertainties of relationships. Saltovskaya (1975) in fact goes even further and places *Chabakovia*, as well as *Renalcis* and many other genera, in synonymy with *Epiphyton*. If this were to be accepted, it would destroy a nomenclatorial framework that, though still imperfect, is essential for description and discussion of these microfossils. The existence of transitional series does not prevent establishment of a workable taxonomy; it simply means that special care is required in drawing distinctions between taxa and that careful comparisons are necessary. It is the lack of these that has helped to create the problems that now exist!

b) Botomaella-Solenopora Series

There is more information concerning the systematic positions of the members of this transitional series than there is for the *Renalcis-Epiphyton* ring, and the special point of interest lies in the difference between the probable affinities of the two ends of the series. *Botomaella* resembles Recent filamentous cyanophytes such as *Scytonema*. *Hedstroemia* is almost certainly a rivulariacean cyanophyte. But solenoporaceans are widely regarded as rhodophytes related to the Corallinaceae. An obvious explanation, especially bearing in mind the comments of the previous paragraph, for this situation would appear to be morphological convergence. However, in this case there is another alternative because the Solenoporaceae is a heterogeneous group, and it is possible that discontinuities exist within it which prevent the *Botomaella-Solenopora* series from being fully transitional. This taxonomic problem is seen most clearly in Silurian material. In specimens from Gotland (see Rothpletz 1913) there can be difficulty in separating *Solenopora compacta* Billings from species of *Hedstroemia*.

Exactly the same difficulty exists in the Cambrian, but in this case specimens have commonly been ascribed to *Bija* and *Rothpletzella* (see Riding and Voronova (1984), and the taxonomy is even more confused (see Bertrand-Sarfati 1979).

Many *Solenopora* species show a delicate meshwork of filaments which contrast with the coarse filaments of *S. compacta*. If the latter were removed from the Solenoporaceae, it would not solve the problem of the relationship hetween *Hedstroemia* and *S. compacta* but it would remove the anomaly of apparent transition between cyanophytes and rhodophytes. If this explanation is correct, then it means that the *Botomaella-Solenopora* series as shown in Fig. 15 is not transitional. This possibility requires further study, and so for the present we retain *Botomaella-Solenopora* as a transitional series. But this example shows the taxonomic value of attempting this type of objective morphological analysis.

c) Girvanella-Proaulopora Series

In contrast to the *Renalcis-Epiphyton* ring and the *Botomaella-Solenopora* series (at least as it is shown in Fig. 15), this is a discontinuous series of unbranched, or rarely branched, tubiform fossils. They are distinguished primarily on the basis of their arrangement, but *Proaulopora* is characterized by the possession of external flange-like projections that give it more complexity than the other genera. Nevertheless, despite their simplicity, even taxa within the *Girvanella*-group fail to show clear mutual transition. Consequently, even though Danielli (1981, p. 96) placed *Batinevia* and *Botominella* in synonymy with *Girvanella*, these genera can be separated by their tube arrangement. We therefore reinstate them as related but mutually discrete taxa.

d) Tuberous and Cup Forms

The *Girvanella-Proaulopora* series could be regarded as the start of a much longer series involving the tuberous and cup forms, but this probably exaggerates the degree of morphological similarity involved. The relationship of the tuberous forms to Ordovician and younger dasycladaleans needs to be examined further, but at present the Cambrian seletonellaceans appear to be unusual, rather isolated, and rare fossils. The cup forms, however, relate to the receptaculitids and archaeocyaths in a long and important series that forms the basis for the group Archeata proposed as a new Kingdom by Zhuravleva and Myagkova (1972). The radiocyathids are especially similar to calathids, and this in turn links them more closely to the receptaculitids than to the archaeocyaths (Nitecki and Debrenne 1979). The basic similarity in all these forms is the cup- or vase-like morphology with a single or double wall and hollow interior. The differences between them lie primarily in the nature of the wall and the skeletal elements forming it. Radiocyathids, calathids, and receptaculitids have rod-like elements, orientated normally to the hollow interior, and these unite distally in a faceted or spiculate surface that forms the external wall. In contrast, archaeocyaths have an integrated wall with pores and have additional skeletal elements that are not present in these other groups. Nevertheless, even the radiocyathid-receptaculitid part of this series is not transitional, because several minor but distinct non-gradational morphological features can be used to distinguish the principal groups within it.

Previous Classifications

Incorporation of poorly based interpretation of affinity into classifications of these fossils has been a cause of instability, with schemes repeatedly undergoing substantial modification as views have changed. Early classifications, such as that by Pia (1927), Maslov (1956), Korde (1961), and Vologdin (1962), show little relationship to the groupings proposed here. Nevertheless, some of the morphological groups we recognize can also be discerned in several more recent classifications, particularly with regard to the tubiform genera. For example, Korde (1973) placed *Batinevia, Bija, Botominella, Nicholsonia,* and *Proaulopora* together in a single order: the Kenellales; and Luchinina (1975) divided this into two parts: *Bija* and *Botomaella,* and *Batinevia, Girvanella, Obruchevella* and *Proaulopora*, which respectively correspond to the fan and non-fan groups recognized here.

However, although these classifications, naturally enough, used morphological information, they always also took account of assumed affinity and there have been very few previous attempts at purely morphological classifications of these fossils. Johnson (1966, Table 3) outlined basic features of Cambrian calcareous algae, but only in a general way and also incorporated characters of doubtful validity, such as the "cells" and "sporangia" of the Epiphytaceae. Pospelov (1974) included some calcareous algae, such as *Epiphyton, Girvanella* and *Renalcis*, in a scheme sub-dividing Precambrian and Early Palaeozoic stromatolites morphologically, but the resulting groups do not resemble those of our classification. Possibly this is in part due to the difficulties of trying to deal with stromatolite microfabrics and calcareous algae together.

Although Voronova's (1976) main classification of calcareous algae near the Precambrian-Cambrian boundary incorporates ideas of affinity, it also presentes a morphological classification of these fossils (Voronova 1976, p. 25, 1979). This scheme is less comprehensive than the one presented here because it deals only with genera present in material Voronova had collected herself, but it still shows some similarities to it. Voronova (1976, p. 25) recognized five major morphological groups characterized by (1) tubes, (2) bushes, (3) bubbles, (4) felted filaments, and (5) compact colonies. The tube-group includes *Girvanella, Proaulopora, Obruchevella, Botominella*, and *Batinevia* and thus corresponds with the non-fan series of the present scheme. The bush-like forms include *Botomaella* and *Epiphyton.* This differs from our current view, but it does express the proximity, which we also recognize, between the *Botomaella*-end of the fan-series and the *Epiphyton* group. However, different groupings can be seen in the separation of *Bija* alone into the "compact colonies" group, *Razumovskia* into the "felted filaments" group, and in the placing together of *Renalcis* and *Chabakovia*, although we do recognize the similarity of the latter by juxtaposing them in the ring-series.

Summary

Many of the calcareous algae described from Lower Cambrian rocks form linked morphological series, several of which are transitional. Previous classifications of these fossils, although based fundamentally on morphological features, have also incorporated interpretations of affinity which have been doubtful. This has caused the classifications to be unstable. Examination of purely morphological groupings provides a more objective taxonomic view of these fossils and also allows the morphological series to be seen clearly. Six criteria are used to establish major groupings: (1) external morphology, (2) internal organization, (3) wall structure, (4) wall thickness, (5) wall layers, and (6) wall surfaces.

Seven principal groups are recognized: (1) spherical (calcispheres, *Uranovia*, etc.); (2) botryoidal (*Renalcis, Tarthinia*, etc.); (3) dendritic (*Angulocellularia, Chabakovia, Epiphyton*, etc.); (4) fans of tubes (*Botomaella, Hedstroemia, Solenopora*, etc.); (5) tangled, coiled, and straight tubes (*Girvanella, Obruchevella, Proaulopora*, etc.); (6) tuberous (Seletonellaceae); and (7) cups (radiocyathids). The spherical group includes internally solid forms that may well be inorganic. The botryoidal and dendritic forms constitute a virtually continuous morhological series: the *Renalcis-Epiphyton* ring. The two groups of tube-forms each constitutes a linear series, one transitional, the other not, both of which originate near the *Epiphyton*-group on the ring. The tuberous forms constitute a discrete group of rare fossils, but the cup-forms show similarities with both receptaculitids and archaeocyaths, and these comprise yet another morphological series.

These similarities may be due to either mutual affinities or to morphological convergence, and the presence of a series should not, alone, be taken as an indication of mutual affinity; nor should taxa juxtaposed within a series necessarily be regarded as good candidates for synonymy. In both cases more information needs to be considered.

Acknowledgments. This work is part of a program of research into calcareous algae near the Precambrian-Cambrian boundary, sponsored by the Natural Environment Research Council with additional support from the Royal Society and the USSR Academy of Sciences. We are grateful to Boris Chuvashov, Nina Drosdova, Veronica Luchinina, and Nora Radionova for advice and helpful discussion.

References

Bassoullet J-P, Bernier P, Deloffre R, Génot P, Jaffrezo M, Vachard D (1979) Essai de classification des Dasycladales en tribus. Bull Cent Rech Explor-Prod Elf-Aquitaine 3 (2):429–442

Bertrand-Sarfati J (1979) Une algue inhabituelle verte, rouge ou bleue dans une formation dolomitique présumée d'âge Précambrien supérieur. Bull Cent Rech Explor-Prod Elf-Aquitaine 3 (2):453–461

Danielli C (1981) The fossil alga *Girvanella* Nicholson and Etheridge. Bull Br Mus Nat Hist (Geol) 35 (2):79–107

Drosdova NA (1980) Algae in Lower Cambrian organic mounds of western Mongolia. Acad Sci USSR, Palaeontol Inst/Acad Sci MPR, Geol Inst, Trans Joint Sov-Mong Pal Exped 10, p 140 (in Russian)

Folk RL, Chafetz HS (1983) Pisoliths (pisoids) in Quaternary travertines of Tivoli, Italy. In: Peryt T (ed) Coated grains. Springer, Berlin Heidelberg New York, pp 474–487

Johnson JH (1966) A review of the Cambrian algae. Colo Sch Mines Q 61(1):162

Korde KB (1961) Cambrian algae of the southeastern Siberian platform. Acad Sci USSR Trans Palaeontol Inst Moscow 89:147 (in Russian)

Korde KB (1973) Cambrian algae. Nauka, Moscow, p 349 (in Russian)

Luchinina V (1975) Palaeoalgological characteristics of the Early Cambrian of the Siberian Platform. Acad Sci USSR Siberian Branch Trans Inst Geol Geophys Novosibirsk 216:99 (in Russian)

Maslov VP (1956) Fossil calcareous algae of the USSR. Acad Sci USSR Trans Inst Geol Sci Moscow 160:301 (in Russian)

Maslov VP (1960) Stromatolites. Acad Sci USSR Trans Geol Inst 41:188 (in Russian)

Nitecki MH, Debrenne F (1979) The nature of radiocyathids and their relationship to receptaculitids and archaeocyathids. Géobios 12(1):5–27

Pia J (1927) Thallophyta. In: Hirmer M (ed) Handbuch der Paläobotanik, vol I. Oldenbourg, München Berlin, pp 31–136

Pospelov AG (1974) On the remains of stromatolite-building algae. In: Biostratigraphy and palaeontology of the Lower Cambrian of Europe and northern Asia. Acad Sci USSR, Siberian Branch, Inst Geol Geophys, Moscow, pp 223–228 (in Russian)

Pratt BR, James NP (1982) *Epiphyton* and *Renalcis* – diagenetic microfossils from calcification of coccoid blue-green algae. Am Assoc Petrol Geol Bull 66:619

Reitlinger EA (1959) Atlas of microscopic organic remains and problematica of Siberia's oldest deposits. Acad Sci USSR Trans Geol Inst Moscow 25:62 (in Russian)

Riding R, Voronova L (1982) Recent oscillatoriacean analogue of the Lower Palaeozoic calcareous alga *Angulocellularia.* Lethaia 15:105–114

Riding R, Voronova L (1984) Assemblages of calcareous algae near the Precambrian/Cambrian boundary in Siberia and Mongolia. Geol Mag 121:205–210

Rothpletz A (1913) Über die Kalkalgen, Spongiostromen und einige andere Fossilien aus dem Obersilur Gotlands. Sver Geol Unders Ca 10:57

Saltovskaya VD (1975) Genus *Epiphyton* Bornemann (probable synonyms and stratigraphic significance). In: Questions of palaeontology of Tadzhikistan. Acad Sci USSR, Donish, Dushanbe, pp 70–88 (in Russian)

Vologdin AG (1939) Middle Cambrian Archaeocyatha and algae of the southern Urals. Prob Palaeontol MGU Moscow 5:210–245 (in Russian)

Vologdin AG (1962) The oldest algae of the USSR. Acad Sci USSR, Moscow, p 656 (in Russian)

Voronova LG (1976) Calcareous algae of the Precambrian-Cambrian boundary deposits of the Siberian platform. In: Voronova LG, Radionova EP. Palaeozoic algae and microphytolites. Acad Sci USSR Geol Inst Moscow 294:3–85, 156–183, 212–214 (in Russian)

Voronova LG (1979) Calcareous algae of the Precambrian and Early Cambrian. Bull Cent Rech Explor-Prod Elf-Aquitaine 3(2):867–871

Walter MR, Awramik SM (1979) *Frutexites* from stromatolites of the Gunflint Iron Formation of Canada, and its biological affinities. Precambrian Res 9:23–33

Zhuravleva IT, Myagkova EI (1972) Archeata – new group of Palaeozoic organisms. In: Palaeontology. Int Geol Congr, Moscow, pp 7–14 (in Russian)

7 Late Ordovician Dasyclad Algae of the Eastern Great Basin

R.E. JOHNSON and P.M. SHEEHAN[1]

Introduction

Dasyclad algae are present and locally abundant in the Upper Ordovician dolomite sequences in eastern Nevada and western Utah. Calcareous algae have been observed in more than 30 measured sections of Upper Ordovician rocks from the carbonate shelf of eastern Nevada and western Utah (Fig. 1). This paper is based on a pilot study in which collections of algae were made at two sections in order to develop a plan for a future comprehensive examination of the flora. The general pattern of algal distribution was examined from both a stratigraphic and paleoenvironmental mode. Silurian calcareous algae were first studied in the Great Basin by Rezak (1959), whereas the Ordovician algae described in this paper (Table 1) have not been studied previously except for a brief note by Johnson (1982). Most algae examined were silicified during diagenisis, and accordingly specimens were prepared by acid etching, a process which allows gross morphology to be examined. However, algal internal structures are often obscured by silicification.

During the Early Paleozoic the margin of the North American continent was in central Nevada. East of the margin was a shallow water carbonate shelf, and open ocean lay to the west (Fig. 1). The region was close to or slightly south of the equator (Ross 1976). The Upper Ordovician Ely Springs and Fish Haven dolomites are about 150 m thick and were deposited on the shallow carbonate shelf.

Environmental Distribution

Depositional environments on the carbonate shelf ranged from intertidal to well below wave base (Fig. 2). Dasyclad algae described in this paper were present when the depositional setting was relatively shallow, and when there was open circulation. The dasyclads were commonly associated with an open marine fauna of tabulate corals, stromatoporoids, crinoids, and brachiopods. Dasyclads are seldom found in deep water deposits, in very shallow restricted environments where algal stromatolites are common, or in carbonates with high terrigenous content.

1 Department of Geology, Milwaukee Public Museum, 800 West Wells Street, Milwaukee, WI 53233, U.S.A.

Paleoalgology: Contemporary Research and Applications
(eds. D.F. Toomey and M.H. Nitecki)

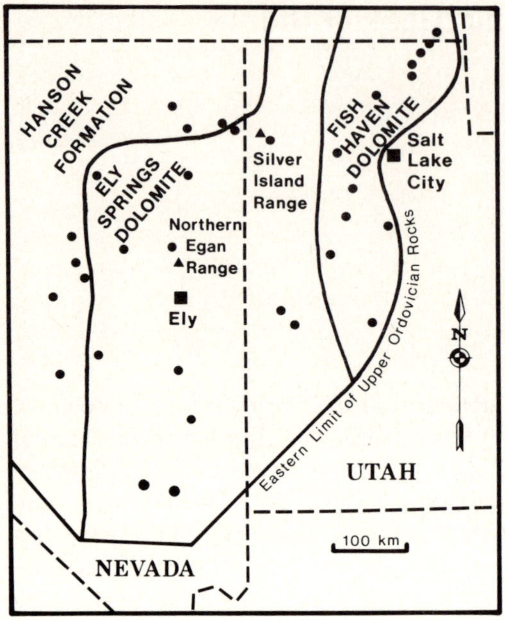

Fig. 1. Location Map. Map of the study areas shows the location of the two sections from which collections were examined for this report (*diamonds*) and other locations where dasyclads have been observed (*dots*) in Ordovician and Silurian rocks. Detailed location of sections is in Sheehan (1980). The distribution of Upper Ordovician formations is also indicated. The Fish Haven and Ely Springs dolomites were deposited on a carbonate shelf and the Hanson Creek Formation in a slope to basinal setting

Table 1. Ordovician algae

M.P.M. locality	3203	3208	3209	3210	3227	3229	3231
Dasyporelleae		3	2	3		4	
Vermiporella sp.		3	1		16	400+	
Rhabdoporella sp.	1	12	8	3	2	37	1
Cyclocrineae		1	2			4	1
Cyclocrinites spp.	6	2	3	2		18	4

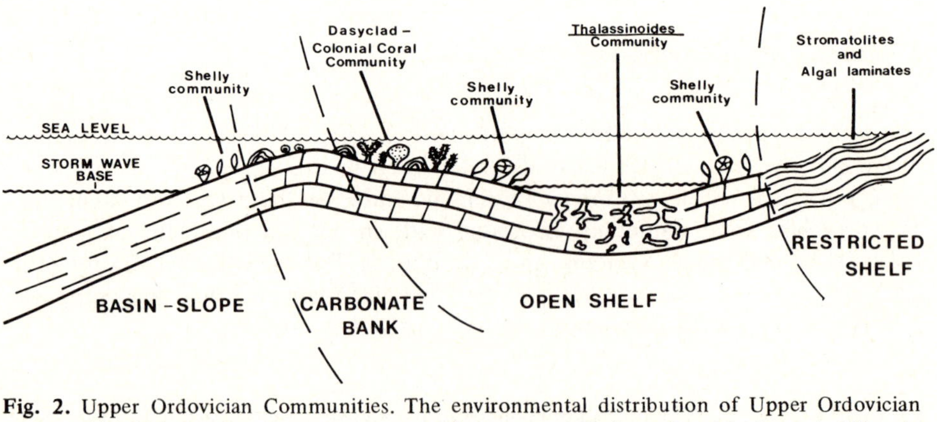

Fig. 2. Upper Ordovician Communities. The environmental distribution of Upper Ordovician communities is shown in an east-west transect from central Utah on the right to central Nevada on the left

Dasyclads occur in greatest abundance in the Dasyclad-Colonial Coral Community which lived in shallow water, protected from strong currents by shoals with oolites and oncolites. Commonly, the biota of the Dasyclad-Colonial Coral Community is preserved in thickly bedded, bioclastic packstones and wackestones. Associated corals include *Calapoecia, Palaeofavosites, Billingsaria, Nyctopora, Agetolites, Catenipora,* and *Palaeophyllum.*

Lateral to the Dasyclad-Colonial Coral Community, but still within wave base, were open marine shelly communities which included some dasyclads but which were dominated by brachiopods, rugose corals, and bryozoans. Below storm wave base, in massive, intensely bioturbated wackestones that lack dasyclads, was the *Thalassinoides* Community, which was dominated by an infaunal burrower (Sheehan and Schiefelbein 1984).

Silurian dasyclads in the Great Basin are common in the Laketown Dolomite and include species of the same genera recognized in the Upper Ordovician. The Upper Ordovician and Silurian algae lived within a depositional setting that was very similar to that described by Riding (1979) for the Devonian System.

Systematic Descriptions

Family DASYCLADACEAE Kuetzing, 1843
Tribe DASYPORELLEAE Pia, 1920
Genus *Rhabdoporella* Stolley, 1893
Rhabdoporella sp. (Fig. 3A–D)

Description. Thallus small, cylindrical, 4.0 mm to 11.0 mm long and 1.8 mm to 2.5 mm wide, straight or slightly curved, wall thickness variable, from 0.2 mm to 0.7 mm. Primary laterals arise radially from the central axis; lateral heads preserved as pores on surface of thallus, approximately 0.1 mm apart.

Remarks. Specimens are fragments of thalli with few terminal ends preserved. The basic, primitive characters of this genus, such as simple, aspondyl laterals are evident. The species is present at all localities, with the greatest abundance at Milwaukee Public Museum (M.P.M.) loc. 3229, where it occurs with *Vermiporella.*

Genus *Vermiporella* Stolley, 1893
Vermiporella sp. (Fig. 3E–H)

Description. Thallus small, cylindrical, ranging from 1.0 mm to 1.8 mm wide, dichotomously ramified and often curved. All thalli recovered are 3.0 mm to 9.0 mm long segments of broken thallus networks. When preserved, surfaces covered with small circular pores representing termini of lateral branches. Wall thickness varies from 0.3 mm to 0.5 mm. Only one order of laterals preserved in thin section.

Remarks. Kozlowski and Kazmierczak (1968) amended the classic descriptions of Stolley (1893) and Pia (1920) with their discovery of three orders of lateral branches

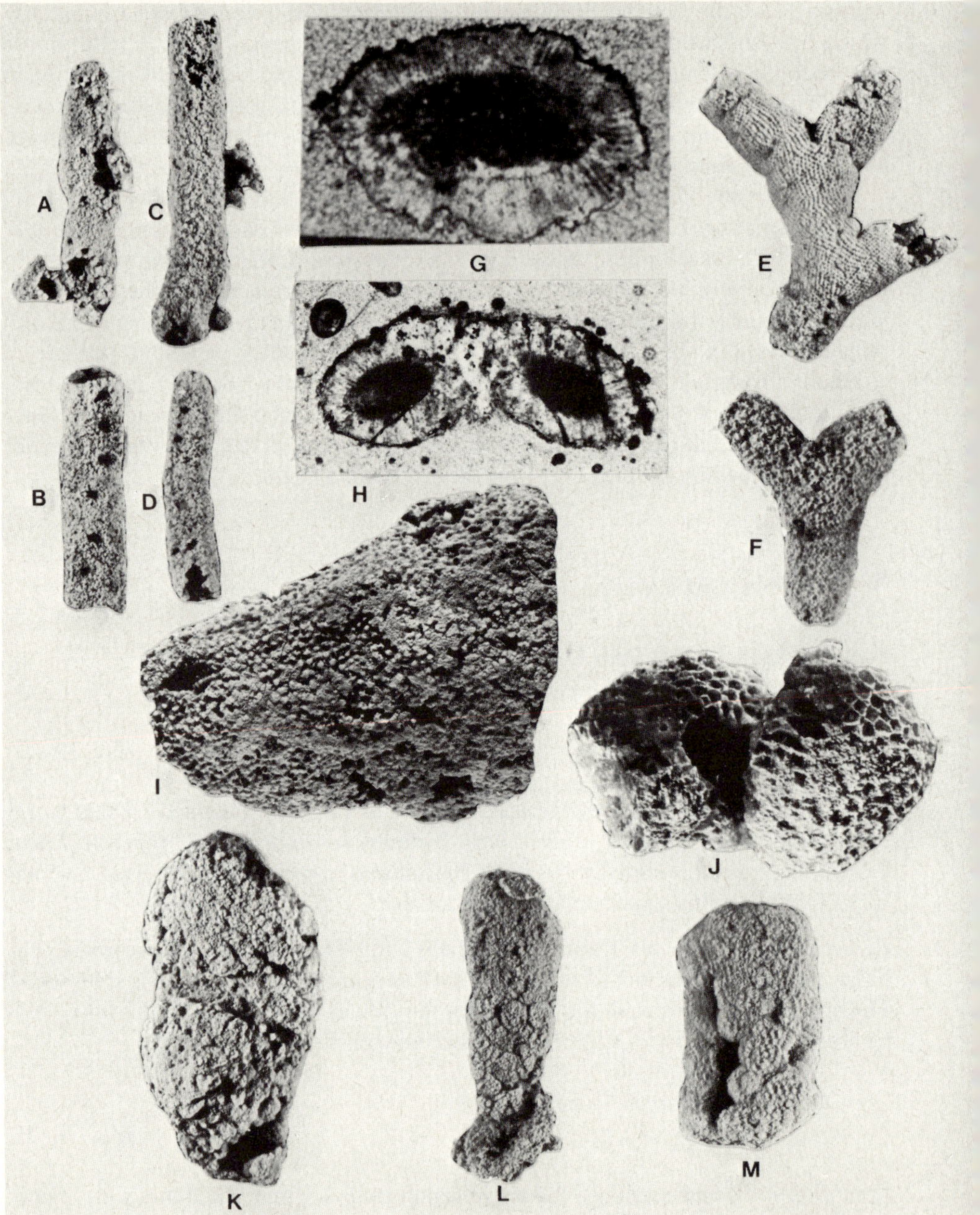

Fig. 3A–M. *Rhabdoporella, Vermiporella, Cyclocrinites* from Ely Springs Dolomite. **A–D** *Rhabdoporella* sp., M.P.M. 2823–2826 respectively, MPM loc. 3229, Floride Member, Ely Springs Dolomite, northern Egan Range, Nevada. Thalli showing mode of preservation and surface pores (**A, B,** and **D** x 3.5; **C** x 4.5). **E–H** *Vermiporella* sp., M.P.M. 2827–2830 respectively, MPM loc. 3229. **E–F** Bifurcated thallus fragments. Note preserved surface pores (x 6). **G–H** Thalli in thin section. Note laterals and nature of bifurcation (**G** x 17; **H** x 14.5). **I** *Cyclocrinites* sp., M.P.M. 2831, MPM loc. 3229. Fragment from a large globullar thallus showing both nipples and facets arranged in whorls. *Vermiporella* is found in four of the seven MPM localities with the greatest number (over 400 thalli) at MPM loc. 3229.

Tribe CYCLOCRINEAE Pia, 1920
Genus *Cyclocrinites* Eichwald, 1840
Cyclocrinites spp. (Fig. 3I–M)

Description. Thallus varies from spheroidal, ovoid, or pyriform to cylindrical or otherwise elongated or globular. Thalli range from 0.27 cm to 3.5 cm in diameter and up to 2 cm long on pyriform and cylindrical individuals. Laterals numerous; lateral heads form a strong cortex with cortical cells, facets, and nipples preserved on different specimens. Cells, facets and nipples range from 0.1 mm to 0.8 mm wide. A number of specimens have both cortical cups and nipple-like lateral heads (Fig. 3I). Main axis rarely preserved. Commonly only the calcareous cortex, which is secondarily silicified, is preserved.

Remarks. This genus, as redescribed by Nitecki (1970), is variable in size and shape of the thallus, extent and degree of calcification, and the shape of the facets, nipples, or pores on the cortex. Many thalli are preserved in a compressed state, often distorting or obscuring surface features. The broad range in size and shape of our sample may be due to environmental parameters or specific differences. The largest specimens (to 3.5 cm across) appear to be pieces of globular thalli which subsequently collapsed and were flattened upon death and burial. Many thalli grade into clavate and cylindrical forms, while others are more spherical. We use Nitecki's synonomy of *Mastopora* Eichwald and *Cyclocrinus* Eichwald with *Cyclocrinites* Eichwald. The genus is found in all but one locality (MPM loc. 3227) with the greatest abundance at MPM loc. 3229, where more than one species is also present.

Summary

The dasyclad algae *Vermiporella, Rhabdoporella,* and *Cyclocrinites,* and associated colonial corals dominate shallow, quiet water, open shelf communities of Upper Ordovician carbonate shelf sediments of the eastern Great Basin. Algal communities are found in clean dolomites, indicating little influx of terrigenous sediment. Commonly, the algae have been replaced by silica.

on surface (x 2). **J** *Cyclocrinites* sp., M.P.M. 2832, MPM loc. 3208, Lost Canyon Member, Ely Springs Dolomite, Silver Island Range, Utah. Two nearly spherical thalli fused together after burial. Note facets (x 6.5). **K** *Cyclocrinites* sp., M.P.M. 2833, MPM loc. 3231, Lost Canyon Member, Ely Springs Dolomite, northern Egan Range, Nevade. Globular thallus (x 5.5). **L** *Cyclocrinites* sp., M.P.M. 2834, MPM loc. 3209, Lost Canyon Member, Ely Springs Dolomite, Silver Island Range, Utah. Clavate thallus (x 5). **M** *Cyclocrinites* sp., M.P.M. 2835, MPM loc. 3210, Lost Canyon Member, Ely Springs Dolomite, Silver Island Range, Utah. Cyclindrical thallus (x 4)

Acknowledgment. This work was supported by the National Science Foundation and the Donors of the Petroleum Research Fund administered by the American Chemical Society.

References

Johnson R (1982) Upper Ordovician dasyclad algae from the eastern Great Basin. Geol Soc Am Abstr 14: no 5, 263

Kozlowski R, Kazmierczak J (1968) On two Ordovician calcareous algae. Acta Palaeontol Pol 13:325–362

Nitecki MH (1970) North American cyclocrinitid algae. Fieldiana Geol 21:182

Pia J (1920) Die Siphoneae Verticillatae von Karbon bis zur Kreide. Abh Zool Bot Ges Wien 11: 263

Rezak R (1959) New Silurian Dasycladaceae from the southwestern United States. Colo Sch Mines Q 54:113–129

Riding R (1979) Devonian calcareous algae. Palaeontol Assoc Spec Pap no 23:141–144

Ross RJ (1976) Ordovician sedimentation in the western United States. In: Bassett MG (ed) The ordovician system. Univ Wales Press, Cardiff, pp 73–105

Sheehan PM (1980) Paleogeography and Marine Communities of the Silurian Carbonate Shelf in the Utah and Nevada. In: Fouch TD, Magathan ER (eds) Rocky Mountain Section. Soc Econ Paleontol Mineral Paleozoic Paleogeogr West-Central United States West Cent US Paleogeogr Symp, no 1, pp 19–37

Sheehan PM, Schiefelbein DRJ (1984) The trace fossil *Thalassinoides* from the Upper Ordovician of the eastern Great Basin: deep burrowing in the Early Paleozoic. J Paleontol 58:440–447

Stolley E (1893) Über silurische Siphoneen. Neues Jahrb Mineral 2:135

8 Masloviporidium, a Cosmopolitan Middle Carboniferous Red Alga

J.R. GROVES[1] and B.L. MAMET[2]

Introduction

The *Atlas of Rock-Building Organisms* was published in Moscow (1973) as an illustrated guide to the recognition of micro- and macrofossils in carbonate rocks. It includes contributions on the stromatoporoids (Yavorsky), foraminifers (Rauzer-Chernousova), and archaeocyathids (Rozanov), as well as extensive treatments by its chief compilator, V.P. Maslov, of the calcareous algae and other major groups. Unfortunately, Maslov's death preceded the publication of the volume.

The chapter on the Rhodophyta (p. 22–27) included the first published report and illustrations of the taxon *Cribroporidium* (Text-Figs. 19a–c and Pl. 21, Figs. 9, 10), which was recovered from Middle Carboniferous rocks of the Donbass (collections no. 196 and 197 of G.D. Kireeva). It is apparent from the text that Maslov considered this form to be morphologically similar to, yet distinguishable from the related genus *Cuneiphycus* Johnson 1960. Further in the text (p. 83), and on Pl. 63 (Figs. 16, 17), under the heading Conodonta and Scolecodonta of Esthonia and Contortopodidia, are two illustrations of specimens referred to as *Contortoporidium.* These are mixed with abundant thalli of *Cribroporidium.* Maslov regarded *Contortoporidium* as a *microproblematicum*, and maintained a distinction between it and *Cribroporidium*, despite their obvious similarity and coincident occurrence in the same sample. Neither *Cribroporidium* nor *Contortoporidium* was proposed formally as a new taxon: i.e., there was no designation of type-species or of type-specimens, and no formal descriptions accompanied the illustrations. Thus, these forms must be considered to be nomina nuda.

We regard the specimens illustrated by Maslov both as *Cribroporidium* and as *Contortoporidium* as representing variously oriented sections through individuals of the same species. The name *Masloviporidium* gen. nov., is proposed here to accommodate these forms.

Since the publication of the "*Atlas*", specimens of *Masloviporidium* been recognized in collections made available to us from elsewhere in the USSR, Algeria, Alaska, and the Midcontinent and western United States (Table 1). These occurrences are Middle Carboniferous in age (Foraminiferal Zones 20 and 21; = Morrowan and Atokan Series of North American terminology, Bashkirian in the Russian sense. Also uppermost Serpukhovian).

1 Department of Geology, University of Iowa, Iowa City, IA 52242, U.S.A.
2 Departément de Géologie, Université de Montréal, Montréal, P.Q., H3C 3J7, Canada

Paleoalgology: Contemporary Research and Applications
(eds. D.F. Toomey and M.H. Nitecki)

Table 1. Worldwide occurrence of *Masloviporidium delicata,* in Middle Carboniferous rocks

Stratigraphic unit	Location	Collector(s)	Associated taxa	Reference
1. C_2, F Limestones	Kruchik Valley, Donetz Basin, USSR	G.D. Kireeva	*Donezella* sp.	Maslov (ed., 1973)
2. Uzunbulaskaya Suite	Uzunbulak Valley, Talasky-Alatau, USSR	B.L. Mamet	*Donezella* sp., *Stacheoides* sp.	Balakhin and others (1975, p. 144)
3. F_2 Limestone	Kruchik Valley, Donetz Basin, USSR	W. Bamber	*Donezella* sp., *Anthracoporellopsis* sp.	Aisenverg et al. (1975, pp. 216 and 310)
4. Tagnana Formation	Colomb-Béchar, Algeria	Y. Lemosquet	*Donezella* sp.	Lys (1979)
5. Peratrovich and Ladron Formations	Peratrovich Island, Alaska	A.K. Armstrong	*Archaeolithophyllum* sp., *Donezella* sp.	Mamet and Pinard this volume
6. Ely Limestone	Granite Mountain, Confusion Range, Utah	A.K. Armstrong, M. Gordon, Jr., R. Hose	*Asphaltina* sp., *Cuneiphycus* sp., *Fasciella* sp., *Sphaeroporella* sp., *Stacheoides* sp.	Hose (1974); Mamet (1982)
7. Prairie Grove Mbr. (Hale Fm.), Brentwood Ls. Mbr. (Bloyd Fm.), Braggs Mbr. (Sausbee Fm.), Chisum Quarry Mbr. (McCully Fm.)	N.W. Arkansas and N.E. Oklahoma	J.R. Groves, B.L. Mamet, P.K. Sutherland	*Archaeolithophyllum* sp., *Cuneiphycus* sp., *Asphaltina* sp., *Stacheoides* sp., *Donezella* sp.	Nodine-Zeller (1977); Groves (1981, 1983a); Mamet (1982; for localities see Sutherland and Manger 1979)
8. Wapanucka Formation	Blanco East, S.E. Oklahoma	B.L. Mamet	*Cuneiphycus* sp., *Donezella* sp.	Grayson, in Sutherland and Manger (eds., 1979, p. 72)
9. Bloom Member (Snaky Canyon Fm.)	Southern Lemhi Range, east-central Idaho	J.R. Groves	*Paraepimastopora* sp., *Osagia* sp., *Stacheoides* sp., *Asphaltina* sp.	Groves (1983b)
10. Type Marble Falls Formation	Northeastern Llano Uplift, central Texas	P.K. Sutherland W.L. Manger	No data	Author's unpublished data (J.R.G.)
11. La Tuna Formation	Hueco Mountains, west Texas	H.R. Lane	*Foliophycus* sp.	Author's unpublished data (J.R.G.)
12. Kearny Formation	West Kansas (subsurface)	Kansas Geological Survey (core)	*Archaeolithophyllum* sp., *Stacheoides* sp., *Donezella* sp.	Author's unpublished data (J.R.G.)

Systematic Paleontology

Rhodophycophyta
Incertae Familiae
Genus *Masloviporidium*, gen. nov.
Type-species: *Donezella delicata* Berchenko 1982 (Fig. 1)

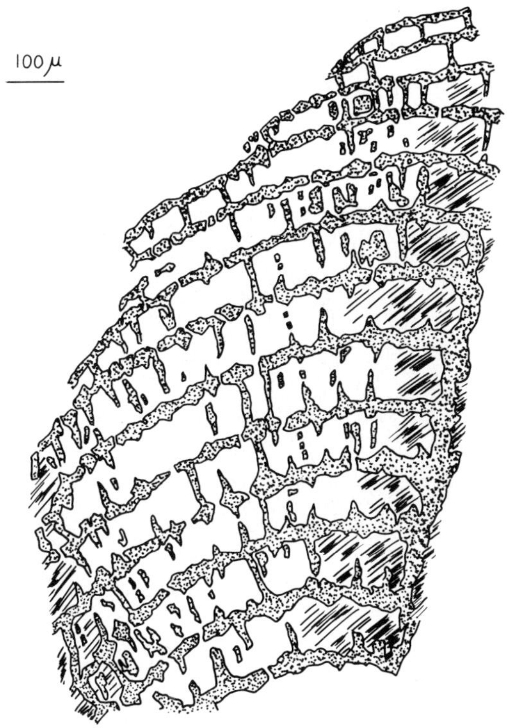

Fig. 1. *Masloviporidium delicata,* gen. nov. (After Maslov 1973, Text-Fig. 19c)

1973 *Cribroporidium* MASLOV. *Atlas of Rock Building Organisms*, Izdat. "Naukova Dumka", Pl. 21, Figs. 9, 10, Pl. 63, Figs. 16, 17, Text-Figs. 19a–c (OBJ, invallid)
1973 *Contortoporidium* MASLOV, *ibid.*, Pl. 63, Figs. 16, 17, Text-Fig. 107 (OBJ, invalid).
1977 *Cuneiphycus* NODINE-ZELLER. Okla. Geol. Sur. Guidebook 18, Pl. 2, Fig. 20.
1982 *Donezella* BERCHENKO. *Systematics and evolution of fossil plants from Ukraine*, "Naukova Dumka", p. 51–52, Pl. II, Figs. 2–5.

Diagnosis. (Text-Fig. 1). Thallus sheet-like, foliate, highly branching, irregular, and composed of arcuate rows of cells, grossly wedge-shaped as in *Cuneiphycus.* Pores in horizontal partitions connect contiguous cells. Wall calcareous secreted, yellowish,

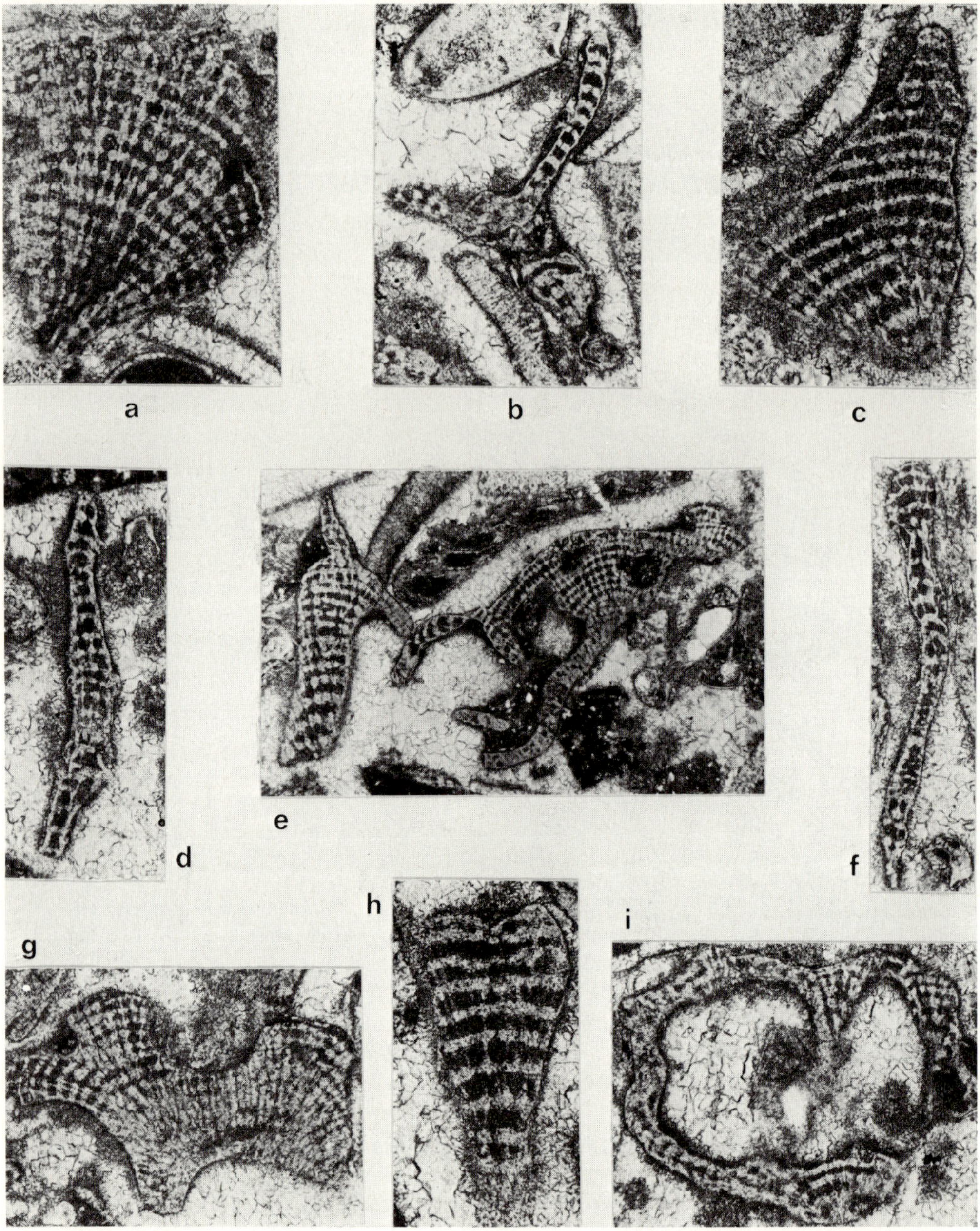

Fig. 2a–i. *Masloviporidium delicata*, gen. nov. All specimens come from the Brentwood Limestone Member of the Bloyd Formation at the Hale Mountain, Arkansas, Zone 20, (sample 10568–14 of Groves 1983a). They are housed in the Paleontological Repository of the University of Iowa (SUI) and the University of Oklahoma Invertebrate Paleontology Repository (OU). **a** SUI 49602c x 32; **b** SUI 49605a x 32; **c** SUI 49604 x 52; **d** SUI 49603e x 32; **e** OU 9911 x 21, showing passage from "*Cribroporidium*" OBJ to "*Contortoporidium*" OBJ; **f** SUI 49606e x 32; **g** SUI 49606d x 32; **h** SUI 49603b x 52; **i** SUI 49605b x 32. All specimens are paratypes

hyaline, probably originally composed of aragonite. Horizontal partitions thick and heavily calcified, whereas vertical ones are much thinner. Reproductive structures unknown.

Geographic and Stratigraphic Distribution (Table 1). Europe (Donets Basin, USSR); North Africa (Algeria); Asia (Kazakhstan, USSR); North America (Alaska, Texas, Oklahoma, Kansas, Idaho, Utah). Middle Carboniferous, uppermost "Serpukhovian", Bashkirian (Foraminiferal Zones 20 and 21; = Morrowan and Atokan Series of North American terminology).

Type-species: *Masloviporidium delicata* (Berchenko 1982) *emend.* (Fig. 2)

1982 *Donezella delicata* BERCHENKO. *Systematics and evolution of fossil plants from Ukraine*, "Nauka Dumka", p. 51–52, Pl. II, Figs. 2–5.
1983 *Donezella delicata* Berchenko. *Upper Serpukhovian Substage in the Donets Basin*, "Nauka Dumka", p. 126, Pl. 85, Figs. 5–7, 9 (not 8), Pl. 86, Fig. 3 (not 2).

Description. *Masloviporidium* composed of irregular cells that vary in height from 20 to 75 μm, and in width from 10 to 50 μm. Diameter of pores in horizontal partitions varies from 5 to 10 μm (higher values may be due to dissolution).

Discussion. Attribution of the species to *Donezella* (Berchenko 1982, 1983) is incorrect as the Donezellinae are tubular, with a cortex perforated by thin blind pores.

The poor "Serpukhovian" material differs from Maslov's material (Kruchick reference-section: see for details Aizenverg et al. 1975) by a less advanced calcification.

Mode of Occurrence. The living plants probably formed wavy, erect palisades that were easily disarticulated. Fossil material is commonly encountered as abraded grains in carbonate packstones and grainstones. The plants do not appear to have encrusted host substrates.

Geographic and Stratigraphic Distribution. See that of the genus.

Summary

Masloviporidium delicata, gen. nov., is erected as type-species of a genus of Middle Carboniferous calcareous red alga. The sheet-like, foliate, and highly branching thallus is composed of wedge-shaped cells defined by heavily calcified horizontal partitions and weakly calcified vertical partitions. Pores in the horizontal partitions connect contiguous cells. The genus is found in carbonate rocks of variable texture in the midcontinental and western United States, Alaska, Algeria, and the Soviet Union, from where it was originally reported.

Acknowledgements. We are grateful to the many people who have made available to us material containing *Masloviporodium.* In particular, K.B. Korde, O.A. Lipina and E.P. Radionova (Geological and Paleontological Institutes of the Academy of Sciences of the USSR) facilitated the examination by Mamet of Maslov's original material. W. Bamber recollected the Kruchick section in 1975. A.K. Armstrong, M. Gordon, R. Hose, R. Lane, Y. Lemosquet, W.L. Manger, and P.K. Sutherland provided useful samples.

References

Aisenberg DE, Babenko AM, Belenko, NG, Belokon VG, Bragin YuN, Dedov VS, Fissunenko OP, Getman VG, Konashov VG, Lagutina VV, Levenstein ML, Maksarov IA, Nagorny YuN, Nesterenko LP, Poletaev VI, Popov VS, Rotai PP, Sharmanova GV, Schegolev AK, Sokolova GU (1975) Field excursion guidebook for the Donetz Basin. 8th Int Congr Carbon Geol Stratigr. Nauka, Moscow, Minist Geol USSR, 359 pp

Armstrong AK, Mamet BL (1977) Carboniferous microfacies, microfossils, and corals. Lisburne Group, Arctic Alaska. US Geol Surv Prof Pap 849:144

Balakhin GV, Bensh FR, Bukharin AK, Dzhenchuraeva AV, Gonchar AD, Iskanderhodzhaev TA, Michno NM, Pitinova AS, Rumyantseva ZS, Savitskaya LI, Sergunkova OI, Sixtel TA, Stankevich YuV, Tsapenko MN, Tyulyandina ZA, Zakharov EF (1975) Field excursion guidebook for the Carboniferous sections of middle Asia. 8th Int Cong Carbon Geol Stratigr. Nauka, Moscow, Minist Geol USSR, 196 pp

Berchenko OI (1982) Systematics and evolution of fossil plants from the Ukraine. Naukova Dumka pp 51–52

Flügel E (1977) Environmental models for Upper Paleozoic benthic calcareous algal communities. In: Flügel E (ed) *Fossil Algae*. Springer, Berlin Heidelberg New York, pp 314–343

Groves JR (1981) Calcareous microfossils from the type Morrowan (Lower Pennsylvanian) region of northwestern Arkansas and northeastern Oklahoma. Geol Soc Am, Abstr With Programs 13: no 7, 465

Groves JR (1983a) Calcareous foraminifers and algae from the type Morrowan (Lower Pennsylvanian) region of northeastern Oklahoma and northwestern Arkansas. Okla Geol Surv Bull 133:65

Groves JR (1983b) Ecologic succession of Lower and Middle Pennsylvanian calcareous algae, southern Lemhi Range, east-central Idaho. Geol Soc Am, Abstr With Programs 15: no 5, 409

Hose R (1974) Geological map of the Granite Mountain S.W. Quadrangle, Juab and Millard counties, Utah. US Geol Surv Misc Rep Ser Map 1–831

Johnson JH (1960) Paleozoic Solenoporaceae and related red algae. Q Colo Sch Mines 55: no 3, 77

Lys M (1979) Micropaléontologie des formations marines du Carbonifère Saharien. 8th Int Congr Carbon Stratigr Geol Moscow 2:37–47

Mamet BL (1982) A report on algal microfacies at the Lower-Middle Carboniferous boundary, North America. In: Ramsbottom WHC, Saunders WB, Owens B (eds) Biostratigraphic data for a mid-carboniferous boundary. Symposium on a mid-carboniferous boundary. Meeting of IUGS subcommission on carboniferous stratigraphy. Leeds, 1981, pp 52–60

Mamet BL, Pinard S (1985) Algal microflora of the Peratrovich Formation (southwestern Alaska, Carboniferous) (this volume)

Mamet BL, Roux A (1977) Algue rouges dévoniennes et carbonifères de la Téthys occicentale. Rev Micropaleontol 19: no 4, 215–266

Maslov VP (ed) (1973) Atlas of rock-building organisms (Calcareous and Siliceous Organisms). Nauka, Moscow, 267 pp (in Russian)

Nodine-Zeller, DE (1977) Microfauna from Chesterian (Mississippian) and Morrowan (Pennsylvanian) rocks in Washington County, Arkansas, and Adair and Muskogee counties, Oklahoma. In: Sutherland PK, Manger WL (eds) Upper Chesterian-Morrowan stratigraphy and the Mississippian-Pennsylvanian boundary in northeastern Oklahoma and northwestern Arkansas. Okla Geol Surv Guideb 18:89–100

Sutherland PK, Manger WL (eds) (1979) Mississippian-Pennsylvanian shelf-to-basin transition. Ozark and Ouachita region, Oklahoma and Arkanasas. Okla Geol Surv Guideb 19:81

Wray JL (1977) Calcareous algae. Elsevier, Amsterdam, 185 pp

9 Carboniferous Algae from the Peratrovich Formation, Southeastern Alaska

B.L. MAMET and S. PINARD[1]

Introduction

The Peratrovich Formation is exposed in the Klawak, Ladrones, Madre de Dios, Shelikof, Toti and Wadleigh Islands off the west side of Prince of Wales Island in southeastern Alaska (Armstrong 1970, Fig. 1). The total thickness of the succession is hard to assess as the base of the formation is extensively faulted. A minimum of 550 meters of carbonates can be measured in nine well-exposed sections. The whole gamut of carbonate facies is observed. In the lower part of the formation, deep-water radiolarites and spiculites are overlain by an extensive carbonate sequence (Vaskey 1982, Pinard 1982) in an island arc setting within the Cordilleran Eugeosyncline (Bamber and Mamet 1985). The succession of Foraminiferal Zones (Mamet and Skipp 1970) ranges from the late Tournaisian to the Bashkirian and indicates a mixture of North American and Tethyan fauna (Mamet and Skipp 1979). This contribution illustrates the algal microflora and its paleobiogeography based on the microfossils. We have also added some references to the Bashkirian Ladrones microflora which overlies the Peratrovich Formation.

Algal Microflora

Algae are ubiquitous throughout the carbonate sequence and are well represented from middle Viséan Zone 13 to Zone 20 (Foraminiferal Zones used herein are described in Mamet 1971, Armstrong and Mamet 1977). *Koninckopora,* characteristic of channel grainstones which are associated with carbonate banks, ranges from middle Viséan to late Viséan. In the upper Viséan, *Issinella* and *Kamaena* form extensive bafflestones trapping and stabilizing sediments. Later carbonates are characterized by a series of red algae boundstones and associated rudstones. Lower Namurian *Ungdarella* boundstones are followed by an upper Namurian *Ungdarella-Komia* association and finally by lower Bashkirian *Ungdarella-Donezella-Masloviporidium* boundstones.

1 Département de Géologie, Université de Montréal, C.P. 6128, Montréal, Québec, Canada H3C 3J7

Paleoalgology: Contemporary Research and Applications
(eds. D.F. Toomey and M.H. Nitecki)

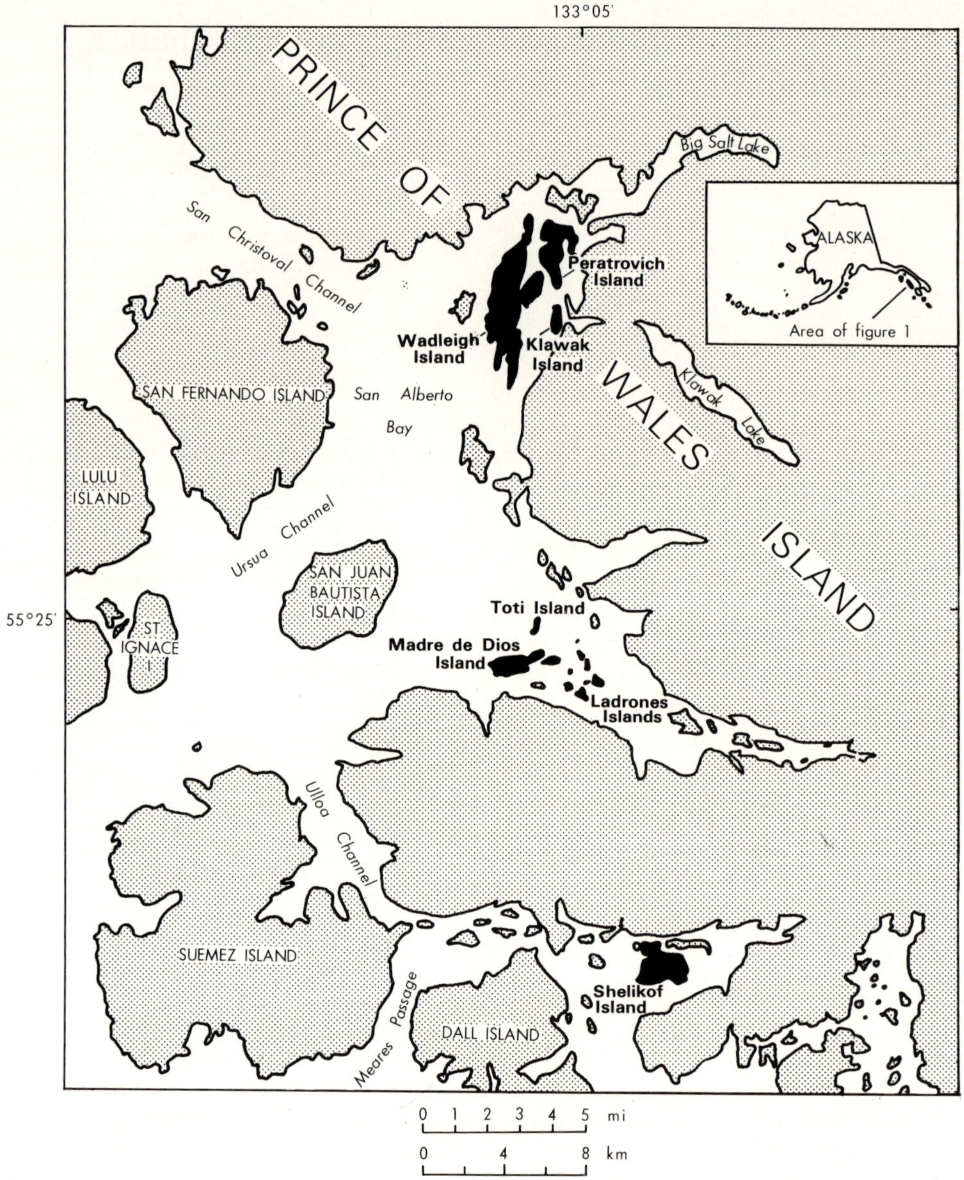

Fig. 1. Peratrovich Formation outcrops at Klawak, Ladrones, Madre de Dios, Peratrovich, Shelikof and Wadleigh Islands. For precise location of the stratigraphic sections, refer to Armstrong (1970)

The microflora with the exception of a few poorly preserved forms is illustrated in Plates 1–3. The abundance of the thirty-four genera, their stratigraphic range in southeastern Alaska, and their distribution in North America and in the Tethys are listed in Table 1.

Plates 1–3. For each taxon, the following information is given: collection number, Department of geology, University of Montreal-locality-position in the section-Foraminiferal Zone-stratigraphic age-magnification.

Plate 1

1 *Girvanella wetheredii* Chapman 1908. Se 22/5, Toti Island, 18.4 m above the base, Zone ⩾13, middle Viséan (or late Viséan). x 83.
2 *Mitcheldeania nicholsoni* Wethered 1886. Se 13/12, Shelikof Island-1, 88.7 m above the base, Zone 15, late Viséan. x 53.
3 Nodule of *Bevocastria conglobata* Garwood 1931, and *Ortonella* sp. Se 12/14, Shelikof Island-2, 39.8 m above the base, Zone 14, late Viséan. x 21.
4 *Atractyliopsis minima* Mamet and Roux 1978. Se 12/21, Shelikof Island-2, 53.6 m above the base, Zone 14, late Viséan. x 62.
5 *Hedstroemia flabellata* (Malakhova 1975). Se 16/23, Ladrones Islands, 33.6 m above the base, L.43, Zone ⩾18, Namurian. x 53.
6 *Pseudohedstroemia polyfurcata* Mamet and Roux 1978. Se 17/6 Ladrones Islands, 33.6 m above the base, L.45, Zone ⩾18, Namurian. x 21.
7 *Asphaltina cordillerensis* Mamet in Petryk and Mamet 1972. Ma 278/9, Shelikof Island-1, 177.6 m above the base, Zone 16s, late Viséan. x 35.
8 *Ortonella* aff. *coloradoensis* Johnson 1945. Se 16/13, Ladrones Islands, 21.4 m above the base, L.33, Zone ⩾18, Namurian. x 53.
9 *Issinella sainsii* Mamet and Roux 1975. Se 12/28, Shelikof Island-1, 34.3 m above the base, Zone 14, late Viséan. x 53.
10 *Kamaena delicata* Antropov 1967. Ma 510/35, Madre de Dios Island, base of the section, Zone 14, late Viséan. x 83.
11 *Calcisphaera laevis* Williamson 1881. Ma 215/20, Shelikof Island-1, 85.6 m above the base, Zone 14/15, late Viséan. x 83.
12 *Calcisphaera pachysphaerica* Pronina 1963). R 30/5, Shelikof Island-2, 38.2 m above the base, Zone ⩾13, middle Viséan (or late Viséan). x 53.
13 *Proninella gracilis* (Vachard in Meissami et al. 1978). Se 22/6, Toti Island, same as 1, but x 83.
14 *Paracaligelloides* sp. Se 11/29, Madre de Dios Island, 45.9 m above the base, L.47, late Viséan. x 53.
15 *Anthracoporellopsis machaevii* Maslov 1956. Se 26/13, Peratrovich Island, 116.3 m above the base, Zone ⩾18, Namurian. x 53.
16 *Aoujgalia richi* Mamet and Roux 1978. Se 19/28, Ladrones Formation, Ladrones Islands, 88.7 m above the base, L.53, Zone 21, late Bashkirian. x 53.
17 *Stacheoides? spissa* Mamet in Petryk and Mamet 1972 encrusting *Fasciella* sp. Ivanova 1973. Se 22/8, Toti Island, 18.4 m above the base, Zone ⩾13, middle Viséan (or late Viséan). x 21.
18,21 *Cuneiphycus johnsoni* Flügel 1966. 18 Se 6/8, Peratrovich Island, 131.5 m above the base, L.4, Zone ⩾ 18, Namurian, x 21. 21 Se 22/24, Toti Island, 31.1 m above the base, Zone ⩾13, middle Viséan (or late Viséan). x 53.
19 *Nostocites vesiculosa* Maslov 1929. Se 26/12, Peratrovich Island, 116.2 m above the base, L.5, Zone ⩾18, Namurian. x 83.
20 *Asphaltinella peratrovichensis* n. sp. Mamet and Pinard 1983. Type of the species. Se 18/27, Ladrones Islands, 55.0 m above the base, Zone 20, early Bashkirian. x 53.

Plate 2

1 Nodular incrustation of *Bevocastria conglobata* Garwood 1931. Se 21/1, Toti Island, 76.5 m above the base, L.45, Zone 14, late Viséan. x 21.
2 *Girvanella wetheredii* Chapman 1908. Se 2/25, Peratrovich Island, 87.2 m above the base, Zone 18, early Namurian. x 53.
3 *Koninckopora tenuiramosa* Wood 1942. R 32/34, Shelikof Island-1, 165.2 m above the base, Zone 16s, late Viséan. x 53.

4,5 *Koninckopora inflata* (de Koninck 1842). 4 R 30/7, Shelikof Island-2, 38.2 m above the base, Zone ⩾ 13, middle Viséan (or late Viséan). x 53. 5 Se 13/2, Shelikof Island-1, 41.3 m above the base, L.56, Zone 14, late Viséan. x 53.

6 *Hedstroemia* sp. Se 2/25, L.7, same as 2.

7 *Orthriosiphonoides*? sp. or *Orthriosiphon*? sp. Se 28/25, Ladrones Islands, 55.0 above the base, Zone 20, early Bashkirian. x 53.

8,9 *Aphralysia capriorae* Mamet and Roux 1975. 8 Se 2/32, Peratrovich Island, 122.4 m above the base, Zone ⩾ 18, Namurian. x 21. 9 Se 2/32, Peratrovich Island, same as 8.

10 *Proninella strigosa* (Vachard in Perret and Vachard 1977). Se 21/12, Toti Island, 4.6 m above the base, L.5A, Zone ⩾ 13, middle Viséan (or late Viséan). x 83.

11 "*Masloviporidium*" sp. Ma 485/33, Ladrones Islands, 54.4 m above the base, Zone 20, early Bashkirian. x 83

12 *Orthriosiphon saskatchewanensis* Johnson and Konishi 1956. Se 10/9, Madre de Dios Island, 33.6 m above the base, L.24, Zone 14, late Viséan. x 53.

13,17 *Fasciella kizilia* Ivanova 1973 (≡ *Shartymophycus fusus* Kulik 1973). 13 Se 22/26, same as 7. Ma 368/27, Shelikof Island-1, 198.9 m above the base. Zone 17, early Namurian. x 53.

14,15 *Cuneiphycus texana* Johnson 1960. 14 Se 18/14, Ladrones Formation, Ladrones Islands, 48.9 m above the base, Zone 21, late Bashkirian. x 53. Ma 507/36, 79.5 m above the base, Zone 21, late Bashkirian. x 83.

16 "*Masloviporidium*" sp. Ma 366/39, Peratrovich Island, 142.6 m above the base, L.1, Zone 20 early Bashkirian. x 83.

Plate 3

1 *Ungdarella uralica* Maslov 1956. Se 5/25, Peratrovich Island, 127.0 m above the base, Zone ⩾ 18, Namurian. x 53.

2 *Stacheoides tenuis* Mamet and Rudloff 1972. Ma 508/15, Ladrones Formation, Ladrones Islands, 79.5 m above the base, Zone 21, late Bashkirian. x 67.

3 *Stacheoides polytrematoides* (Brady 1876). Se 19/16, Ladrones Formation, Ladrones Islands, 82.6 m above the base, Zone 21, late Bashkirian. x 53.

4 *Ungdarella peratrovichensis* Mamet and Rudloff 1972. Se 19/9, same as 2, but x 21.

5 *Epistacheoides connorensis* Mamet and Rudloff 1972. Se 8/30, Peratrovich Island, 143.8 m above the base, Zone 20, early Bashkirian. x 21.

6 *Epistacheoides nephroformis* Mamet in Petryk and Mamet 1972. Se 19/24, Ladrones Formation, Ladrones Islands, 85.7 m above the base, Zone 21, late Bashkirian. x 21.

7 *Archaeolitophyllum missouriensum* Johnson 1956. Ma 495/35, Ladrones Islands, 52.0 m above the base, Zone 20, early Bashkirian. x 67.

8,10 *Asphaltinella peratrovichensis* n. sp. Mamet and Pinard. 8 Ma 507/26, same as 2, but x 53 10 Se 19/12, same as 2, but x 53.

9 *Stacheoides* aff. *tenuis* Mamet and Rudloff 1972. Se 15/30, Ladrones Islands, 15.3 m above the base, L.23, Zone 20, early Bashkirian. x 53.

11 *Cuneiphycus johnsoni* Flügel 1966. Se 19/8, Peratrovich Island, same as 2, but x 83.

12 *Komia abundans* Korde 1951. Se 7/1, Ladrones Formation, Ladrones Islands, 134.6 m above the base, Zone 21, late Bashkirian. x 53.

13 *Stacheoides meandriformis* Mamet and Rudolf 1972. Ma 507/24, Ladrones Formation, Ladrones Islands, 78.0 m above the base, Zone 21, late Bashkirian. x 53.

The microflora has a low specific diversity (species/genus ratio: 1,3) and the majority of the taxa has very wide geographic dispersion. It is quite surprising not to recognize a single characteristic American genus such as *Albertaporella, Columbiapora, Pekiskopora, Pseudoissinella, Sphinctoporella, Windsoporella* and *Yukonella.* Schizophyta are represented by *Girvanella.* Spongiostromids are very rare. Among the Chlorophycophytes, the five recognizable Codiaceans are scarce, and the charac-

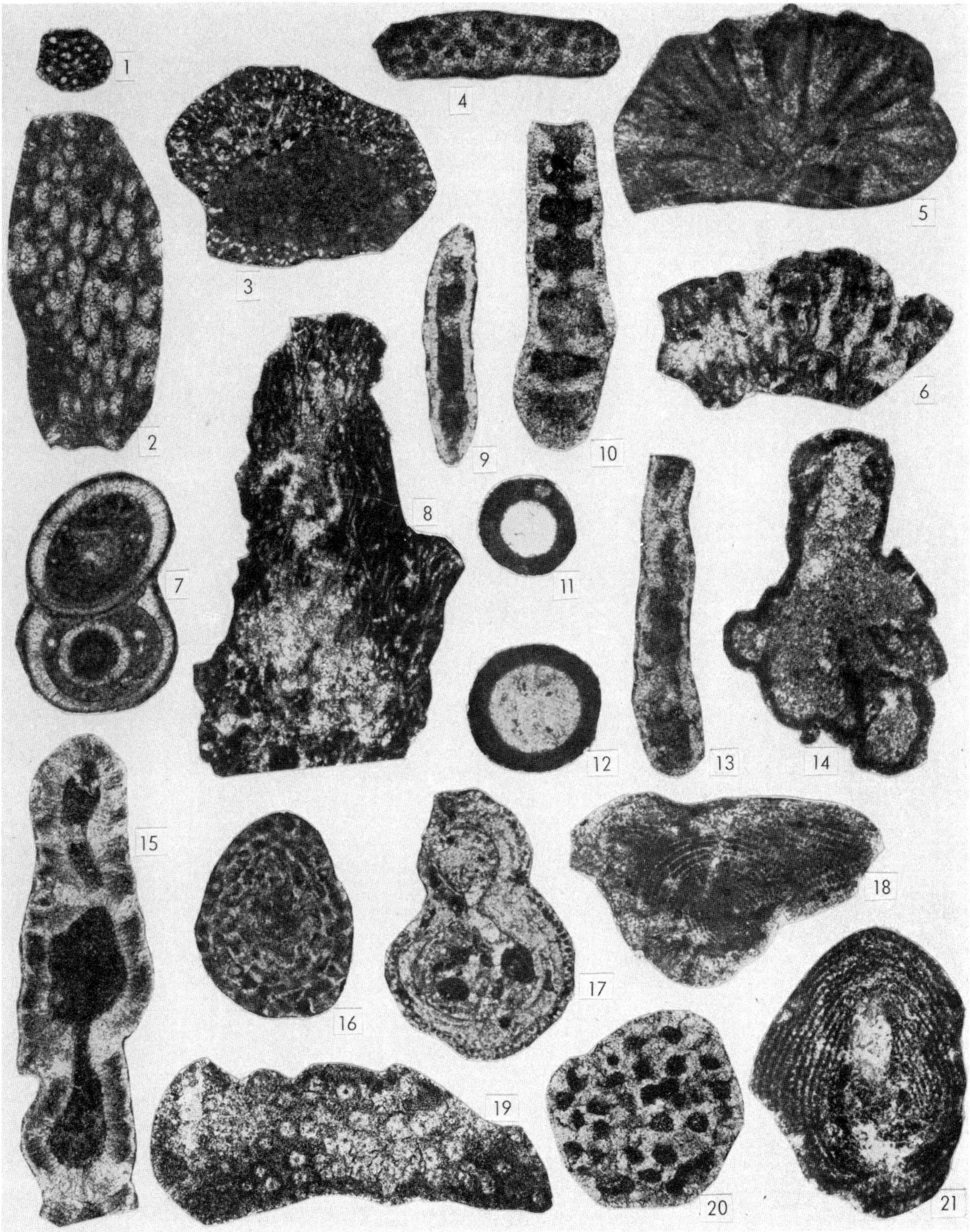

Plate 1 (legend see p. 93)

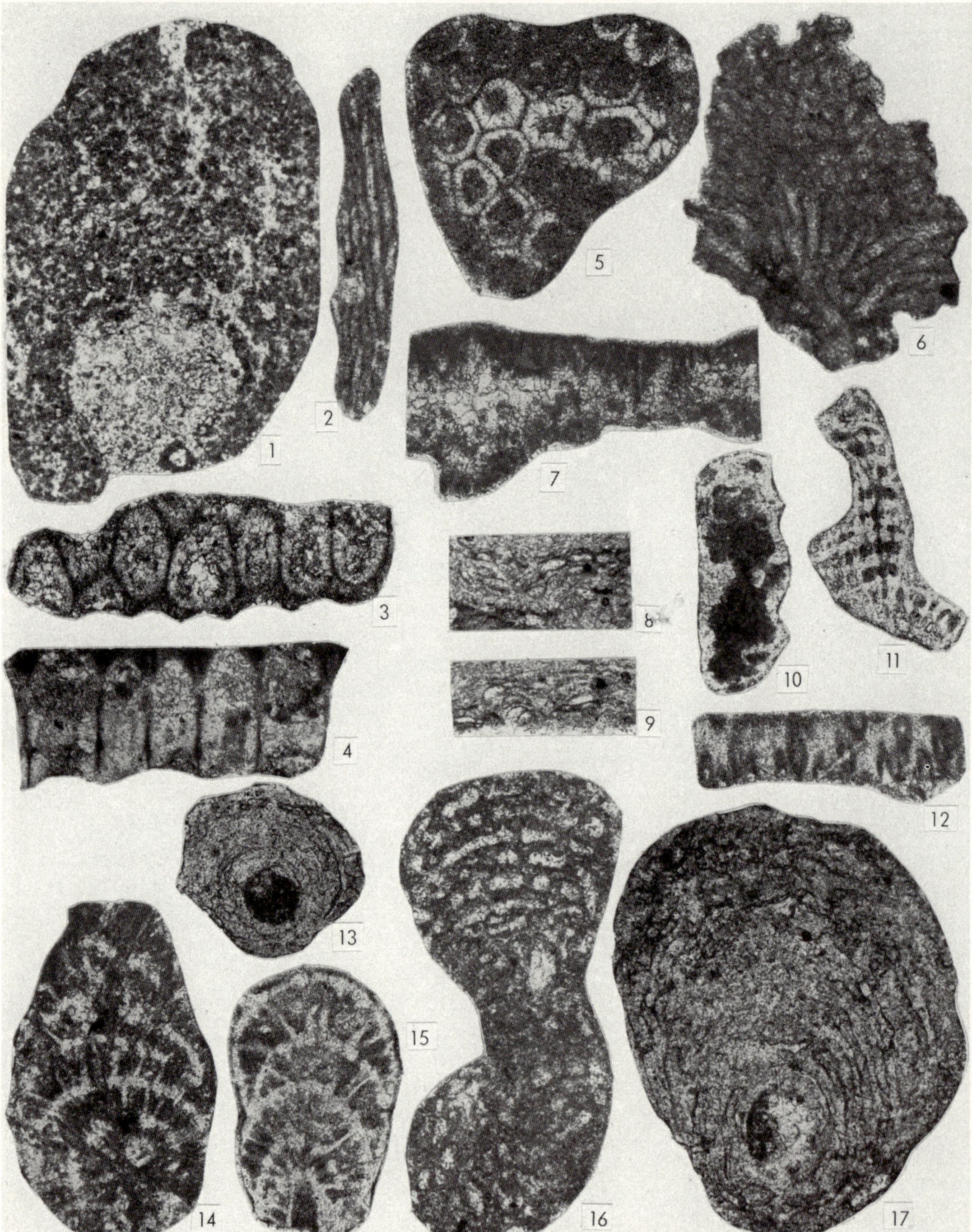

Plate 2 (legend see p. 93–94)

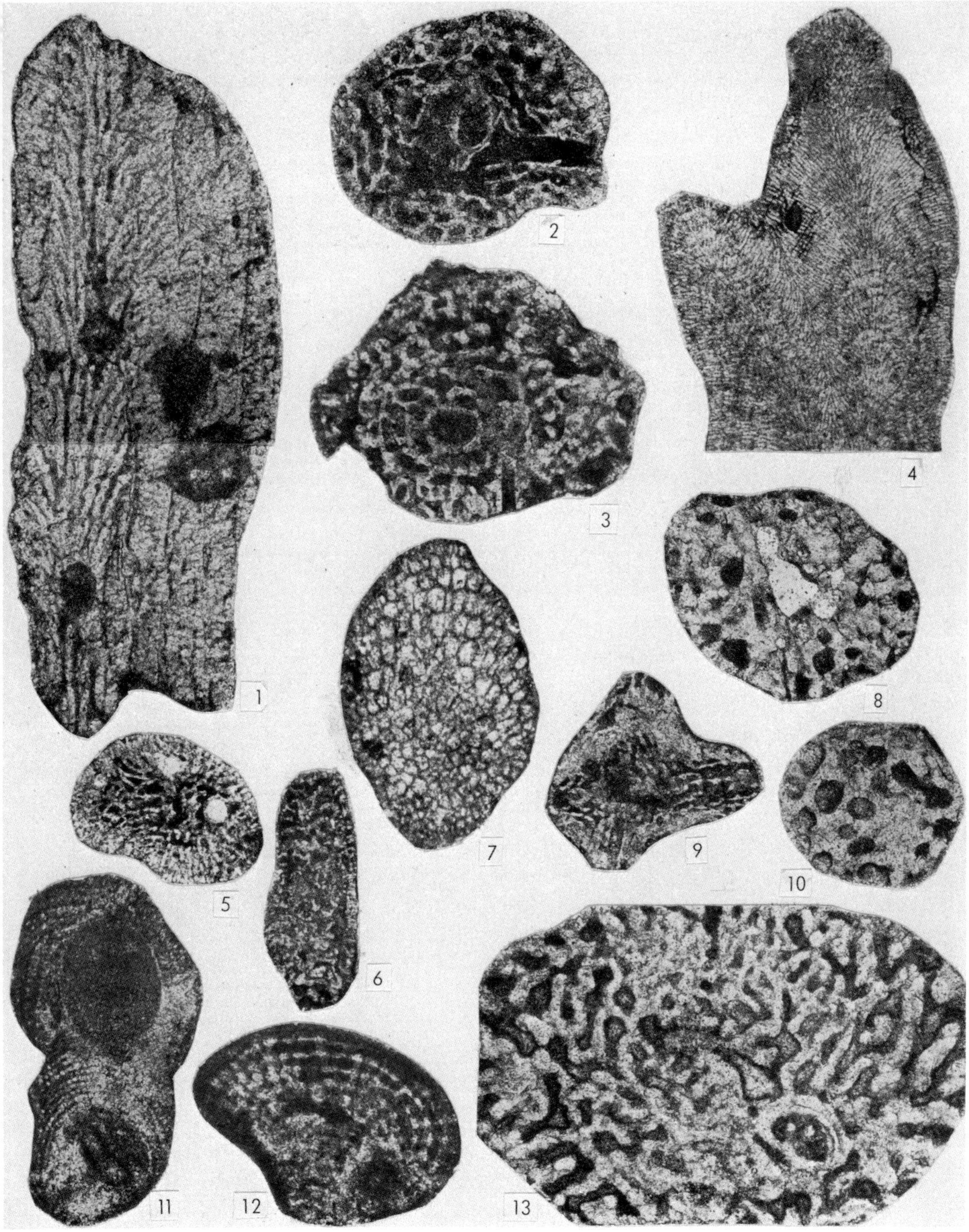

Plate 3 (legend see p. 94)

Table 1. Peratrovich microflora, showing stratigraphic range and abundance of taxa within Peratrovich Formation. Foraminiferal Zones, here varying from 10 to 20, are described in Armstrong and Mamet (1977). The table also indicates paleogeographic distribution and relative abundance of the flora in North America and in the Tethys. Cosmopolitan distribution are modified from Mamet and Roux (1977)

Group	Taxon	Peratrovich Formation (zones)	Peratrovich Formation (abundance)	North America	Tethys	COS°
• SCHIZOPHYTA	*Girvanella wetheredii*	13-20	rare	rare/present	very abundant	X
• CHLOROPHYCOPHYTA	*Aphralysia capriorae*	16s-18	rare	rare	present	X
• CODIACEAE	*Bevocastria conglobata*	13-20	rare	present	common	X
	Hedstroemia flabellata	18	rare		rare	
	Mitcheldeănia nicholsoni	15	rare	present	present	X
	Ortonella aff. *coloradoensis*	12	present	present	present	X
	Pseudohedstroemia polyfurcata	15-18	rare	present		
• DASYCLADACEAE and DASYCLADACEAE?	*Actractyliopsis minima*	14	rare	present		
	Anthracoporellopsis machaevii	14-20	present	present	present	X
	Donezella lutugini *	18-20	variable	variable	variable	X
	Issinella devonica *	13-20	variable	variable	variable	X
	Issinella sainsii	14-16i	rare	rare	rare	X
	Kamaena delicata	13-20	rare	rare	present	X
	Koninckopora inflata	13-16s	variable	variable	very abundant	X
	Koninckopora tenuiramosa	13-16i	variable	variable	very abundant	X
	Orthriosiphon saskatchewanensis	14	present	common		
	Orthriosiphonoides tenuiramosa	14	rare	present		
	Paleoberesella lahuseni *	10-16i	variable	variable	variable	X
• RHODOPHYCOPHYTA	*Aoujgalia richi*	16i-20	present	present		
	Archaeolithophyllum missouriensum	20	present	variable	variable	X
	"Cribroporidium" sp.	20	rare	present	variable	X
	Cuneiphycus cf. *aliquantulus*	20?	rare	present	present	X
	Cuneiphycus johnsoni	18	rare	present	present	X
	Cuneiphycus texana	18?-20	rare	present	?	?
	Epistacheoides connorensis	14-20	present	present	rare	X
	Epistacheoides nephroformis	14-20	present	common	rare	X
	Komia abundans	18-20	common	common	abundant	X
	Pseudosolenopora sp. *		rare	present	present	X
	Pseudostacheoides loomisi *	13-15	rare	present	present	X
	Stacheoides meandriformis	13-20	common	common	common	X
	Stacheoides polytrematoides	18-20	present	present	present	X
	Stacheoides? spissa	13-18	rare	present		
	Stacheoides aff. *tenuis*	13-20	rare	present		X
	Stacheoides tenuis	13-20	present	common	common	X
	Ungdarella peratrovichensis	18	abundant	rare		
	Ungdarella uralica	16i-20	variable	common	very abundant	X
• INCERTAE SEDIS	*Asphaltina cordillerensis*	16s-20	present	variable	present	X
	Asphaltinella peratrovichensis	15-20	present			
	Fasciella kizilia	16s-20	common	present	very abundant	X
	Nostocites vesiculosa	13-18	rare	present	present	X
	"Paracaligelloides" sp.	14-16i	present	present	present	X
	Proninella gracilis	13-17	present	present	common	X
	Proninella strigosa	13-18	present	present	common	X
• ALGAL CYSTS	*Calcisphaera laevis*	9-18	present	abundant	abundant	X
	Calcisphaera pachysphaerica	10-20	present	abundant	abundant	X
	Parathurammina sp. *	12	present	common	common	X

° COSMOPOLITAN — — — — RARE ——— COMMON ▬▬ VERY ABUNDANT
* NOT ILLUSTRATED ——— PRESENT ▬ ABUNDANT ■■■■ VARIABLE

teristic Tethyan *Calcifolium* is conspicuously absent. Dasyclads and related forms are represented mainly by *Koninckopora* and again they are of low diversity. *Issinella* in the Viséan and *Donezella* in the Namurian play an important role in sedimentation as they form extensive bafflestones. The Rhodophycophytes are represented by Ungadarellaceae and Stacheiinae, while Solenoporaceae are scarce.

Taxonomy

All taxa have been previously described, including *Ungdarella peratrovichensis* Mamet and Rudloff (1972), which is a very common fossil in the Namurian. To this microflora can be added a new taxon, whose description follows.

INCERTAE SEDIS
Asphaltinella peratrovichensis sp. nov.
(Plate 1, Fig. 20, Plate 3, Figs. 8 and 10)

Diagnosis. Thallus encrusting, composed of closely packed, unsegmented, cylindrical to subcylindrical tubes, with regular concentric arrangement.

Description. Thallus encrusting, composed of closely packed, regular cylindrical tubes, with concentric arrangement. The type has a width of 520 μm for 30 tubes; internal diameter of the tubes, 60–75 μm. Wall is single layered, pseudofibrous, and perforated by thin pores, 25 to 40 μm.

Comparison. Asphaltinella peratrovichensis is similar in general dimensions to *A. horowitzi* Mamet and Roux 1978, but differs from it by a much greater regularity of the arrangement of the tubes.

Material. The richest assemblages come from interval 55–80 m above the base of the Ladrones section and at the junction of the Peratrovich and Ladrones Formations. Also known from the late Viséan Peratrovich Formation in the Shelikof section.

Holotype. Université de Montréal Se 18/27, 66X-12+180C. Plate 1, Fig. 20.

Stratigraphic Range. Late Viséan-Bashkirian.

Paleobiogeography

The algal microflora is neither typically Tethyan, nor characteristically North American, but is essentially cosmopolitan. Hence it is interesting to compare this flora with the associated foraminiferal microfauna.

Peratrovich foraminifers have many Tethyan genera hitherto unrecorded or very scarce in North America: e.g., *Cribrospira, Howchinia, Janischewskina, Mediocris, Omphalotis, Spinothyra* and *Viseidiscus* mixed with typical North American forms such as *Banffella* and *Zellerinella.* However such Tethyan markers as *Endostaffella, Forschiella, Haplophragmella, Loeblichia, Permodiscus* and *Valvulinella* are absent. While quite rich, the microfauna is rather and undiversified, and the species/genus ratio is much lower than that observed in Eurasia.

In conclusion, Peratrovich microfossils constitute a link between Eurasia and North America. Foraminifers have a strong Tethyan affinity, although they certainly cannot be considered typically Tethyan. The microflora is remarkably cosmopolitan. There are only two species restricted to the formation while 36 of 46 taxa have a north hemispheric distribution.

Summary

The well-preserved Peratrovich algal microflora (Tournaisian?-basal Pennsylvanian) is a rich assemblage of low specific diversity and is characteristic of a cosmopolitan flora. Thirty-four genera of Schizophyta, Chlorophycophyta, Rhodophycophyta and some incertae sedis have been identified. However, there are few representatives of the Schizophyta, which reflects the scarcity of lagoonal facies within the carbonate sequence. Viséan Codiaceae and Dasycladaceae are well documented. Dasyclads(?) such as *Issinella* and *Kamaena* stabilize the sediment and form bafflestones. *Koninckopora* are observed in channelized grainstones within banks. Rhodophycophyta dominate the Namurian and Bashkirian microflora. *Ungdarella* and *Komia* flourish and build boundstones in high energy environments. *Asphaltinella peratrovichensis* n. sp. is placed among the small encrusting incertae sedis.

Acknowledgments. The authors wish to express their gratitude to Dr. A.K. Armstrong of the U.S. Geological Survey, Menlo Park, who collected the Peratrovich material. Support from the Natural Sciences and Engineering Research Council of Canada is also kindly acknowledged.

References

Armstrong AK (1970) Mississippian Rugose Corals, Peratrovich Formation, West Coast, Prince of Wales, Southeastern Alaska. US Geol Surv Prof Pap 534:44

Armstrong AK, Mamet B (1977) Carboniferous Microfacies, Microfossils, and Corals, Lisburne Group, Arctic Alaska. US Geol Surv Prof Pap 849:129

Bamber EW, Mamet B (1985) Carboniferous stratigraphy and depositional history of Canada. 10th Int Congr Carbon Stratigr and Geol, Madrid, 1983, in press

Mamet B (1971) Une Zonation par Foraminifères du Carbonifère Inférieur de la Téthys Occidentale. 7th Congr Int Stratigr Geol Carbon, Krefeld. C-R 3:391–408

Mamet B, Roux A (1977) Algues rouges dévoniennes et carbonifères de la Téthys occidentale. Quatrième partie. Rev Micropaleont 19: no 4, 215–266

Mamet B, Roux A (1978) Algues viséennes et namuriennes du Tennessee (Etats-Unis). Rev Micropaleontol 21: no 2, 68–97

Mamet B, Skipp B (1979) Lower Carboniferous Foraminifera–Paleogeographic implications. 8th Congr Int Stratigr Geol Carbon, Moscow. C-R 2:48–66

Pinard S (1982) Zonation micropaléontologique (Foraminifères, Algues) de la Formation de Peratrovich (Carbonifère Inférieur), Alaska. Unpubl Mem M Sc, Univ Montreal, 174 p

Vaskey GT (1982) Carboniferous facies and some Pennsylvanian brachiopods, Western Prince of Wales Island, Southeast Alaska. Unpubl M Sc thesis, Univ Oregon, 255 p

10 Review of Tethyan Mesozoic Algae of Romania

O. DRAGASTAN[1]

Introduction

Tethyan Mesozoic algae have been the subject of several studies, which have often reviewed long forgotten taxa and inspired critical systematic reviews. Lately the taxonomic discussions have focused on section Porostromata Pia, 1927 (Cyanophyta) and on Trevisan family Codiaceae Zanardini 1843 (Chlorophyta).

The Paleozoic Tethyan algae have been revised recently by Mamet and Roux (1974, 1975), Mamet et al. (1977), Toomey (1979), Riding (1977a), and Flügel (1977, 1979, 1981), while the Mesozoic and Cenozoic were revised by Elliott (1964, 1975), Ott (1972), Masse (1976, 1979), Flügel (1975, 1977), and Schäfer (1979).

The present study is concerned with the taxonomic criteria and possible nomenclature of Porostromata and the Mesozoic and Cenozoic genera.

Criteria for Distinguishing Between Porostromata Genera and Species. The criteria for distinguishing the Porostromata taxa may be stated by referring to the study of Richter et al. (1979) regarding *Rivularia haematites,* based on the dimensions of pseudobifurcate filaments. Three dimensional types have been recognized: type A with diameter between 27–72 μ reported from structures with hemispherical, zonal morphology, and erect growth which represents 20%; type B with diameter between 8–37 μ and a frequency which exceeds 30% in simple, isolated, zonal hemispherical structures; type C with diameter between 2–12 μ and a frequency exceeding 75% may be encountered in crusts and nodules. Type C dimensions corresponds to those reported for the original specimens of *R. dura* and *R. haematites*, while type B resembles other living species of the genus *Rivularia.* The authors mentioned above show that the Cyanophyceae algae are rather problematic with respect to the classification and nomenclature of Mesozoic and Cenozoic genera; *Rivularia* is still omitted from fossil types, although several authors have compared their Meso-Cenozoic specimens to the present-day *Rivularia.*

According to Pia (1927), the criteria for recognizing Porostromata are as follows: the branching or pseudo-branching of cells or filaments, the shape of filaments (cylindrical, straight, sinuous, and constricted), length of filaments, diameter of filaments, thickness of filament walls, and the shape of the thallus and its zonal growth.

1 University of Bucharest, Laboratory of Palaeontology, Bd. N.Bălcescu nr.1, 70111 Bucharest, Romania

Paleoalgology: Contemporary Research and Applications
(eds. D.F. Toomey and M.H. Nitecki)

For Maslov (1949) the only objective measures for distinguishing between the *Girvanella* species are a diameter along the filament but he does not specify whether he means the inner or outer diameter.

Mamet and Roux (1974, 1975) introduce two elements for Porostromata: the inner diameter of the filament at the base of the branching (= pseudobifurcation) and the wall thickness. According to these authors, wall thickness is a convincing criterion because the precipitation of $CaCO_3$ on the filament surface is extremely variable.

Mitcheldeania, Garwoodia, Ortonella, Hedstroemia, and *Bevocastria* are assigned to the Codiaceae. It is believed that most Paleozoic genera and families have no living representatives. On the other hand, for the post-Jurassic time, it is easy to establish analogies with living algae, especially for Cyanophyta; however, any classification attempts are speculative. Nevertheless, I believe that there may be modern analogues even for the Paleozoic Cyanophyta. For example, *Renalcis, Bija*, and some species of *Ortonella* may be compared with the recent algae.

Using Pia's criteria as a starting point, Flügel (1975, 1977) states that the branching type and diameter of filaments (lumen) are important for delimiting the Porostromata species. The criterion of inner diameter, or lumen, had been used by Richter et al. (1979) for the study of *Rivularia haematites.* Riding (1977a) and Wray (1977) discuss the affinities of *Ortonella, Garwoodia, Cayeuxia,* and *Girvanella*, and consider these to be skeletal stromatolites (Cyanophyta).

Masse (1979) introduced the term "schizophytoidal" and distinguishes several basic structures: (1) cellular or filamentlike types, e.g., Porostromata, to which he assigns *Bacinella sterni* Radoicic; (2) "*Cayeuxia*" with diverging filaments and simple fanlike branching assigned by Shuyskiy (1973) to Rivulariaceae or Garwoodiaceae; (3) "*Ortonella*" with Y-shaped filaments; (4) *Girvanella* with flexuous, interwoven filaments; (5) *Epiphyton* with diverging, successively branched filaments; and (6) *Marinella lugeoni* with fasciculate filaments.

He assigns to the Cyanophyta *Marinella lugeoni* Pfender and *Pycnoporidium sinuosum* Johnson and Konishi, in which the filaments show partitions or "septa". *Marinella* has diameter of filaments reduced (10–20 μ), which is characteristic of Cyanophyta, and *Pycnoporidium* and *Stenoporidium* have filaments with greater diameter, not exceeding 100 μ, which is the accepted limit for Cyanophyta filaments. Thus, the occurrence of partitions is rather unusual for these algae, since in the Cyanophyta the "partitions" correspond to adjoining cell walls.

Schäfer (1979), also concerned with the nomenclature of Porostromata, on the basis of Upper Rhaetian material recognized the following three groups: (1) *Cayeuxia/Garwoodia* with short filaments and simple or complicated filaments; (2) *Zonotrichites/Apophoretella* with long and thin filaments; and (3) *Ortonella*, which is deeply and divergently branched. Schäfer also recognize types with small dimensions, in which the angle between filaments (bifurcation) is considered for the first time, and she recognizes two groups – one with the angle of 90°, and the other with the angle below 90°. However, this distinction is not important, since genera have not been classified by angles of bifurcation. Schäfer's study could have been more valuable if all his types were included, if the taxonomic values of the types were evaluated, and if possible synonymies between genera and species were established.

The classification of genera and species of Cyanophyta (Porostromata) should be based on the type of pseudobifurcation of filaments or cells. It is known that most of living Cyanophyta are characterized by V-shaped or Y-shaped sympodial pseudobifurcation. Types with polyfurcate filaments are not known. In my opinion, all pseudobifurcate algae should be assigned to Cyanophyta. Another important taxonomic character is the inner diameter of the filament, or lumen, measured at the base of the branching. The maximum size of the filament diameter of living and fossil Cyanophyta does not exceed 100 μ, a value that must be considered while classifying cyanophytes with filamentlike thalli.

Another essential taxonomic character is the angle of divergence of pseudobifurcate filaments. There are clearly three angular classes. The first class, between 2° and 20°, is characteristic of *Rivularia* and its synonyms: *Zonotrichites* (3°–8°, 10°–15°, and rarely 20° in Bornemann and Elliott and Ott), *Cayeuxia* (10°–20°, Frollo 1938), *Stenoporidium* (10°–15°), *Diversocallis* (3°–5°), *Dimorphostroma* (5°–10°), *Dendractis* (5°–7°), *Brachydactylus* (10°), and *Rivularialithus* (10°) – all according to the publications of different authors. The second (25°–30°) I call an interference, or transition class, e.g., *Rivularia* and *Ortonella*, and *Paraortonella* n.gen. The third class includes *Ortonella*, and its values range between 30°–60° (50°–80°). The *Ortonella furcata* of Garwood shows two angular values of 25°–40° and 35°–45°, but according to Mamet and Roux (1975), its angular values exceed 45°–50°.

The genus *Mitcheldeania* is characterized by the shape of the pseudobifurcate filaments with constrictions, and by the angle which in the original specimen is 10°–15° and in other specimens 20°–30°. *Apophoretella* exhibits fine, long filaments with large modulations which are characteristic of this genus; the angle between the pseudobifurcate filaments measured in the original specimen is 2°–5°. The arrangement of pseudobifurcate filaments (strongly interwoven in the central zone and vertical-erect in the marginal zone) is a characteristic feature of *Pseudogirvanella* n.gen.

Other features, such as the morphology of the thallus or its dimensions, may be used when defining the species, although many cyanophytes exhibit a large morphological variability and numerous transitional types. The occurrence of compact, semilax and lax inner structures is characteristic of certain species without being a distinctive criterion for their definitions.

Among the unimportant characters are the presence or absence of growth zones, the space between filaments, and the thickness of filament walls.

Systematic Paleontology

CYANOPHYTA

Order Nostocales

Family Rivulariaceae Rabenhorst 1865

Rivularia (Roth 1802) Agardh 1824

Type-species: *Rivularia dura* Roth 1802

Zonotrichites Bornemann 1887
Stenoporidium Yabe and Toyama 1928
Cayeuxia Frollo 1938
Diversocallis Dragastan 1967, emend. Dragastan 1978
Pycnoporidium Yabe and Toyama 1928 (pro parte?)
?*Ortonella* (pro parte – all Paleozoic species with the angle of less than 20° between the pseudobifurcate filaments)
?*Dimorphostroma* Reis 1923
?*Dendractis* Reis 1923
?*Rivularialithus* Maslov 1955

Diagnosis: Globular or hemispherical thallus with erect growth, mono-, bi- or multidigitate, crossed by trichoms or pseudobifurcate filaments of sympodial type. Inner diameter of filaments (lumen) of type species is 4–9 μ and usually varies between 2–70 μ. The angle of divergence between filaments varies between 2°–20°. It may or may not show zonal structure.
Discussion: *Rivularia* (Roth 1802) Agardh 1824, includes more than ten living species in lacustrin or littoral wave areas. The species have been described by Tilden (1920), Geitler (1932), Fritsch (1945), Bourelly (1970), Golubic (1973, 1976), and Monty (1976).

Special attention was given by geologists to the *Rivularia haematites* (De Candolle) Agardh 1824, a living algae with a strongly calcified thallus, forming large bioherms in littoral areas (gulfs) (Schneider 1977, Richter et al. 1979).

The morphology of the thallus, hemispherical or with a tendency to form branched columns, with or without growth zones crossed by pseudobifurcate filaments, favoured the comparison with the Paleozoic taxa (*Palaeorivularia* Korde 1958, *Primorivularia* Edhorn 1973), Mesozoic (*Zonotrichites* Bornemann 1887, *Cayeuxia* Frollo 1938) and Cenozoic (*Dimorphostroma* Reis 1923, *Dendractis* Reis 1923 and *Rivularialithus* Maslov 1955). In spite of the extremely similar, sometimes identical features, none of those authors failed to introduce new taxa, some of which are junior synonyms of *Rivularia* (Roth 1802) Agardh 1824.

The first comparison of *Rivularia* was made by Bornemann (1887) to the new genus *Zonotrichites* with nodular thallus of 8 x 4 cm, which he described from the Upper Triassic (Rhaetian) lagoonal-brackish facies of Upper Silesia in Poland. *Zonotrichia* Agardh (considered by some authors a subgenus of *Rivularia*) is synonymous with *Rivularia*, and the striking resemblance mentioned by Bornemann (pp. 127, 128, Pl. VI, Fig. 4) between *Zonotrichia (Euactis) heeriana* Naegeli and *Zonotrichites* further supports the opinion that *Zonotrichites* is a synonym of *Rivularia*. Both genera, however, occur in different formations. The original specimens of *Zonotrichites lissaviensis* were lost from the Bornemann Collection during World War II (Dr. T. Peryt of the Geological Institute in Warsaw, pers. comm.). According to the rules of the ICBN, the loss of original specimens allows for the designation of a neotype for *Rivularia lissaviensis* (Bornemann), from the specimens described from Upper Triassic or Liassic. Of all the specimens of *Zonotrichites* in the collections of Elliott 1964, 1975, Ott 1972, Abate et al. 1977, only the fossils shown by Ott on p. 275, Pl. II, and associated with *Oligoporella chia* Herak, may be designated a neotype of *R. lissaviensis* (Bornemann). This specimen is from the Upper Ladinian Nohutalan (Karaburun Peninsula in Turkey). The assignment of this specimen as a type species for Bornemann's taxon should also be approved of by Ott (Munich). Only by this means will *R. lissaviensis* have a valid type.

Reis (1923) describes *Dimorphostroma* as a new Codiacean, and *Dendractis* as a new Chlorophycean from the fresh water Miocene (Seesinterkalke) in the Rheinpfalz region (Donnersberg sheet). On p. 119 he mentions *Rivularia (Zonotrichia)*, comparing it to *Dimorphostroma*, and showing that there is a great similarity between the two genera. However, according to Bornemann, the assignment and description in comparison with living *Zonotrichia* are not sufficiently obvious, the main problem being which type of branching is prevalent or more important. In my opinion, *Dendractis* with pseudobifurcate filaments and a diameter of 56–90 μ is also synonymous with *Rivularia*. In order to elucidate their systematic position and those of other taxa (e.g., *Ternithrix* Reis 1923, possibly synonymous with *Girvanella* Nicholson and Etheridge 1878, *Chlorellopsis* Reis 1923, *Microchorton* Reis 1923, *Brachydactylus* Reis 1923, and *Cladophorites* Reis 1923), it is necessary to revise Reis's work and to identify the types. Since Pia's (1927) work, Reis's study has been forgotten; these taxa, therefore, should now be restudied and reintroduced to paleontological literature.

Yabe and Toyama (1928) describe the Upper Jurassic *Stenoporidium* with nodular zonal thallus and pseudobifurcate filaments with diameter of 41–81 μ, also a synonym of *Rivularia*.

Frollo (1938), without mentioning Bornemann's study, introduces the genus *Cayeuxia* and two new species, *C. piae* and *C. moldavica*, codiaceans from the Tithonian of the Bicaz Gorges, East Carpathians. The species have been recognized by the variable diameter of filaments bifurcating at a sharp angle. Frollo (1938, p. 270, Fig. 1) provides the reconstruction of the branching type, corresponding to a sinistral angle of about 45°, which is characteristic of *Cayeuxia*; on the other hand, it is in disagreement with the pictures of the two species which show branching angles of less than 20°. The description of the genus demonstrates clearly that *Cayeuxia* has pseudobifurcate filaments at an open angle of 20°, which corresponds to the picture of new species but contradicts the reconstruction drawing of *Garwoodia* Wood 1941. Because Frollo's original specimen is lost, it is necessary that the species not included in the synonymy – *Rivularia piae* (Frollo) – gets a neotype from the type locality, the Bicaz Gorges in East Carpathians, and thus becomes valid.

Cayeux moldavica Frollo and the other species of *Cayeuxia* reported from Mesozoic formations are synonyms of *Rivularia lissaviensis* (Bornemann). The validity of the synonymy is discussed in the section on description.

Elliott (1964, 1975) is the first to revive *Zonotrichites*, which he describes by considering the diameter of filaments; the genus is Triassic from the Arabian Peninsula, and from the Middle Jurassic of the Middle East. The diameter of pseudobifurcate filaments is 18–25 μ and not 30 μ (35 μ) as shown by Bornemann. Elliott compares *Zonotrichites* Bornemann 1887 with *Cayeuxia* Frollo 1938, which he considers very similar but not synonymous. As regards the facies, Elliott shows that *Zonotrichites* and other cyanophytes may tolerate considerable range in temperature and salinity and occur in fresh, brackish, lagoonal, and marine facies.

Freytet and Plaziat (1965) reported Upper Cretaceous and Eocene *Rivularia haematites* crusts from southern France.

Dragastan (1967) reported a new Barremian-Lower Aptian gymnocodiacean *Diversocallis* from Pădurea Craiului, Apuseni Mountains, with crustlike thallus, mesothallus, and lower and upper perithallus. The genus is a synonym of *Ethelia* Weber van Boose; and *Diversocallis undulatus* Dragastan 1967 is a synonym of *Ethelia* (= *Pseudolithothamnium*) *alba* (Pfender) (Bakalova 1975 and Masse 1976, who cautiously assign *Diversocallis* Dragastan to *Stenoporidium* Yabe and Toyama 1928). In 1978, Dragastan and Bucur Ioan emended *Diversocallis* and show that the thallus is semi-elliptical or fanlike, with a central "medullary" zone surrounded by concentric growth beds crossed by non-bifurcate filamentous cells. The Tithonian and Neocomian *D. moesicus* and *D. dianae* have been reported from the Carpathians and the Moesian Platform. These two species with hemispherical thalli and growth zones crossed by pseudobifurcate filaments and with coarsely crystallized walls, belong to *Rivularia* (Roth 1802) Agardh 1824, and according to the 1967 and 1978 descriptions, this genus is a synonym of *Ethelia* and of *Rivularia* pro parte.

Ott (1972) mentions *Zonotrichites* from the Ladinian lagoonal facies of the calcareous Northern Alps and the Karaburun Peninsula in Turkey, and notes that the codiacean or solenoporacean *Ortonella, Cayeuxia,* and *Dimorphostroma* were reported from the Mesozoic and Cenozoic. The oldest name of the genus is *Zonotrichites* described by Bornemann in 1887, and according to the law of priority the other genera are synonyms.

D'Argenio et al. (1975) notes that the algal nodules consist of thalli with branched tubes displayed radially, and are similar to the fossil *Cayeuxia* and the living *Rivularia*, both in *Mitcheldeaniaceae.*

Rivularia lissaviensis (Bornemann 1887)

(Text-Figs. 1–2; Pl. I, Figs. 1–3; Pl. III, Figs. 1–2; Pl. IV, Figs. 1–6; Pl. V, Figs. 1–5; Pl. VI, Figs. 1–2; Pl. VII, Figs. 1–4; Pl. VIII, Figs. 1–5).[1]

Cayeuxia alpina
1975, Flügel, p. 312, Taf. 1, Fig. 5; Taf. 2, Fig. 4; Taf. 4, Fig. 1
Cayeuxia austriaca
1967, Fenninger and Hötzl, p. 30, Taf. 1
Cayeuxia doerflesiana
1951, Kamptner, pp. 182–184, Pl. 1, Figs. 1–5
1974, Flügel, Taf. 3/1
Cayeuxia kurdistanensis
1967, Dragastan, p. 449, Pl. VII, Fig. 47
1970, Bouroullec and Deloffre, p. 87, Pl. 4, Figs. 1–2, 4
1972, Deloffre, p. 119, Pl. 3, Fig. 1
1978, Bakalova, p. 11, Tab. IV, Fig. 2

1 Plates I–XXVIII appear on pp. 128–155

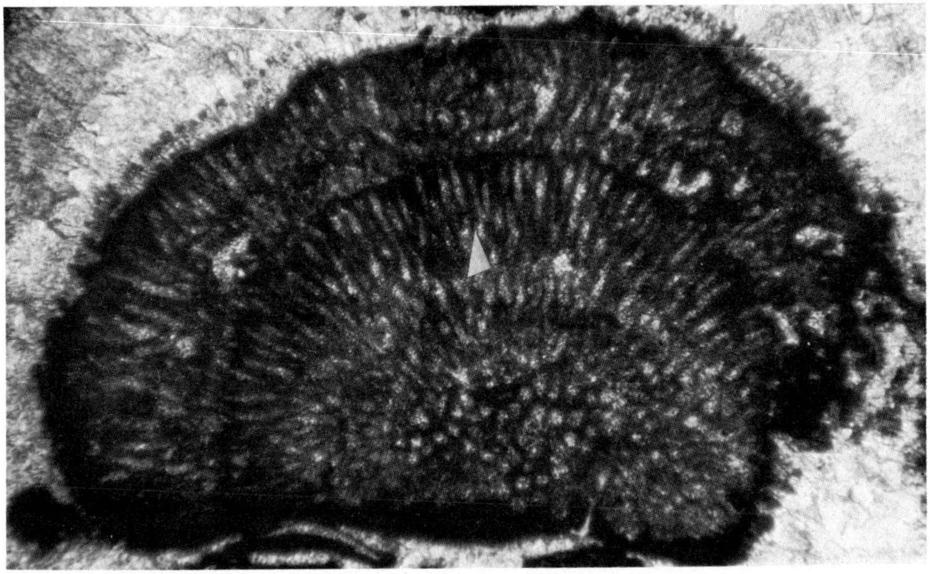

Text-Fig. 1. *Rivularia lissaviensis* (Bornemann), vertical section with typical V pseudobifurcations, compact inner structure and zones of growth. *Hypotype*: Coll. L.P.B. – V no. 0089. Tithonian, Făgetul Ciucului, Bicaz Gorges, Eastern Carpathians. (x 12)

Cayeuxia mediterranea
1978, Düzbastilar, p. 362, Pl. I, Figs. 1–2
Cayeuxia moldavica
1938, Frollo, p. 269, Pl. 18
1967, Dragastan, p. 449, Pl. V, Fig. 44; Pl. VI, Figs. 45–46
1968a, Johnson, p. 10, Pl. 1, Figs. 3–4; p. 64, Pl. 10, Fig. 2
1975, Dragastan, Pl. XVIII, Fig. 1: Pl. XLVI, Fig. 4: Pl. LV, Fig. 1
1978, Bucur, Pl. IV, Fig. 7; Pl. V, Fig. 2; Pl. X, Fig. 2
1978, Düzbastilar, p. 363, Pl. I, Figs. 3–6
1978, Bakalova, p. 10, Tab. II, Fig. 2; Tab. IV, Fig. 3
1978, Wnendt, p. 46, Fig. 15
1980, Steiger and Wurm, Taf. 26, Fig. 8
1980/1981, Bucur, p. 59, Pl. VI, Fig. 2
1981, Steiger, p. 328, Taf. 15, Fig. 4
1982, Flügel, p. 577, Pl. 52, Fig. 4
Cayeuxia piae
1940, Frollo, Pl. VIII, Fig. 4
1965, Maslov, p. 1154, Fig. 2v
1966, Praturlon, p. 173, Fig. 4a
1968, Bouroullec and Deloffre, p. 220, Pl. 2, Figs. 1,3
1978, Bakalova, p. 11, Tab. IV, Fig. 5

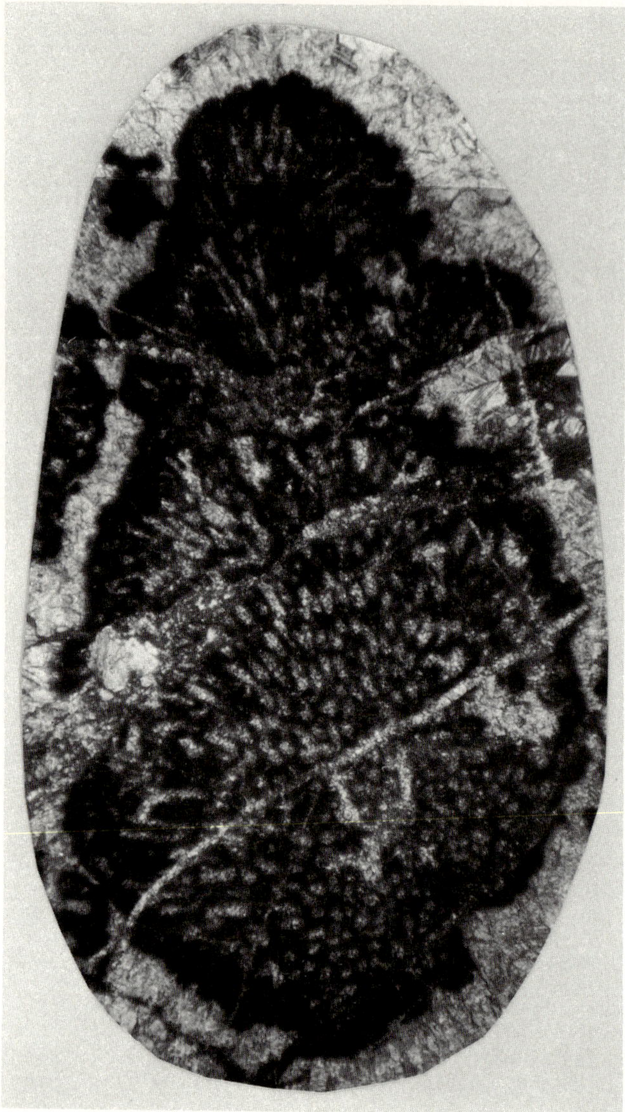

Text-Fig. 2. *Rivularia lissaviensis* (Bornemann), vertical oblique section. *Hypotype*: Coll. L.P.B. – V no. 0090. Tithonian, Piatra Craiului Massif, Eastern Carpathians. (x 12)

Cayeuxia sp.

1958, Dufaure, p. 109, Fig. 19
1964, Johnson, p. 77, Fig. 5
1966, Misik, p. 103, Pl. 29, Fig. 1
1966, Derin and Reiss, p. 21, Fig. 59
1974, Catalano et al., p. 15, Fig. 6: p. 22, Fig. 10
1975, Flügel, p. 314, Taf. 4, Fig. 9
1977, Abate et al., Pl. 3/6
1982, Wurm, p. 272, Taf. 30, Fig. 2; p. 278, Taf. 33, Fig. 1

Cayeuxia/Garwoodia
1979, Schäfer, Taf. 17/3
Codiacées
1958, Dufaure, p. 109, Pl. 4, Figs. 6–7
1959, Radoicic, Tab. VII, Fig. 3
1960, Radoicic, p. 45, Tab. V, Fig. 1
1966, Radoicic, p. 53, Pl. XIV, Fig. 2; p. 72, Fig. 2; p. 117, Fig. 2
Cyanophycean
1982, Flügel, p. 565, Pl. 47, Fig. 5
Mitcheldeania africana
1935, Le Maitre, p. 48, Pl. 12
Ortonella sp.
1977, Flügel, p. 133, Fig. 2
Ortonella/Cayeuxia
1981, Steiger, p. 332, Pl. 17, Fig. 6
1982, Flügel, p. 327, Pl. 28, Fig. 3
Pycnoporidium sp.
1966, Misik, p. 213, Pl. 84, Fig. 2
Stenoporidium chaetetiformis
1928, Yabe and Toyama, Pl. XXII, Figs. 2–4
1961, Endo, p. 66, Pl. 4, Fig. 5: Pl. 9, Figs. 1–3; Pl. 10, Figs. 2–3
Zonotrichites lissaviensis
1887, Bornemann, Taf. V, Figs. 1–2, Taf. VI, Figs. 1–2
1964, Elliott, Pl. I, Figs. 1–2, Pl. II, Figs. 1–3
Zonotrichites sp.
1972, Ott, p. 275, Taf. II
1975, Elliott, p. 352, Pl. 48, Fig. 3
1981, Brandner and Resch, p. 227, Fig. 24
Zonotrichites/Apophoretella
1979, Schäfer, Taf. 17, Fig. 1

Neotype: *Zonotrichites* sp. Ott (1972, p. 275, Taf. II). Upper Ladinian, Nohutalan, Karaburun Pen.
Diagnosis: Thallus with wide morphological variability, varying in size from 2.2 x 8 mm to 1 x 6 cm. Subspherical, hemispherical, columnarerect or fanlike nodules with more or less flaring borders. Thallus crossed by filamentlike pseudobifurcate cells, sympodially V-shaped, with the angles of divergence 2–10° and 15–20°. The inner structure of the thallus is compact and the filaments tightly joined and parallel. The inner diameter of filaments at base of branches varies between 20–50 μ, only rarely as much as 60 μ. Growth zones may or may not occur in variable amount depending on the size of the thallus. The small, hemispherical or subspherical thalli do not exhibit growth zones, but the fanlike or columnar-erect do (Pl. IV, Figs. 1, 3, 5; Text-Fig. 1; Pl. V, Figs. 1–5; Text-Fig. 2; Pl. VII, Figs. 1–2).
Remarks: The Liassic *Mitcheldeania africana* Le Maîtrae 1935 from Morocco exhibits V-shaped, pseudobifurcate filaments, with diameter of 15–60 μ and an angle of 8–10°. On the basis of the morphology of the thallus, and the branching type and

its size, *M. africana* is synonymous with *Rivularia lissaviensis* (Bornemann). *Cayeuxia moldavica* Frollo 1938 is also synonymous with *R. lissaviensis*, based on its morphology, diameter of filaments which in the original specimen is 25–50 μ, and on the angle of divergence which is less than 20°. The following are also synonyms: the Tithonian *C. doerflesiana* Kamptner, 1951, from Austria, with bifurcate filaments at an angle 5–10° and diameter 30–50 μ; the Tithonian *C. austriaca* Fenninger and Hötzl 1967, with the angle 10–15° and diameter of filaments 40 μ; and the Upper Triassic *C. alpina* Flügel (1975) with diameter of filaments 25–50 μ and the angle of divergence 5–8°, which are just like *Cayeuxia* sp. If one agrees with Masse (1979) that some cyanophytic structures may show filaments with partitions, then *Pycnoporidium lobatum* Yabe and Toyama 1928, *P. sinuosum* Johnson and Konishi with diameter of filaments 50–94 μ and *Stenoporidium chaetetiformis* Yabe and Toyama 1928 (which, according to Endo, has a diameter of filaments of 41–81 μ and angle of divergence 5–10°) may also be considered junior synonyms of *R. lissaviensis* (Bornemann).

I conclude that *R. lissaviensis* (Bornemann) shows a varied morphology of the thallus and the diameter of filaments 20–50 μ (60 μ), and V-shaped branches with angles of divergence 2–10° to 15–20°. These parameters have been obtained by measuring the photographs of original specimens (photographs by Bornemann, Frollo, and Ott).

Stratigraphic Distribution: Mesozoic, Middle and Upper Triassic interval (Anisian-Rhaetian), Jurassic (Liassic, Bajocian-Bathonian, Oxfordian-Tithonian), and Lower Cretaceous (Valanginian-Aptian).

Rivularia lemaitrae (Dragastan 1969)
(Pl. I, Figs. 4–10)

Holotype: *Cayeuxia lemaitrae* Dragastan 1969, Pl. I, Fig. 7, Coll. L.P.B V – no. 9216, Glimeia Hill, Ladinian-Lower Carnian, Pădurea Craiului Mountains.

Diagnosis: Thallus small, 1.05 x 2.7 mm, ovoidal or fanlike, crossed by pseudobifurcate filaments with diameter 75–90 μ, branching, V-shaped with the angle of 4–10°.

Stratigraphic Distribution: Triassic, Ladinian-Carnian, Pădurea Craiului Mountains and Norian-Rhaetian, Vaşcău Plateau, Apuseni Mountains.

Rivularia sp.
(Pl. II, Fig. 1)

Specimens: Glimeia Hill, Apuseni Mountains, Ladinian-Lower Carnian, Coll. L.P.B. V – no. 0103.

Diagnosis: Thallus fanlike with irregular shape, 2.8 x 4.5 mm, and lax inner structure. Filaments pseudobifurcate with V-shaped branching, with angle of divergence between 5–10° and diameter of filaments 75–90 μ.

Remarks: Lax inner structure and irregularly disposed subparallel filaments make this form transitional between *Ortonella* and *Paraortonella*.

Stratigraphic Distribution: Triassic, Ladinian-Lower Carnian.

Rivularia piae (Frollo 1938)
(Pl. III, Figs. 5–6; Pl. IX, Figs. 1–5; Pl. X, Figs. 1–2)

Cayeuxia piae
1938, Frollo, p. 264, Pl. 18
1968a, Johnson, p. 9, Pl. 2, Fig. 2
1970, Golonka, p. 89, photo 14
1970, Bouroullec and Deloffre, p. 86, Pl. 3, Figs. 1–4, 6
1975, Dragastan, Pl. LXXXII, Fig. 4
1976, Luperto Sinni, p. 408, Taf. 56, Fig. 4
1978, Dragastan et al., Pl. IV, Fig. 1; Pl. VI, Fig. 6
1978, Düzbastilar, p. 360, Pl. I, Figs. 7–8
Cayeuxia sp.
1958, Dufaure, p. 109, Pl. 4, Fig. 3
1966, Derin and Reiss, p. 24, Fig. 89
1974, Jaffrezo, p. 28, Pl. 1, Fig. 2
Cayeuxia/Garwoodia
1979, Schäfer, Taf. 17, Fig. 2
Codiacées
1959, Radoicic, Tab. VII, Sl. 2
1966, Radoicic, p. 117, Fig. 1; p. 281, Fig. 1
Pycnoporidium lobatum
1964, Johnson in Pfender, p. 53, Pl. 8, Fig. 2

Neotype: *Cayeuxia piae* Frollo 1938, Pl. IX, Fig. 5, Coll. L.P.B. V – no. 0135, Tithonian, Bicaz Gorges, East Carpathians.
Diagnosis: Thallus hemispherical, subspherical, or ellipsoidal, 2.1 x 4.5 mm, crossed by V-shaped pseudobifurcate filaments, with diameter of 75–90 μ and angle of divergence 5–10°. Filaments coarsely crystallized, subparallel, generating a semi-lax inner structure.
Stratigraphic Distribution: Lower Jurassic, Karaburun, Turkey; Middle Jurassic, Sărăţii Valley, Persani Mountains, Romania; Haifa boreholes, Israel; Upper Jurassic, Oxfordian-Tithonian; Lower Cretaceous, Berriasian-Aptian, and Upper Cretaceous, Senonian-Altamura limestones, Italy.

Rivularia fruticulosa (Johnson and Kaska 1965)
(Pl. X, Figs. 3–7; Pl. XI, Figs. 1–5)

Cayeuxia fruticulosa
1968a, Johnson, p. 64, Pl. 10, Fig. 3
1969, Johnson, p. 137, Pl. 25, Figs. 1–2
1975, Dragastan, Pl. LXXXV, Fig. 4
1978, Dragastan et al., Pl. VII, Fig. 2
Codiacées
1966, Radoicic, p. 59, Fig. 2

Type: *Cayeuxia fruticulosa* Johnson and Kaska 1965, Lower Cretaceous, Guatemala.
Hypotypes: Pl. X, Figs. 3–7; Pl. XI, Figs. 1–5, Coll. L.P.B. V – nos. 0138–0147.
Diagnosis: Thallus hemispherical, nodular, subspherical, or irregular ellipsoidal, sometimes very flaring (Pl. X, Fig. 7), height 3.0 and width 9.0 mm. Inner structure is lax with wide micritic zones between filaments. Filaments pseudobifurcate, fine with straight, curved, or sinuous-irregular line. Diameter of filaments 10–45 μ and angle of divergence 5–10° (rarely 20°). False furcate filaments with marked angles (Pl. X, Figs. 5–7). Some filaments show wide, coarsely recrystallized areas (Pl. X, Fig. 3) or "ampoules" (Pl. XI, Figs. 1–2) on the exterior of the thallus and corresponding to the heterocysts (Pl. XI, Figs. 1–2).
Stratigraphic Distribution: Upper Jurassic, Tithonian, Lower Cretaceous, Valanginian-Lower Aptian.

Rivularia atanasiui (Dragastan 1971)
(Pl. XII, Figs. 1–4)

Cayeuxia atanasiui
1971, Dragastan, p. 184, Pl. 9, Figs. 5–6, *not* Figs. 4, 7–10

Holotype: *Cayeuxia atanasiui* Dragastan 1971, Pl. 9, Fig. 5, Tithonian, Bicaz Gorges, East Carpathians.
Hypotypes: Pl. XII, Figs. 2–4, Coll. L.P.B. V – nos. 0148–0150.
Diagnosis: Thallus vertical-erect, 3.0 mm high and 2.2 mm wide, consisting of superposed hemispherical structures and with constant vertical growth (hemisphere of 0.6 x 0.9 mm). Filaments pseudobifurcate, with diameter of 30–75 μ, V-shaped branching with angle of divergence 15–20°.
Remarks: The main features of this species are the bidigital shape of the thallus and the superposed hemispheroids.
Stratigraphic Distribution: Upper Jurassic, Tithonian, and Lower Cretaceous, Barremian-Lower Aptian.

Rivularia tadeuszi n.sp.
(Pl. II, Fig. 2; Pl. XIII, Figs. 1–8)

Cayeuxia atanasiui
1971, Dragastan, Pl. 9, Figs. 4, 7–10
1975, Dragastan, Pl. XCI, Fig. 1
1978, Dragastan et al., Pl. III, Fig. 4; Pl. V, Fig. 5; Pl. VI, Figs. 1, 4: Pl. XV, Fig. 5
1978, Bucur, Pl. X, Fig. 1
Cayeuxia kurdistanensis
1958, Dufaure, p. 109, Pl. 4, Fig. 13
1970, Bouroullec and Deloffre, p. 87, Pl. 4, Figs. 5–6
Cayeuxia sp.
1968a, Johnson, p. 12, Pl. 4, Fig. 8
1975, D'Argenio et al., p. 528, Figs. 4B–C
1976, Masse, Pl. 8, Fig. 6

Codiacées
1960, Radoicic, p. 45, Tab. V, Fig. 1
1966, Radoicic, p. 59, Fig. 2 (left of photograph)
Pila ?
1972, Bignot, Pl. IX, Figs. 4–5

Holotype: Pl. XIII, Fig. 3, Tithonian, Bicǎjel, East Carpathians, Coll. L.P.B. V – no. 0153.
Syntypes: Pl. II, Fig. 3; Pl. XIII, Figs. 1–2, 4–8, Ladinian-Carnian, Pǎdurea Craiului, Tithonian, Cheile Bicazului, East Carpathians, Coll. L.P.B. V – nos. 0104; 0151–0152; 0154–0158.
Derivatio nominis: Dedicated to Dr. Tadeusz M. Peryt, Institute of Geology, Warsaw, for his contribution to the study of Cyanophyta.
Diagnosis: Thallus consisting of several, 2–6, "bushes" superposed with erect or fanlike growth (Pl. XIII, Fig. 3). Thallus is 1.5 mm high and 6.0 mm wide. Each "hemispheroid" is 0.9 x 1.5 mm. Thallus is crossed by pseudobifurcate filaments with diameter of 30–45 μ (sometimes 60 μ) and with angle of divergence 5–10°. Sedimentation causes the bushes to break, and they occur in isolation (Pl. XIII, Figs. 2, 4, 5).
Remarks: *Rivularia tadeuszi* n.sp. differs from other *Rivularia* by the morphology of thallus, the growth of superposed, erect, monodigitated, or fanlike bushes, small diameter of filaments, and the angle of divergence from 5° to 10°.
Stratigraphic Distribution: Middle and Upper Triassic, Ladinian-Lower Carnian; Upper Jurassic, Oxfordian-Tithonian; Lower Cretaceous, Neocomian, Barremian-Aptian; and Upper Cretaceous, lacustrian Senonian, Jadran region, Yugoslavia.

Rivularia theodori n.sp.
(Pl. III, Figs. 3–4; Pl. XIV, Figs. 1–7)

Cayeuxia sp.
1958, Dufaure, p. 109, Pl. 14, Fig. 16

Holotype: Pl. XIV, Fig. 1, Neocomian, Fǎgetul Ciucului, East Carpathians, Coll. L.P.B. V – no. 0159.
Syntypes: Pl. III, Figs. 3–4, Oxfordian, Sǎrǎţii Valley, Perşani Mountains, Pl. XIV, Figs. 2–7, Tithonian, Bicaz Gorges, Haţeg-Pui, Vinturariţa and Barremian-Aptian, Fǎgetul Ciucului, Coll. L.P.B. V – nos. 0108–0109; 0160–0165.
Derivatio nominis: Dedicated to Prof. Theodor Neagu, University of Bucharest, for his contribution to palaeontology.
Diagnosis: Thallus small, fanlike, width 1.2–2.4 mm and height 0.8–1.2 mm, crossed by large coarsely crystallized filaments in groups of two (Pl. XIV, Figs. 1, 3–4, 7). The diameter of the filaments is 60–75 μ, and the angle of divergence is 3–10°.
Remarks: *Rivularia theodori* n.sp. is characterized by the small size of the thallus, filaments in groups of two, and by a constant diameter.
Stratigraphic Distribution: Upper Jurassic, Oxfordian-Tithonian, and Lower Cretaceous, Neocomian and Barremian-Aptian.

Rivularia kurdistanensis (Elliott 1957)
(Pl. XIV, Figs. 8–14)

Cayeuxia kurdistanensis
1957, Elliott, p. 790–791, Pl. 25, Figs. 8–10
1968a, Johnson, p. 9, Pl. 2, Fig. 1
non 1968a, Johnson, p. 64, Pl. 10, Fig. 1 is synonymous with *Pseudolithocodium carpathicum* 1979, Misik
1968, Bouroullec and Deloffre, p. 223, Pl. 2, Fig. 2
1970, Bouroullec and Deloffre, p. 87, Pl. 2, Figs. 5–6; Pl. 4
Cayeuxia sp.
1976, Masse, Pl. 8, Fig. 7

Hypotypes: Pl. XIV, Figs. 8–14, Tithonian, Bicaz Gorges, East Carpathians; Cetea, Southern Apuseni, Tithonian-Neocomian, Vinturariţa, South Carpathians, Coll. L.P. B. V – nos. 0166-0172.
Diagnosis: Thallus nodular, hemispherical, fanlike, size small, width 1.5–3.3 mm, height 1.4–3.0 mm. Compact inner structure crossed by pseudobifurcate filaments with diameter 15–30 μ and angle of divergence 3–5°.
Stratigraphic Distribution: Upper Jurassic, Upper Oxfordian-Tithonian, and Lower Cretaceous, Neocomian and Barremian-Lower Aptian.

Rivularia undet.
(Pl. XV, Fig. 1)

Specimens: A single Neocomian specimen, Făgetul Ciucului, East Carpathians, Coll. L.P.B. V – no. 0173.
Diagnosis: Thallus erect, elongated, and very narrow, width 0.3 mm and height 2.5 mm. Pseudobifurcate filaments with diameter 30–45 μ and angle of divergence 10–15°. Because of its characteristic shape it may belong to a new taxon.
Stratigraphic Distribution: Neocomian, Bicaz Gorges, East Carpathians.

Rivularia dianae (Dragastan and Bucur 1978)
(Pl. XV, Figs. 2–3)

Diversocallis dianae
1978, Dragastan and Bucur, p. 187, Pl. II, Figs. 1–2; Pl. III, Figs. 1–2
Diversocallis sp.
1978, Bucur, Pl. IV, Figs. 4, 6
Lithophyllum ? *maslovi*
1978, Wnendt, p. 43, Fig. 12

Holotype: in paper of 1978, Coll. L.P.B. V – no. 0071, Tithonian-Berriasian, Piatra Craiului, East Carpathians.

Hypotypes: Pl. XV, Figs. 2–3, Piatra Craiului, Coll. L.P.B. V – nos. 0174–0175.
Diagnosis: Thallus semielliptical or fanlike, 3.2–6.0 mm wide and 2.1–5.0 mm high, with 4 to 6 growth zones, 0.30, 0.60, 0.90 mm thick. Pseudobifurcate filaments with coarsely crystallized walls and diameter 15–20 μ. Small angle of divergence 3–5°.
Stratigraphic Distribution: Upper Jurassic, Kimmeridgian-Tithonian; and Lower Cretaceous, Berriasian.

Rivularia moesica (Dragastan and Bucur 1978)
(Pl. XV, Figs. 4–6)

Diversocallis moesicus
1978, Dragastan and Bucur, p. 186, Pl. I, Figs. 1–4
Diversocallis-
1975, Dragastan, Pl. XXV, Fig. 3; Pl. XXXI, Fig. 4
Myxophyte koloni
1975, Bursuk, Tab. XXIII, Figs. 2–4
Stenoporidium
1976, Masse, Pl. 8, Fig. 2

Holotype: Pl. XV, Fig. 4, Coll. L.P.B. V – no. 0017, Tithonian, as *Clypeina jurassica*, Făgetul Ciucului, East Carpathians.
Hypotypes: Pl. XV, Figs. 5–6, Coll. L.P.B. V – nos. 0176–0177
Diagnosis: Thallus hemispherical, 1.5–3.0 mm wide and 1.3–1.8 mm high, with 4 to 8 growth zones 0.075–0.15 mm thick. Filaments pseudobifurcate with coarsely crystallized walls and diameter 10–15 μ. Inner structure of the thallus is compact, with angle of divergence 2–5°.
Stratigraphic Distribution: Upper Jurassic, Tithonian, and Lower Cretaceous, Barremian.

Rivularia carpathica n.sp.
(Pl. XV, Fig. 7, Pl. XVI, Figs. 1–2)

Holotype: Pl. XVI, Fig. 7, Tithonian, Bicaz Gorges, East Carpathians, Coll. L.P.B. V – no. 0179.
Syntypes: Pl. XV, Fig. 1, Tithonian, Cetea, Southern Apuseni, Coll. L.P.B. V – no. 0178; Pl. XVI, Fig. 2, Coll. L.P.B. V – no. 0180
Derivatio nominis: Derived from the name of the Carpathian Mountains.
Diagnosis: Thallus hemispherical, height 4.5 mm and width 3.0 mm, with 5 to 8 growth zones, 0.30 mm thick. Filaments pseudobifurcate with diameter of 20–30 μ. Angle of divergence less than 3°, inner structure compact and very coarsely crystallized.
Stratigraphic Distribution: Upper Jurassic, very frequent in the Tithonian.

Apophoretella Elliott 1975
Apophoretella dobunnorum Elliott 1975
(Pl. XVI, Fig. 4)

Apophoretella dobunnorum
1975, Elliott, p. 354, Pl. 49, Fig. 3

Hypotype: Pl. XVI, Fig. 4, Tithonian, Bicăjel, Bicaz Gorges, East Carpathians. Coll. L.P.B. V – no. 0181.
Diagnosis: Thallus hemispherical, width up to about 7.0 mm; height, however, only 3.0 mm. It exhibits six growth zones with pseudobifurcate, slightly sinuous filaments. The diameter of filaments is 10 μ and the angle of divergence is 3–4°. Inner structure compact.
Stratigraphic Distribution: Middle Jurassic, Bathonian, England, and Upper Jurassic, Tithonian, Bicaz Gorges, East Carpathians.

Family Scytonemataceae
Ortonella Garwood 1914
Ortonella lemoineae Dragastan 1971
(Pl. XVI, Fig. 3)

Ortonella lemoineae
1969, Pugh, p. 519, Fig. 1
1978, Bucur, Pl. VI, Fig. 1

Holotype: Dragastan 1971, p. 187, Pl. X, Fig. 5, Coll. L.P.B. V – no. 9345, Tithonian, Ghilcoş, East Carpathians.
Diagnosis: Thallus nodular, hemispherical, 3.1 mm wide and 2.1 mm high. Pseudobifurcate filaments with diameter 45–75 μ and angle of divergence 40–50°. Lax inner structure.
Remarks: Dufaure (1958) reports *Orthonella* sp. (p. 109, Pl. 4, Figs. 1–2) from the Bathonian, but without description. Without revising the Dufaure Collection, at present, together with the Derin and Reiss (1966) Collections, one of the most important collections of Porostromata, these algae cannot be assigned to specific taxa. Bouroullec and Deloffre (1970, p. 86, Pl. 3, Fig. 5) describe the taxon as *Cayeuxia piae* Frollo, from the Sinemurian, which is, in fact, *Ortonella* with the angle of divergence 40°. I propose to transfer the Upper Ladinian *Cladogirvanella cipitensis* Ott, 1966 (from Southern Tyrol) to *Ortonella cipitensis* (Ott) because of the morphology of the thallus and the angles of divergence of filaments which are between 50–80°. It is possible that some specimens of *Microtubus communis* Flügel also belong to the genus *Ortonella*.

Stratigraphic Distribution: Jurassic, Sinemurian, Tithonian.

Paraortonella n.gen.

Type Species: *Paraortonella getica* n.gen., n.sp.,
Tithonian, Ghilcoş, East Carpathians.
Diagnosis: Thallus small, subspherical, sometimes much developed in width, with pseudobifurcate filaments and the angle of divergence 25–30°.

Remarks: *Paraortonella* is a transition genus between *Rivularia* and *Ortonella*, having an angle of divergence of 25–30°.

Paraortonella getica n.sp.
(Pl. XVII, Figs. 1–3)

Holotype: Pl. XVII, Figs. 1–2, Tithonian, Ghilcoş, East Carpathians. Coll. L.P.B. V – no. 0182.
Syntypes: Pl. XVII, Fig. 3, Barremian-Aptian, Bardos Massif, East Carpathians. Coll. L.P.B. V – nos. 0183, 0184.
Derivatio nominis: "Getica" derived from the ancient inhabitants of Romania.
Diagnosis: Thallus small, hemispherical, 1.5–2.2 mm wide and 0.9–1.3 mm high. Inner structure of the thallus lax, with pseudobifurcate filaments and diameter 30 μ, and angle of divergence 25–30°.
Remarks: The angles of divergence of 25–30° of different *Rivularia* and *Ortonella* species show clearly an intermediate class between these two genera, which makes the assignment of certain algae to either one of these genera very difficult.
Stratigraphic Distribution: Upper Jurassic, Tithonian, and Lower Cretaceous, Barremian-Lower Aptian.

Paraortonella richteri n.sp.
(Pl. XVII, Figs. 4–5)

Cayeuxia sp.
1969, Pugh, p. 519, Pl. 17B

Holotype: Pl. XVII, Fig. 4, Tithonian, Ghilcoş, East Carpathians. Coll. L.P.B. V – no. 0185.
Syntype: Pl. XVII, Fig. 5, Tithonian, Suhard, East Carpathians. Coll. L.P.B. V – no. 0186.
Derivatio nominis: Dedicated to Dr. Detlev Richter, Institute of Geology of the Ruhr University in Bochum, for his contribution to the study of cyanophycean algae.
Diagnosis: Thallus small, fanlike, 1.5–1.6 mm wide and 1.4–1.5 mm high. Lax inner structure of the thallus. Coarsely crystallized pseudobifurcate filaments with diameter of 30–45 μ (rarely 60 μ) and angle of divergence 25–30°.
Stratigraphic Distribution: Upper Jurassic, Tithonian.

Family Mitcheldeaniaceae
Mitcheldeania Wethered 1886
Mitcheldeania americana (Johnson 1961)
(Pl. XVIII, Figs. 1–3)

Cayeuxia americana
1961, Johnson, p. 148, Pl. 31, Figs. 3–5

Hypotypes: Pl. XVIII, Figs. 1–2, Tithonian, Bicăjel, East Carpathians. Coll. L.P.B. V – nos. 0188, 0189.

Diagnosis: Thallus subnodular, 6.6 mm wide and 5.6 mm high, crossed by pseudobifurcate constricted filaments with diameter 60–90 μ. Angle of divergence 10–20°.
Remarks.: The branching type of filaments and the presence of constrictions (Pl. XVIII, Fig. 3) warrant its assignment to *Mitcheldeania.*
Stratigraphic Distribution: Upper Jurassic, Tithonian; Texas, and East Carpathians, Romania.

Order Oscillatoriales
Pseudogirvanella n. gen.

Type species: *Pseudogirvanella cretacea* n. gen. n.sp., Barremian-Lower Aptian, Vîrciorog-Fişca, Pădurea Craiului, Apuseni Mountains.
Diagnosis: Thallus nodular with bifurcate filaments, rotarypseudoconcentric in its central area turning into vertical-erect towards its marginal areas.
Remarks: The arrangement of filaments in the central and marginal areas allows one to recognize and distinguish it from *Girvanella.*

Pseudogirvanella cretacea n.sp.
(Pl. XVII, Fig. 6)

Holotype: Pl. XVII, Fig. 6, Barremian-Lower Aptian, Vîrciorog-Fişca, Pădurea Craiului, Apuseni Mountains. Coll. L.P.B. V – no. 0187.
Diagnosis: Thallus nodular, 2.7–3.1 mm high. Inner structure of the thallus compact, crossed by flexuous, pseudobifurcate filaments interwoven in the central area, with rotary-pseudoconcentric arrangement. Laterally the filaments tend to be arranged vertically, with erect growth and regular bifurcation. The diameter of filaments is 45–60 μ, and the thickness of the walls 10–15 μ. The angle of divergence is reduced 2–5°.
Remarks: Differs from the other species of *Girvanella* in the inner structure of the thallus and in having two distinct arrangements of filaments.
Stratigraphic Distribution: Lower Cretaceous, Barremian-Lower Aptian.

Order Entophysalidales
Family Entophysalidaceae Geitler
Hormathonema Ercegovic 1929
Hormathonema massei n.sp.
(Pl. II, Fig. 3)

Hormathonema
1979, Masse, p. 698, Pl. 1, Fig. 7

Holotype: Ladinian-Lower Carnian, Someş Graben, Apuseni Mountains. Coll. L.P.B. V – no. 0105.
Derivatio nominis: Dedicated to Dr. J.P. Masse of the University of Marseilles.
Diagnosis: Thallus attachment consists of columnar-thrombolitic aggregates and crescentiform elements. The aggregates are crossed by fine filaments without parti-

tions. Each columnar or domelike aggregate is made up of a sequence of micritic and microsparitic laminae, similar to algal laminites. The morphostructure may be compared with the living *Hormathonema* Ercegovic.
Stratigraphic Distribution: Middle and Upper Triassic, Ladinian-Lower Carnian, and Lower Cretaceous, Barremian.

Addendum

The Upper Jurassic *Cayeuxia mediterranea* Herak 1957 from Kios Island may be transferred to *Rivularia mediterranea* (Herak 1967) because of filament diameter of 30–92 μ, and the angle of divergence of 10–20°.

CHLOROPHYTA
Family Codiaceae (Trevisan) Zanardini 1843
Garwoodia Wood 1941
Garwoodia fluegeli n.sp.
(Text-Fig. 3; Pl. XIX, Figs. 1–8)

Cayeuxia kurdistanensis
1970, Bouroullec and Deloffre, Pl. 4, Fig. 3
Cayeuxia aff. *moldavica*
1958, Dufaure, p. 109, Pl.4, Fig.11

Text-Fig. 3. *Garwoodia fluegeli* n.sp., horizontal axial section with typical dextral ramifications. *Syntype:* Coll. L.P.B. V–no. 0091 Tithonian Lapos Valley Bardos Massif, Bicaz Gorges, Eastern Carpathians. (x 12)

Cayeuxia piae
1958, Dufaure, p. 109, Pl. 4, Fig. 9
Cayeuxia septata
1982, Bucur et al., p. 49, Pl. X, Fig. 11

Holotype: Pl. XIX, Fig. 1, Tithonian, Bicăjel, East Carpathians. Coll. L.P.B. V – no. 0190.
Syntypes: Pl. XIX, Figs. 2–8, Tithonian-Neocomian, Bicaz Gorges, East Carpathians; Tithonian, V. Cornilor, Remeţi, Apuseni Mountains. Coll. L.P.B. V – nos. 0091, 0191–0197.
Derivatio nominis: Dedicated to Prof. Erik Flügel from the University of Erlangen, Nürnberg, for his contribution to the study of algae.
Diagnosis: Thallus hemispherical 2.7–4.0 mm wide and 1.7–4.5 mm high. Semi-lax inner structure of the thallus, crossed by dichotomous filaments (Pl. XIX, Figs. 3, 4, and 8) and polychotomous with three filaments (Pl. XIX, Figs. 1, 2, 6, 7). In dichotomous filaments the angle of divergence is about 45–50°, and in the polychotomous filaments the angle is 90°. The diameter of filaments is 50–75 μ, and they are always dextrally branched.
Remarks: This species resembles the Paleozoic *Garwoodia gregaria* but differs from it by the diameter of filaments and by the angles of divergence of 45–50° and 90°, respectively, as well as by its dextral branching.
Stratigraphic Distribution: Upper Jurassic, Oxfordian-Tithonian, and Lower Cretaceous, Neocomian and Aptian.

Garwoodia bardosi n.sp.
(Pl. XX, Figs. 1–5)

Cayeuxia moldavica
1970, Golonka, p. 85, photo 13
Cayeuxia aff. *moldavica*
1958, Dufaure, p. 109, Pl. 4, Fig. 10

Holotype: Pl. XX, Fig. 4, Tithonian, Ghilcoş, East Carpathians. Coll. L.P.B. V – no. 0201.
Syntypes: Pl. XX, Figs. 1–3, 5, Tithonian, Neocomian, Barremian-Lower Aptian, Bicaz Gorges, East Carpathians. Coll. L.P.B. V – nos. 0198–0200, 0202.
Derivatio nominis: "Bardosi" derived from the Bardos Massif in the Bicaz Gorges, East Carpathians.
Diagnosis: Thallus vertical or hemispherical, 4.8–5.2 mm wide and 3.0–4.5 mm high. Semi-lax inner structure crossed by dichotomous filaments with diameter of 45–60 μ, and angle of divergence about 90°. Sinistral branching (Pl. XX, Figs. 1, 4).
Remarks: *G. bardosi* n.sp. differs from other *Garwoodia* species by its sinistral, branched diochtomic filaments and the angle of divergence of 90°.
Stratigraphic Distribution: Upper Jurassic, Oxfordian-Tithonian, and Lower Cretaceous, Neocomian and Barremian-Lower Aptian.

Hedstroemia Rothpletz 1913
Hedstroemia moldavica n.sp.
(Pl. XXI, Figs. 1–9; Pl. XXV, Fig. 1)

Cayeuxia sp.
1958, Dufaure, p. 109, Pl. 4, Figs. 3–4
1964, Colacicchi, p. 111, Fig. 17
1966, Derin and Reiss, p. 21, Fig. 56

Holotype: Pl. XXI, Figs. 1–2, Tithonian, Făgetul Ciucului, East Carpathians. Coll. L.P.B. V – no. 0203.
Syntypes: Pl. XXI, Figs. 3–9, Tithonian, Bicaz Gorges and Barremian-Aptian, borehole Şopotul Nou, Reşiţa Zone. Coll. L.P.B. V – nos. 0204–0210.
Derivatio nominis: "Moldavica" derived from the name of the northeastern Romanian province.
Diagnosis: Thallus hemispherical or fanlike, 1.5–3.2 mm wide and 1.5–2.4 mm high. Semi-lax inner structure in all 4-filament types and compact in 3-filament types. The filaments are pitchfork shaped, axialpolychotomically branched. The angle of divergence is 4–5°. The diameter of filaments is 50–60 μ.
Remarks: Differs from Paleozoic species by its axialpolychotomic branchings in 3- or 4-filament types, and by constant diameter.
Stratigraphic Distribution: Middle Jurassic, Bajocian, Bathonian-Callovian; Upper Jurassic, Tithonian; Lower Cretaceous, Barremian-Aptian; and Upper Cretaceous, Cenomanian, Marsica Nord Orientale, Italy.

Carpathocodium n.gen.

Type Species: *Cayeuxia anae* Dragastan 1971, p. 183, Pl. 9, Fig. 1. Coll. L.P.B. V – no. 9339. Tithonian, Făgetul Ciucului, East Carpathians.
Diagnosis: Codiacean of cylindrical segments with a wide hollow medullar zone or with longitudinal sinuous filaments separated by large spaces, and a cortical region of dichotomous, flexuous tube fascicles which, along the wall, are either vertical or parallel to the longitudinal axis of the thallus. The almost vertical tube network is of the 2+1 type and constitutes an ellipsoidal (fusiform) fascicle which contacts another tube fascicle of 2+1 or 2+2 type. The fascicles occur at different levels and are slightly aligned on the longitudinal axis. The thallus fragments from the cortical zone exhibit successive longitudinal alternating fascicles, parallel to the axis of the thallus (Pl. XXII, Fig. 3).
Remarks: *Carpathocodium* n. gen. resembles, by its inner structure and the arrangement of dichotomous filaments, the Dinantian *Malakhovella* Mamet and Roux 1977, and differs from *Arabicodium* by the complex, fasciculate structure of the cortical zone with tubes parallel to the longitudinal axis of the thallus.

Carpathocodium anae (Dragastan 1971)
(Pl. XXII, Figs. 1–6; Pl. XXIII, Fig. 1)

Arabicodium anae
1978, Bakalova, p. 9, Tab. II, Figs. 1, 3, 5
Cayeuxia anae
1971, Dragastan, p. 183, Pl. VIII, Figs. 5–6; Pl. IX, Figs. 1–3
1975, Dragastan, Pl. XXXII, Fig. 1; Pl. LIV, Fig. 3; Pl. XXIV, Fig. 2
1976, Masse, Pl. 8, Fig. 10
1978, Bucur, Pl. VI, Fig. 7
1980, Jaffrezo, p. 343, Pl. XIII, Figs. 7–8
1981, Misik and Sykora, p. 106, Pl. VI, Fig. 3
1982, Bucur et al., p. 49, Pl. X, Figs. 8–9
Cayeuxia sp.
1958, Dufaure, p. 109, Pl. 4, Fig. 15
Codiacées
1960, Radoicic, p. 119, Tab. XLII, Fig. 1
Girvanella sp.
1976, Luperto Sinni, p. 410, Tab. 57, Figs. 4–5, 7
Organism B
1974, Jaffrezo, p. 30, Pl. 2, Figs. 4–6
Technitella legumen
1938, Pfender, Pl. XIV, Fig. 7

Specimens: Pl. XXII, Figs. 1–6, Pl. XXIII, Fig. 1, Tithonian, Bicaz Gorges, East Carpathians; Dealul Dreptului, Haţeg-Pui, South Carpathians. Coll. L.P.B. V – nos. 0211–0216.

Diagnosis: Thallus consisting of cylindrical segments with wide, hollow medullar zone, sometimes crossed by longitudinal sinuous filaments. The cortical zone is crossed by dichotomic filaments of the 2+1 or 2+2 type which consitute ellipsoidal or fusiform fascicles, alternatively aligned. Because of the pronounced vertical arrangement, the thallus breaks up easily into fragments which preserve the characteristic structure of cortical filaments. Length 4.4 mm, outer diameter 3.2, diameter of medullar cavity 1.5, thickness of cortical walls 0.75, diameter of filaments 20–60 μ at the base and 75–90 μ apically.

Remarks: The taxon was initially assigned to *Cayeuxia* by Dragastan (1971) and by Masse (1976) to Caulerpal genus *Arabicodium* or *Halimeda* because of the morphologic features of filaments. Bakalova (1978), based on inconclusive and poorly represented specimens, assigned it to *Arabicodium.* The most representative morphologic features of *Carpathocodium anae* (Dragastan) have been illustrated by Radoicic (1960) from the Santonian, by Misik (1979) and by Misik and Sykora (1981), p. 106, Pl. VI, Fig. 3), from the Tithonian-Neocomian.

Scotlandella n.gen.

Type Species: *Cayeuxia nodosa* Anderson 1948, p. 138–140, Great Estuarine Series of Skye, Middle and Upper Jurassic, Scotland.

Diagnosis: Thallus with erect-columnar or nodular-spheroidal growth up to 1 cm, crossed by radial polychotomous filaments, generally branched on one side. The

branching system is characteristic, with a main filament and larger diameter from which several tubes (4–5) of smaller diameter with generally sinistral, and rarely dextral, successive branching occur.
Remarks: *Scotlandella* resembles *Ortonella* and *Garwoodia* and differs from them by its polychotomous, successive, mainly sinistral branching.

Scotlandella nodosa (Anderson 1948)
(Pl. XXIII, Fig. 6)

Cayeuxia nodosa
1948, Anderson, p. 110
1970, Hudson, p. 21, Pl. 22, Figs. A–C
Hedstroemia sp.
1969, Pugh, p. 520, Pl. 18A

Holotype: Coll. Anderson, Purbeckian, Skye, Scotland.
Diagnosis: Thallus columnar-erect, up to 1 cm high, crossed by radially arranged filaments. The filaments are polychotomous and their inner diameter is 25–50 μ (frequently 30 μ). The branching consists of a main filament with diameter of 70 μ, from which 4–5 successive sinistral filaments originate.
Remarks: Anderson (1948) compared this species to *Cayeuxia* and *Ortonella*, while Hudson (1970) recognized the special features of this alga, requiring a new taxonomic name.
Stratigraphic Distribution: Middle and Upper Jurassic, Purbeckian

Scotlandella alpina n.sp.
(Pl. XXIII, Figs. 2–5)

Holotype: Pl. XXIII, Fig. 3, Neocomian, Ghilços, East Carpathians. Coll. L.P.B. V – no. 0217.
Syntypes: Pl. XXIII, Figs. 2, 4–5, Tithonian and Barremian-Lower Aptian, Bicaz Gorges, East Carpathians. Coll. L.P.B. V – no. 0217A.
Diagnosis: Thallus nodular, subspherical, 1.5 x 1.8 mm, crossed by polychotomous filaments separated by wide spaces filled with micrite, the branching consisting of main filament with diameter 75–90 μ, dividing into two or three sinistral filaments with diameter 45–60 μ. The main filament is 0.8–1.0 mm long. Sometimes the filaments are dextrally divided (Pl. XXIII, Fig. 5).
Remarks: *S. alpina* n.sp. differs from *S. nodosa* by its small thallus and the 2(3) sinistral branching type.
Stratigraphic Distribution: Upper Jurassic, Tithonian, and Lower Cretaceous, Neocomian, Barremian-Aptian.

Johnsonicodium n.gen.

Type Species: *Johnsonicodium stratiforme* n.gen., n.sp., Neocomian, Ghilços, East Carpathians.

Diagnosis: Thallus of bladed crust type with irregular contour, sometimes attached. The length of the thallus is 10–15 mm. The thallus exhibits a central, hollow, and narrow medullar zone and, depending on the attachment, thick or thin cortical zones. The branching type consists of a tubular filament of large diameter dividing into dichotomous or trichotomous short filaments distally widely opened.

Johnsonicodium stratiforme n.sp.
(Pl. XXIV, Figs. 1–2)

Holotype: Pl. XXIV, Fig. 1, Neocomian, Ghilcoş, East Carpathians. Coll. L.P.B. V – no. 0221.
Syntype: Pl. XXIV, Fig. 2, Neocomian, Ghilcoş, East Carpathians. Coll. L.P.B. V – no. 0222.
Diagnosis: Thallus crustlike-bladded with irregular shape and a medullar hollow, 0.40–0.50 mm. Thickness of the cortical zones differs according to the attachment type: basally 0.90 mm and apically 2.1 mm. Cortical zones crossed by branched tubes with fibro-radial walls and thickness of 20–30 μ. The branching system consists of long tubes with diameters of 0.10–0.14 mm from which di- or trichotomous tubes, with diameters of 0.045–0.075 mm, branch further.
Stratigraphic Distribution: Lower Cretaceous, Neocomian.

Arabicodium Elliott 1957

Arabicodium bicazensis n.sp.
(Pl. XXV, Figs. 2–4)

Arabicodium sp.
1958, Dufaure, p. 109, Pl. 4, Fig. 5
Codiacées
1966, Radoicic, p. 183, Fig. 2
Permocalculus sp.
1975, Dragastan, Pl. LII, Fig. 1

Holotype: Pl. XXV, Fig. 2, Tithonian, Suhard, East Carpathians. Coll. L.P.B. V – no. 0223.
Syntypes: Pl. XXV, Figs. 3–4, Tithonian-Neocomian, Bicaz Gorges, East Carpathians. Coll. L.P.B. V – nos. 0224–0225.
Derivatio nominis: "Bicazensis" derived from the Bicaz Valley, East Carpathians.
Diagnosis: Digitiform thallus, 4.5 mm long and diameter 1.5 mm. Its central medullar zone is narrow, with diameter of 0.30 mm. The thick cortical zones are crossed by small filaments with diameter of 10–15 μ.
Remarks: Differs from other *Arabicodium* species by the narrow medullar zone with large cortical zones.
Stratigraphic Distribution: Middle Jurassic (Bathonian), Upper Jurassic (Oxfordian-Tithonian), and Lower Cretaceous (Neocomian).

Arabicodium hansi n.sp.
(Pl. XXV, Figs. 5–6)

Holotype: Pl. XXV, Fig. 6, Barremian-Lower Aptian, Bardos, East Carpathians. Coll. L.P.B. V – no. 0227.
Syntype: Pl. XXV, Fig. 5, Barremian-Lower Aptian, Bardos, East Carpathians. Coll. L.P.B. V – no. 0226.
Derivatio nominis: Species dedicated to Prof. Hans Mensink, Institute of Geology of Ruhr University in Bochum.
Diagnosis: Thallus with subspherical segments and diameter of 3.0–4.0 mm. Medullar zone of 0.30–1.5 mm and narrow cortical zones crossed by filaments with diameter 20–30 μ.
Remarks: Differs from other *Arabicodium* by the almost spherical shape of the segments.
Stratigraphic Distribution: Lower Cretaceous, Barremian-Lower Aptian.

Tethysicodium n.gen.

Type Species: *Tethysicodium wrayi* n.gen., n.sp., Tithonian, Făgetul Ciucului, East Carpathians.
Diagnosis: Thallus shoelike, consisting of segments with a narrow medullar channel, sometimes branched. The thick cortical zone is crossed by a network of channels from the central zone. There is a main long filament, with or without constrictions (Pl. XXVI, Fig. 1, upper left), continuing for several dichotomous channels.

Tethysicodium wrayi n.sp.
(Pl. XXVI, Figs. 1–4)

Holotype: Pl. XXVI, Fig. 1, Tithonian, Făgetul Ciucului, East Carpathians. Coll. L.P.B. V –no. 0228.
Syntypes: Pl. XXVI, Figs. 2–4, Tithonian, Vinturariţa, South Carpathians, Barremian–Aptian, Bardos, East Carparthians. Coll. L.P.B. V – nos. 0229–0231.
Derivatio nominis: Dedicated to Dr. John L. Wray for his contribution to the study of algae.
Diagnosis: Thallus elongated, shoelike, 6.0–8.0 mm long and 2.0–3.0 mm in diameter. A central, narrow medullar channel of 0.15–0.30 mm continues with primary cortical filaments 75–90 μ in diameter, and several bifurcating filaments of 30 μ. The angles between primary filaments are 20–30°, and those between secondary filaments are 10°. The thallus consists of segments 0.9–1.5 mm high.
Stratigraphic Distribution: Upper Jurassic, Tithonian, and Lower Cretaceous, Barremian-Lower Aptian.

Bacinellacodium n.gen.

Type Species: *Bacinellacodium calcareus* n.gen., n.sp., Tithonian, Făgetul Ciucului, East Carpathians.

Diagnosis: Thallus subcylindrical with a large, coarsely crystallized medullar zone of *Bacinella* type. The cortical walls are thin and crossed by simple or bifurcating filaments. The thallus may be attached to *Bacinella* or *Lithocodium*.

Bacinellacodium calcareus n.sp.
(Pl. XXVII, Figs. 1–3)

Holotype: Pl. XXVII, Fig. 1, Tithonian, Făgetul Ciucului, East Carpathians. Coll. L.P.B. V – no. 0232.
Syntypes: Pl. XXVII, Figs. 2–3, Neocomian, Barremian-Lower Aptian, Ghilcoş, East Carpathians. Coll. L.B.P. V – nos. 0233–0234.
Diagnosis: Thallus subcylindrical, 4.5–6.0 mm long and 2.8–3.0 mm in diameter. The coarsely crystallized medullar zone is 0.6–1.5 mm. The cortical walls are 0.30–0.60 mm thick and crossed by simple or bifurcating filaments 0.015–0.030 mm in diameter.
Remarks: This alga occurs frequently with *Bacinella irregularis* or *Lithocodium aggregatum* with which it forms a calcareous "crust".
Stratigraphic Distribution: Upper Jurassic, Tithonian, and Lower Cretaceous, Neocomian, Barremian-Lower Aptian.

Bevocastria Garwood 1931
Bevocastria elliotti (Dragastan 1971)
(Pl. XXVII, Fig. 4: Pl. XXVIII, Figs. 1–4)

Cayeuxia elliotti
1971, Dragastan, p. 187, Pl. X, Fig. 3, *not* Figs. 1, 2, 4
1978, Bucur, Pl. V, Fig. 4: Pl. VI, Fig. 3

Holotype: Pl. XXVIII, Fig. 1, Tithonian, Făgetul Ciucului, East Carpathians. Coll. L.P.B V – no. 0235.
Syntypes: Pl. XXVIII, Figs. 2–4, Tithonian, Făgetul Ciucului, East Carpathians. Coll. L.P.B. V – nos. 0236–0238.
Diagnosis: Thallus ovoidal, 4.8 mm long and 2.7–3.5 mm in diameter. It has a central medullar hollow, coarsely recrystallized, 0.60–0.90 mm in diameter. The cortical zone is crossed by fine, simple or bifurcating filaments with diameter 0.030 mm. The angle of divergence is 10–20°.
Remarks: Differs from the Paleozoic *Bevocastria* species by the occurrence of the medullar hollow and simple or bifurcate filaments which are not intermingled.
Stratigraphic Distribution: Upper Jurassic, Tithonian.

Bevocastria toomeyi n.sp.
(Text-Figs. 4–5)

Holotype: Text-Fig. 4, Tithonian, Cetea, Southern Apuseni. Coll. L.P.B. V – no. 0092.
Syntype: Text-Fig. 5, Tithonian, Presaca Ampoiului, Southern Apuseni. Coll. L.P.B. V – no. 0093.

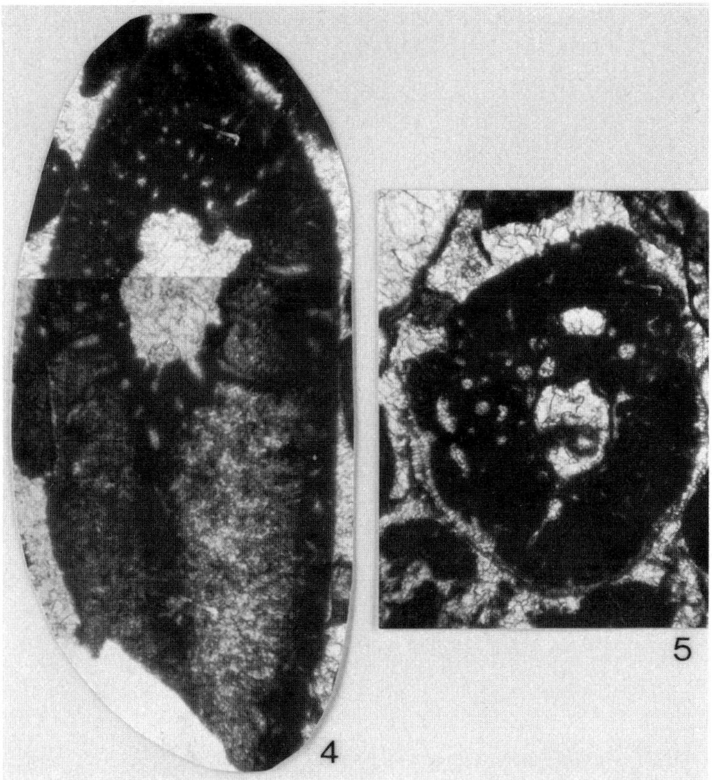

Text-Figs. 4–5. *Bevocastria toomeyi* n.sp. **4** Vertical oblique section. *Holotype*: Coll. L.P.B. – V no. 0092. **5** Horizontal section. *Syntype*: Coll. L.P.B. – V no. 0093. Tithonian, Presaca Ampoiului, Southern Apuseni. (x 8)

Derivatio nominis: Dedicated to Dr. Donald F. Toomey for his contribution to the study of algae.
Diagnosis: Very elongated thallus, up to 6.0 mm long and diameter 3.0 mm. It exhibits a large medullar cavity of 0.90 mm, and the walls of the thallus are micritized. The walls are crossed by long, thin filaments with diameter 0.075–0.090 mm.
Remarks: Differs from *B. elliotti* by the inner structure of the walls.
Stratigraphic Distribution: Upper Jurassic, Tithonian.

Addendum

In order to avoid any confusion and to observe the ICBN Rules, paragraph 45, we introduce *Appeninocodium* n.gen. for *Pseudocodium* Praturlon 1964, which is a junior homonym of the Recent *Pseudocodium* Weber van Bosse 1896. Thus, *Pseudocodium convolens* Praturlon 1964 is transferred to a codiacean alga, *Appeninocodium convolens* (Praturlon 1964).

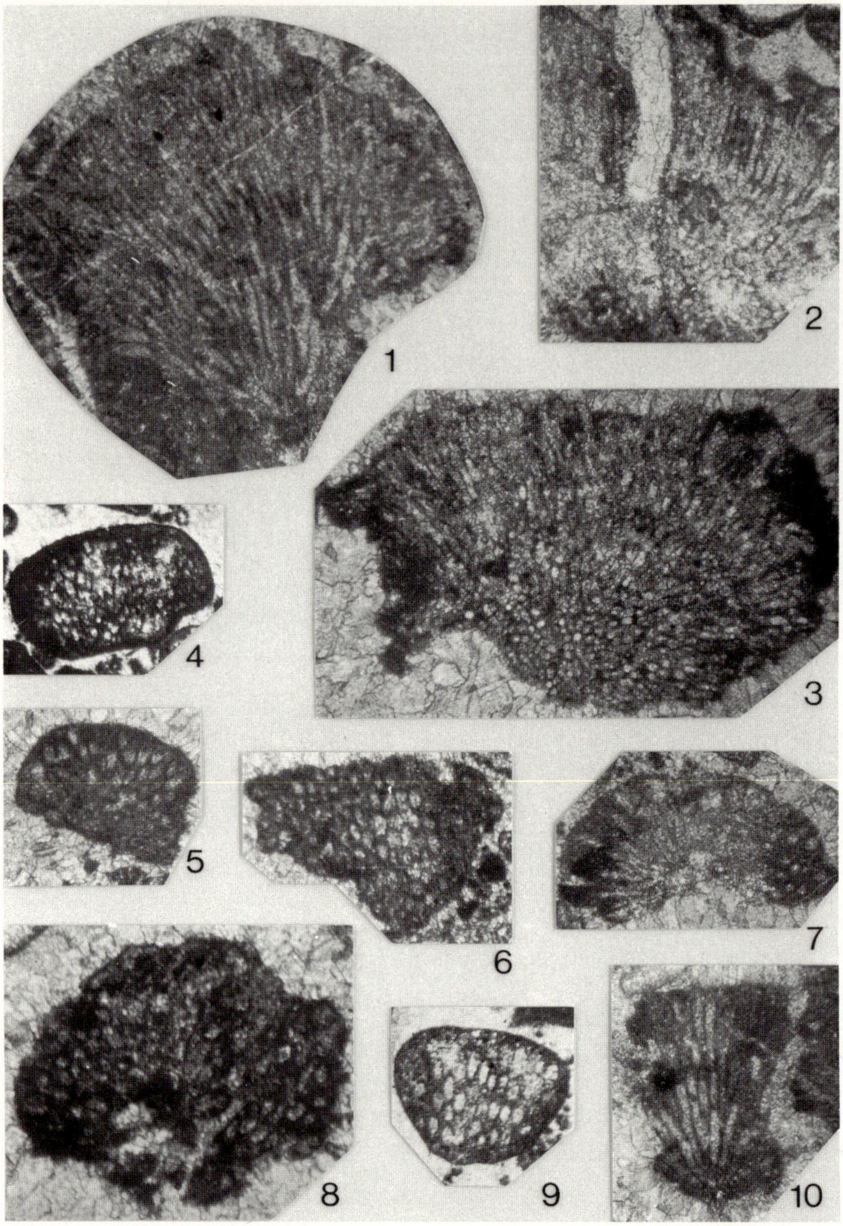

Plate I

Figs. 1–3. *Rivularia lissaviensis* (Bornemann). **1** Norian-Rhaetian with *Diplopora tubispora*. *Hypotype*: Coll. L.P.B. – V no. 0094. **2–3** Ladinian-Lower Carnian. *Hypotypes*: Coll. L.P.B. – V nos. 0095–0096, Vaşcău, Northern Apuseni and Glimeia Hill, Pădurea Craiului Mountains, Northern Apuseni. (x 8)

Figs. 4–10. *Rivularia lemaitrae* (Dragastan). **8** *Holotype* of Diaconu and Dragastan 1969, Ladinian-Lower Carnian, Glimeia Hill, Pădurea Craiului Mountains, Northern Apuseni. **4–7, 9–10** *Syntypes*: Coll. L.P.B. – V nos. 0097–0102, Glimeia Hill, Ladinian Lower Carnian, Pădurea Craiului Mountains, Northern Apuseni. **10** Carnian, Cîmp Moţi, Vaşcău, Northern Apuseni. (x 8)

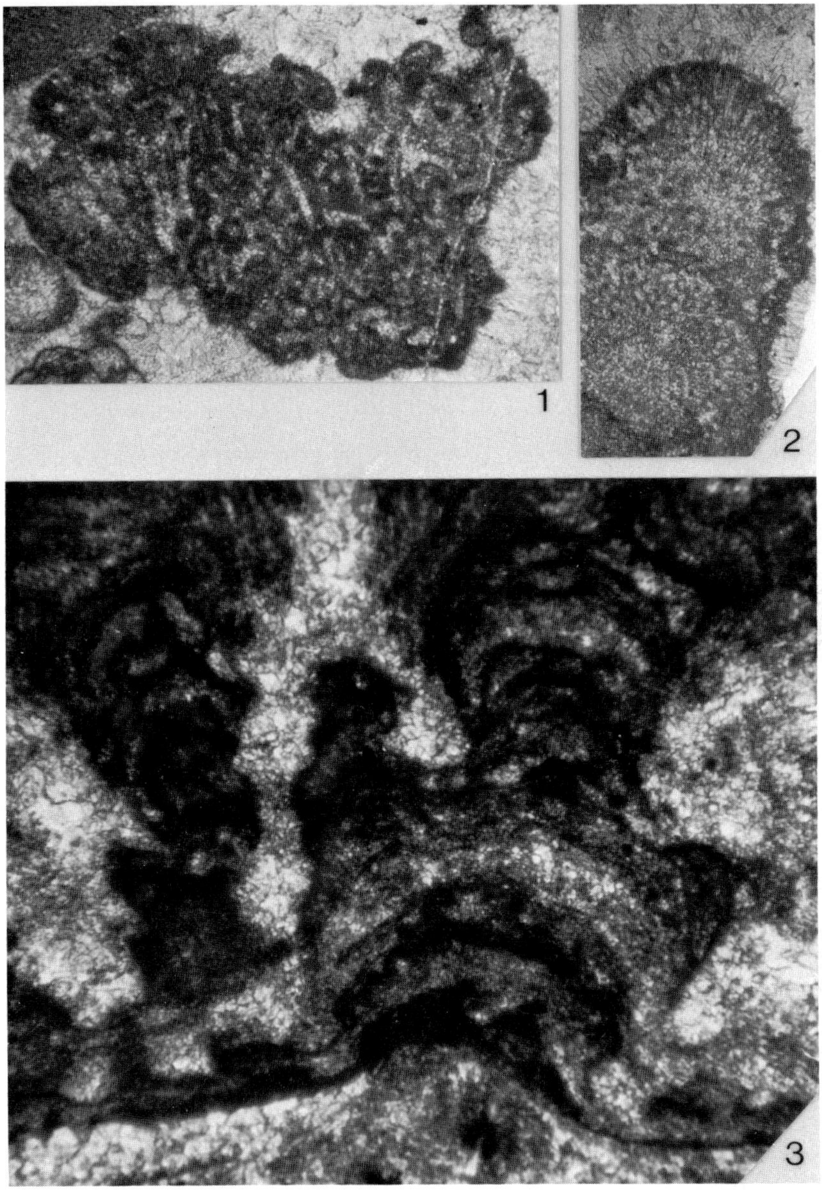

Plate II

Fig. 1. *Rivularia* sp., lax inner structure. Coll. L.P.B. – V no. 0103, Ladinian-Lower Carnian, Glimeia Hill, Pădurea Craiului Mountains, Northern Apuseni. (x 8)

Fig. 2. *Rivularia tadeuszi* n.sp. *Syntype*: Coll. L.P.B. – V no. 0104, Ladinian-Lower Carnian, Glimeia Hill, Pădurea Craiului Mountains, Northern Apuseni. (x 8)

Fig. 3. *Hormathonema massei* n.sp. *Holotype*: Coll. L.P.B. – V no. 0105, Ladinian, Someş Graben, Northern Apuseni. (x 8)

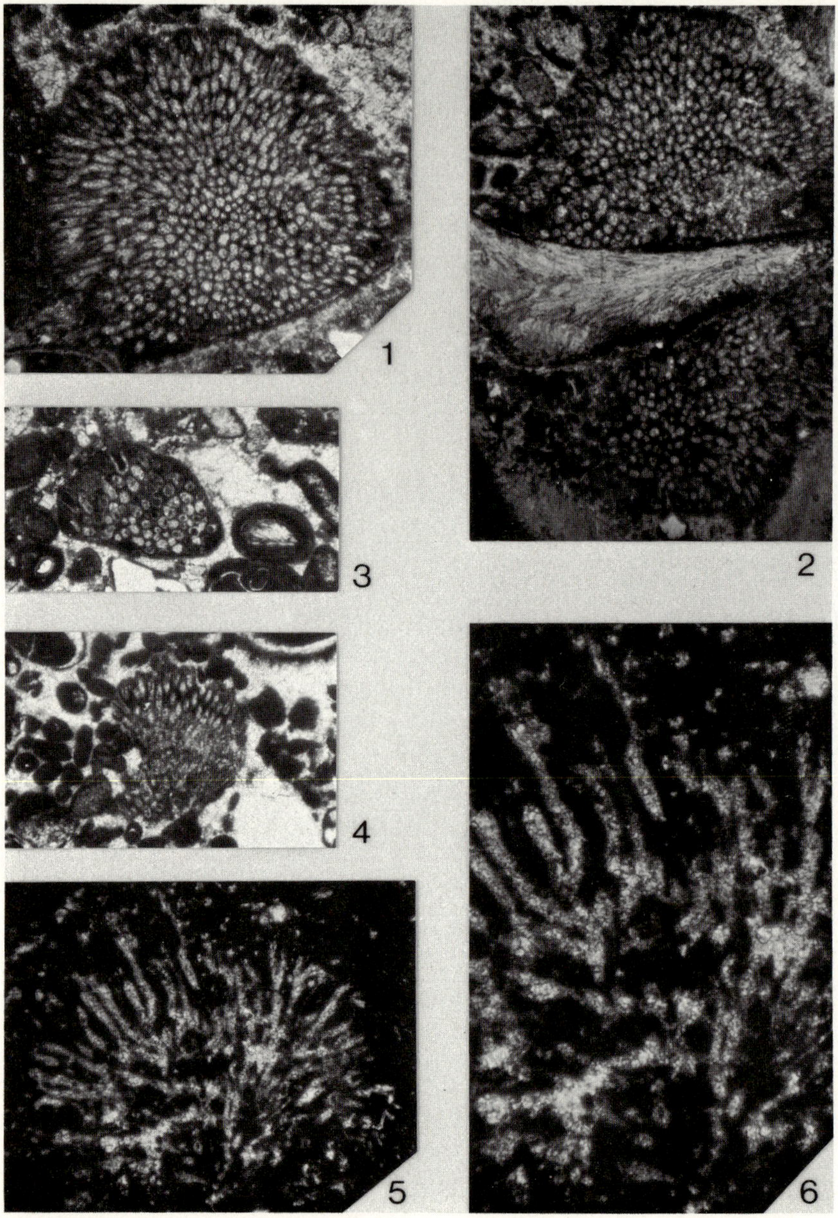

Plate III

Figs. 1–2. *Rivularia lissaviensis* (Bornemann). *Hypotypes*: Coll. L.P.B. – V nos. 0106–0107, Bajocian-Lower Bathonian, Sărătii Valley, Perşani Mountains, Eastern Carpathians. (x 8)
Figs. 3–4. *Rivularia theodori* n.sp. *Syntypes*: Coll. L.P.B. – V nos. 0108–0109, Oxfordian-Lower Kimmeridgian, Sărătii Valley, Perşani Mountains, Eastern Carpathians. (x 8)
Figs. 5–6. *Rivularia piae* (Frollo). *Syntypes*: Coll. L.P.B. – V nos. 0110–0111, Bajocian-Lower Bathonian, Sărătii Valley, Persani Mountains, Eastern Carpathians. **5** (x 8); **6** (x 16)

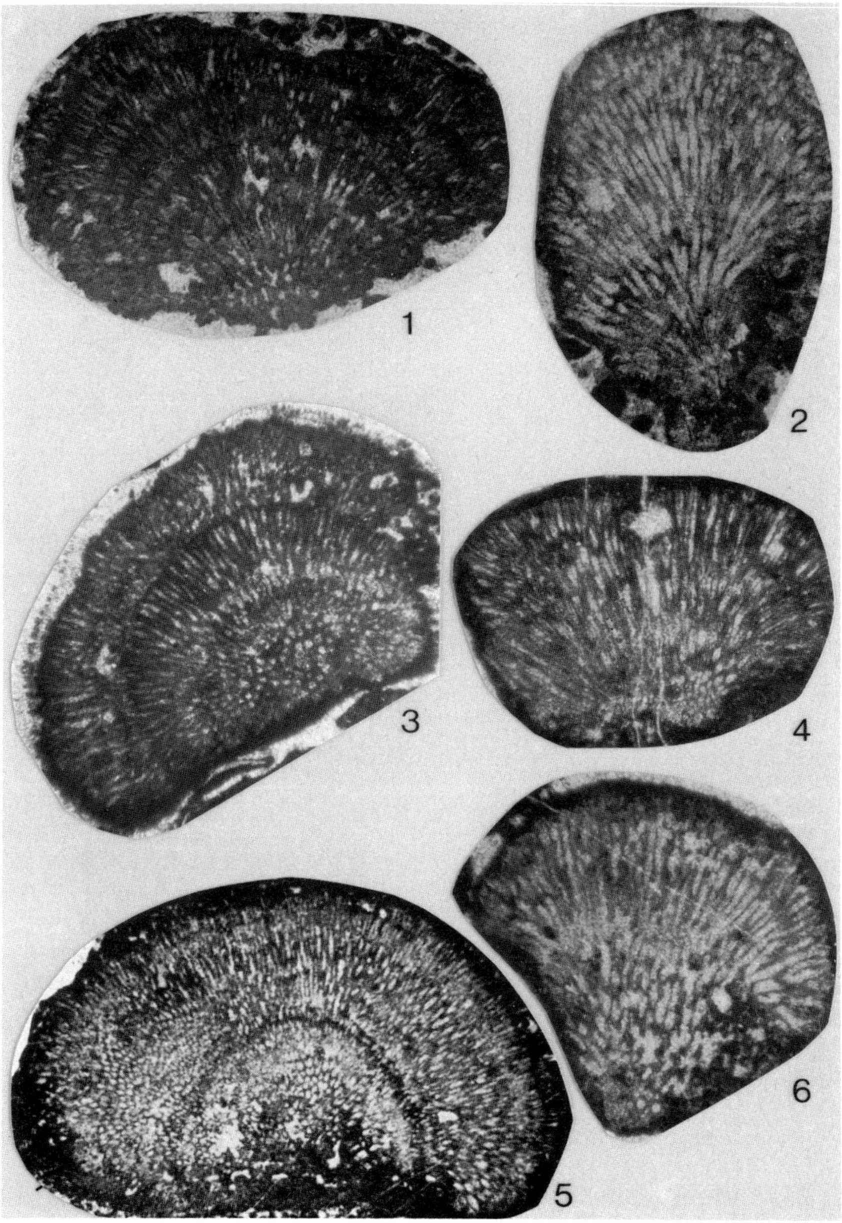

Plate IV

Figs. 1–6. *Rivularia lissaviensis* (Bornemann), compact inner structure with or without zones of growth. *Hypotypes*: Coll. L.P.B. – V nos. 0112–0117. **1** Tithonian, Frăsinet core Moesian Platform. **2, 4–6** Tithonian, Bicăjel, Bicaz Gorges. **3** Făgetul Ciucului, Eastern Carpathians. (x 8)

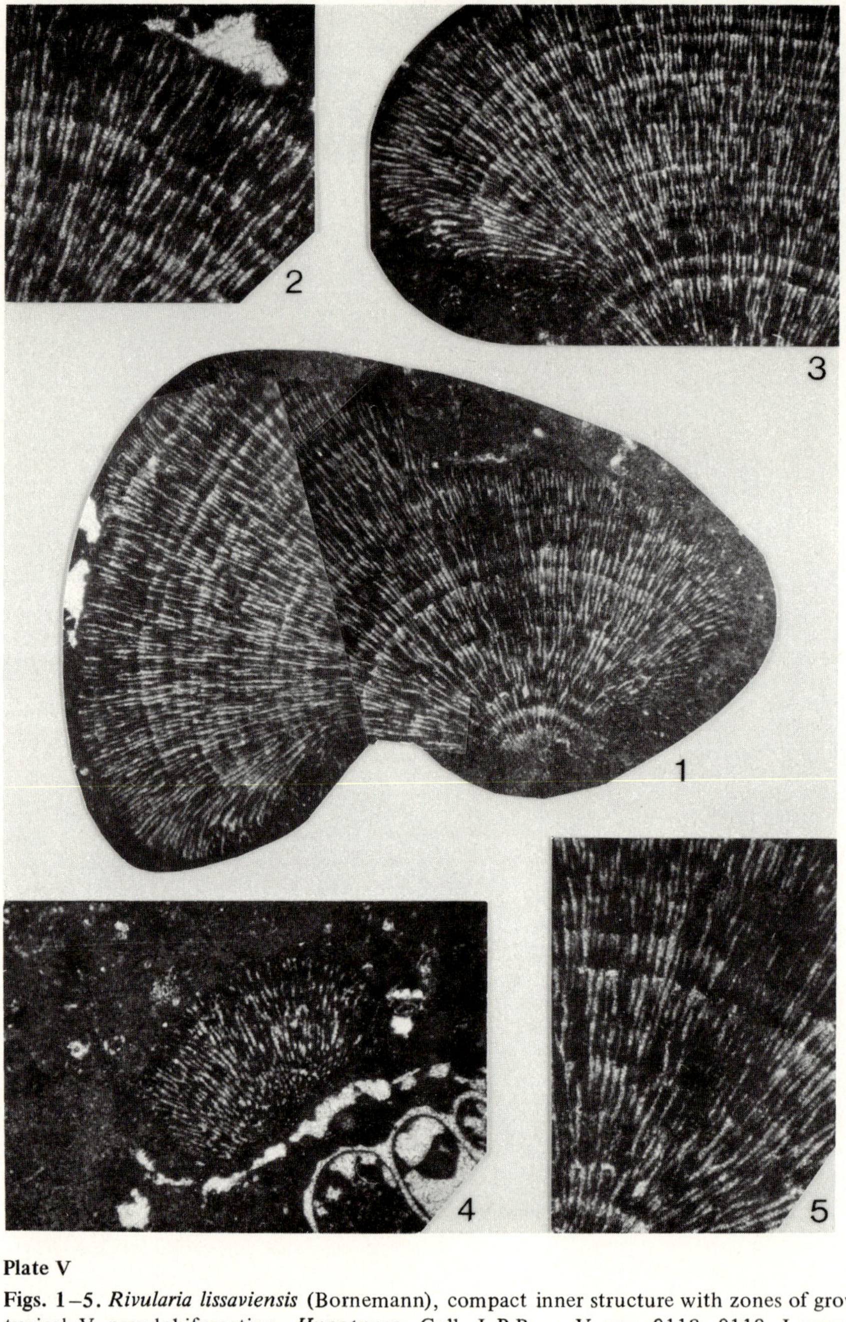

Plate V

Figs. 1–5. *Rivularia lissaviensis* (Bornemann), compact inner structure with zones of growth and typical V pseudobifurcation. *Hypotypes*: Coll. L.P.B. – V nos. 0118–0119, Lower Aptian, Petroşani, Southern Dobrogea. (x 8)

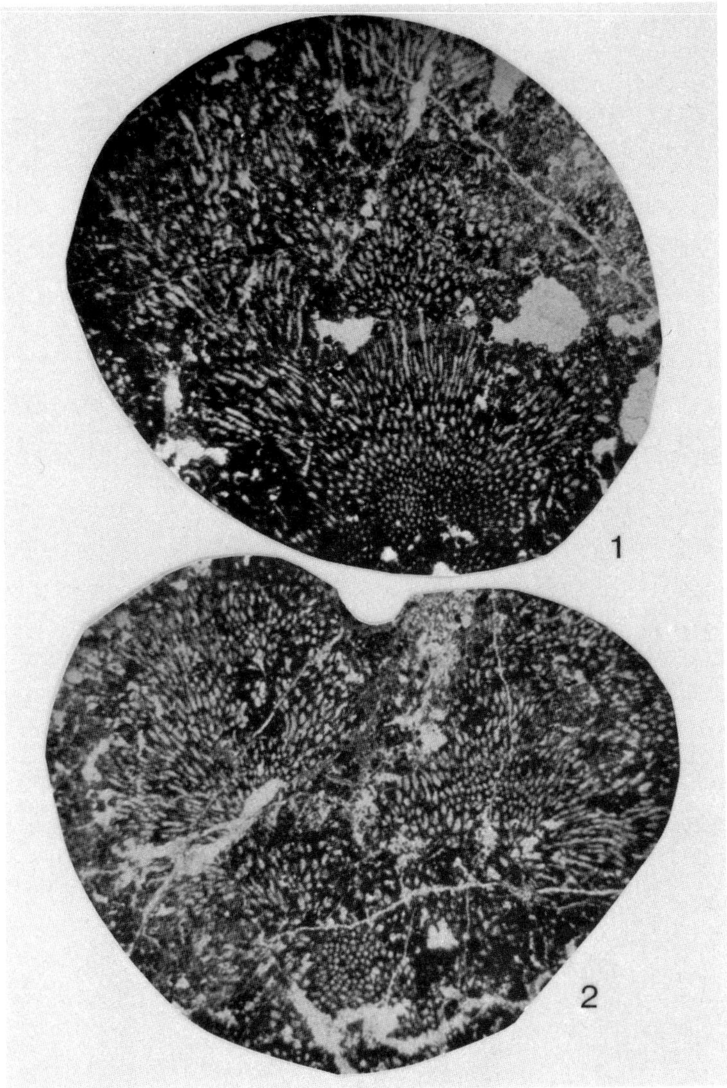

Plate VI

Figs. 1–2. *Rivularia lissaviensis* (Bornemann), similar to type of *Zonotrichites lissaviensis* of Bornemann 1887. *Hypotypes*: Coll. L.P.B. – V nos. 0120–0121, Tithonian, Făgetul Ciucului, Bicaz Gorges, Eastern Carpthians. (x 8)

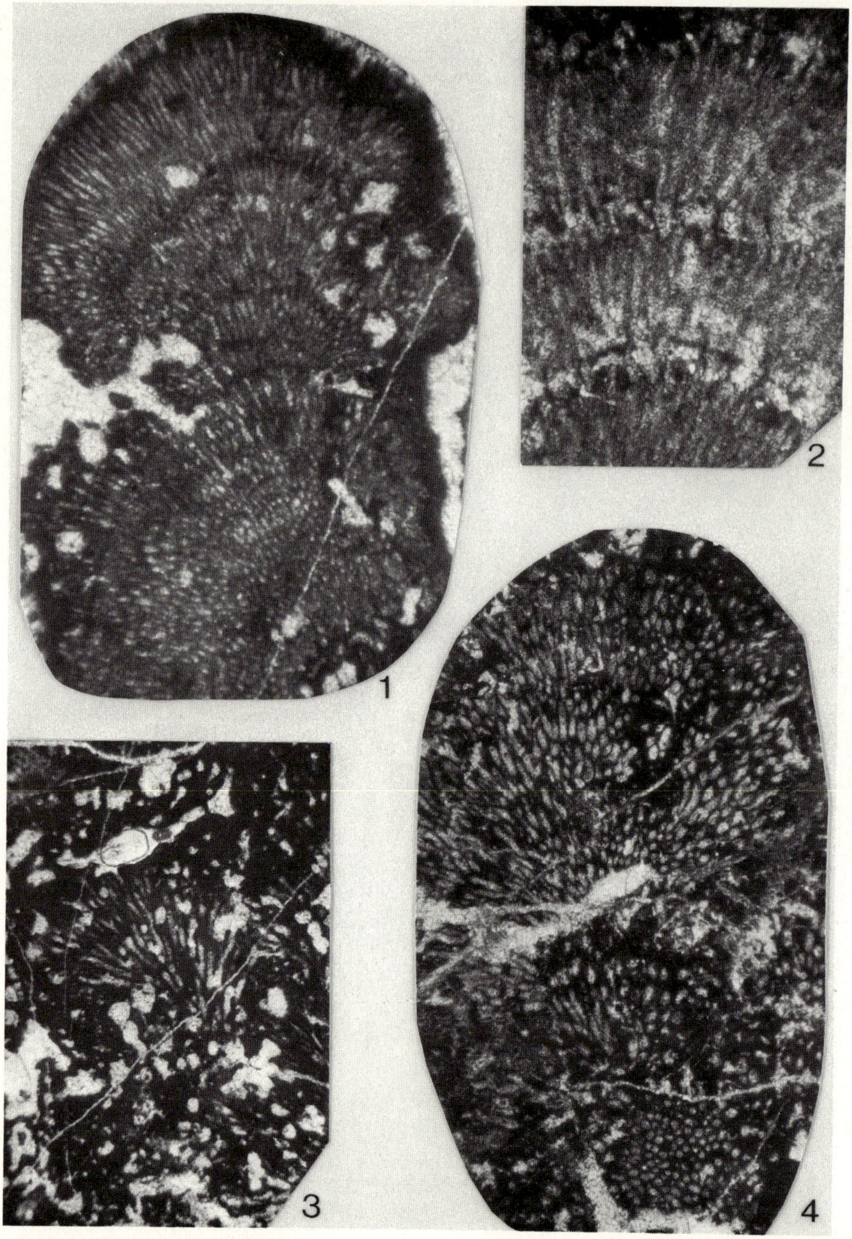

Plate VII

Figs. 1–4. *Rivularia lissaviensis* (Bornemann). *Hypotypes*: **1** Tithonian, Făgetul Ciucului, Coll. L.P.B. – V no. 0122. **2** Lower Tithonian, Kapfelberg, Franconia, Coll. L.P.B. – V no. 0123. (x 16). **3–4** Thallus with crusts of *Lithocodium aggregatum*. Coll. L.P.B. – V nos. 0124–0125, Tithonian, Bicăjel, Bicaz Gorges, Eastern Carpathians. (x 8)

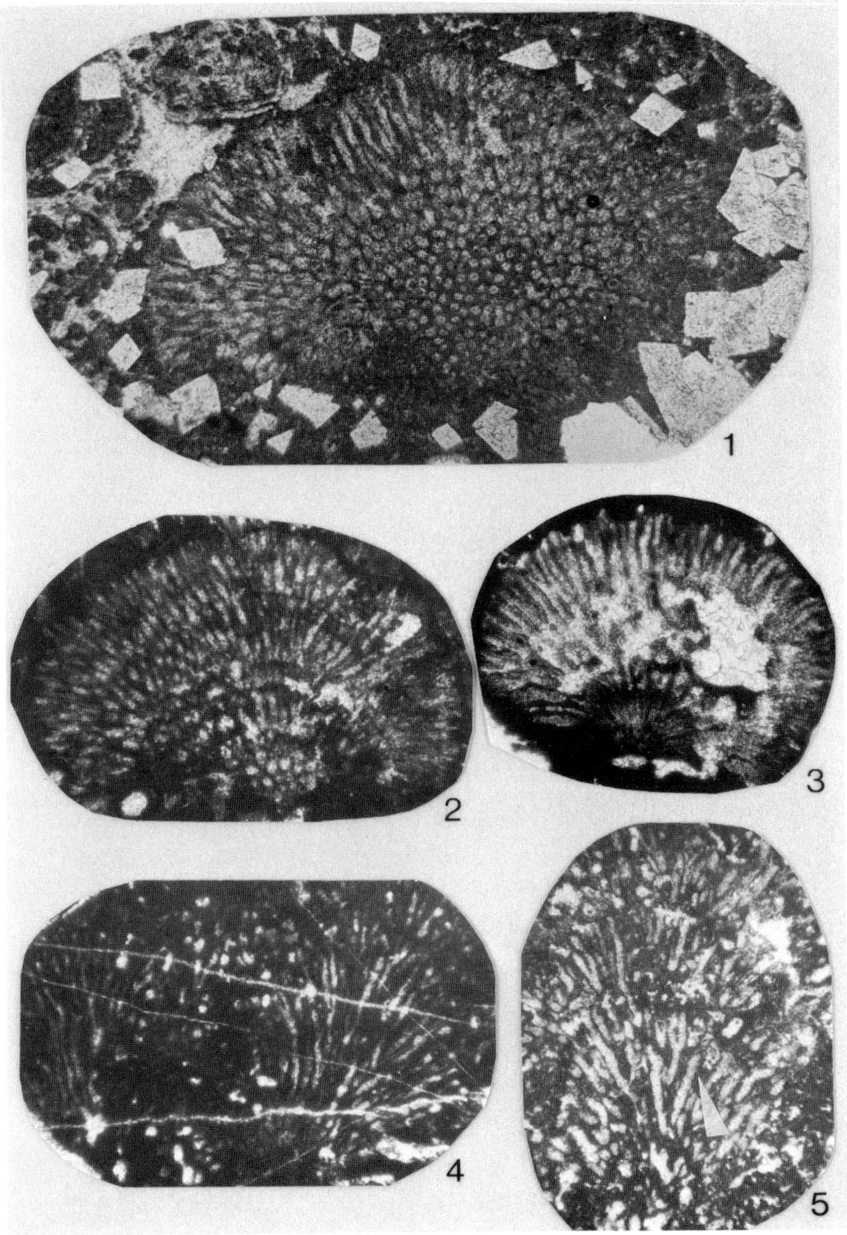

Plate VIII

Figs. 1–5. *Rivularia lissaviensis* (Bornemann). **1–3** Compact inner structure. **4–5** Semi-lax inner structure. *Hypotypes*: **1** Coll L.P.B. – V no. 0126, Neocomian, Bîldana and Călugăreni core, Moesian Platform. **2** Coll. L.P.B. – V no. 0127, Berriasian, Bicăjel, Bicaz Gorges. **3–5** Coll. L.P.B. – V nos. 0128, 0129, 0130, Barremian-Lower Aptian, Bardos, Bicaz Gorges, Eastern Carpathians. (x 8)

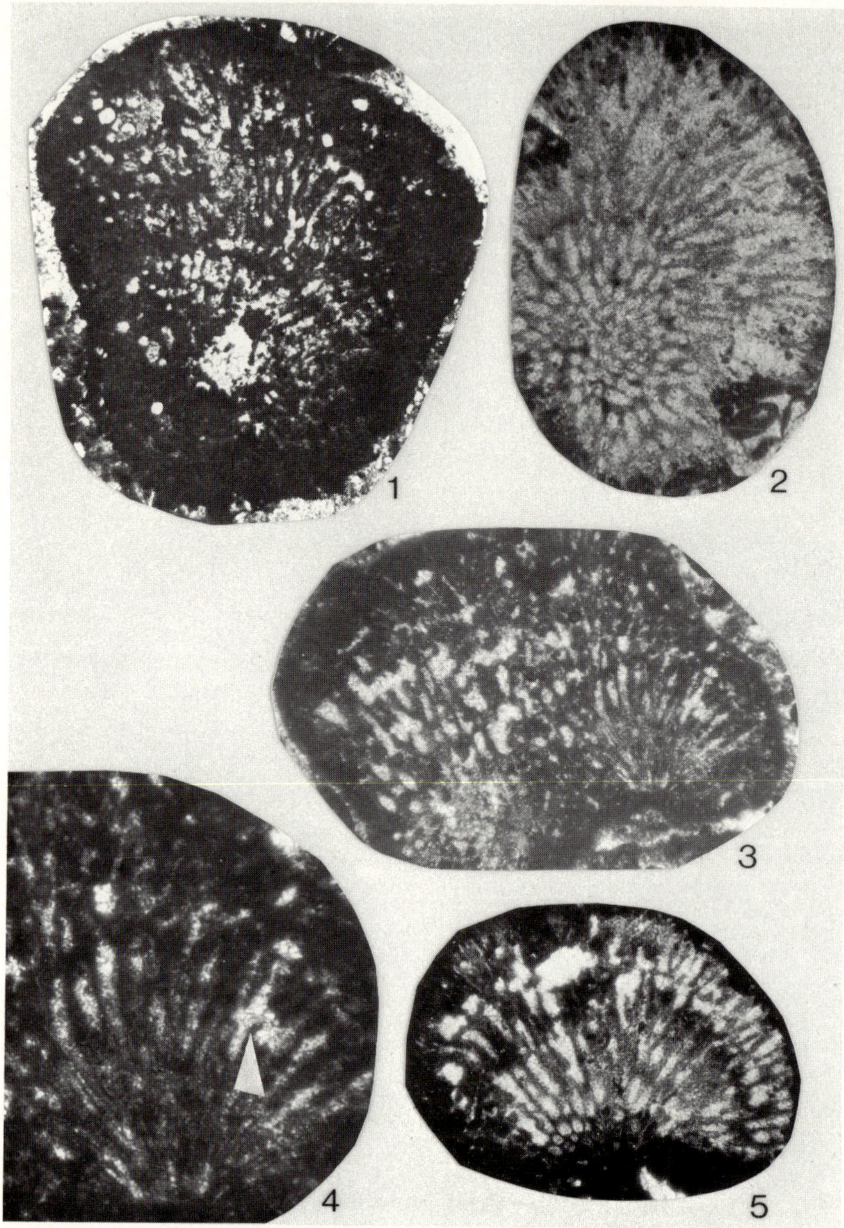

Plate IX

Figs. 1–5. *Rivularia piae* (Frollo). Compact inner structure. **1** *Hypotype*: Coll. L.P.B. – V no. 0131, Tithonian, Ghilcoş, Bicaz Gorges. **2** Coll. L.P.B. – V no. 0132, Neocomian, Suhard. **3–4** Coll. L.P.B. – V nos. 0133–0134, Lower Tithonian, Kapfelberg, Franconia (*arrow* indicates type of ramification). **5** Neotype, Coll. L.P.B. – V no. 0135, Tithonian, Făgetul Ciucului, Bicaz Gorges. **1–3, 5.** (x 8); **4**. (x 16)

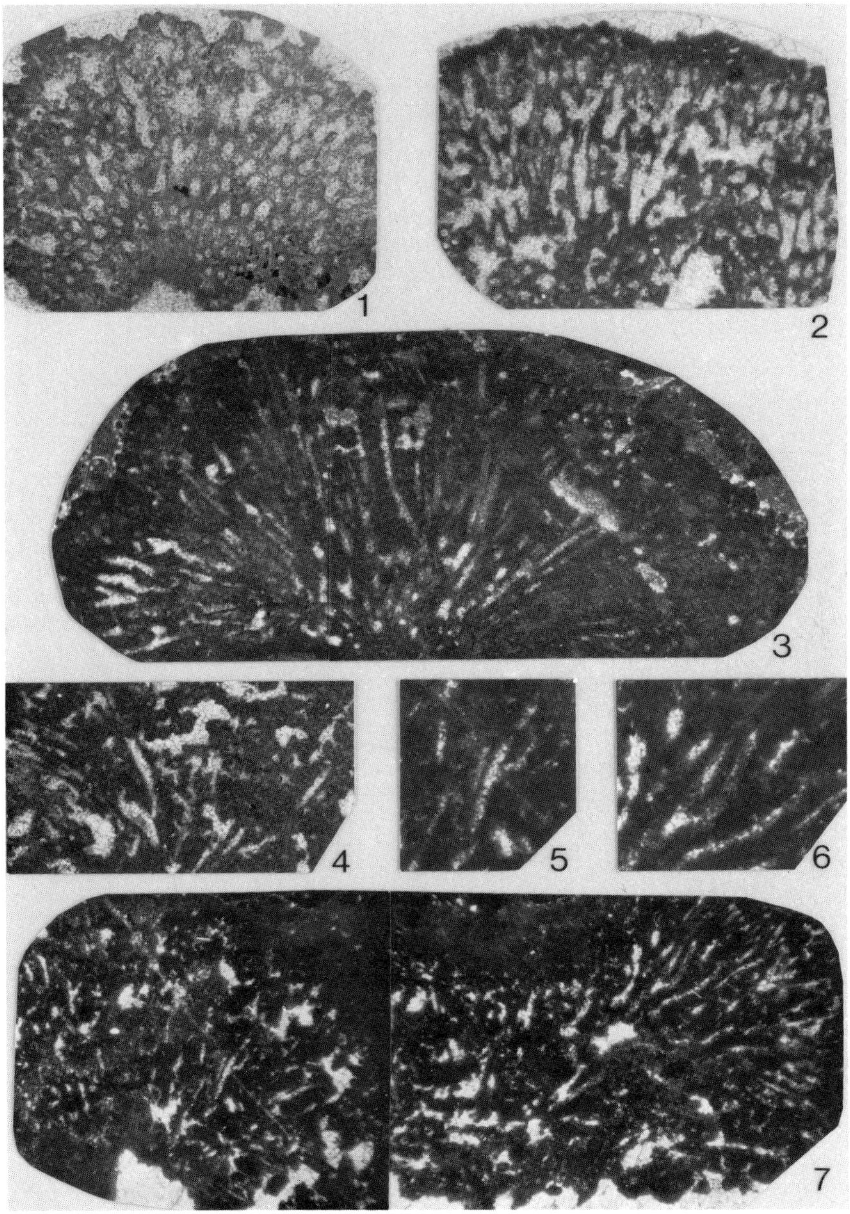

Plate X

Figs. 1–2. *Rivularia piae* (Frollo), compact inner structure. *Hypotypes*: Coll. L.P.B. – V nos. 0136–0137, Tithonian, Frăsinet core, Moesian Platform. (x 8)

Figs. 3–7. *Rivularia fruticulosa* (Johnson and Kaska), lax inner structure, large coarsely recrystallized area probably heterocysts (**3**), types of ramification (**5–6**). *Hypotypes*: Coll. L.P.B. – V nos. 0138–0142, Tithonian, Ghilcoş, Bicaz Gorges, Eastern Carpathians. **3–5, 7** (x 8); **6** (x 16)

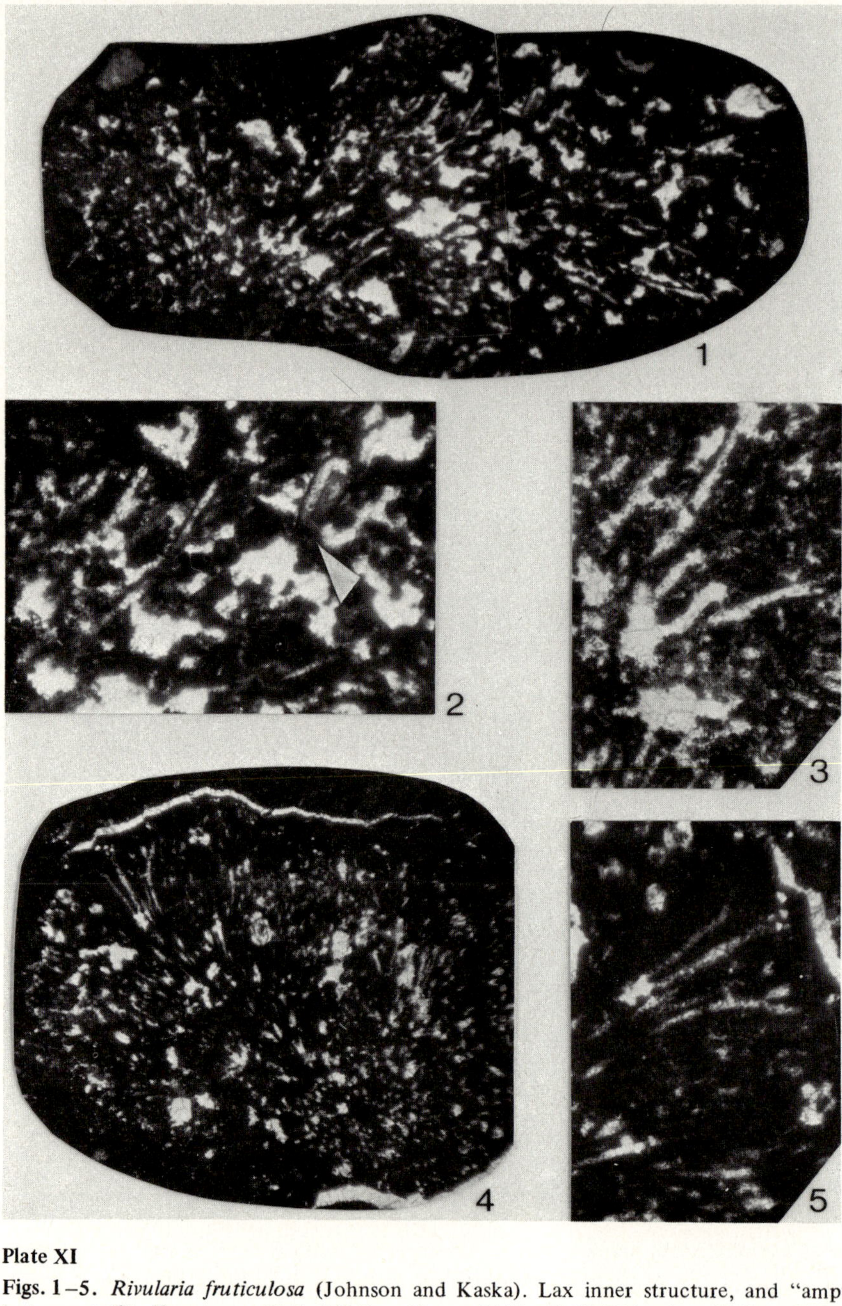

Plate XI

Figs. 1–5. *Rivularia fruticulosa* (Johnson and Kaska). Lax inner structure, and "ampulae" as heterocyst (**2**). *Hypotypes*: Coll. L.P.B. – V nos. 0143–0147; Tithonian, Ghilcoş, Bicaz Gorges, Eastern Carpathians. **1, 4** (x 8); **2, 3, 5** (x 16)

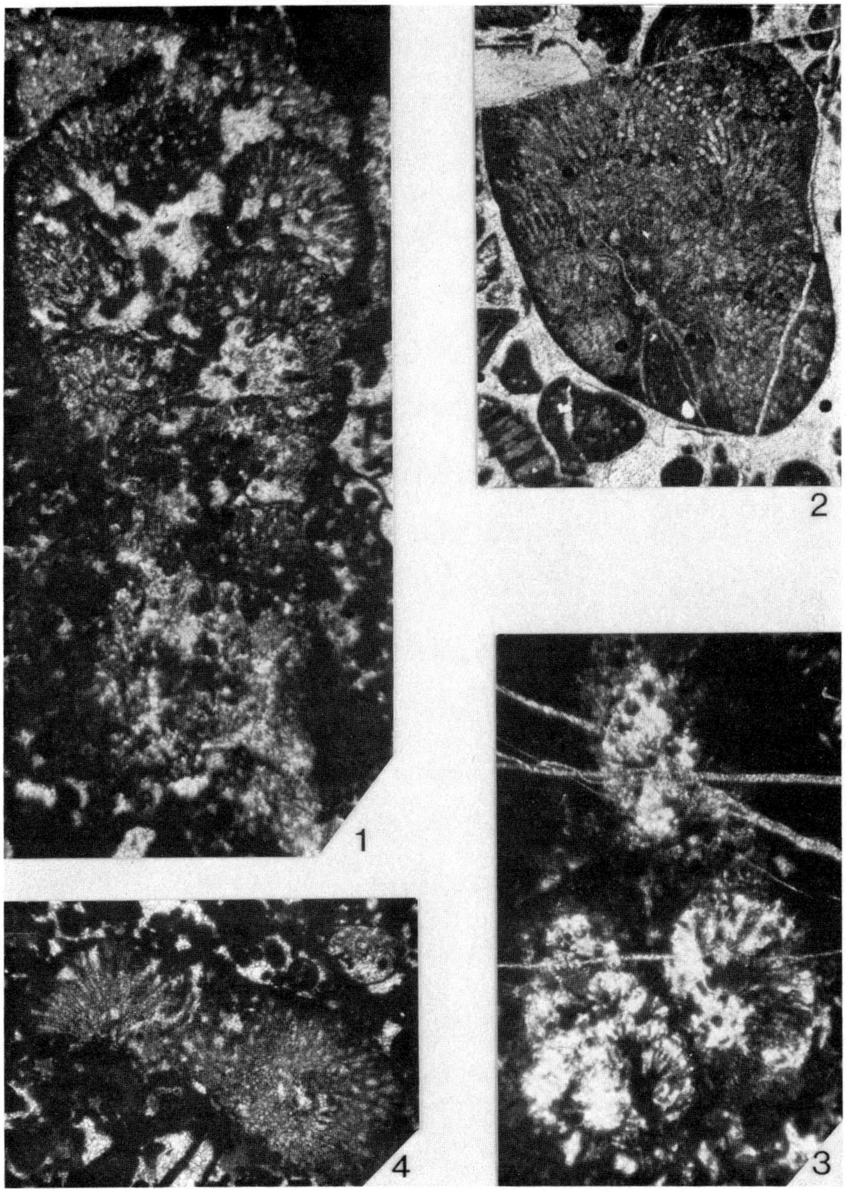

Plate XII

Figs. 1–4. *Rivularia atanasiui* (Dragastan). **1** Holotype of Dragastan 1971. **2–4** *Hypotypes*: Coll. L.P.B. – V nos. 0148–0150, Făgetul Ciucului. **4** Barremian-Lower Aptian, Ghilcoş, Bicaz Gorges, Eastern Carpathians. **1** (x 12); **2–4** (x 8)

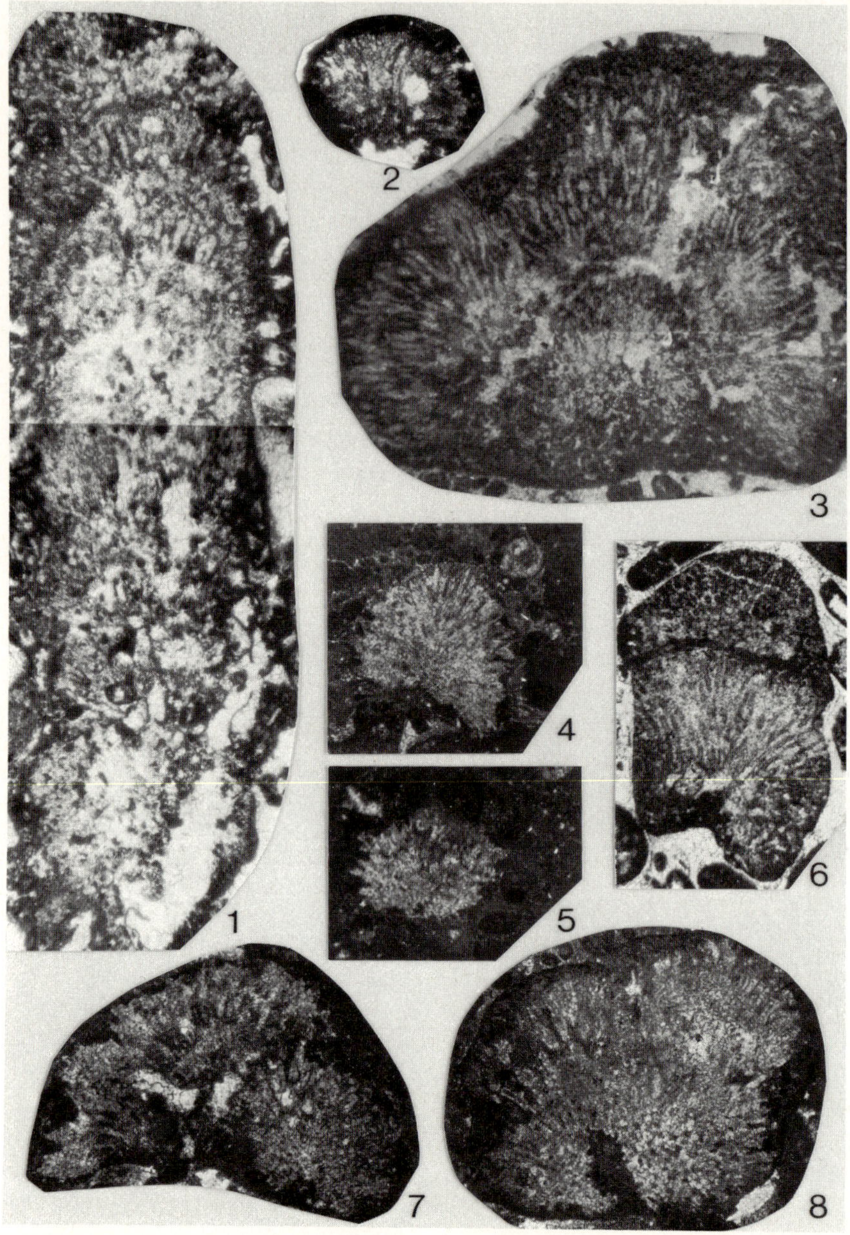

Plate XIII

Figs. 1–8. *Rivularia tadeuszi* n.sp. *Holotype*: **3** Coll. L.P.B. – V no. 0153, Tithonian, Bicăjel; *Syntypes*: **1–2** Coll. L.P.B. – V nos. 0151–0152. **4–8** Coll. L.P.B. – V nos. 0154–0158. Tithonian, Făgetul Ciucului, Ghilcoş. **8** Neocomian, Făgetul Ciucului, Bicaz Gorges, Eastern Carpathians. (x 8)

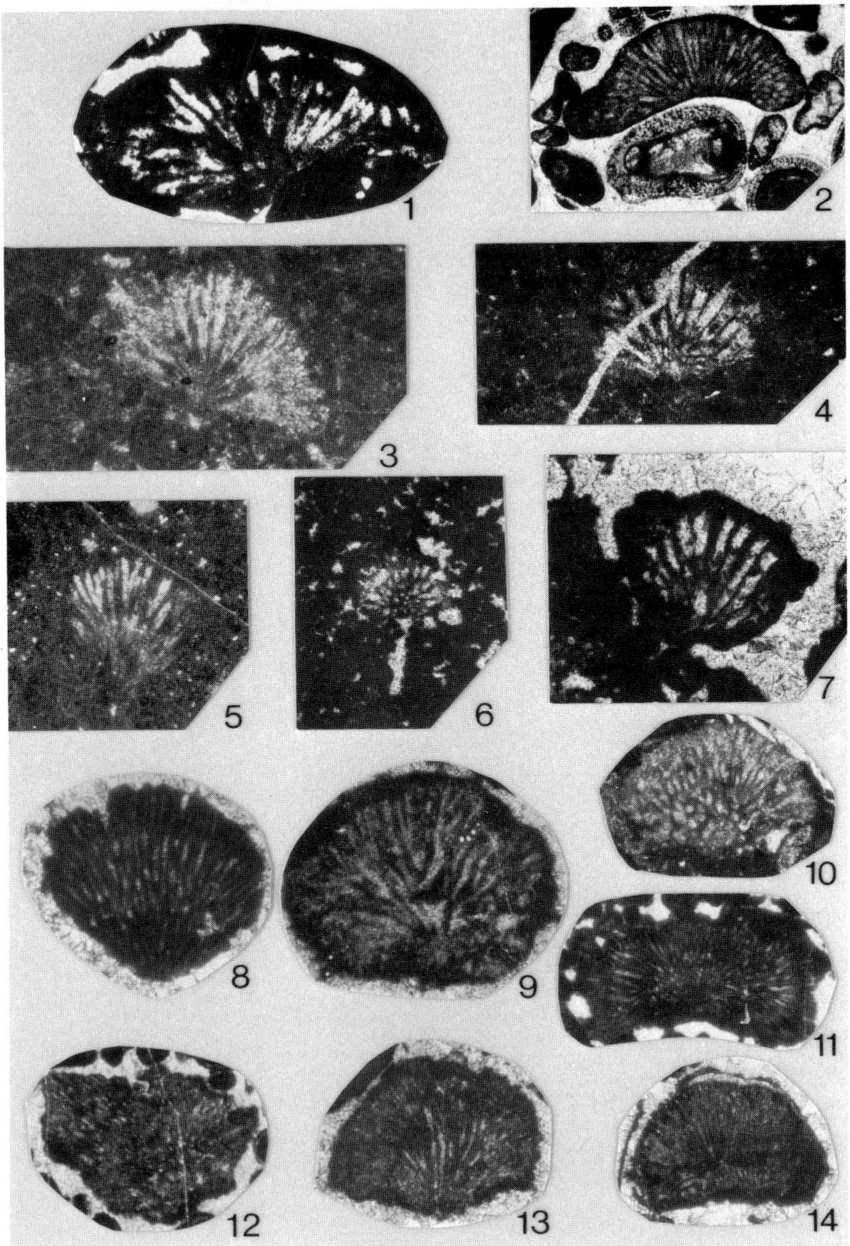

Plate XIV

Figs. 1–7. *Rivularia theodori* n.sp. **1** *Holotype*: Coll. L.P.B. – V no. 0159, Neocomian, Făgetul Ciucului. **2–7** *Syntypes*: Coll. L.P.B. – V no. 0160 (**4–5, 7**). Coll. L.P.B. – V nos. 0162–0163, 0165, Tithonian, Făgetul Ciucului, Bicăjel, Bicaz Gorges. **3** Coll. L.P.B. – V no. 0161, Tithonian-Neocomian, Pui-Haţeg, Southern Carpathians. **5** Barremian-Lower Aptian, Făgetul Ciucului, Bicaz Gorges, Eastern Carpathians. **6** Coll. L.P.B. – V no. 0164, Tithonian, Vînturariţa, Southern Carpathians. (x 8)

Figs. 8–14. *Rivularia kurdistanensis* (Elliott). *Hypotypes*: **9–12.** Coll. L.P.B. – V nos. 0167–0170, Tithonian, Bicăjel, Bicaz Gorges, Eastern Carpathians. **8, 14** Coll. L.P.B. – V nos. 0166, 0172, Tithonian, Cetea, Southern Apuseni. **13** Coll. L.P.B. – V no. 0171, Tithonian, Vînturariţa, Southern Carpathians. (x 8)

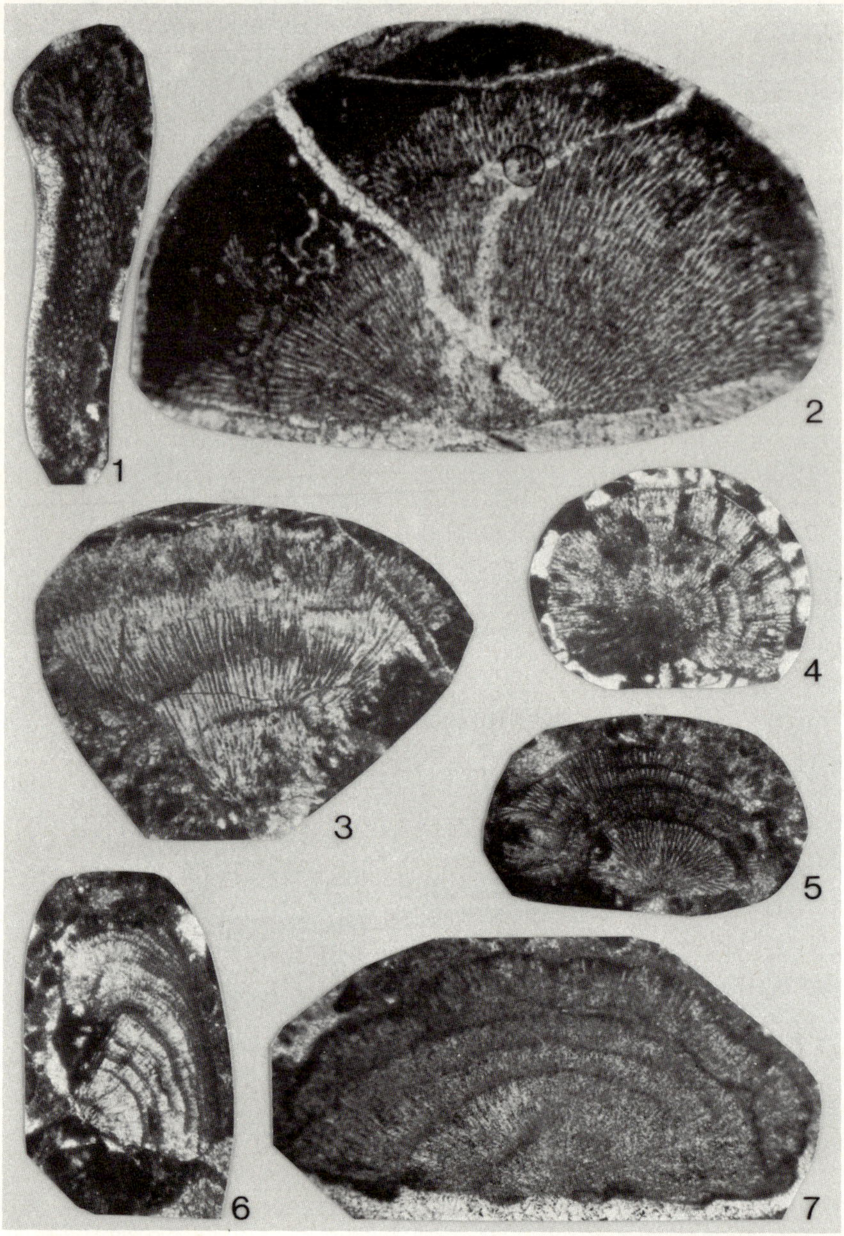

Plate XV

Fig. 1. *Rivularia* undet. Coll. L.P.B. – V no. 0173, Neocomian, Făgetul Ciucului, Bicaz Gorges, Eastern Carpathians. (x 8)

Figs. 2–3. *Rivularia dianae* (Dragastan and Bucur). *Hypotypes*: Coll. L.P.B. – V nos. 0174, 0175, Tithonian, Călineţ, altitude 1950 m, Piatra Craiului Mountains, Eastern Carpathians. **4–6** *Rivularia moesica* (Dragastan and Bucur). **4** *Holotype*: **5–6** *Hypotypes*: Coll. L.P.B. – V no. 0176–0177. **5** Tithonian, Presaca Ampoiului, **6** Tithonian, Poiana Ampoiului, Southern Apuseni. (x 8)

Fig. 7. *Rivularia carpathica* n.sp. *Syntype*: Coll. L.P.B. – V no. 0178, Tithonian, Cetea, Trascău Mountains, Southern Apuseni. (x 8)

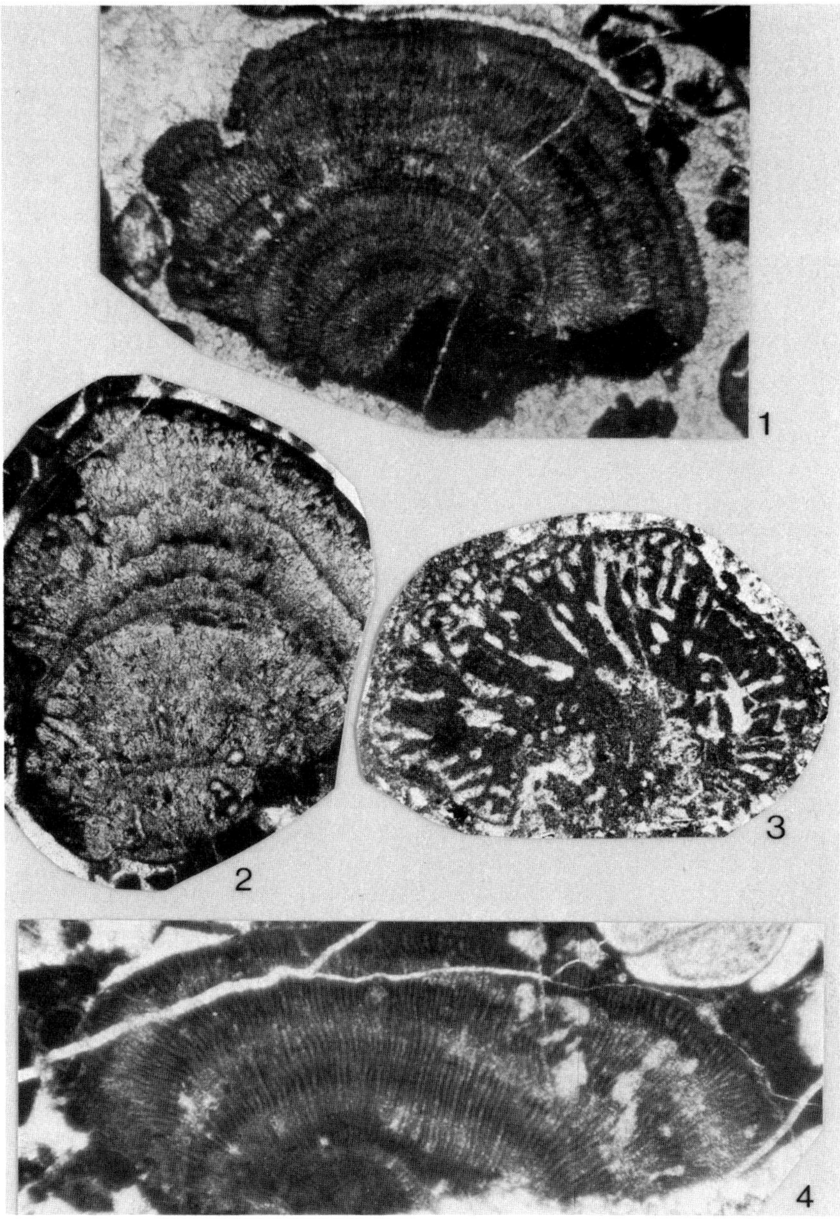

Plate XVI

Figs. 1–2. *Rivularia carpathica* n.sp. **1** *Holotype*: Coll. L.P.B. – V no. 0179, Tithonian, Făgetul Ciucului. **2** *Syntype*: Coll. L.P.B. – V no. 0180, Tithonian, Făgetul Ciucului, Bicaz Gorges, Eastern Carpathians. (x 8)

Fig. 3. *Ortonella lemoineae* Dragastan. *Holotype*: Ghilcoş, Bicaz Gorges, Eastern Carpathians. (x 8)

Fig. 4. *Apophoretella dobunnorum* Elliott. *Hypotype*: Coll. L.P.B. – V no. 0181, Tithonian, Bicăjel, Bicaz Gorges, Eastern Carpathians. **1–2, 4** (x 8); **3** (x 16)

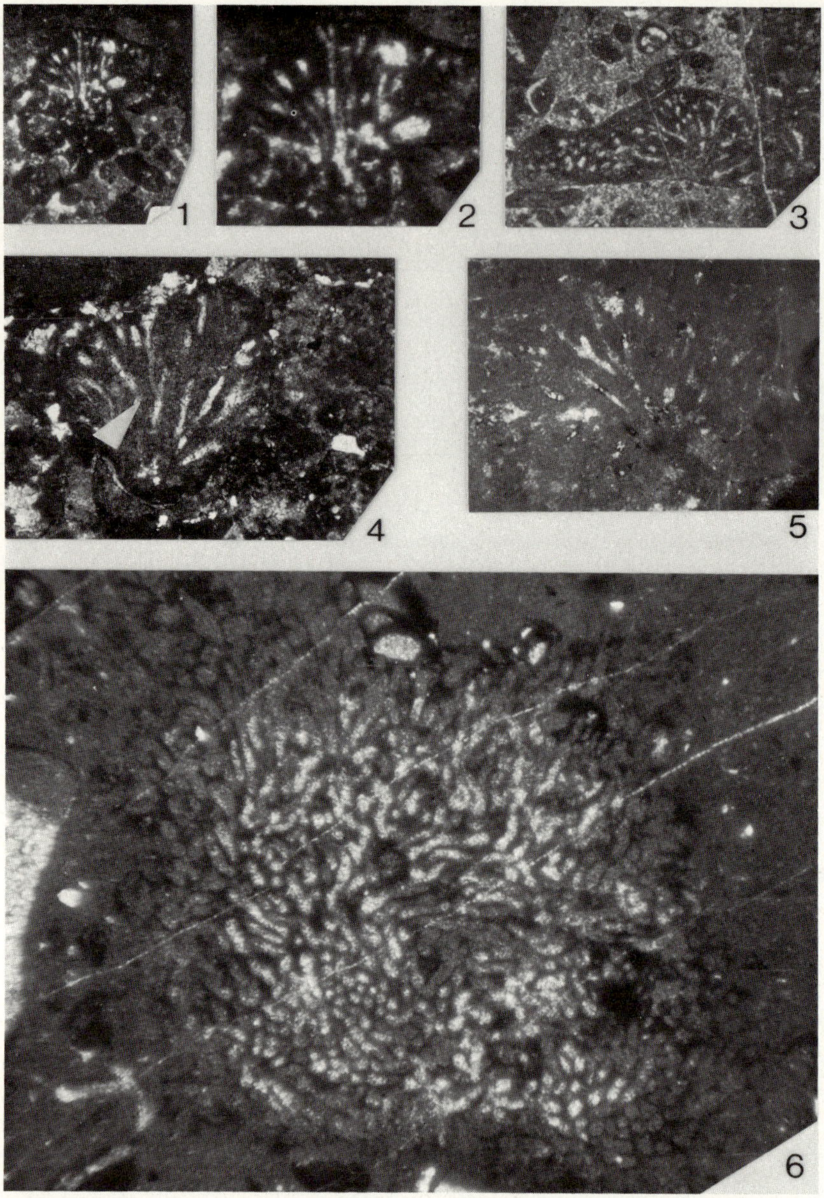

Plate XVII

Figs. 1–3. *Paraortonella getica* n.gen., n.sp. **1–2** *Holotype*: Coll. L.P.B. – V no. 0182, Tithonian, Ghilcoş. **3** *Syntype*: Coll. L.P.B. – V no. 0183, 0184, Barremian-Lower Aptian, Ghilcoş, Bicaz Gorges, Eastern Carpathians. **1, 3** (x 8); **2** (x 16)

Figs. 4–5. *Paraortonella richteri* n.gen., n.sp. **4** *Holotype*: Coll. L.P.B. – V no. 0185, Tithonian, Ghiloş, Bicaz Gorges (*arrow* indicates type of ramification). **5** *Syntype*: Coll. L.P.B. – V no. 0186, Tithonian, Suhard, Bicaz Gorges, Eastern Carpathians. (x 16)

Fig. 6. *Pseudogirvanella cretacea* n.gen., n.sp. *Holotype*: Coll. L.P.B. – V no. 0187, Barremian-Lower Aptian, Vîrciorog-Fîşca, Pădurea Craiului Mountains, Northern Apuseni. (x 16)

Plate XVIII

Figs. 1–3. *Mitcheldeania americana* (Johnson). **3** *Holotype*: from Johnson 1961. **1–2** *Hypotypes*: Coll. L.P.B. – V nos. 0188, 0189, Tithonian, Bicăjel, Bicaz Gorges, Eastern Carpathians. **1** (x 8); **2** (x 16)

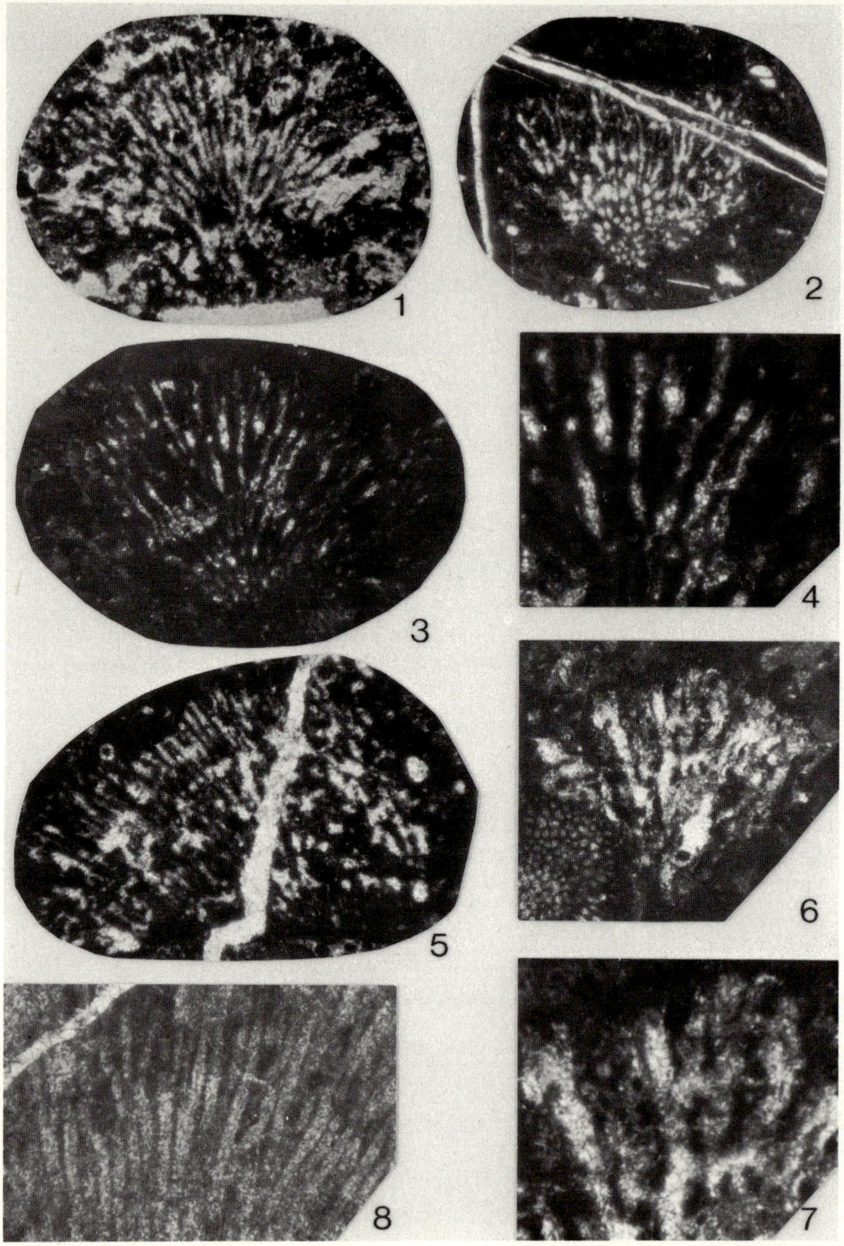

Plate XIX

Figs. 1–8. *Garwoodia fluegeli* n.sp. **1** *Holotype*: Coll. L.P.B. – V no. 0190, Tithonian, Bicăjel, Bicaz Gorges. **2–8** *Syntypes*: Coll. L.P.B. – V nos. 0191–0197. **2, 3–4, 6–7** Tithonian, Bicăjel. **5** Coll. L.P.B. – V no. 0194, Neocomian, Bicaz Gorges. **8** Tithonian, Cornilor Valley, Remeţi, Northern Apuseni. **4, 7–8** (x 16); others (x 8)

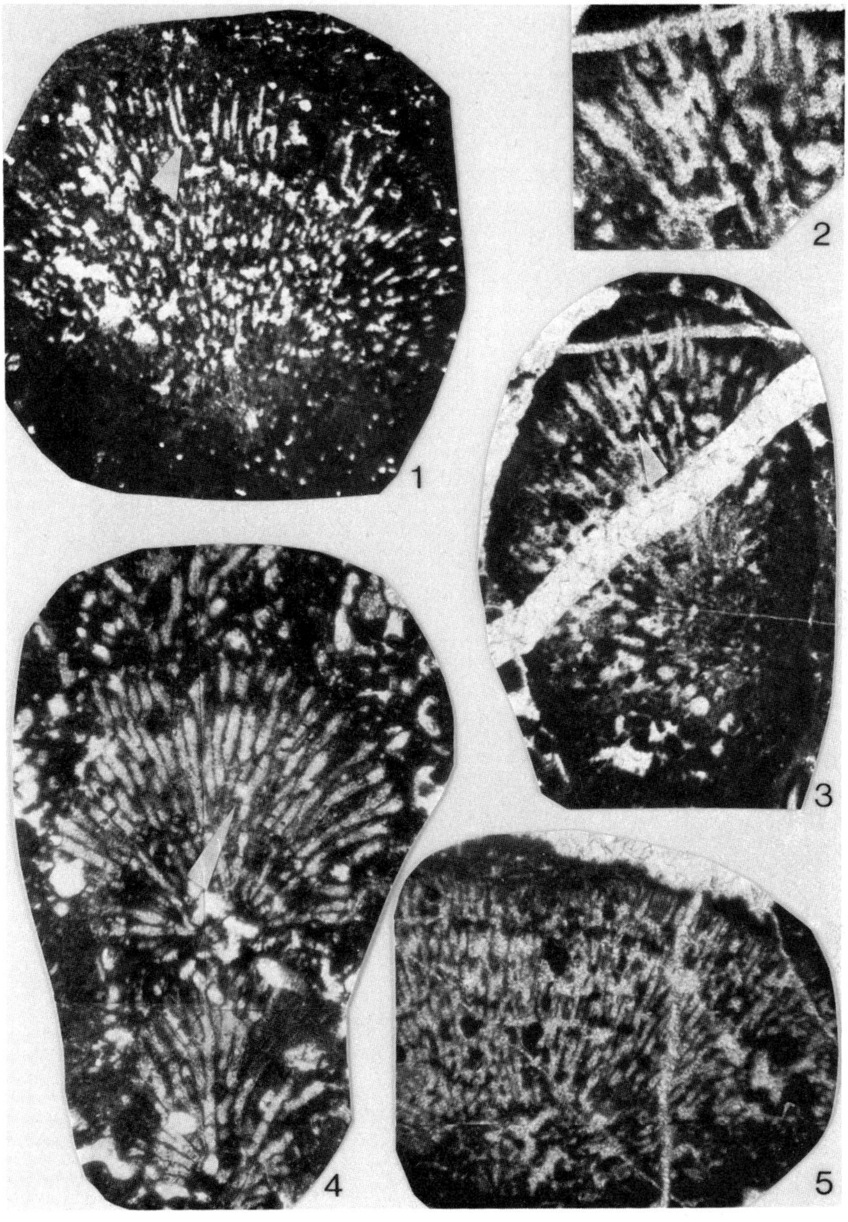

Plate XX

Figs. 1–5. *Garwoodia bardosi* n.sp. **4** *Holotype*: Coll. L.P.B. – V nos. 0201, Tithonian, Ghilcoş, Bicaz Gorges. **1–3, 5** *Syntypes*: Coll. L.P.B. – V nos. 0198, 0200,0202. **1** Neocomian, Ghilcos. **2–3** Tithonian, Făgetul Ciucului. **5** Barremian-Lower Aptian, Bardos, Bicaz Gorges, Eastern Carpathians (*arrow* indicates type of ramification). **1, 3, 5** (x 8); **2** (x 16); **4** (x 12)

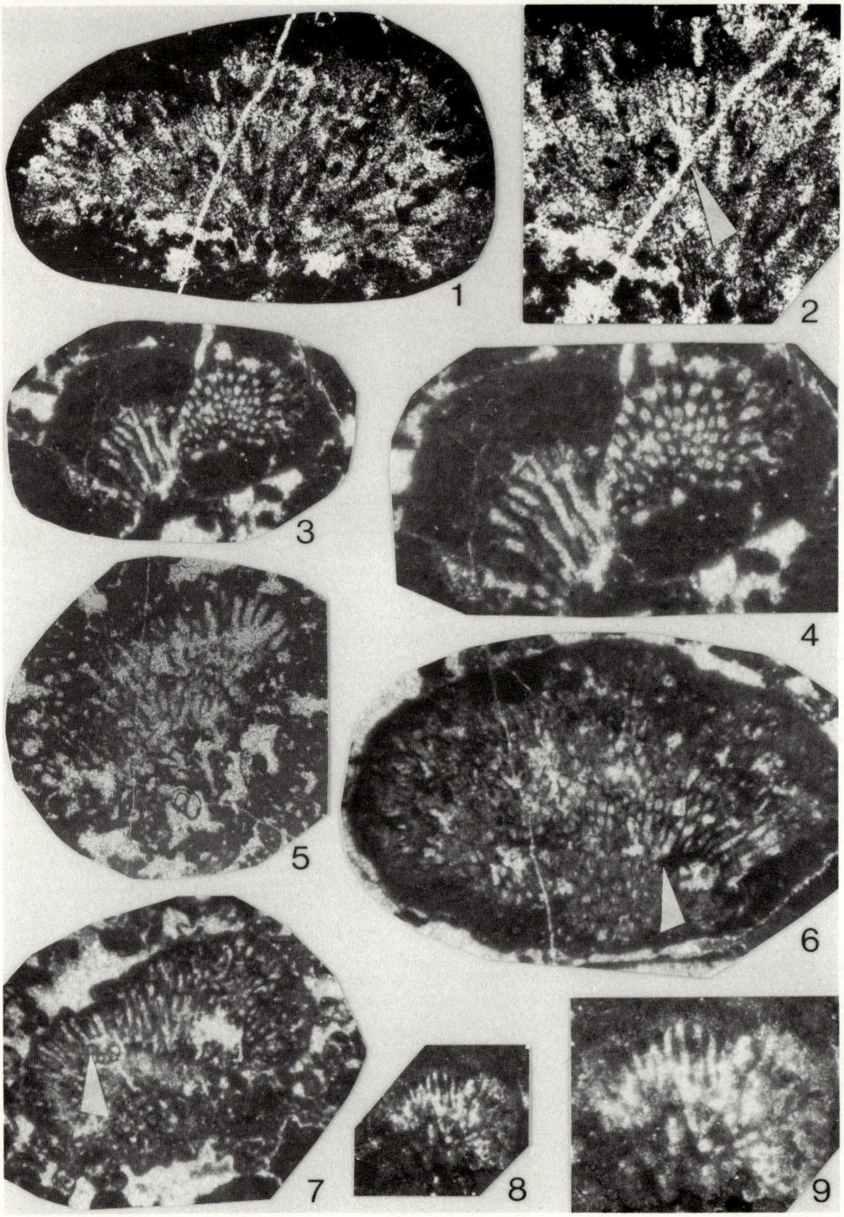

Plate XXI

Figs. 1–9. *Hedstroemia moldavica* n.sp. **1–2** *Holotype*: Coll. L.P.B. – V no. 0203, Tithonian, Făgetul Ciucului, Bicaz Gorges, Eastern Carpathians. **3–9** *Syntypes*: Coll. L.P.B. – V nos. 0204–0210. **3–4** Tithonian, Bicăjel. **6–7** Tithonian, Ghilcoş. **8–9** Tithonian, Ghilcoş. **5** Upper Barremian-Lower Aptian, Şopotul Nou core, Reşiţa-Moldova Nouă Zone (Banat), Southern Carpathians (*arrow* indicates type of ramification). **2, 4, 9** (x 16); others (x 8)

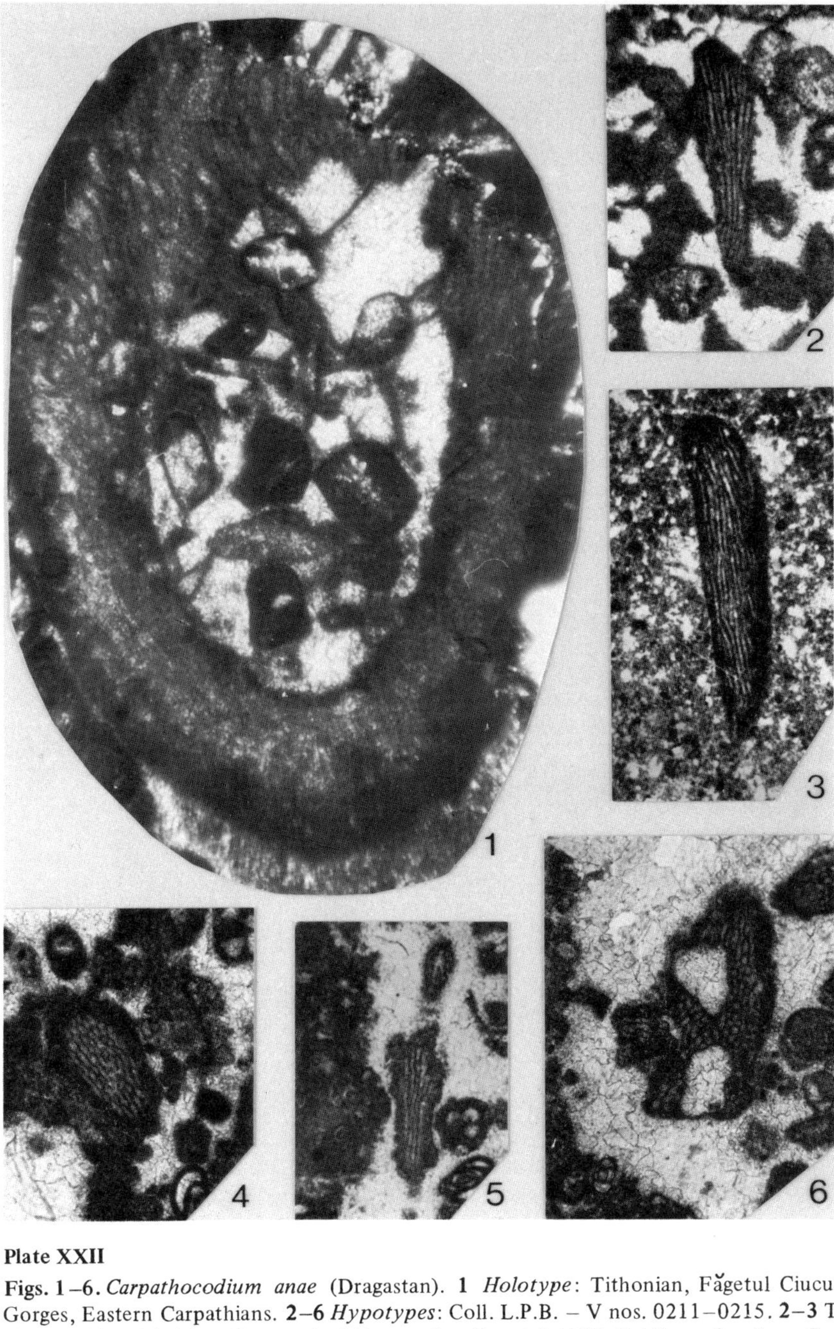

Plate XXII

Figs. 1–6. *Carpathocodium anae* (Dragastan). **1** *Holotype*: Tithonian, Făgetul Ciucului, Bicaz Gorges, Eastern Carpathians. **2–6** *Hypotypes*: Coll. L.P.B. – V nos. 0211–0215. **2–3** Tithonian, Făgetul Ciucului, Bicaz Gorges. **4–6** Neocomian. Dreptului Hill, Pui Zone. Southern Carpathians. (x 12)

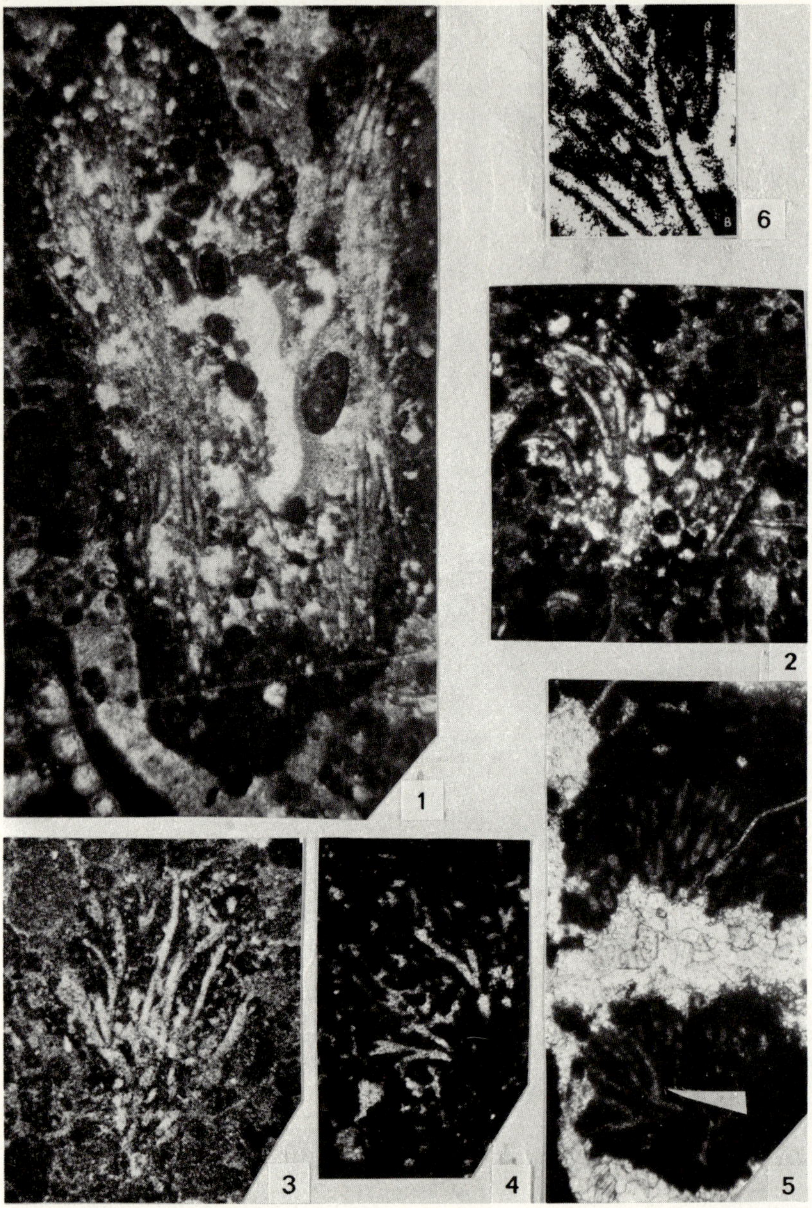

Plate XXIII

Fig. 1. *Carpathocodium anae* (Dragastan). *Hypotype*: Coll. L.P.B. – V no. 0216, Tithonian, Bicăjel, Bicaz Gorges, Eastern Carpathians. (x 12)

Figs. 2–5. *Scotlandella alpina* n.gen., n.sp. **3** *Holotype*: Coll. L.P.B. – V no. 0217, Neocomian, Ghilcoş, Bicaz Gorges, Eastern Carpathians. **2, 4–5** *Syntypes*: Tithonian, Bicăjel. **4** Barremian-Lower Aptian, Ghilcos. **5** Tithonian, Lapoş Valley, Bardos, Bicaz Gorges, Eastern Carpathians (*arrow* indicates type of ramification. (x 8)

Fig. 6. *Scotlandella nodosa* (Anderson). From Anderson in Hudson 1970, Purbeckian, Series of Skye, Scotland. (x 16)

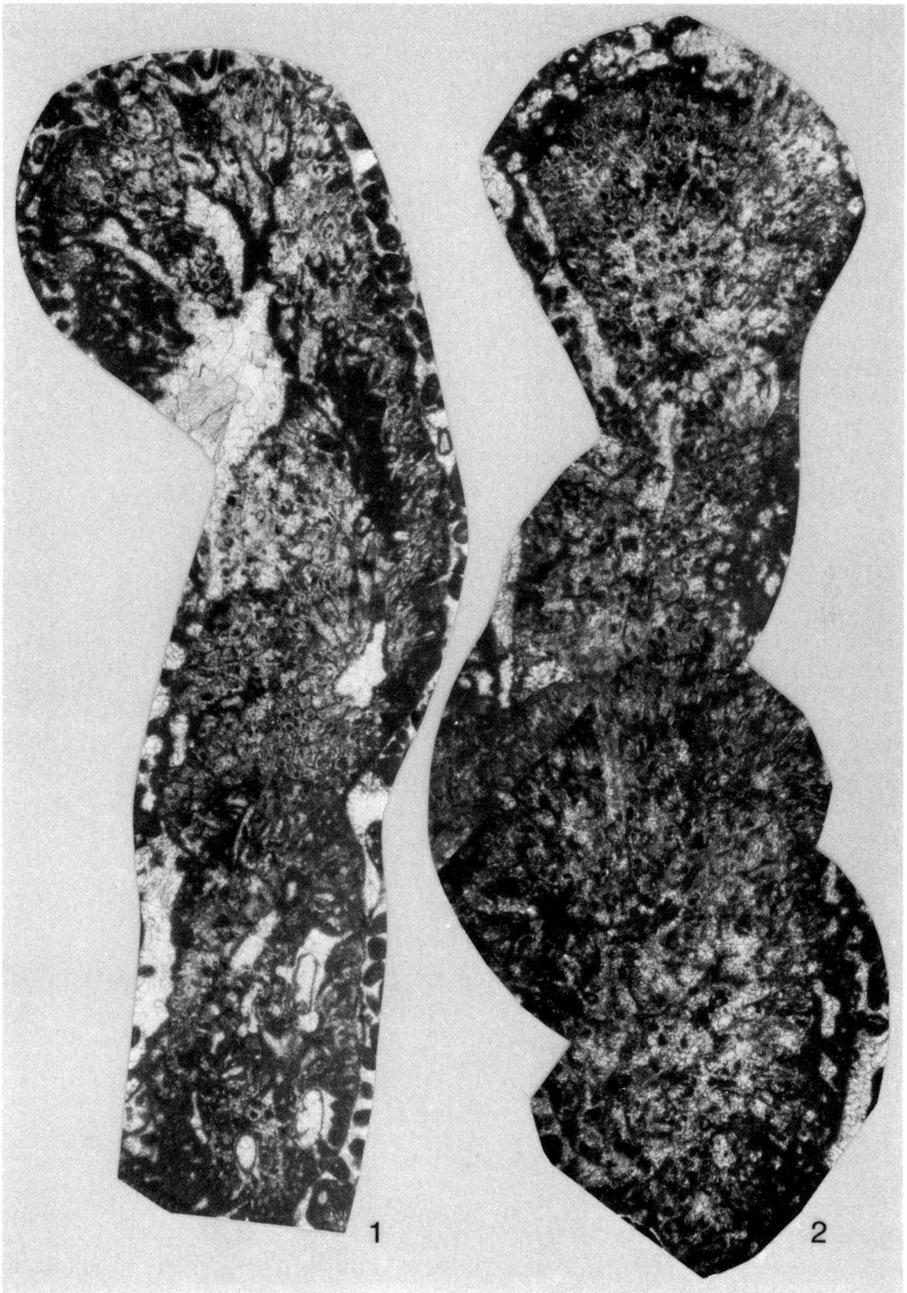

Plate XXIV

Figs. 1–2. *Johnsonicodium stratiforme* n.gen., n.sp. **1** *Holotype*: Coll. L.P.B. – V no. 0221, Neocomian, Ghilcoş, Bicaz Gorges, Eastern Carapathians. **2** *Syntype*: Coll. L.P.B. – V no. 0222, Neocomian, Ghilcoş, Bicaz Gorges, Eastern Carpathians. (x 4,8)

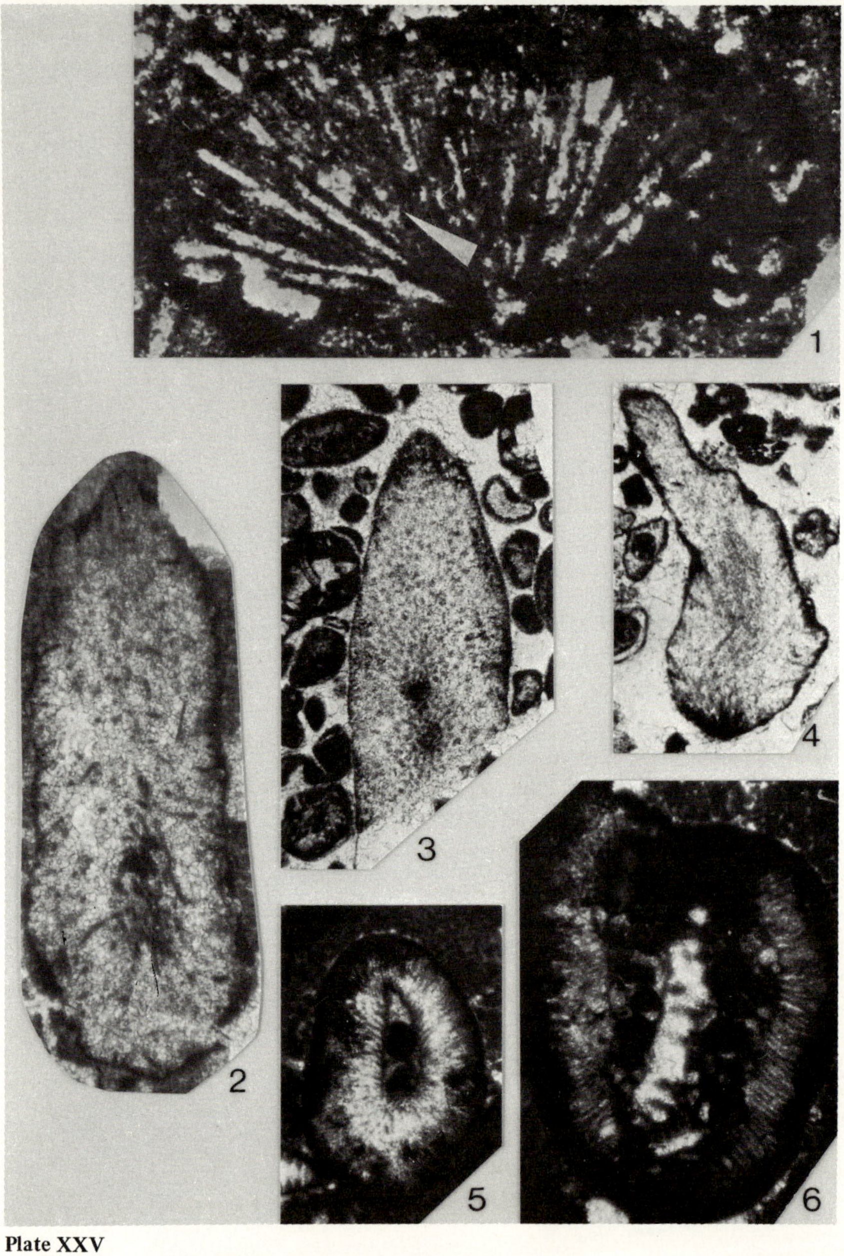

Plate XXV

Fig. 1. *Hedstroemia moldavica* n.sp. *Syntype*: from R. Colacicchi 1964, Cenomanian, Chiarano, Marsica Nord–Orientale, Italy. (x 40)

Figs. 2–4. *Arabicodium bicazensis* n.sp. **2** *Holotype*: Coll. L.P.B. – V no. 0223, Tithonian, Suhard, Bicaz Gorges, Eastern Carpathians. **3–4** *Syntypes*: Coll. L.P.B. – V nos. 0224, 0225, Neocomian, Fagetul Ciucului, Bicaz Gorges. (x 8)

Figs. 5–6. *Arabicodium hansi* n.sp. **5** *Syntype*: Coll. L.P.B. – V no. 0226, Barremian-Lower Aptian, Bardos, Bicaz Gorges, Eastern Carpathians. **6** *Holotype*: Coll. L.P.B. – V no. 0227, Barremian-Lower Aptian, Bardos, Bicaz Gorges. (x 8)

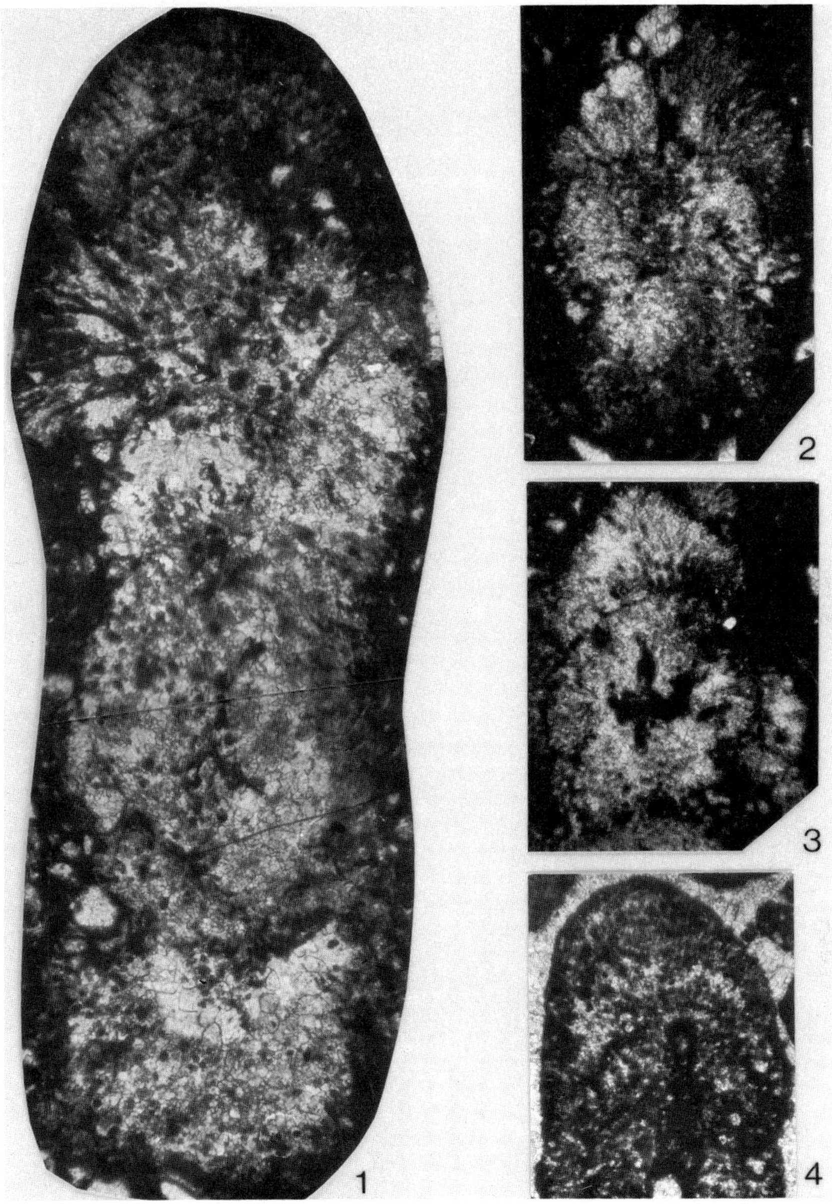

Plate XXVI

Figs. 1–4. *Tethysicodium wrayi* n.gen., n.sp. **1** *Holotype*: Coll. L.P.B. – V no. 0228, Tithonian, Făgetul Ciucului, Bicaz Groges, Eastern Carpathians; **2–3** Coll. L.P.B. – V nos. 0229,0230, Barremian-Lower Aptian, Bardos, Bicaz Gorges, Eastern Carpathians. **4** Coll. L.P.B. – V no. 0231, Tithonian with *Nerinea*, Vinturariţa, Southern Carpathians. (x 8)

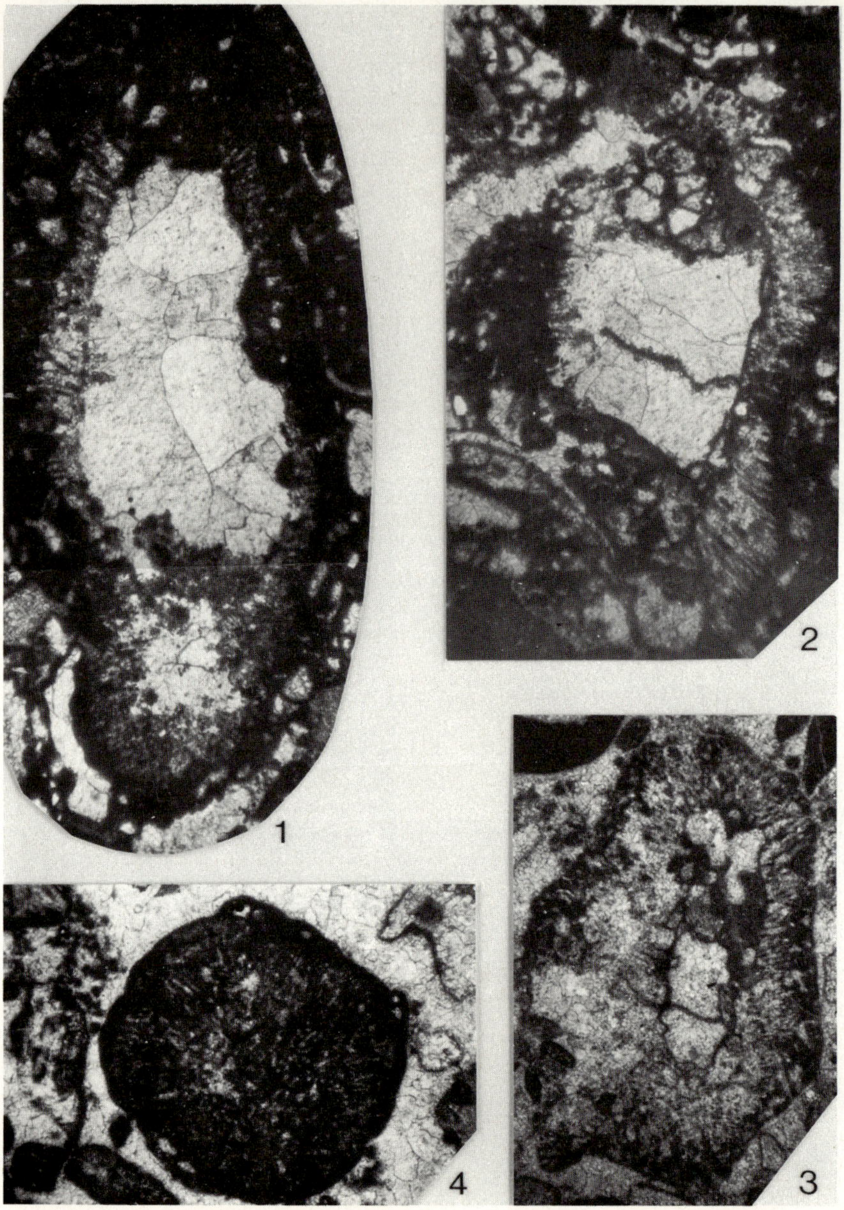

Plate XXVII

Figs. 1–3. *Bacinellacodium calcareus* n.gen., n.sp. **1** *Holotype*: Coll. L.P.B. – V no. 0232, Tithonian, Făgetul Ciucului, Bicaz Gorges, Eastern Carpathians. **2–3** *Syntypes*: Coll. L.P.B. – V nos. 0233, 0234, Barremian-Lower Aptian, Ghilcoş, Bicaz Gorges, Eastern Carpathians. (x 8)
Fig. 4. *Bevocastria elliotti* (Dragastan). *Hypotypes*: Coll. L.P.B. – V no. 0235, Upper Tithonian, Cetea, Trascău Mountains, Southern Apuseni. (x 8)

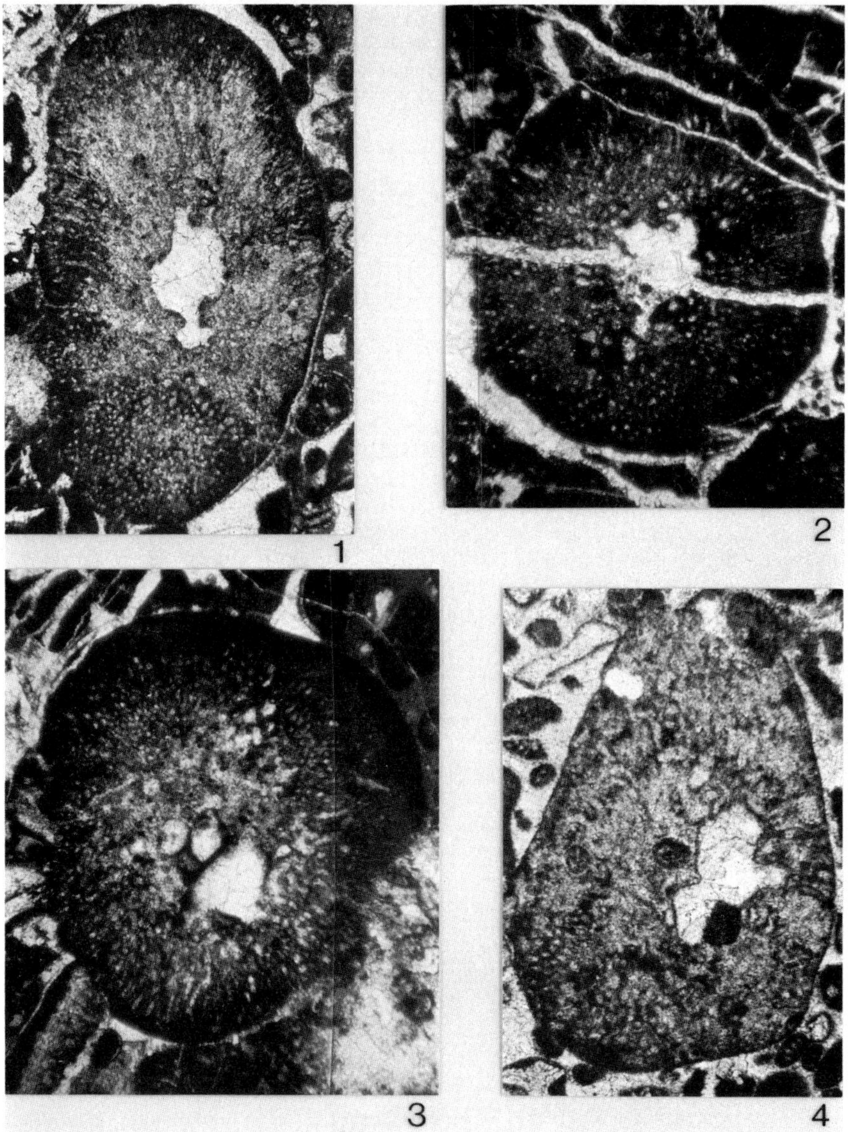

Plate XXVIII

Figs. 1–4. *Bevocastria elliotti* (Dragastan). **1** *Holotype*: Dragastan 1971. Coll. L.P.B. V – no. 0235. **2–4** *Hypotypes*: Coll. L.P.B. – V nos. 0236–0238, Tithonian, Făgetul Ciucului, Bicaz Gorges, Eastern Carpathians. (x 8)

Acknowledgments. I sincerely thank Prof. Dr. Erik Flügel (Erlangen) for his interest and discussion, Dr. D. Richter (Ruhr, Bochum) for his suggestions, Dr. T. Steiger (München) for his information on the status of the Reis Collection, Dr. T. Peryt (Warsaw) for his information on the status of the Bornemann Collection, and the Alexander von Humboldt Foundation for financial assistance. My deep gratitude goes to Prof. Donald F. Toomey (Conoco Inc., Ponca City) and to Prof. Matthew Nitecki (Chicago) for the revision of the manuscript.

References

Abate B, Catalano R, d'Argenio B, Di Stefano P, Riccobono R (1977) Relationships of algae with depositional environments and faunal assemblages of the Panormide carbonate platform, Upper Triassic, Northwestern Sicily. In: Flügel E (ed) Fossil algae. Recent results and developments. Springer, Berlin Heidelberg New York, pp 301–313

Anderson FW (1948) Algal beds in the great estuarine series of Skye. Proc R Phys Soc Edinburgh, XXIII, 2,pp 123–142

Azema J, Bourrouilh R, Champetier Y, Fourcade E, Rangheard Y (1974) Rapports stratigraphiques, paléogéographiques et structuraux entre la Chain Ibérique, les Cordillières bétiques et les Baléares. Bull Soc Geol Fr 7:XVI, 2, 140–160

Bakalova D (1975) Algue calcaires des sédiments urgoniens dans la montagne de Prébalkan central (Bulgarie centrale). Paleontology, stratigraphy and lithology, vol 2. Bulg Acad Sci, pp 49–56

Bakalova D (1978) Algues calcaires de la Formation de Brestnica (Tithonique supérieur – Barremien supérieur). Rev Bulg Geol Soc XXXIX:1, pp 3–16

Bassoullet JP, Bergougnan H (1981) Faune et faciès typiques du domaine sud-tethysien: le Lias du Munzur-Dag (Anatolia Orientale). Bull Soc Geol Fr 7:XXII, 1, 83–93

Benke K, Dürkoop A, Errenst Ch, Mensink H (1981) Die Korallenkalke im Ober-Jura der nordwestlichen Iberischen Ketten (Spanien). Facies 4, Erlangen, pp 27–94

Bignot G (1972) Recherches stratigraphiques sur les calcaires du Crétacé supérieur et de l'Eocène d'Istrie et des régions voisines. Essai de révision du Liburnien. Trav Lab Micropalaeontol 2. Thèse, Paris, 353 pp

Bornemann JG (1887) Geologische Algenstudien. Jahrb K Preuss Geol Landesanst Bergakad 1866:116–134

Bouroullec J, Deloffre R (1968) Les algues du Néocomien d'Aquitaine. Bull Centre Rech Pau. SNPA 2(2):213–261

Bouroullec J, Deloffre R (1970) Les algues du Jurassique aquitaine. Bull Centre Rech Pau. SNPA 4(1):79–127

Bourelly P (1970) Les algues d'eau douce. Initiation à la Systématique, 3. Les algues bleues et rouges. Paris, 512 pp

Brandner R, Resch W (1981) Reef development in the Middle Triassic (Ladinian and Cordevolian) of the Northern Limestones Alps near Innsbruck, Austria. Soc Econ Paleontol Mineral Spec Publ 30:203–231

Brinkmann R, Flügel E, Jacobshagen V, Lechner H, Rendel B, Trick P (1972) Trias, Jura und Unterkreide der Halbinsel Karaburun (West-Anatolien). Beitr Geol Anatolien 13. Geol Paleontol 6:139–150

Brown PR (1964) Petrography and origin of some Upper Jurassic beds from Dorset, England. J Sediment Petrol 34:254–269

Bucur I (1978) Microfaciesurile calcarelor albe din partea nordică a masivului Piatra Craiului. Consideraţii biostratigrafice, D.S. ale Sedinţelor. Inst Geol Geofiz LXIV (1976–1977), Bucuresti, pp 89–105

Bucur I (1980–1981) Algues calcaires du Crétacé inférieur des Monts Padurea Craiului. Nymphaea VIII–IX, pp 53–68

Bucur I, Bădăluţa A, Popescu O (1982) Date noi privind biostratigrafia depozitelor jurasice şi cretacice din partea mediană a Zonei Reşiţa (Banat), D.S. Inst Geol Geofiz, LXVI, 1979, Bucuresti, pp 21–51

Bursuk A (1975) Byburt yoresinian mikropaleontolojik ve stratigrafik irdelemesi. These, Trabzon, 196 pp

Catalano R, d'Argenio B, De Castro P (1974) Rapporti tra le facies di piattaforma carbonatica del Giurassico e del Cretacico delle Madonie orientali (Sicilia). Boll Soc Nat Napoli, LXXXIII, pp 3–37

Chiocchini M, Manganelli V, Pannuzi L (1977) Ricerche geologhiche sul Mesozoico della Sabina (Lazio). I Micropaleontologia, stratigrafia ed evoluzione paleoambientale dell'area del Monte Calvo. Boll Soc Geol Ital 94:1753–1790

Chiriac M, Barbulescu A, Neagu Th, Dragastan O (1977) La Dobrogea centrale et du sud pendant le Jurassique et le Crétacé. Rev Roum Géol Géophys Géogr Géol 21:145–153

Colacicchi R (1964) La facies di transizione della Marsica nordorientale. Geol Romana III:93–124

Colacicchi R(1967) La facies di transizione a livello del Lias nelle Marsica orientale. Sezione della Serra Rufigno. Riv Ital Paleontol 73(3):759–808

D'Argenio B, De Castro P, Emiliani C, Simone L (1975) Bahamian and Appenninic Limestones of identical lithofacies and age. Am Assoc Petrol Geol 59(3):524–530

Deloffre R (1972) *Palaeodasycladus elongatulus* Praturlon, 1966 dans le Lias inférieur d'Andalousie (Espagne). Bull Centre Rech Pau. SNPA 6(1):107–119

Derin B, Reiss Z (1966) Jurassic Microfacies of Israel. Isr Inst Petrol Spec Publ, 43 pp

Diaconu M, Dragastan O (1969) Triassic calcareous algae from the Apuseni Mountains (Rumania). Rev Palaeobot Palynol 8:63–101

Diaconu M, Dragastan O (1970) Date noi asupra depozitelor triasice din Pădurea Craiului (Munţii Apuseni). DS Inst Geol LVI, 3, Bucuresti, pp 43–50

Dragastan O (1967) Alge calcaroase în Jurasicul superior şi Cretacicul inferior din Munţii Apuseni. St. Şi Cerc Geol Geofiz Geogr Ser Geol Bucuresti 12(2):441–454

Dragastan O (1969) Algues calcaires du Jurassique supérieur et du Crétacé inférieur de Roumanie. Rev Micropalaeontol 12(1):53–62

Dragastan O (1970) New species of Dasycladaceae (calcareous algae) in the Lower Cretaceous of the Eastern Carpathians (Rumania). Rev Palaeobot Palynol 10:117–129

Dragastan O (1971) New Algae in the Upper Jurassic and Lower Cretaceous in the Bicaz Valley, East Carpathians (Romania). Rev Esp Micropalaeontol III(2):155–192

Dragastan O (1975) Upper Jurassic and Lower Cretaceous microfacies from the Bicaz Valley Basin (East Carpathians). Mem XXI Inst Geol Geophys, Bucuresti, 87 pp

Dragastan O (1980) Alge calcroase din Mesozoicul şi Terţiarul României. Acad RS Romania, Bucuresti, 169 pp

Dragastan O (1981) Mesozoic Dasycladaceae from Romania: Distribution and Biostratigraphical Importance, Facies, 4, Erlangen, pp 165–196

Dragastan O, Bucur I (1978) New species of the genus *Diversocallis* in the Jurassic and Cretaceous from Romania. Rev Roum Geol Geophys Geogr Geol Bucuresti 22:185–187

Dragastan O, Bucur I, Demeter I (1978) Date noi privind biostratigrafia depozitelor barremian-albiene din partea central-estică a Zonei Resita-Moldova Nouă (Banat), obţinute prin forajul de referintă de la Sopotul Nou. D.S. ale Sed, LXIV, Bucareşti, pp 17–36

Dragastan O, Diaconu M, Popa E, Damian R (1982) Biostratigraphy of the Triassic Formations in the East of the Pădurea Craiului Mountains. D.S. Inst Geol Geofiz, LXVII, 4 (1979–1980), Bucureşti, pp 29–61

Dufaure Ph (1958) Contribution a l'étude stratigraphique et micropaléontologique du Jurrasique et du Néocomien de l'Aquitaine a la Provence. Rev Micropalaeontol 1(2):87–115

Düzbastilar MK (1978) Notes on the Triassic-Jurassic Calcareous Algae of the Karaburun Peninsula (II). Codiaeae, E.U. Fen Fak Dergisi, Ser A, C II, S.U., Izmir, pp 359–369

Elliott GF (1957) Algues calcaires Codiacée fossiles d'Iraq nouvelles et peu connues. Bull Soc Geol Fr VI(6):789–795

Elliott GF (1964) Zonotrichites (Calcareous Algae) from the Arabian Triassic. Ecl Geol Helv 57(2):567–570

Elliott GF (1970) New and little-known Permian and Cretaceous Codiaceae (Calcareous Algae) from the Middle East. Palaeontology 13(2):327–333

Elliott GF (1975) Transported algae as indicator of different marine habitats in the English Middle Jurassic. Palaeontology 18(2):351–366

Endo R (1961a) Phylogenetic relationships among the calcareous algae. Spec Publ Saitama Univ, 52 pp

Endo R (1961b) Calcareous algae from the Jurassic Torinosu limestone in Japan. Sci Rep Saitama Univ Commem Vol Urawa, pp 53–75

Fenninger A, Hötzl H (1967) Die Mikrofauna und -flora des Plassen- und Tressensteinkalkes der Typuslokalite (Nördliche Kalkalpen). N Jb Geol, Paläont 128(1):1–37

Flügel E (1964) Mikroproblematika aus rhätischen Riffkalken der Nordalpen. Palaeontol Z 38(1, 2):74–78

Flügel E (1972) Mikroproblematika in Dünnschliffen von Trias-Kalken. Mitt Ges Geol Bergbaustud Innsbruck 21:957–988

Flügel E (1974) Fazies. Interpretation der Cladocoropsis-Kalke (Malm) auf Karaburun, W.-Anatolien. Archiv für Lagerstattfor in den Ostalpen, Sonderband 2, Leoben, pp 79–94

Flügel E (1975) Kalkalgen aus Riffkomplexen der alpin-mediterranen Obertrias. Verh Geol BA Wien 2–4:297–346

Flügel E (1977) Verkalkungsmuster porostromater Algen aus dem Malm der Südlichen Frankenalb. Geol Bl NOBayern 27(3–4), München, pp 131–140

Flügel E (1979) Paleoecology and Microfacies of Permian, Triassic and Jurassic algal communities of Platform and Reef Carbonates from the Alps. Bull Cent Rech Explor Prod, Elf-Aquitaine. 3,2 (2nd International Symposium on Fossil Algae) pp 569–586

Flügel E (1981) Palaeoecology and facies of Upper Triassic Reefs in the Northern Calcareous Alps. Soc Econ Paleontol Mineral Spec Publ 3:291–359

Flügel E (1982) Microfacies analysis of limestones. Springer, Berlin Heidelberg New York, 633 pp

Flügel E, Lein R, Senowbari-Daryan B (1978) Kalkschwämme, Hydrozoen, Algen und Mikroproblematika aus Cidarisschisten (Karn, Ober Trias) der Mürztaler Alpen (Oberösterreich). Mitt Ges Geol Bergbaustud Österr 25:153–195

Fourcade E, Jerez L, Rodriguez T, Jaffrezo M (1972) El Jurassico terminal y el Cretacico inferior de la Sierra de la Muela (Provincia Murcia). Consideraciones sobre las biozonas con foraminiferos de la Albense-Aptense del Sureste de Espana. Rev Esp Micropalaeontol, num extraord XXX Aniver E N Adaro, Madrid pp 215–248

Freytet P, Plaziat JC (1965) Importance des constructions algaires dues à des Cyanophycées dans les formations continentales du Crétacé supérieur et de L'Eocène du Languedoc. Bull Soc Geol Fr 7(VII):679–694

Fritsch FE (1945/1961) The structure and reproduction of the Algae. 2nd edn, vol I, 791 pp; vol II, 842 pp, Cambridge Univ Press, Cambridge

Frollo MM(1938) Sur un nouveau genre de Codiacée du Jurassique supérieur des Carpates Orientales. Bull Soc Geol Fr 8(3–4):269–271

Frollo MM (1940) Etude pétrographique des séries sédimentaires de la Vallée du Bicaz, Carpates Orientales, Thèse, Paris, 179 pp

Geitler L (1932) Cyanophyceae. In: Rabenhorsts Kryptogamen-Flora, vol 14. Akad Verlagges, Leipzig, p 1196

Golonka J (1970) Calcareous Algae from the Upper Jurassic of the Southern periphery of the Swiętokrzyckie Mountains, part II. Chlorophyta. Bull Acad Pol Sci, XVIII, 2, Kraków, pp 85–93

Golubić S (1973) The relationship between blue-green algae and carbonates deposits. In: Carr NG, Whitton BA (eds) The biology of blue-green algae. Blackwell Scientific Publ, Oxford London Edinburgh Melbourne, pp 434–472

Golubic S (1976) Taxonomy of extant stromatolite-building Cyanophytes. In: Walter MR (ed) Stromatolites. Elsevier, Amsterdam, pp 127–140

Groiss JTh, Keupp H, Zeiss A (1981) Exkursion A_2 – Faziesentwicklung im Malm der Südlichen Frankenalb. Palaeontol Ges 51 Jahresvers, Programm u Exkursionführen, Erlangen, pp 55–87

Herak M (1967) Mesozoische Kalkalgen der Insel Chios (Griechenland). N Jb Geol Paläont, Abh 129(1):97–104

Hudson DJ (1970) Algal limestones with pseudomorphs after gypsum from the Middle Jurassic of Scotland. Lethaia 3(1):11–40

Jaffrezo M (1974) Les algues calcaires du Jurassique supérieur et du Crétacé inférieur des Corbières (2ème partie). Rev de Micropal 17(1):23–32

Jaffrezo M (1980) Les formations carbonatées des Corbières (France) du Dogger à l'Aptien: micropaléontologie, stratigraphique, biozonation, paléoécologie, extension des résultats a la Mésogée. Thèse, Paris, 614 pp

Johnson JH (1961) Jurassic algae from the subsurface of the Gulf, Coast. J Paleontol 35(1):147–151

Johnson JH (1964) The Jurassic algae. Colo Sch Mines Q 59(2):129

Johnson JH (1968a) Lower Cretaceous algae from Texas. Colo Sch Mines Prof Contrib 4:71

Johnson JH (1968b) Lower Cretaceous algae from the Blake Escarpment, Atlantic Ocean and from Israel. Colo Sch Mines Prof Contrib 5:46

Johnson JH (1969) A review of the Lower Cretaceous algae. Colo Sch Mines Prof Contrib 6:180

Kamptner E (1951) Über das Auftreten der Codiaceen-Gattung Cayeuxia Frollo im Ober-Jura von Ernstbrunn (Niederösterreich) Österreich. Akad Wiss Math Naturh Sitzungsber Abt I 160(3–4):177–197

Korde KB (1973) Vodoroslii Kembria, Moskva: Izdatel Nauka, Moscow, 349 pp

Le Maitre D (1935) Etudes paléontologiques sur le Lias du Maroc-Spongiomorphides et Algues. Maroc Serv Mines. Notes 34:19–58

Le Maitre D (1946) Sur la présence d'Algues solénoporacées et mitcheldeniacées dans le Trias de Slovaquie. C R Soc Geol Fr, pp 106–108

Luperto Sinni E (1976) Microfossili senoniani delle Murge. Riv Ital Paleontol 82(2):293–416

Mamet B, Roux A (1974) Sur quelques algues tubulaire scalariformes de la Téthys paléozoique. Rev de Micropal 17(3):134–156

Mamet B, Roux A (1975) Algues devoniennes et carbonifères de la Téthys Occidentale (troisième partie). Rev Micropalaeontol 18(3):134–187

Mamet B, Roux A (1977) Algues rouges dévoniennes et carbonifères de la Téthys occidentale (4ème partie). Rev de Micropal 19(4):215–266

Maslov VP (1949) Vodoroslî Girvanella ee ekologhia i stratigraficeskoe znacenie. Bull Moskv ispît, Prirodî Moskva 24:89–110

Maslov VP (1965) Kopetdgarii-novaia triba mutovciatih sifonei (Zelenie vodorosli). Dokl Akad Nauk USSR 164(5):1154–1157

Masse JP (1976) Les calcaires urgoniens de Provence. Stratigraphie, paléontologie, les paléoenvironments et leur évolution. Thèse, Fac Sci Univ Marseille, p 445

Masse JP (1979) Schizophytoides du Crétacé inférieur. Caractéristiques et signification écologique. Bull Centre Rech Explor. Prod Elf-Aquitaine, 3, 2 (2nd Int Symp Fossil Algae) pp 686–703

Masse JP, Poignant AF (1971) Contribution à l'étude des algues du Crétacé inférieur provencal. Intérêt stratigraphique. Rev Micropalaeontol 13(4):258–266

Mercier J (1973) Etude géologiques des zones des Hellénides en Macedonie centrale (Grèce). Ann Geol Pays Hellen 20:596

Misik M (1966) Microfacies of the Mesozoic and Tertiary Limestones of the West Carpathians. Slov Akad Vied, Bratislava, 269 pp

Misik M (1979) Jurassic and Cretaceaous Algae (Dasycladales excepted) from the West Carpathians. Bull Centre Rech Explor Prod Elf-Aquitaine 3(2):705–712

Misik M, Sykora M (1981) Pieninsky exoticky chrbat rekonstrnovany z valunov karbonatovych hornin kriedovych zlepencov bradloveho pasma a maninskey jednotky. Zapadne Karpaty Ser Geol 7, Geol Ust D Stura, Bratislava, pp 7–111

Monty LVC (1976) The origin and development of cryptalgal fabrics. In: Walter MR (ed) Stromatolites. Elsevier, Amsterdam, pp 193–249

Neagu Th, Pana I, Dragastan O (1977) Biostratigrpahie de la Série des calcaires Eocrétacés de l'aire Cernavoda-Alimanu-Ostrov. Rev Roum Géol Géogr Géologie 2:137–144

Ott E (1966) Zwei neue Kalkalgen aus den Cassianer Schichten Südtirols (Ober ladin, mittlere Trias). Mitt Bayer, Staatssamml Palaeontol Hist Geol 6:155–166

Ott E (1972) Mitteltriadische Riffe der Nördlichen Kalkalpen und altersgleiche Bildungen auf Karaburum und Chios (Ägäis). Mitt Ges Geol Bergbaustud 21:251–276

Patrulius D (1965) Inventar sumar al algelor calcaroase Neojurasice şi Eocretacice din Carpaţii româneşti şi Platforma Precapatică. D S ale şedinţelor Inst Geol LI(2):31–42

Peryt MT (1981) Phanerozoic Oncoids - An Overview. Facies, 4, Erlangen, pp 197–214

Pfender J (1938) Les Foraminifères du Valanginien Provencal. Bull Soc Geol Fr 8(3–4):231–239

Pfender J (1939) Sur la repartition d'une Codiacée jurassique: Cayeuxia Frollo. C R Geol Fr, pp 122–123

Pia J (1927) Abteilung Thallophyta. In: Hirmer M (ed) Handbuch der Paläobotanik, vol I. Oldenbourg, München Berlin, 134 pp

Praturlon A (1964) Calcareous algae from Jurassic-Cretaceous limestones of central Apennines (Southern Latium-Abruzzi). Geol Romana III:171–202

Praturlon A (1966) Algal assemblages from Lias to Paleocene in southern Latium-Abruzzi: a review. Bull Soc Geol Ital 85:167–194

Pugh EM (1969) Algae from the Lower Purbeck Limestones of Dorset. Proc Geol Assoc 79(4): 513–523

Radoicic R (1959) Paleontolosko odlike sedimenata okoline Bacinskog jezero (Gradac-Ploce). Bull Geol III:55–63

Radoicic R (1960) Microfaciès du Crétacé et du Paléogene des Dinarides externes de Yougoslavie. Inst Rech Geol R P Crna Gora 4(1):172

Radoicic R (1966) Microfaciès du Jurassique des Dinarides externes de la Yougoslavie. Geol (Razpr in Porocila) 9:377

Ramalho M (1971) Contribution à l'étude micropaléontologique et stratigraphique du Jurassique supérieur et du Crétacé inférieur des environs de Lisbonne (Portugal). Serv Geol Port Mem 19: (Nov Ser), 212

Reis MO (1923) Kalkalgen und Seesinterkalke aus dem rheinpfälzischen Tertiär. Geogn Jahresh 36:103–130

Riche Ph, Prestat B (1980) Paléogeographie du Crétacé moyen du proche et moyen-orient et sa signification pètroliere. Proc World Petrol Congr, vol II. Heydem, Bucharest, pp 57–75

Richter KD, Herforth A, Ott E (1979) Pleistozäne, brackische Blaugrünalgenriffe mit Ruvularia haematites auf der Perachorahalbinsel bei Korinth (Griechenland). Neues Jahrb Geol Palaeontol Abh 159(1):14–40

Riding R (1977a) Problems of affinity in Paleozoic calcareous algae. In: Flügel (ed) Fossil algae. Recent results and developments. Springer, Berlin Heidelberg New York, pp 202–211

Riding R (1977b) Skeletal Stromatolites. In: Flügel E (ed) Fossil algae. Recent results and developments. Springer, Berlin Heidelberg New York, pp 57–60

Schäfer P (1977) Fazielle Entwicklung und palökologische Zonierung zweier Obertriadischer Riffstrukturen in den nördlichen Kalkalpen (Öberrhät-Riff-Kalke, Salzburg). Facies 1, Erlangen, 245 pp

Schneider J (1977) Carbonate construction and decomposition by epilithic and endolithic microorganisms in slat and freshwater. In: Flügel E (ed) Fossil algae. Recent results and developments. Springer, Berlin Heidelberg New York, pp 248–260

Senowbani-Daryan R, Schäfer P (1979) Distributional patterns of calcareous algae within Upper Triassic Patch Reef of the Northern Calcareous Alps (Salzburg) (2nd Intern Symp Fossil Alpae). Bull Centre Rech Explor Prod Eld-Aquitaine 3(2):811–820

Shuysky VP (1973) Algues calcaires récifales du Dévonien inférieur de l'Oural (in Russian). Akad Nauk SSSR, Centre Sci Oural, Inst Geol Geochim, AN Zavaritsky, p 155

Stafleu FA (ed) (1972) International code of botanical nomenclature, Utrecht, 426 pp

Steiger T (1981) Kalkturbidite im Oberjura der Nördlichen Kalkalpen (Barmsteinkalke, Salzburg, Austria) Facies 4, Erlangen, pp 215–348

Steiger T, Wurm D (1980) Faciesmuster oberjurassischer Platform-Karbonate (Plassen-Kalke, Nördliche Kalkalpen Steirisches Salzkammergut, Österreich). Facies 2, Erlangen, pp 241–284

Tilden JE (1920) Calcareous algae. In: Index algarum universalis. London, 314 pp
Toomey DF (1979) Role of archaeolithophyllid algae within a late Carboniferous algal-Sponge Community, South-western United States. Bull Cent Rech Explor Prod Elf-Aquitaine, 3,2 (2nd Int Symp Fossil Algae), pp 943–853
Toomey DF (1981) Organic buildup constructional capability in Lower Ordovican and Late Paleozoic Mounds. In: Gray J et al (eds) Communities of the past. Hutchinson Ross, Stroudsburg, pp 35–68
Vinogradov C, Dragastan O (1975) Microfacial study of the Upper Jurassic and Lower Cretaceous deposits from the central part of the Moesic Platform (Romania). Rev Roum Géol Géophys Géogr Géol 19:105–118
Weber van Bosse J (1896) On a new genus of Siphonean Algae "Pseudocodium". J Linn Soc Bot 32:209–212
Wnendt EJ (1978) Die Geologie und Mikrofazies des Jura-Korallenkalkes von Olvega (NW-Iberische Ketten, Spanien), Diplomkartierung und Arbeit. Bochum, Univ Ruhr, 78 pp
Wray JL (1977) Calcareous algae. Dev Paleontol Strat 4. Elsevier, Amsterdam, 185 pp
Wurm D (1982) Mikrofazies, Paläontologie und Paläokologie der Dachsteinriffkalke (Nor) des Gosaukammes. Österreich. Facies 6, Erlangen, pp 203–296
Yabe H, Toyama S (1928) On some rock-forming algae from the younger Mesozoic of Japan. Sci Rep Tohuku Univ 2(12):141–152

11 Key for Specific Determinations of Cretaceous *Archaeolithothamnium*

A.F. POIGNANT[1]

Specific determinations of Cretaceous *Archaeolithothanium* are made with the help of several criteria of different hierarchic values. Only practical criteria for geologist are retained here and are based upon the qualitative aspects and not upon quantitative.

The key includes:
- the external morphology of the thallus
- the texture of the tissue
- the structure of the reproductive organs

CORALLINACEA

● Articulated thallus. Differentiated tissues (Hypothallus and perithallus).
I. *Corallinoideae*

● Non articulated thallus. Differentiated tissues (Hypothallus and perithallus).
II. *Melobesioidea*

MELOBESIOIDEAE

● Conceptacles perforated by a single pore 1
● Conceptacles perforated by a few or many pores. 2
● Sporangia isolated or in sori 3
Archaeolithothamnium Rothpletz 1891

I. Incrusting thallus, superimposed

1. Zoned tissue.
P. Layers of variable height. 1S or nS.
Thick transversal partitions. Small cells.
H. Reduced or absent; narrow cells; passage P/H with a right angle.
Sp. Small, separated or joined, non-merged. 1...7 rows.
A. brevium Lemoine

1 Centre d'Etude des Algues Fossiles, Université P. & M. Curie – 4, place Jussieu, 75230 Paris Cédex 05, France

Paleoalgology: Contemporary Research and Applications
(eds. D.F. Toomey and M.H. Nitecki)

2. Non-zoned tissue.
   P. Layers of variable height; presence of short layers.
   Transversal partitions absent or pronounced every other or fourth layer.
   Calcitic edge.
   Great and more or less ondulated cells.
   H. Long and narrow cells. Passage P/H abrupt.
   Sp. Great size, of ten merged (D/d = 1,2). 1...4 rows.
   A. rude Lemoine

   P. Layers (or grating).
   Continuous and well pronounced transversal partitions.
   Small cells, nearly cubic.
   H. Small cells.
   Sp. Separated or joined non-merged. Trend to the sphericity.
   A. mamillosum (Gümbel) Lemoine

   P. Layers (or grating).
   Transversal partitions more continuous than longitudinal ones.
   Variable sized cells.
   H. Thick; variable sized cells.
   Sp. Serried, non-merged, great size (D/d = 2,1). 1...9 rows.
   A. nummuliticum (Gümbel) Rothpletz

   P. More on layers than in threads.
   Flexuous transversal partitions more pronounced than longitudinal ones.
   S, dark, grouped or isolated.
   H. Reduced, narrow cells.
   Sp. Separated or joined, non-merged, small size.
   A *A. feuillei* Lemoine

   P. Regular; threads or layers.
   Irregular transversal partitions – often cubic cells.
   H. Narrow and variable sized cells.
   Sp. Separated or joined, non-merged; great size. 1...10 rows.
   A. lugeoni Pfender

   P. More in threads than in layers (or grating). nS, dark and compressed.
   Transversal partitions less pronounced than longitudinal ones.
   Cubic small sized cells.
   H. Variable sized cells.
   Sp. Serried or joined, non-merged. 1...19 rows.
   A. gosaviense Rothpletz

   P. Threads (and layers).
   Longitudinal partitions often pronounced.
   Continuous and thin traversal partitions, or in alignements.
   H. Thin, short cells, rather narrow. Passage P/H with right angle.
   Sp. Sori. *A. batalleri* Lemoine

P. Threads.
Irregular transversal partitions.
Pronounced longitudinal partitions. Small sized cells.
H. Rare, variable sized cells.
Sp. Separated, non-merged.
A. aschersoni (Schwager) Foslie

II. Free thallus, spherical (or incrusting)
P. Non zone if spherical; in grating.
Zone if incrusting; layers of variable height.
Continuous transversal and longitudinal partitions.
H. Rarely recurrent, great cells.
Sp. Separated or joined, non-merged, rather small. 1...20 rows.
A. lycoperdioide (Michelin) Lemoine

III. Branching bulkies
1. Alternations in medullary tissue.
a) Simple and regular alternations 1L/1S.
Right longitudinal partitions
Continuous transversal partitions.
P. Thin. Longitudinal partitions more pronounded than transversal ones.
H. Unknown.
Sp. Spherical and separated (D/d) = 1,2). 1 row.
A. alternatum Poignant

b) Irregular alternations – 2-3 L/1S
Flexuous longitudinal partitions
Transversal partitions on zig-zag.
P. Thin. Transversal partitions more pronounced than longitudinal ones.
H. Unknown.
Sp. Unknown.
A. alternum Conti

2. Non-zoned medullary tissu.
a) n irregular layers;
1 to 8 S. and middle sized cells.
Continuous longitudinal partitions more pronounced than transversal ones.
Transversal partitions not at the same level; 1 on 2 or 3 more pronounced; aspect of threads. Narrow and small cells.
P. n regular layers and nS.
H. Reduced.
Sp. Separated, sometimes joined. 1...2 rows.
A. anastomosans Pfender

b) Homogenous and regular.
Flexuous longitudinal partitions. Transversal partitions in zig-zag.
P. Rather thin; passage P/H without transition.
n regular layers and 1 to nS.
H. Thick, great cells.
Sp. Small, joined.
A. penicillum Pfender

c) Heterogenous and disorganized.
Undulating or flexuous longitudinal partitions.
Transversal partitions on zig-zag.
P. Thin; passage P/H without transition.
More in layers than in threads.
H. Thick, great and narrow cells.
Sp. Spherical. 2-3 desorganized rows.
A. incompositum Conii

3. Little zoned medullary tissue (nL/1S or rarely nL/nS).
a) Homogenous zones, often weakly pronounced; possibly narrowing cells.
Flexuous or inclined longitudinal partitions.
Normal or weakly inclined transversal partitions.
P. Very thick, cells in layers, n.S.
H. Reduced, great and narrow cells.
Sp. Rather wide. (D/d = 2). 1...4 rows.
A. turonicum Rothpletz

b) Homogenous zones and short branches.
Longitudinal partitions neither inclined nor flexuous
Well pronounced transversal partitions. Wide cells.
P. Thin. n.S. Isolated or gathered, wide cells.
H. Thin, variable sized cells, rigid threads.
Sp. Wide.
A. parisiense (Gümbel) Lemoine

4. Zoned medullary tissue: regularly narrowing layers.
a) Zones: ((1)2...8 L clear, to 1...5 S dark), to (1L to 1...3S).
Inclined longitudinal partitions.
Irregular but not flexuous transversal partitions.
Almost horizontal layers in the axis of branches.
P. Thin, non-zoned, in layers; calcitic edge.
H. Rather thin, long and narrow cells.
Sp. Serried or joined. 1 row.
A. latifoliaceum Pfender

b) Zones: (1)2...6 L to 1... n S. Irregular layers.
Longitudinal partitions neither flexuous nor inclined.
Irregular and flexuous transversal partition.
Very curved layers.

P. Thick, non zoned, in grating with sometimes a large and distinct layer.
H. Thin, recurrent, long and narrow cells.
Sp. Great, joined or separated. 1...3 rows.
A. hippuritorium (Munier-Chalmas) Pfender

c) Zones: (1...3 L to 1...3 S), to (1 L to 1 S).
Flexuous or inclined longitudinal partitions.
Transversal partitions neither inclined nor flexuous, sometimes irregular.
P. Non zoned, rather thin, in layers; presence of short dark, wide cells.
H. Thin and recurrent.
Sp. 1 row.
A. haugi Pfender

d) Zones: (2...5 L to -1-2 S), to (nL to nS).
Longitudinal partitions neither flexuous nor inclined.
Flexuous transversal partitions, sometimes thin. Very curved layers.
P. Non-zoned, distinct layers; small cells.
H. Rather thin, long and narrow cells.
Sp. Rather wide, separated or serried, abundant. 1...3 rows.
A. digitatum Pfender

5. Zoned medullary tissue: layers non-regular.
Narrowing cells.
a) Zones: 2...4 L clear and 1...5 S dark.
Often continuous and radiating longitudinal partitions, sometimes flexuous.
Irregular and very weakly staggered transversal partitions.
P. Homogenous, without short layers, thick. Larger than higher cells.
H. Variable sized cells, rather narrow; recurrent.
Sp. Serried, non-joined, variable size, little abundant. 1...4 rows.
A. provinciale Pfender

b) Zones: 2 L clear and 2...6 dark. Important basal crusts, short branches or annular aspect.
Rigid longitudinal partitions.
Irregular transversal partitions. Imperceptible passage P/H.
P. Very thick, with irregular, small layers.
H. Reduced, recurrent, narrow cells.
Sp. Joined or separated, occasional wide, often absent.
1...2 rows.
A. cretaceum (Munier-Chalmas) Pfender

c) Zones: nL and nS.
Strong longitudinal partitions, trend to threads.
Thin transversal partitions in zig-zag.
Short layers, irregular, isolated or gathered; narrow cells.
P. Rather thin, non-zoned, regular layers; short cells.
H. Rare, small cells.
Sp. Rather large. 1...2 rows.
A. dehornae Pfender

d) Zones: nL and nS.
Flexuous and inclinated longitudinal partitions.
Thin transversal partitions in zig-zag. Short and gathered layers.
P. Regular, horizontal partitions thicker than vertical ones.
Sometimes threads in crusts, sometimes passage P/H with a line and a dark zone of small cells.
H. Narrow cells.
Sp. Separated, rarely joined, high and narrow (D/d = 2,7).
1-2 rows.
A. antenorense Pfender

Summary

The specific determinations are based upon several criteria, but it is obvious that no one criterion for Cretaceous *Archaeolithothanium* allows to identify a species in the certainty; it is absolutely necessary to use an assemblage of features according to a well chosen hierarchy.

Qualificative and quantitative asepcts can be considered. The first includes:
- the external morphology of the thallus,
- the texture of the tissue,
- the structure of reproductive organs.

The second, not included here, deals with the sizes of the different elements noted above.

References

Lemoine M (1960) Les Algues Floridées calcaires du Crétacé du Sud de la France. Mém MNHN Paris 7, X:127–240

Poignant AF (1977) Sur une clé de détermination générique des Algues rouges mésozoiques; implications phylogénétiques. CR Acad Sci 285:1229–1231

Poignant AF (1979) Détermination générique des Corallinacées mésozoiques et cénozoiques. IIème Sym Int Algues Fossiles, Paris. Bull Cent Rech Explor Prod Elf-Aquitaine 3–2:757–765

Poignant AF (1980) Critères de détermination spécifique des Corallinacées fossiles. CR Acad Sci 291:389–391

Reefs, Buildups and Other Frameworks

12 Devonian Reef-Associated Articulate Red Algae from Western Canada

E.R. MAGATHAN[1]

Introduction

The existence of articulate red algae, which are widespread in shallow tropical and temperate marine habitats today, has not been positively established from pre-Cretaceous rocks. Accordingly, the Upper Devonian algae documented in this paper demonstrate a much earlier appearance in the geologic record of these plants than previously confirmed.

The fossils in thin section, referred to here informally as *Archaeoamphiroa*, exhibit the gridlike arrangement of calcite cell walls that is diagnostic of calcareous red algae, and they occur in various shapes and sizes as discrete subcylindrical bead-like grains which evidently represent the calcified segments, or intergenicula, of a flexible, branching plant thallus. Dolomite rhombs have selectively crystallized in and around many of the segments, suggesting an original mineralogy of high magnesian calcite similar to that secreted by modern coralline algae, and the alternating length of cell tiers within the segments suggests close affinities with the Amphiroideae, a subfamily of the articulate coralline algae whose most conspicuous diagnostic feature is a regular pattern of variation in cell length.

The distribution of *Archaeoamphiroa* is narrowly confined to reef-margin facies in the Frasnian Alexandra Reef-Complex of northwest Canada, where it was first identified. In this reef the fragments occurs as components of grainstones in crevices of the stromatoporoid-built reef frame, and in packstones and wackestones which accumulated in its sheltered lee. The ecologic preference indicated by this distribution is remarkably like that of its modern counterpart, *Amphiroa.*

It is recommended that, because the parallels between these Devonian algae and modern articulate corallines are so obvious, they merit classification as ancestral corallines, an interpretation that has been tentatively supported by Johansen (pers comm, August 1983). Recognition of *Archaeoamphiroa* in at least one other Upper Devonian reef-complex in western Canada by the author suggests that articulate corallines may have been widespread during the Devonian, but have been overlooked in other reef studies, or altered beyond recognition by diagenesis.

---

1 Consulting Geologist, 535 Garfield Street, Denver, CO 80206, U.S.A.

Paleoalgology: Contemporary Research and Applications
(eds. D.F. Toomey and M.H. Nitecki)

Description

Thin Section Identification. Thin section examination of more than 20 specimens (which included both longitudinal and transverse sections) indicates that the plant segments are subcylindrical (terete) in shape (Fig. 1A–C). Individual fragments range

Fig. 1A–C. Photomicrographs illustrating typical segments of the articulate red alga *Archaeoamphiroa* surrounded by the coarse carbonate sands with which they are commonly found. All specimens collected from reef-flat facies of the Alexandra Reef-Complex, Northwest Territories, Canada. **A** Longitudinal section, showing tiered gridlike structure (emphasized by dark micrite infill of many individual cells) and apparent apical flaring suggestive of the cell expansion which is a prelude to branching in articulate coralline algae (Sample 84JE-M). **B** Transverse section, illustrating subcircular shape of the segment and "honeycomb" appearance of polygonal cells in cross-section view (Sample 11JE-S). **C** Longitudinal section, showing the alternation of long and short cell tiers which is characteristic of *Archaeoamphiroa*, and demonstrating preferential seeding of dolomite rhombs (*arrows*) around the plant segment. Clear calcite cells walls are exceptionally well-preserved in this specimen (Sample 85JE-T)

in length from 1.5 to 5 mm in length, and in breadth from 0.9 to 3.9 mm, and average 2.5 mm in length and 1.5 mm across. They are large enough to be recognized on polished surfaces at low magnification by a trained eye, but because they appear as porcellanous white nodules, larger segments might easily be mistaken for crinoids in a cursory examination.

Cells are stacked in regularly alternating long and short tiers, and long tiers generally range from 290–500 microns in length, while short tiers range from 60 to 170 microns in length. They are polygonal in cross-section, and range in diameter from 80–140 microns. Cell walls are 4 microns thick and are of clear, colorless calcite.

Inferred Mineralogy. The excellent preservation of cell walls in many segments suggests that they were originally of calcite, rather than aragonite. It also suggests that this calcite must have been a high-magnesian form, for, in many rocks where dolomite does not otherwise exist, dolomite rhombs have selectively crystallized in and around the *Archaeoamphiroa* segments (Fig. 1C). I infer that these crystals were seeded by an original mineralogy of high magnesian calcite. That dolomite can be seeded in this manner is supported by many observations of selective dolomitization in Tertiary and younger coralline algae (Schlanger 1957, Johansen 1981). Proof that the cells were originally of high magnesian calcite would support affinities of *Archaeoamphiroa* with the coralline algae, which today secrete higher amounts of magnesium in their skeletons than any other lime-secreting organisms.

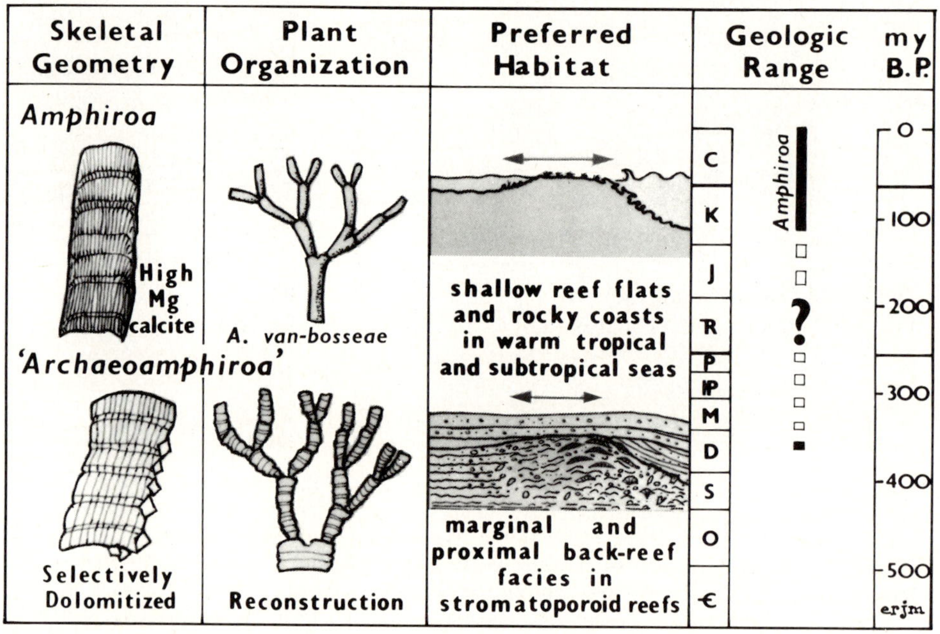

Fig. 2. Comparative chart emphasizes similarities between Upper Devonian *Archaeoamphiroa* and modern articulate coralline algal genus *Amphiroa*, shows their respective geologic ranges, and suggests that apparent gap in the record may be documented in future research. (Range of the genus *Amphiroa* is from Wray 1977)

Reconstruction. Observations collectively suggest the reconstruction illustrated in Fig. 2. The living plant thallus consisted of subcylindrical calcified intergenicula of variable length between 2 and 6 mm, connected by flexible genicular organic tissue. The intergenicula were externally ringed by microscopic grooves or furrows demarcating tier junctions. The plant fronds undoubtedly branched and, although there is no unequivocal evidence of this branching, apical flaring in several segments suggests the cell expansion that is normally associated with branching in modern coralline segments (Fig. 1A). It is speculated that this branching was dichotomous and similar to that in the modern *Amphiroa van-bosseae* (Norris and Johansen 1981, p. 23).

Lacking evidence for perithallic tissue, the writer assumes that the present fossil outlines represent the external surfaces of the original fragments. Constrained by that assumption, one must conclude that the cell tier junctions, which express irregular 'stepped' profiles along the outer margins of the segments (Fig. 1A), must have been expressed externally as a system of minute ridges and grooves around the segments, as reconstructed in Fig. 2.

Habitat

Distribution in the Alexandra Reef-Complex. *Archaeoamphiroa*'s distribution has been well documented by the author (as Jamieson) in a paleoecologic evaluation of the Alexandra Reef-Complex, a Frasnian carbonate reef which outcrops discontinuously along a 30-mile long (48 km) escarpment southwest of Great Slave Lake in Canada's Northwest Territories (Jamieson 1967a, 1971), and this discussion is based on results of this evaluation.

Figure 3 shows all occurrences of *Archaeoamphiroa* documented in the Alexandra Reef-Complex. Each occurrence is plotted as a closed back circle on a panel reconstruction of the lithostratigraphy along the escarpment, and the extrapolated distribution pattern is indicated by a light gray screen. This distribution pattern clearly shows that this articulate alga was confined exclusively to facies identified as shallow-water marginal reefs, and was excluded from protected mudbanks and mounds leeward of the reef margin and from shoal-water sands and lagoon deposits which accumulated after the demise of active reef growth.

Facies Associations. Within the narrowly circumscribed marginal reef habitat, lithofacies and fossil associates of *Archaeoamphiroa* document the characteristics of its habitat (Fig. 4):

1. Horizontal attitudes of associated sessile organisms indicate that the surface upon which it grew must have been essentially horizontal.
2. Stromatolites and encrusting coralline-like algae with which it is usually associated indicate that it grew as part of a thick algal turf which veneered the horizontal reef surface.
3. Coarse sands with which they are associated indicate that the plants grew in waters agitated enough to concentrate corase sediments and winnow out fines.

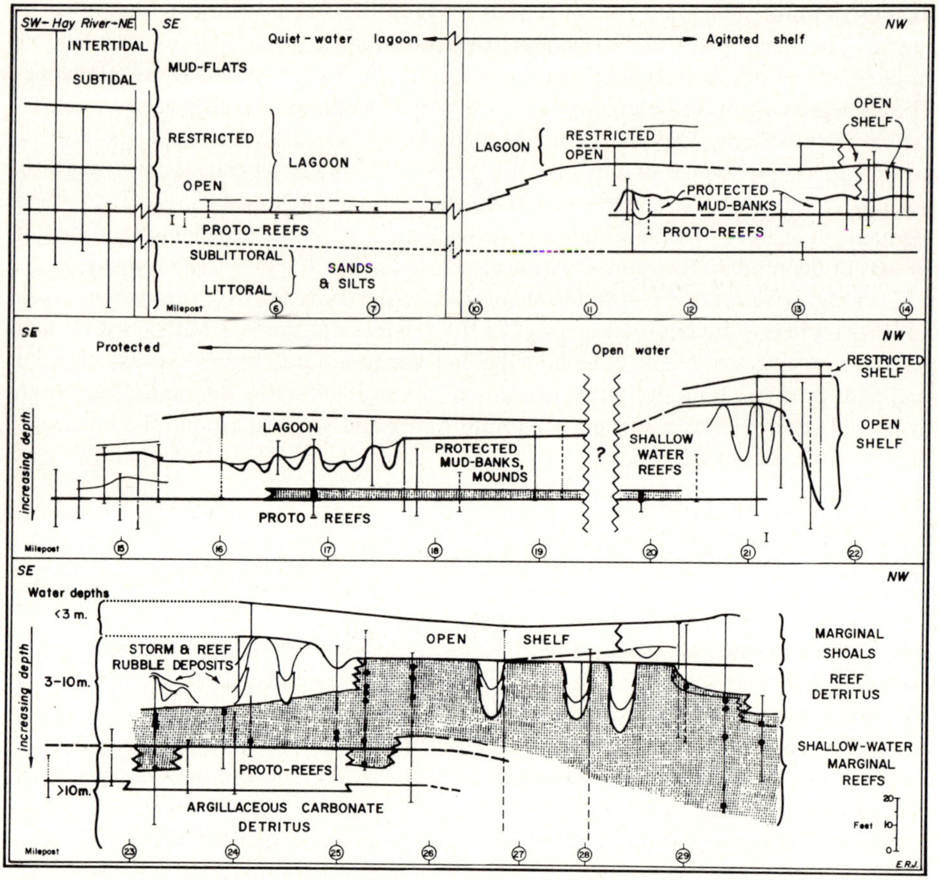

Fig. 3. Distribution of *Archaeoamphiroa* in panel reconstruction of Alexandra Reef-Complex facies. Reconstruction was based on study of specimens collected at 2-foot intervals along measured sections which are indicated by *vertical lines.* Each occurrence of *Archaeoamphiroa* is plotted as a *closed black circle*, and the extrapolated distribution pattern is screened in *light gray.* The distribution pattern clearly shows that the alga was confined to shallow-water marginal reef facies

4. Coexistence with *Actinostroma*, a stromatoporoid which dominated the sloping outer spurs of the reef-margin in the Alexandra Reef-Complex, indicates close proximity to the outer reef margin.
5. Occurrence with *Renalcis*-encrusted cavities, many of which are floored with internal sediment, suggests that the growth surfaces must have been cavernous like those of modern-day reef flats.

Reconstruction. Collectively, the distribution pattern and facies associations indicate that *Archaeoamphiroa* must have been confined to a shallow, surf-swept reef flat immediately leeward of the *Actinostroma*-dominated reef margin (Fig. 5). In this habitat, it evidently flourished in association with a carpet of soft filamentous algae continually flushed by strong wave surges.

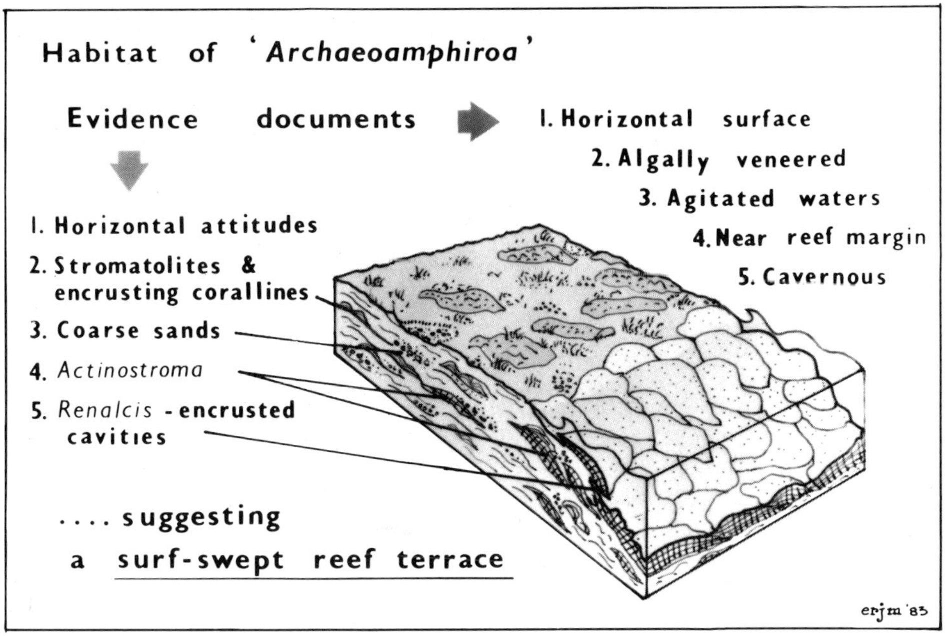

Fig. 4. Habitat of *Archaeoamphiroa*: evidence of facies and fossil associations suggests a surf-swept reef terrace

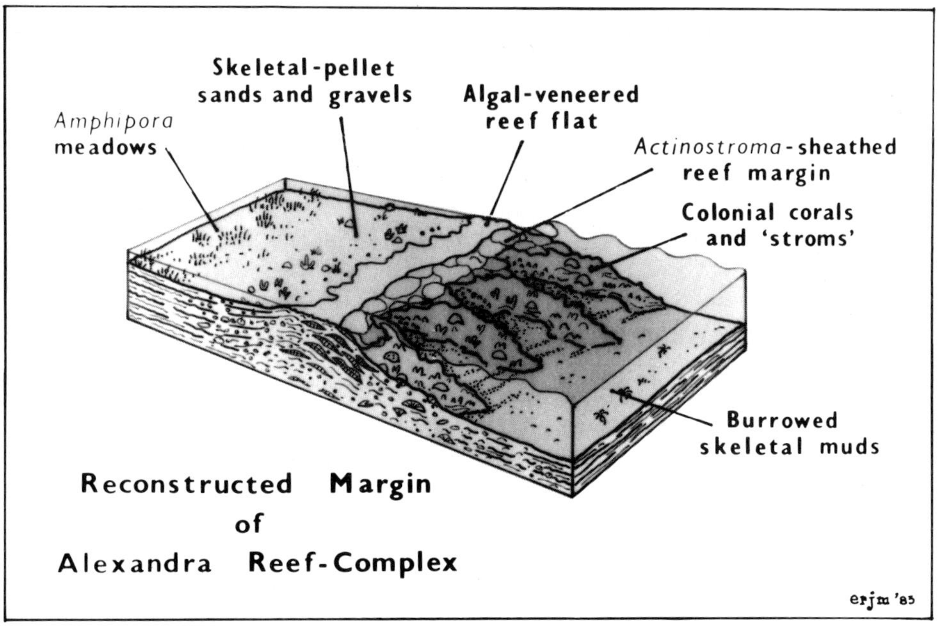

Fig. 5. Facies reconstruction of Alexandra Reef-Complex indicates that *Archaeoamphiroa*'s habitat – the algal-veneered reef flat – was a broad terrace immediately leeward of the growing reef margin

Phylogenetic Affinity

Classification as Calcareous Red Algae. The gridlike arrangement of cellular calcite diagnoses *Archaeoamphiroa* positively as a calcareous red alga. The excellent preservation of this cellular calcite in many specimens excludes affinities with the aragonite-secreting Squamariacean algae, indicating classification with either the Solenoporacean or Corallinacean families.

Because large cell size is frequently cited as a distinguishing characteristic of the Solenoporaceae, the relatively large size of cells in *Archaeoamphiroa* might suggest affinity with this family. However, for reasons cited below, this author favors ancestral affinity with the Corallinaceae:

1. The articulate growth mode characterizes a major subfamily of the Corallinaceae, but is unknown in Solenoporacean forms.
2. Cells as long as 500 microns have been described from certain coralline species (Johansen 1981, p. 225), indicating that cell lengths in *Archaeoamphiroa* do not preclude a Corallinacean affinity.
3. Evidence previously cited in this paper for an original mineralogy of high-magnesian calcite supports affinity with the Corallinaceans, which characteristically secrete high-magnesian calcite, rather than the Solenoporaceans, for which no evidence has yet been documented to suggest a mineralogy of high-magnesian calcite.
4. Similarities with the modern coralline genus *Amphiroa* are so pronounced as to suggest a close affinity with the articulate Subfamily Amphiroideae.

Similarity to Amphiroa. Many observations suggest close similarities between *Archaeoamphiroa* and the modern *Amphiroa*, most particularly the species *A. van-bosseae*, including:

1. cell tiers of alternating cell length, a feature which is apparently unique amongst modern coralline algae to the genus *Amphiroa*;
2. the size and shape of the intergenicula, which are remarkably akin to those in *Amphiroa van-bosseae* specimens which have been described by Norris and Johansen (1981, p. 23);
3. the habitat preference of *Amphiroa* today for rocky coasts and reef prominences in shallow, surf-swept tropical to subtropical marine waters such as the algal-veneered reef flat rimming the eastern and southern coasts of Saipan, which has been vividly described by Cloud (1959).

All evidence – cell organization, flexible branched growth mode, inferred high-magnesian calcite mineralogy, segment size and shape, and preferred habitat – overwhelmingly suggest classification of this Late Devonian algae with the coralline Subfamily Amphiroideae.

Implied Coexistence of Related Crustose Forms

Interpretation of *Archaeoamphiroa* as an ancestral coralline alga indicates coexistence during the Late Devonian of related encrusting forms, for the ontologic history of modern articulate corallines shows that they clearly evolved from encrusting *Lithothamnium*-like precursors (Johansen 1981). If one can assume that *Archaeoamphiroa*, like its modern counterparts, evolved from an encrusting forbear, then one cannot ignore the likelihood that its appearance in the fossil record *must have been preceded* by encrusting precursors.

Conventionally, the lineage of coralline algae is traced back only as far as the mid-Mesozoic (Wray 1977, p. 48), although the notion that ancestral coralline forms existed in the Paleozoic has been speculated on over several decades. Johnson first raised the possibility in 1956, noting significant structural similarities between the Pennsylvanian *Archaeolithophyllum* and modern crustose species; but lacking evidence for direct linkage in the fossil record, he conservatively assigned this and similar coralline-like forms to a group of red algae "of uncertain affinities" (Johnson 1961). Subsequently there have been numerous reports of Upper Devonian crustose coralline algae from regions as far apart as western Canada (Sonnenfeld 1964, Jamieson 1967b, Wray and Playford 1970, Machielse 1972) and western Australia (Wray 1967); and although Riding (1974) has disputed the interpretations of some of these forms as algal, the fact that these reports emanated independently from many different localities argues strongly for the collective interpretation that they represent.

The implications of this study are that ancestral encrusting coralline algae must have been in existence by Upper Devonian times, thus reinforcing the conclusions of earlier workers who have reported ancestral crustose coralline algae from rocks of Upper Devonian age in North America and Australia. Furthermore, it is suggested that future research will more fully document the existence of encrusting coralline algae in Devonian rocks throughout the world and may verify the comparison of presently unclassified algae like *Katavella* (Chuvashov 1966) with modern articulate coralline forms.

Summary

This paper documents the existence in Upper Devonian reefs of articulate calcareous red algae homologous with the modern articulate coralline genus *Amphiroa.* The habitat range of these algae has been profiled across a 30-mile (48 km) outcrop transect of the Frasnian Alexandra Reef-Complex southwest of Hay River in Canada's Northwest Territories.

Distribution of the beadlike algal fragments in this reef-complex indicate that they are confined exclusively to marginal organic shoal facies, where they are most commonly associated with the reef-building stromatoporoid *Actinostroma*, encrusting coralline-like algae, crinkly-surfaced stromatolites, and *Renalcis.* Significantly, their occurrences are inversely related to lagoon-dwelling organisms such as dasycladacean and codiacean algae, calcispheres, and *Amphipora.*

Dolomite rhombs preferentially replacing many of the fragments suggest an original mineralogy of high-magnesian calcite such as characterizes modern coralline algae.

Close correspondence in structure, inferred mineralogy and habitat between this alga, informally referred to as *Archaeoamphiroa*, and the living *Amphiroa* suggests a phylogenetic linkage, and it is recommended that *Archaeoamphiroa* be classified with the articulate coralline subfamily Amphiroideae. Furthermore, because the ontologic history of articulate coralline algae indicates an encrusting precursor, documentation of these algae supports evidence previously cited for the existence of crustose coralline algae in the Devonian.

Acknowledgments. The writer acknowledges support for field research for this project by the Geological Survey of Canada, capable field assistance by Gudula von Schoenberg and H. Miriam Steele, supporting scholarships by the British Council, Canadian Federation of University Women, and Marty Memorial fund of Queen's University, Canada, and stimulating guidance during the study that led to this paper from the penetrating critiques of Alan Lees at Reading University and supportive confidence of Digby J. McLaren of the Geological Survey of Canada.

References

Chuvashov BI (1966) *Katavella*, a new genus of fossil red algae. Int Geol Rev 8:89–93

Cloud PE, Jr (1959) Geology of Saipan, Mariana Islands, part 4: Submarine topography and shoal-water ecology. U S Geol Surv Prof Pap 280-K:361–445

Jamieson ER (1967a) The Alexandra Reef-complex (Frasnian), Hay River area, NWT, Canada: stratigraphy, sedimentology, and palaeoecology. Unpubl Ph D thesis, Reading Univ Reading, England, 220 p

Jamieson ER (1967b) Discovery of reef-building coralline algae in the Canadian Devonian. Abstr Int Symp Devonian Syst, Calgary, Alberta, p 74

Jamieson ER (1971) Paleoecology of Devonian reefs in western Canada. Proc North Am Paleontol Convent, part J:1300–1340

Johansen HW (1981) Coralline algae, a first synthesis. Boca Raton, Florida. CRC Press, 239 p

Johnson JH (1956) *Archaeolithophyllum*, a new genus of Paleozoic coralline algae. J Paleontol 30:53–55

Johnson JH (1961) Limestone-building algae and algal limestones. Color Sch Mines, 297 p

Machielse S (1972) Devonian algae and their contribution to the Western Canada Sedimentary Basin. Bull Can Pet Geol 18:187–237

Norris JN, Johansen HW (1981) Articulated coralline algae of the Gulf of California, Mexico, I: *Amphiroa* Lamouroux. Smithson Contrib Mar Sci 9:1–29

Riding R (1974) The Devonian genus *Keega* (Algae) reinterpreted as a stromatoporoid basal layer. Palaeontology 17:565–577

Schlanger SO (1957) Dolomite growth in coralline algae. J Sediment Petrol 27:181–186

Sonnenfeld P (1964) Famennian reefs in Alberta, Canada. Bull Am Assoc Pet Geol 48:548

Wray JL (1967) Upper Devonian calcareous algae from the Canning Basin, Western Australia. Colo Sch Mines Prof Contrib 3:76

Wray JL (1977) Calcareous algae. Elsevier Sci Publ, New York, 185 p

Wray JL, Playford PE (1970) Some occurrences of Devonian reef-building algae in Alberta. Bull Can Pet Geol 18:544–555

13 Lower Permian Phylloid Algal Mounds, Southern Tatum Basin, Southeastern New Mexico, U.S.A.

J.M. CYS[1]

Introduction

The Tatum Basin (Fig. 1) is located in southeastern New Mexico on the Northwest Shelf of the prolific hydrocarbon-productive Permian Basin region. It is a local subsurface basin that originated in Early Pennsylvanian time, with maximum development during Middle Pennsylvanian. As a result of depositional infilling, the area ceased to be a basin by Early Permian (Wolfcamp) time and became a shallow open shelf that was the site of cyclic carbonate deposition. Phylloid algal mounds are an integral part of these cycles and are the subject of this paper.

This paper is based on a core study of Lower Permian phylloid algal mounds and associated facies in the southern Tatum Basin. The fields from which cores were studied are shown in Fig. 1 and the cored wells studied are listed in Table 1. Appreciation is expressed to Union Oil Company of California, Dorchester Exploration, Inc., Coastal Oil & Gas Corporation, Mapco Production Company, and Shell Oil Company, all of Midland, Texas, for permission to examine their respective cores. Thanks are also due Mapco Production Company, Midland, Texas, for support of this study, which was done while the author was employed by Mapco. D.F. Toomey, Midland, Texas, and Susan Longacre, Getty Research, Houston, Texas, very kindly reviewed an earlier draft of this paper.

Stratigraphy

The Lower Permian sediments of the southern Tatum Basin are a cyclic sequence of interbedded limestones and shales that is assigned to the Hueco Formation. At the base of the formation are four shoaling cyclic carbonate sequences that have been informally designated the Bough member. Hydrocarbon-productive phylloid algal mounds are common within the Bough and are the subject of this paper. Fusulinids, especially the occurrence of *Leptotriticites* (Wilde 1971), date the informal A, B, and C units of the Bough member in the study area as Early Permian (earliest Wolfcamp) in age. The Bough D appears to be of Late Pennsylvanian (Virgil) age.

1 602 West Kansas, Midland, TX 79701, U.S.A.

Paleoalgology: Contemporary Research and Applications
(eds. D.F. Toomey and M.H. Nitecki)

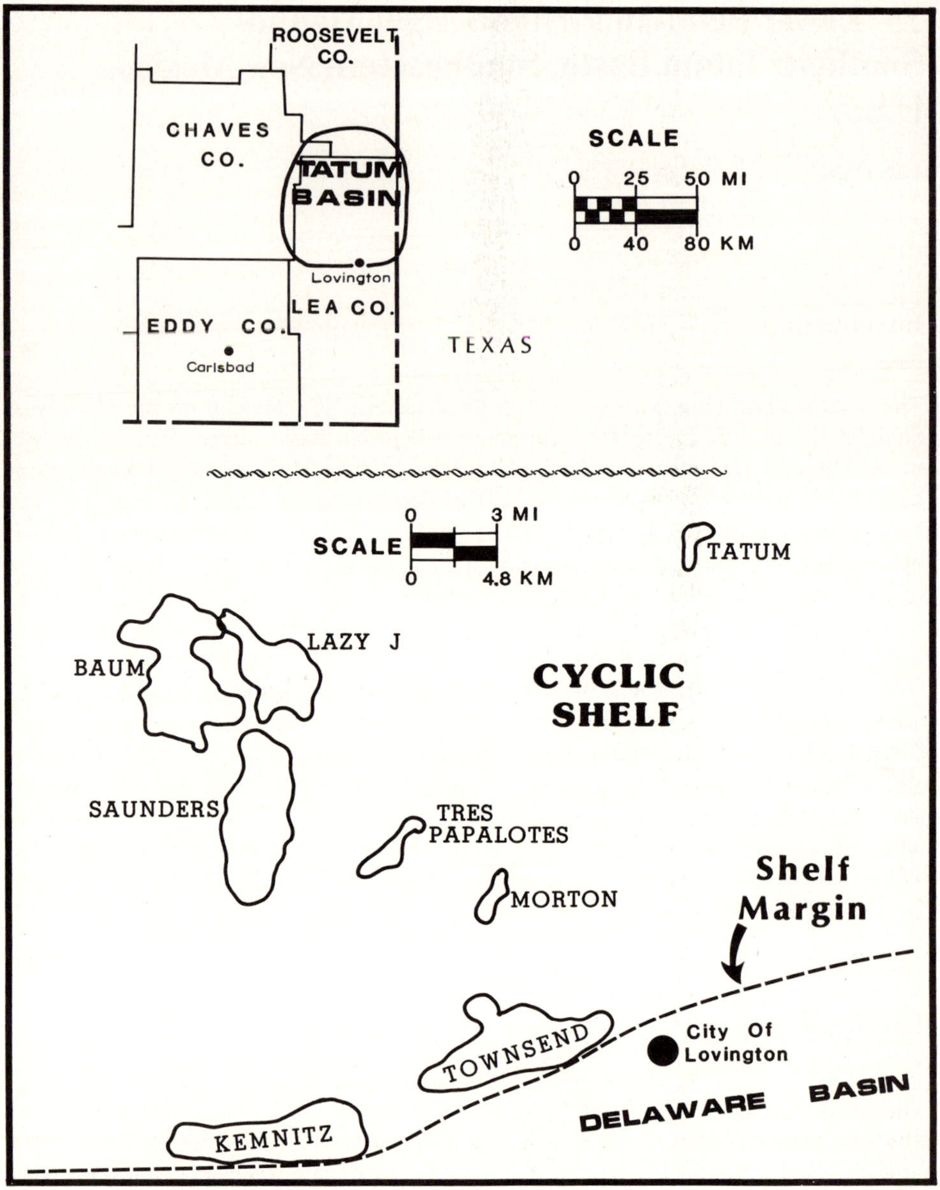

Fig. 1. Index maps showing the general location of the Tatum Basin in southeastern New Mexico and the outline of the Tatum Basin study area. Cores were studied from Baum, Lazy J, Saunders, Tatum, Tres Papalotes, and Morton Fields. Kemnitz and Townsend Fields, located on the shelf-edge between the Tatum Basin shelf and deep Delaware Basin, are also shown for reference

Table 1. Wells from which cores were studied for this paper

Well name	Field	Location
Great Western No. 1 Town	Morton	Section 18, T-15-S, R-35-E
Union of California No. 1 State "7"	Morton	Section 07, T-15-S, R-35-E
Union of California No. 1–12 Gulf-Federal	Morton	Section 12, T-15-S, R-34-E
Dorchester No. 1 Morton Solid State Unit	Tres Papalotes	Section 04, T-15-S, R-34-E
Magnolia No. 1 Read-State	Lazy J	Section 34, T-13-S, R-33-E
Shell No. 1-A State	Saunders	Section 34, T-14-S, R-33-E
Shell No. 2 State "A"	Saunders	Section 34, T-14-S, R-33-E
Coastal Oil and Gas Corp. No. 3 State "8"	Baum	Section 08, T-14-S, R-33-E
Union of California No. 1–6 Duncan	Tatum	Section 06, T-13-S, R-36-E
Union of California No. 1–6 Duncan "B"	Tatum	Section 06, T-13-S, R-36-E
Union of California No. 1–5 Atwood Bradley	Tatum	Section 05, T-13-S, R-36-E

Note: All wells are located in Lea County, New Mexico

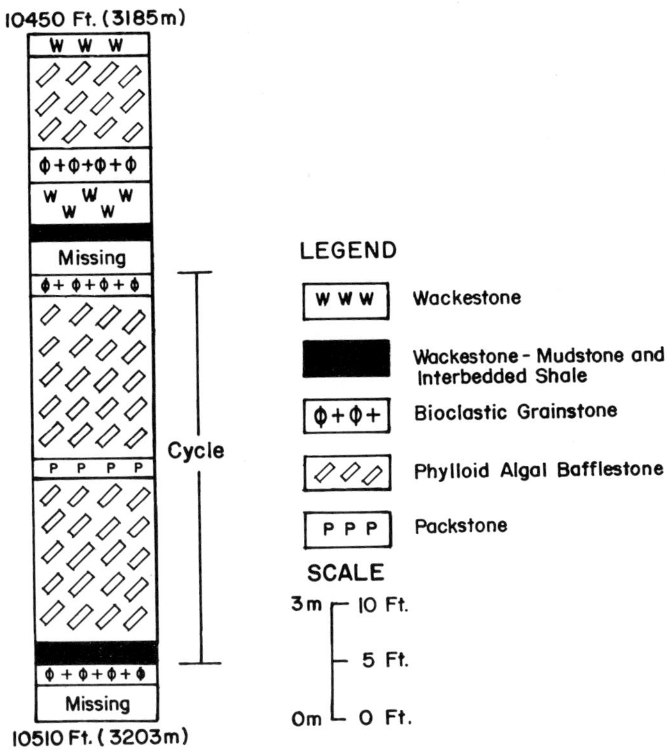

Fig. 2. Schematic core lithology diagram showing shoaling carbonate cycles present in the informally named Bough member of the Hueco Formation, from the Dorchester No. 1 Morton Solid State Unit, Tres Papalotes Field. Packstone and wackestones on the diagram are all bioclastic

Each Bough interval consists of shoaling carbonate cycles separated by interbedded shale and argillaceous limestone. A typical cycle within a phylloid algal mound complex (Fig. 2) consists of the following superposed units:

1. A basal transgressive marine unit, 2 to 30 feet (0.6–9.1 m) thick, of black shale and interbedded dark-gray argillaceous lime mudstone and wackestone with scattered crinoids and fusulinids (Fig. 3A).
2. Phylloid algal bafflestone (Fig. 3B, E) described more fully below.
3. Crestal high-energy grainstones (Fig. 3C), 2 to 10 feet (0.2–3.0 m) thick with an average of 4 to 6 feet (1.2–2 m), consisting of tubular foraminifers, fusulinids, crinoids, molluscs, *Tubiphytes*, ooids, and minor amounts of algae and intraclasts.

Although the typical ideal cycle has the above three elements in sequence, some cycles have missing or repeated units, as shown in the upper part of the core in Fig. 2.

Away from areas of mound development phylloid algal bafflestones and crestal grainstones are generally replaced in the cycle by a unit of fossiliferous lime wackstone (Fig. 3D). Fusulinids and crinoids are the most abundant fossils; brachiopods, molluscs, and bryozoans occur in minor numbers. Contacts between various lithologic units appear to be conformable.

Phylloid Algal Mounds

Description. As discussed above, the phylloid algal mounds in the informal Bough member are found in shoaling carbonate cycles on a shallow, open carbonate shelf. They occur as complexes rather than large individual mounds. Most complexes have an average size of 1 to 2 square miles (2.59–5.18 km^2), although a few attain a size as large as Baum and Saunders Fields (Fig. 1), which have respective areas of 14 and 12 square miles (36.26 and 31.08 km^2). Thickness of mound complexes within the study area averages from 200 to 400 feet (70–122 m). Individual mounds range in thickness from 6 to 31 feet (1.8–9.4 m) with an average of 17 feet (5.2 m). These are thinner than the shelf-edge mounds of the southern Sacramento Mountains of southern New Mexico, which attain a thickness of 70 feet (21 m) (Toomey et al.

Fig. 3A–E. Photographs of polished core slabs of typical Bough carbonate cycle lithologies. **A** Deeper water transgressive sequence from a depth of 10,503 through 10,504 feet (3201.3–3201.6 m), Dorchester No. 1 Morton Solid State Unit, Tres Papalotes Field. Six-inch (15 cm) ruler is on very calcareous fissle black shale. Below ruler is fusulinid wackestone with irregular, wavy black shale laminae. **B** Phylloid algal bafflestone with good phylloid algal blade moldic and fabric-nonselective porosity. Note solution-channel porosity at top center and gastropod cross-section near center left edge. Scale is in cm. Depth of 10,339 feet (3151.3 m), Union of California No. 1–6 Duncan well, Tatum Field. **C** Tubular foraminifer-crinoid grainstone with scattered intraclasts at a depth of 10,471 feet (3191.6 m), Dorchester No. 1 Morton Solid State Unit, Tres Papalotes Field; scale in cm. **D** Fusulinid-crinoid wackestone of the open-marine shelf environment; depth, 10,395 feet (3168.4 m), Union of California No. 1 State 7 well, Morton Field. Diameter of coin is 1.85 cm. **E** Phylloid algal bafflestone with crinoids; depth, 10,451 feet (3185.5 m), Union of California No. 1 State 7 well, Morton Field. Diameter of coin is 1.85 cm

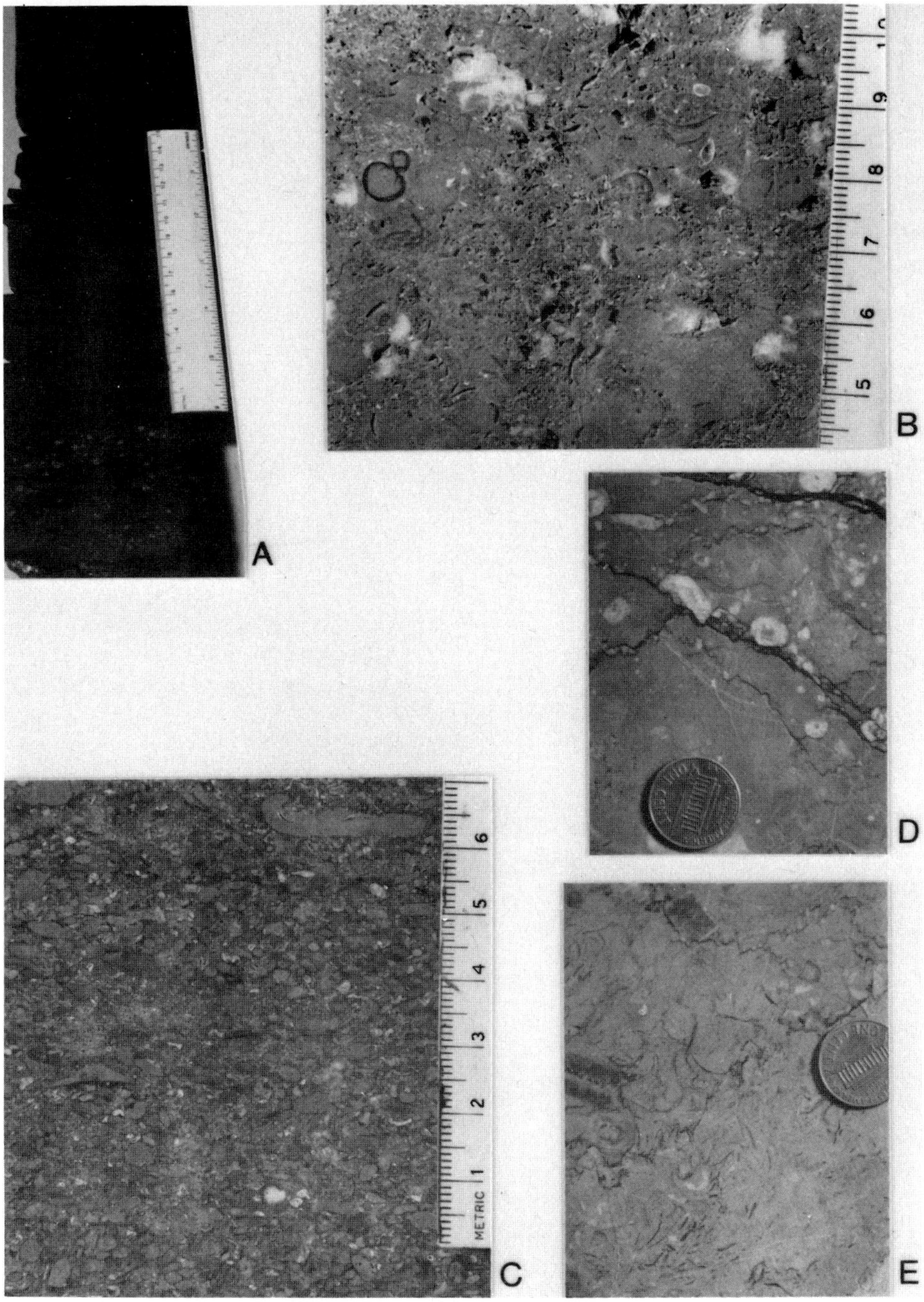

Fig. 3A–E

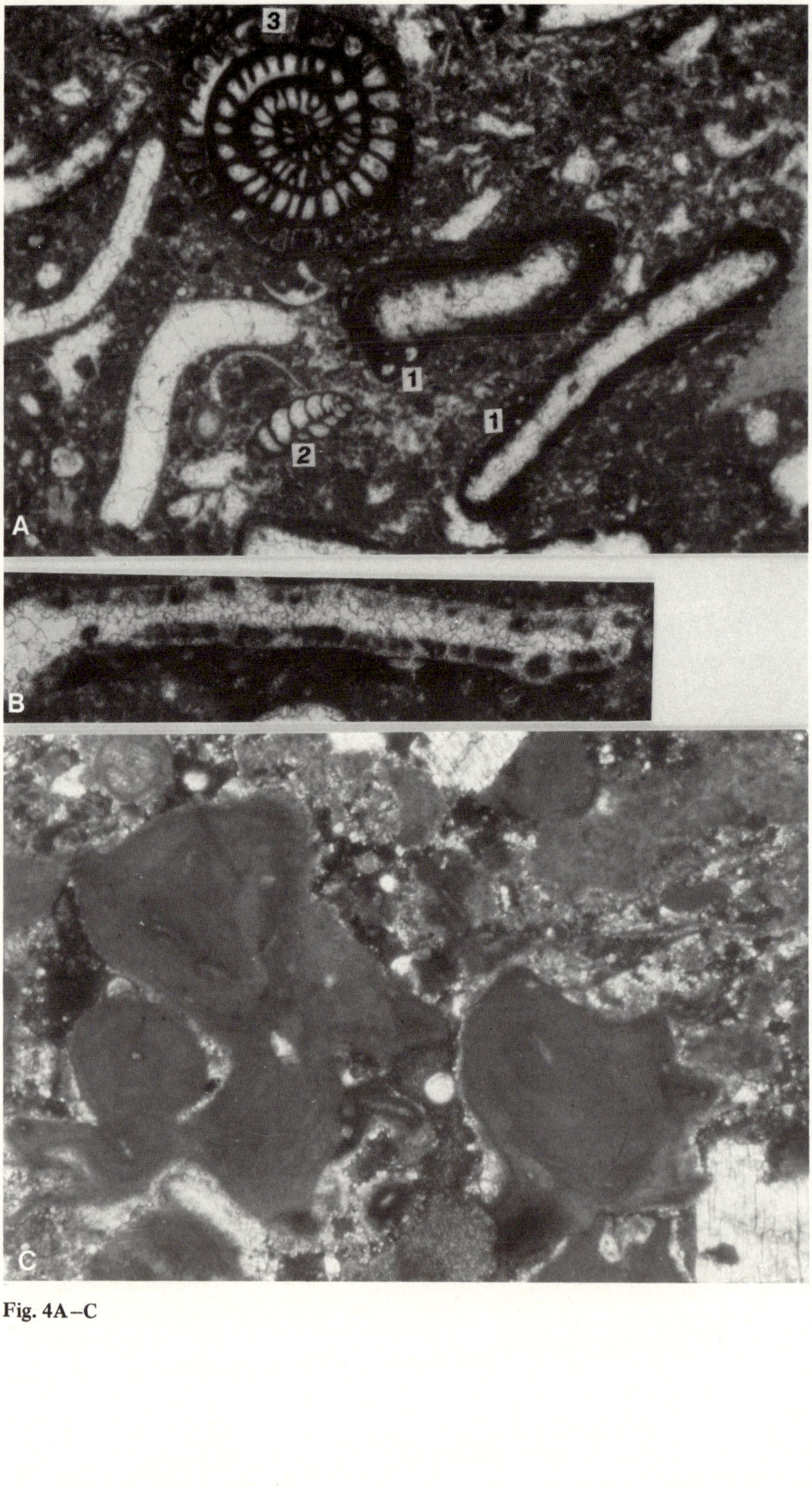

Fig. 4A–C

1977). Areal extent of individual mounds is small, probably in the range of 0.25 to 0.5 square mile (0.65–1.3 km^2). The mounds have both biohermal and biostromal shapes, the former being the most common. Petrographically, the mound rock is a phylloid algal bafflestone (Fig. 4A).

Phylloid algae (Fig. 4A), principally of the genus *Eugonophyllum* (Fig. 4B), are the primary mound builders. Encrusting tubular foraminifers are the most common accessory organisms (Fig. 4A) within the mounds. They are especially common in beds adjacent to the mounds. Crinoids (Fig. 3E), bryozoans, gastropods (Fig. 3B), fusulinids (Fig. 4A), and the problematical micro-organism *Tubiphytes* are less common accessory organisms. *Tubiphytes*, usually in association with encrusting tubular foraminifers, frequently forms local pockets of boundstone (Fig. 4C) within the phylloid algal mounds. This is in direct contrast to Kemnitz and Townsend Fields (Fig. 1), situated on the Tatum Basin shelf-edge, where *Tubiphytes*-tubular foraminifer boundstone is the primary mound builder and phylloid algae have a minor secondary role (Malek-Aslani 1970). In contrast to the well-known shelf-edge phylloid algal mounds of the Sacramento Mountains of southern New Mexico (Toomey et al. 1977), the shelf mounds of the southern Tatum Basin do not exhibit an observable vertical biotic zonation. However, the mounds in both areas exhibit the same basic shoaling facies sequence.

Porosity. The mounds and associated grainstones are important hydrocarbon reservoirs in the southern Tatum Basin area. Reserves of the more common smaller fields, such as Morton and Tres Papalotes (Fig. 1), range between 1.2 to 2 million barrels of oil. With this quantity of reserves, the mound complexes are potential lucrative exploration targets.

The mound complexes form both stratigraphic-structural and stratigraphic traps. Porosity is best developed in the phylloid algal bafflestone and consists primarily of molds of blades of phylloid algae and fabric-nonselective types (Fig. 3B) in almost equal abundance. Primary porosity of shelter cavities beneath phylloid algal blades is of minor importance. The dissolution of aragonitic fossil fragments, the presence of solution fractures, breccia pockets, internal vadose silt, red staining, and local red-shale infill of breccia pockets and solution fractures are all supporting the contention that the porosity originated in the vadose zone during subaerial exposure at the end of each regressive carbonate cycle. A typical mound complex field in the study area has recently been described in detail by Cys and Mazzullo (in press).

Fig. 4A–C. Thin section photomicrographs (plain light) of phylloid algal mound rock. **A** Typical phylloid algal bafflestone. Note foraminifers (1) encrusting phylloid algae, palaeotextularid foraminifer (2), and fusulinid (3); depth 10,459 feet (3187.9 m), Union of California No. 1 State 7 well, Morton Field. Width of photographic field is 8 mm. **B** *Eugonophyllum*; note floating utricles which are diagnostic of this genus; depth 10,459 feet (3187.9 m), Union of California No. 1 State 7 well, Morton Field. Width of photographic field is 3.4 mm. **C** Local pocket of *Tubiphytes* boundstone; depth 10,439 feet (3181.3 m), Union of California No. 1–6 Duncan "B" well, Tatum Field. Width of photographic field is 4.25 mm

Porosity in the tubular foraminifer-*Tubiphytes* boundstone pockets is both primary constructional and fabric-nonselective. Biomoldic, oomoldic, and primary interparticle are the major porosity types in the grainstones. Primary intraskeletal porosity is a relatively minor fourth type.

Paleotectonics

In the southern Tatum Basin, control of location of the mound complexes is apparently paleostructural. Seismic (common depth point) data and deep subsurface well-control demonstrate that mound complexes are situated on flanks or crests of deep-seated paleostructures, that probably originated during Early Pennsylvanian time. This is shown by pronounced thinning of the Early Pennsylvanian sequence across the structures. These paleostructures created subtle paleobathymetric high areas on the sea floor which apparently had optimum paleoenvironmental conditions for the prolific growth of phylloid algae. Wilson (1975) stated that the complexes preferentially occur on the downthrown sides of growth faults. This statement needs clarification. The faults were active during Early Pennsylvanian time but were not active during deposition of the "Bough" member, and on seismic records these cannot be identified above the Lower Pennsylvanian sequence. They are also high-angle reverse in their orientation. Hence, they are not growth faults in the restricted sense of the term. Although several phylloid algal mound complexes are on the downthrown sides of faults (see Fig. VI-19 in Wilson 1975), it can be shown seismically that they are also located on paleostructures.

Summary

Phylloid algal mounds of Early Permian age formed on a shallow carbonate shelf associated with shoaling cycles, in the Tatum Basin of southeastern New Mexico. These mounds are similar to the well-known shelf-edge phylloid algal mounds of the Sacramento Mountains described by Wilson (1975), and Toomey et al. (1977). Mounds in both areas developed as integral units of regressive carbonate cycles, exhibit shoaling facies sequences, and are intimately associated with paleostructure. The major difference is that the southern Tatum Basin shelf mounds are much thinner.

References

Cys JM, Mazzullo SJ (in press) Depositional and diagenetic history of Morton Field (Lower Permian), southeastern New Mexico. In: Roehl PO, Choquette PW (eds) Carbonate petroleum reservoirs; a casebook. Springer, Berlin Heidelberg New York

Malek-Aslani M (1970) Lower Wolfcampian reef in Kemnitz Field, Lea County, New Mexico. Am Assoc Pet Geol Bull 54:2317–2335

Toomey DF, Wilson JL, Rezak R (1977) Evolution of Yucca Mound complex, Late Pennsylvanian phylloid-algal buildup, Sacramento Mountains, New Mexico. Am Assoc Pet Geol Bull 61:2115–2133

Wilde GL (1971) Phylogeny of *Pseudofusulinella* and its bearing on Early Permian stratigraphy. In: Dutro JT Jr (ed) Paleozoic perspectives: A paleontological tribute to G. Arthur Cooper. Smithson Contrib Paleobiol 3:363–379

Wilson JL (1975) Carbonate facies in geologic history. Springer, Berlin Heidelberg New York, 471 p

14 Sedimentation of an Upper Pennsylvanian (Virgilian) Phylloid Algal Mound Complex, Hueco Mountains, El Paso Country, West Texas

J.C. POL[1]

Introduction

Phylloid algal mounds are common in Late Paleozoic carbonates of west Texas and southern New Mexico (Wray 1968, Wilson 1975, 1977). The accumulated literature suggests a close relationship between carbonate sedimentation and progradational clastic sequences as well as the strong influence of local tectonics (Wilson 1967, 1972). Studies by various workers indicate that distinct community succession and characteristic sedimentological processes are common to algal mound deposition. These tend to produce similar vertical and lateral facies relationships (Wilson 1975, Toomey 1976, Toomey et al. 1977, Toomey and Cys 1979). However, minor differences in environmental conditions (i.e., climate, water circulation, local tectonics, etc.) ensure that no two algal mound complexes are identical.

The purpose of this paper is to describe the sedimentological history of a Late Pennsylvanian (Virgilian) phylloid algal mound complex which outcrops on the western flank of the Hueco Mountains, El Paso County, west Texas (Fig. 1). This paper is based on a portion of the writer's work at the University of Texas, Austin. The outcrop, named "Reef" Hill by King et al. (1945), is located about 25 miles (40 km) east of the city of El Paso and is approximately two miles south of Hueco Tanks State Historical Park. The outcrop is an erosional remnant with nearly 350 feet (107 m) of topographic relief covering an area of about one-half square mile (Figs. 2 and 3). The outcrop section on "Reef" Hill has been divided into three informal units: (1) lower clastic unit, (2) middle carbonate unit, and (3) upper clastic unit (Lower Permian Hueco Group). This paper will focus only on the phylloid algal biohermal middle carbonate unit (Fig. 3).

Geological Setting

Stratigraphy. The Hueco Mountains are a north-south trending escarpment that constitutes the eastern edge of the Basin and Range province. The range has a gently dipping eastern margin and a sharp, fault-bounded western margin. A major angular unconformity places progressively older strata beneath the Lower Permian Hueco

1 ARCO Exploration Company, P.O. Box 5540, Denver, CO 80217, U.S.A.

Paleoalgology: Contemporary Research and Applications
(eds. D.F. Toomey and M.H. Nitecki)

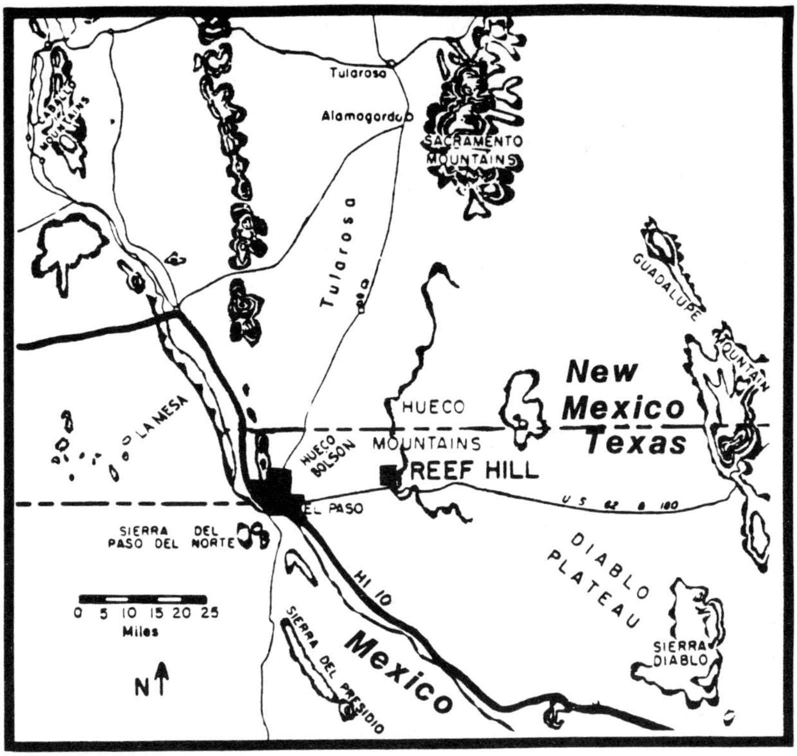

Fig. 1. Index map of the study area showing major mountain ranges of West Texas and south-central New Mexico. (After Nelson and Haig 1958)

Fig. 2. Aerial photograph of "Reef" Hill, Hueco Mountains, West Texas; view is toward the east-southeast. The *two arrows* indicate the upper and lower boundaries of the middle carbonate unit. (Photo by D.F. Toomey)

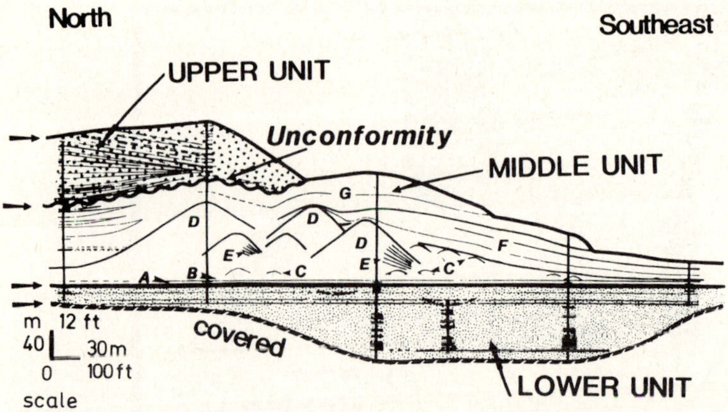

Fig. 3. Schematic cross-section of "Reef" Hill based on six measured sections. Interpreted subenvironments of the middle unit include *A* rugose coral bed; *B* dasycladacean algal flat zone; *C* "small mound" zone; *D* massive mound core; *E* accretionary bedding; *F* flank beds; *G* drape beds

Group in a southerly direction. A nearly complete Paleozoic section (Cambrian interval not present) and a crystalline Precambrian basement are exposed in the Hueco Mountains. Above the Permian beds, the section includes Lower Cretaceous (Comanchean) units, Tertiary intrusive igneous rocks, and Quaternary terrigenous basin fill clastics (King et al. 1945).

The Pennsylvanian section of the Hueco Mountains was first defined by King et al. (1945). They named the beds of the entire Pennsylvanian System, consisting of over 1500 feet (450 m) of limestone and shale, the Magdalena Group. They further subdivided the Pennsylvanian into a lower division of Morrow-Atoka age, a middle division of Desmoines age, and an upper division of Missouri-Virgil age. The "Reef"Hill outcrop consists of the upper division limestone and shale, Late Pennsylvanian (Magdalena Group), and the Lower Permian Pow Wow Conglomerate (Hueco Group). Equivalent strata in the Sacramento Mountains are known as the Holder and Beeman Formations (Pray 1961, Seewald 1968); it may be appropriate to extend this terminology into the Hueco Mountains area.

Paleogeography and Paleotectonics. The Late Paleozoic was a time of major tectonic collision that resulted in fragmentation of the North American continent into a series of adjoining landmasses and basins (McKee and Crosby 1975). The Late Pennsylvanian "Reef" Hill carbonates were deposited on the Pennsylvanian Sacramento shelf, west of the Pedernal Landmass and east of the Oro Grande Basin (Fig. 4). The Sacramento Shelf (Wilson 1967) was variable in width, being very narrow near present-day Alamogordo, and widening towards the north and south.

Persistent movement of the La Luz structure during the Late Paleozoic was responsible for rapidly changing depositional patterns of the Sacramento Shelf (Wilson 1972). During times of uplift, clastic sedimentation spread far into the basinal area, bypassing the emergent, subaerially exposed shelf platform. Rising sea level conditions, resulting from subsidence, produced clear shallow-water environments condu-

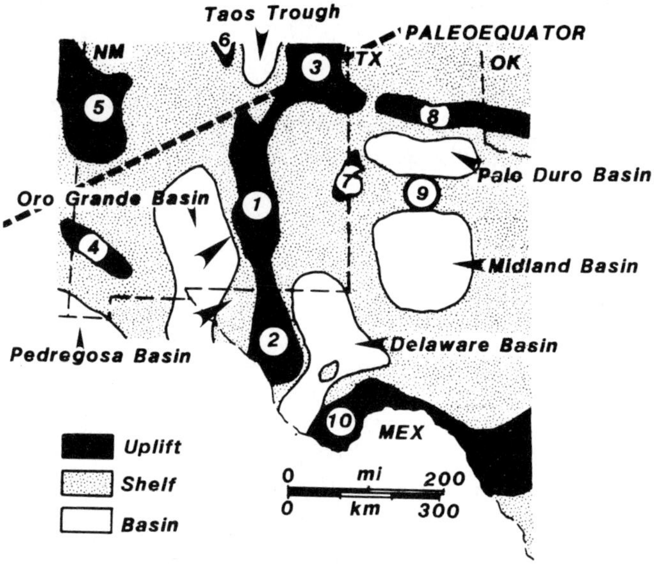

Fig. 4. Regional Late Paleozoic paleogeography and tectonic elements of a portion of southern New Mexico and West Texas. Tectonically negative areas (basins) are shown *blank*, positive areas are *black*, and platforms are *stippled*. Key: *1* Pedernal Landmass; *2* Diablo Platform; *3* Sierra Grande Uplift; *4* Florida Platform; *5* Defiance-Zuni Uplift; *6* Uncompahgre Uplift; *7* Roosevelt Positive Element; *8* Amarillo-Wichita Uplift; *9* Matador Uplift; *10* Marathon Salient of the Quachita Foldbelt (McKee and Crosby 1975, Smith 1981)

cive to carbonate production and accumulation. Shelf margins surrounding deeper basins became sites of localized biohermal deposition during the deepening-water phase of the transgression-regression cycle (Wilson 1967, 1977, Toomey et al. 1977).

Uplift of the Diablo Platform was not as pronounced as that of the Pedernal Landmass (McKee and Crosby 1975). It is believed that the Diablo Platform was a stable, gently subsiding feature throughout Lower and Middle Pennsylvanian time. During Middle Pennsylvanian (Desmoines) time it is thought to have been emerging; by Upper Pennsylvanian (Missouri), it became subaerially exposed and appears to have been connected with the Pedernal Landmass. A brief period of subsidence produced a relatively minor marine transgression, which led to deposition of the "Reef" Hill phylloid algal mound complex. Renewed uplift at the close of the Pennsylvanian produced a major angular unconformity that resulted from the resumption of clastic deposition (Pol 1982).

Facies Relationships

Clastic Facies. The deposition of the lower clastic unit had an important influence in the initiation and development of the phylloid algal mounds of the middle carbonate unit. The details of the sedimentation and diagenesis of this unit, however, are beyond the scope of this paper and will be treated separately.

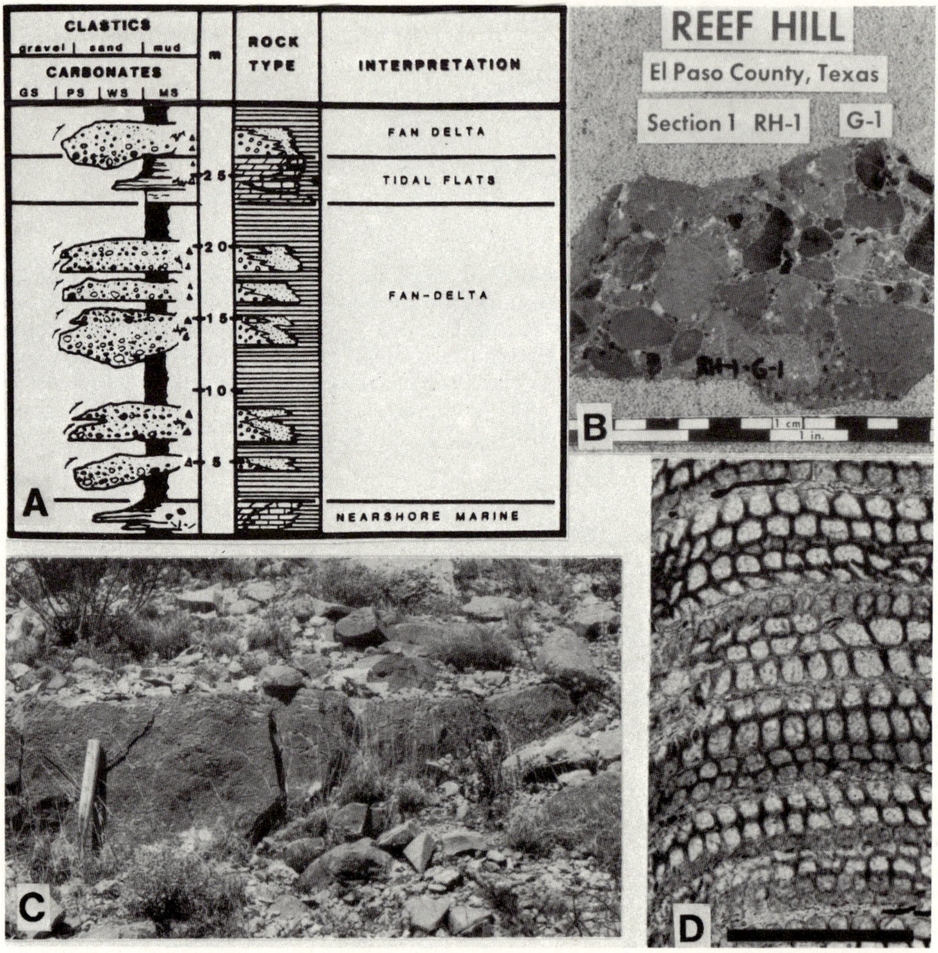

Fig. 5A–D. Representative examples of clastic facies from the lower unit exposed on "Reef" Hill. **A** Generalized section of the lower unit, composed primarily of coarse calclithite and shale, with minor occurrences of carbonates. **B** Polished slab of poorly sorted coarse calclithite, showing variety of clast types and degree of rounding. **C** Outcrop photograph of a coarse calclithite channel; note hammer for scale. **D** Thin-section photomicrograph of silicified wood from a 13-foot (4 m) petrified tree found in place on the outcrop; scale bar is 0.5 mm; (sample, RH-3-F1-1b)

Fig. 6A–E. Representative examples of carbonate facies from the lower unit exposed on "Reef" Hill. **A** Thin-section photomicrograph of a dasycladacean? algal packstone/wackestone; (sample, RH-1-A-2; plane polarized light; scale bar = 1 mm). **B** Thin-section photomicrograph of a fossiliferous packstone; (sample, RH-2-B-1, crossed-nicols, scale bar = 1 mm). **C** Polished slab of a bioturbated sponge spicule packstone. **D** Thin-section photomicrograph of a sponge spicule packstone; (sample, RH-1-M-1, plane light, scale bar = 0.5 mm). **E** Outcrop photograph of the dolomitic mudstone zone [lens cap for scale = 2 inches (5 cm) in diameter]

The lower clastic unit exposed on "Reef" Hill (Fig. 5A) has been interpreted to have been deposited as a fan-delta system. This is associated with minor interbedded shallow-water carbonates and spiculites, and marginal carbonate tidal flat-lagoonal complexes (Pol 1982, 1983, Toomey and Babcock 1983). The clastic facies of this unit consist of limestone conglomerates composed of coarse clasts of limestone, dolomite, shale, chert, and silicified reworked fossils. The abundance of such coarse lithic clasts is indicative of a source area with high topographic relief situated in close proximity to the depositional area (Creaser 1977). It is thought that the existence of exaggerated topography near a depositional basin may be related to contemporaneous movement along a fault plane, dividing the uplifted platform from the subsiding basin. Figures 5 and 6 show some of the more important textural and compositional features of this unit. A local marine transgression caused a sharp eastward shift in clastic deposition away from the study area and initiated overlying carbonate deposition.

Carbonate Facies. Drowning of the delta platform produced an area with a few meters of bathymetric relief above the surrounding sea bottom, which became the site of carbonate production and deposition (Fig. 7). The middle unit carbonate records a shift in the ecologic community structure, and resultant facies pattern change with time. The relationship to water depth is of critical importance to the community evolution and depositional facies. One of the most important biotic groups in the middle unit are the calcareous algae. Phylloid algae (*Eugonophyllum*, see Fig. 8), various types of dasycladacean algae, as well as the blue-green alga *Girvanella* were found. The recent contribution by Toomey and Babcock (1983) contains a detailed and concise set of photographs and descriptions of these and other fossils from the "Reef" Hill outcrop. In that volume two unique, undescribed algae are introduced for the first time.

The depositional sequence includes the following: (1) colonization phase, (2) algal flat deposition, (3) phylloid algal mound initiation, (4) phylloid algal mound dominance, and (5) shallowing upward phase.

Colonization Phase. The deposition of the "Reef" Hill limestone essentially began with colonization of the drowned delta platform by rugose corals (Fig. 9A,B). This biostromal bed represents the first clear-water environment after marine transgression.

Compositionally, this bed is a rugose coral packstone (following usage of Dunham 1962) or rugose coral biomicrite (Folk 1959, 1962, terminology). This rock consists of phylloid algae (*Eugonophyllum*), dasycladacean algae (*Epimastopora*), micrite, rugose corals, crinoids, pelecypods, brachiopods, ostracodes, and rare fusulinids (Fig. 9B). This fossil grouping indicates normal marine conditions in relatively shallow water, on the order of less than 10 to 30 m deep. Stevens (1965) states that optimum depth for rugose corals is between 10 and 30 m. More recently, Waters and Holland (1983) have shown that there is a variety of water depths in which rugose corals may be found.

Large, curved (ceratoid) solitary rugose corals were found in this zone (Fig. 9A). The twisted or contorted morphology of these corallites may be related to relatively

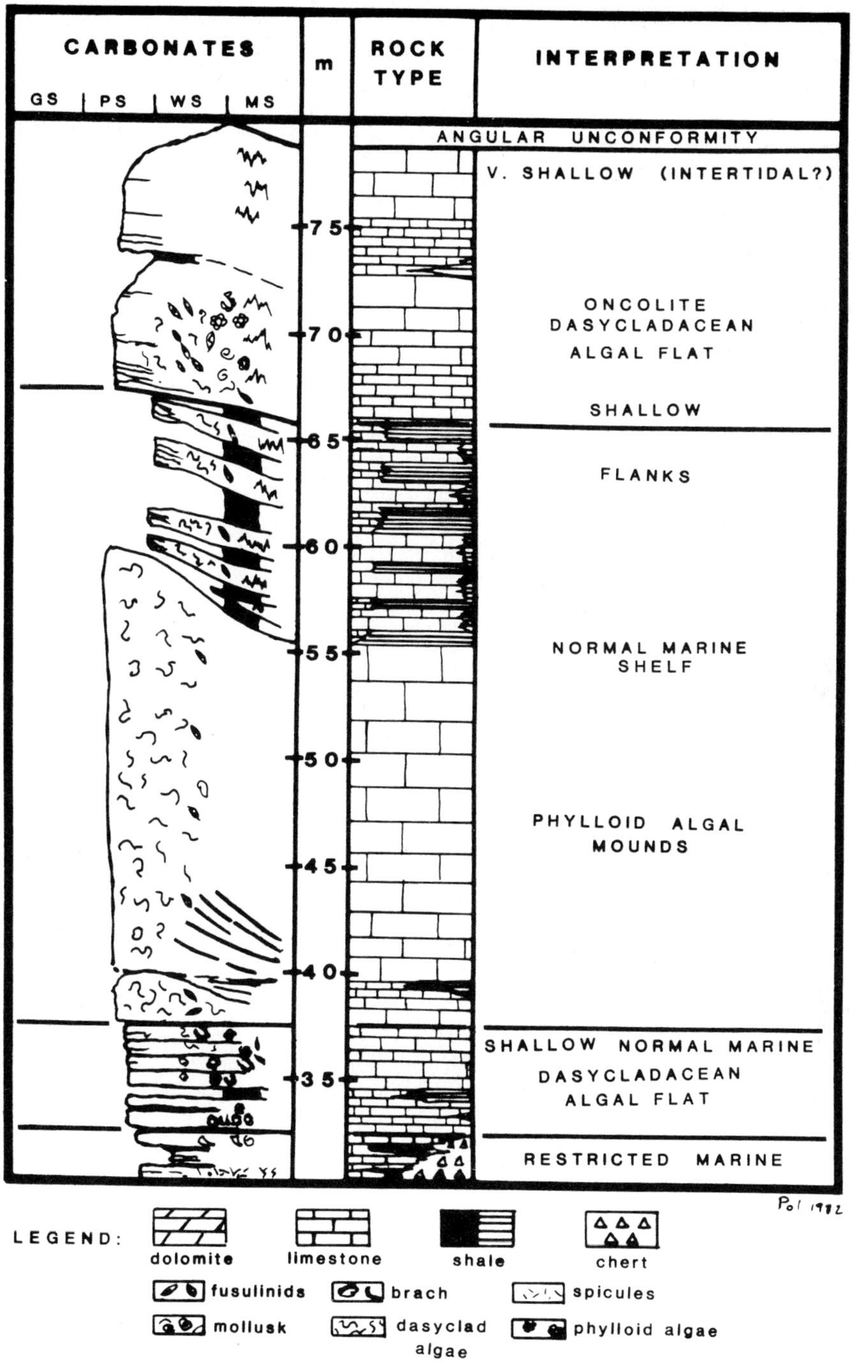

Fig. 7. Schematic measured section of the middle carbonate unit exposed on "Reef" Hill, Hueco Mountains, West Texas

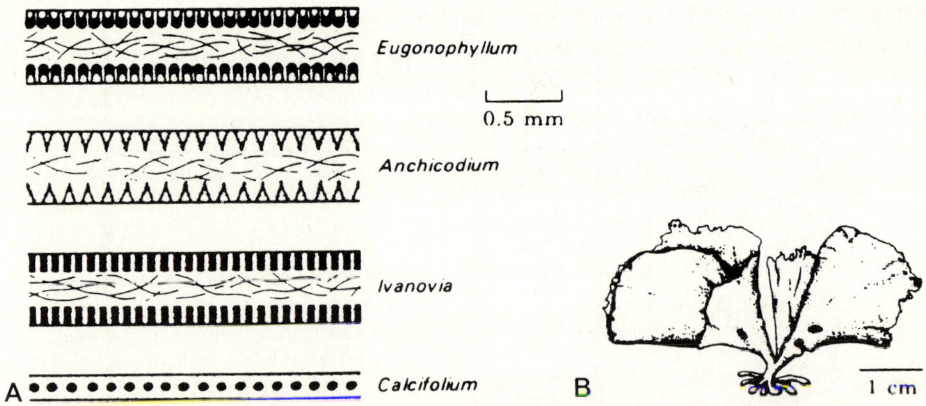

Fig. 8. A Thin-section subdivision of the phylloid algae based on internal morphology (utricles in *solid black*). **B** Probable growth morphology of *Eugonophyllum* (Wray 1977)

high energy conditions (Tasch 1980). Wave and current agitation of the substrate probably caused toppling of the corallites and renewed upward growth, resulting in twisted corallites.

The rugose coral bed has been interpreted to have been deposited as a biostromal unit within a relatively normal marine, high energy, shallow-water regime. This is based on vertical facies relationships as well as the fossil evidence noted above.

Dasycladacean Algal Flats. Overlying the rugose coral bed is a zone of light gray, uniformly bedded limestone (Fig. 9C), consisting of dasycladacean algal packstone (or biomicrite). These sediments contain foraminifers, dasycladacean algae, phylloid algae, crinoids, brachiopods, molluscs, fusulinids, and ostracodes (Fig. 9D), suggesting shallow-water, normal marine conditions. Most of the carbonate grains of this zone are unbroken and unabraded and contain thick micrite coatings, presumably made by endolithic boring organisms (Fig. 9E).

The abundance of micrite and fine bioclastic grains suggests a low-energy environment. The origin of the micrite can be attributed to the bioerosion of calcareous grains, with a minor contribution from abrasion and direct carbonate precipitation. The apparent low-energy conditions (i.e., abundant micrite) may have been more a function of local baffling by dense algal growth and sediment-binding organic mats (Bathurst 1967, Neumann et al. 1970) than an overall low-energy setting. Bioturbation is pervasive and has obliterated any primary sedimentary structures that may have been present. The apparent bedding in this zone is a result of stylolitization (Fig. 9C).

This interval is interpreted to have been deposited within a clear, shallow-water lagoonal setting. Evidence for this includes the abundance of dasycladacean algae (Ginsburg et al. 1971, Flügel 1977, Wray 1977). The high faunal diversity would tend to indicate an environment devoid of ecologic stress, and thus favorable to a variety of organisms.

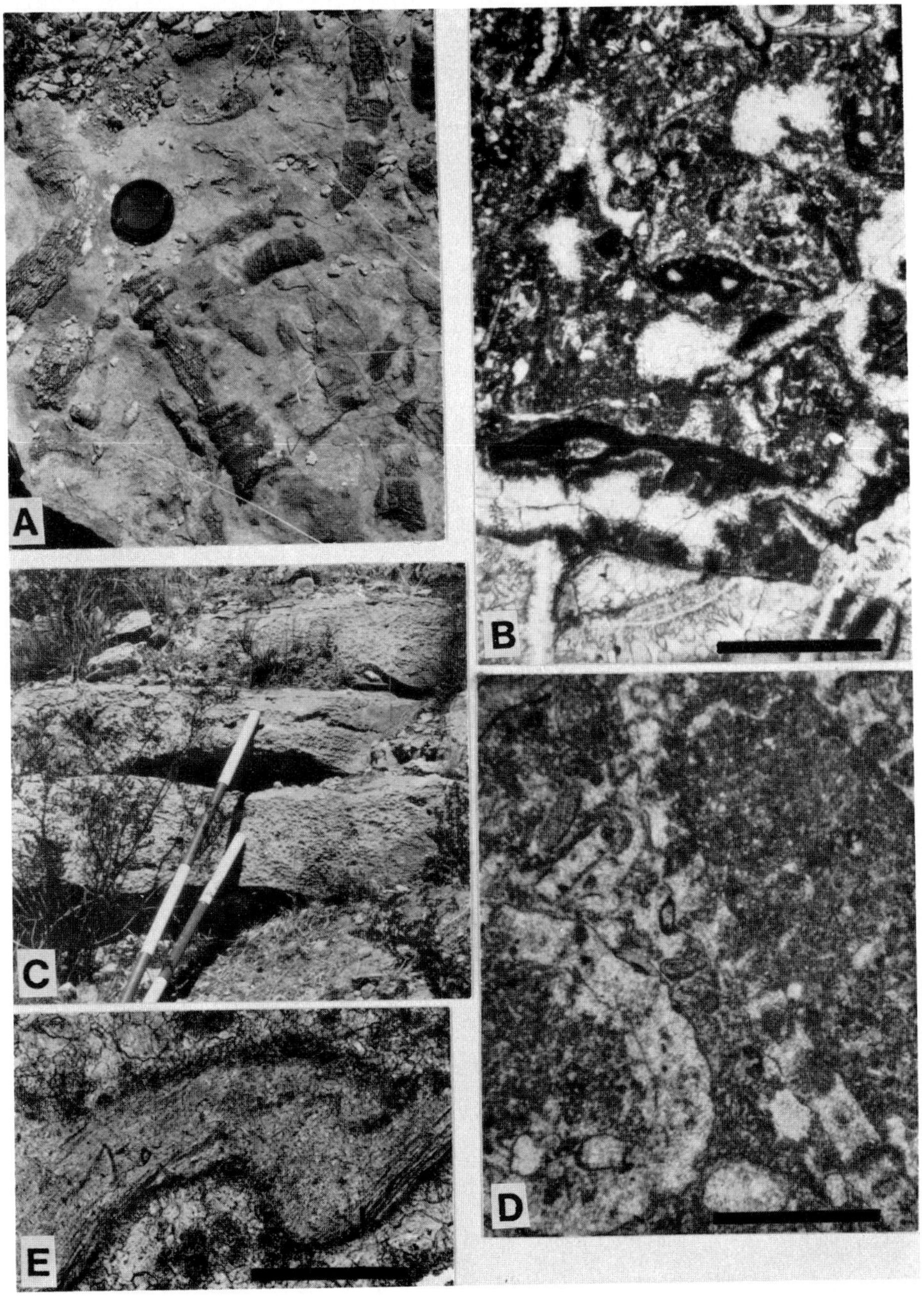

Fig. 9A–E. Characteristic features observed in the basal zones of the middle carbonate unit. **A** Outcrop photograph, rugose coral with large ceratoid coral. Note lens cap for scale, 2 inches (5 cm) diameter. **B** Thin-section photomicrograph of rugose coral bed; rugose coral (*r*) partially silicified pore spaces filled by coarse calcite cement; (sample, RH-5-F-1, plane light, bar = 1 mm). C Outcrop photograph of dasycladacean algal flat, Jacob's staff in feet. **D** Thin-section photomicrograph of dasycladacean algal wackestone; (sample, RH-1-0-5, plane light, bar = 1 mm). **E** Thin-section photomicrograph of brachiopod fragment with micrite coating produced by endolithic microborings; (sample, RH-1-0-2, plane light, bar = 0.5 mm)

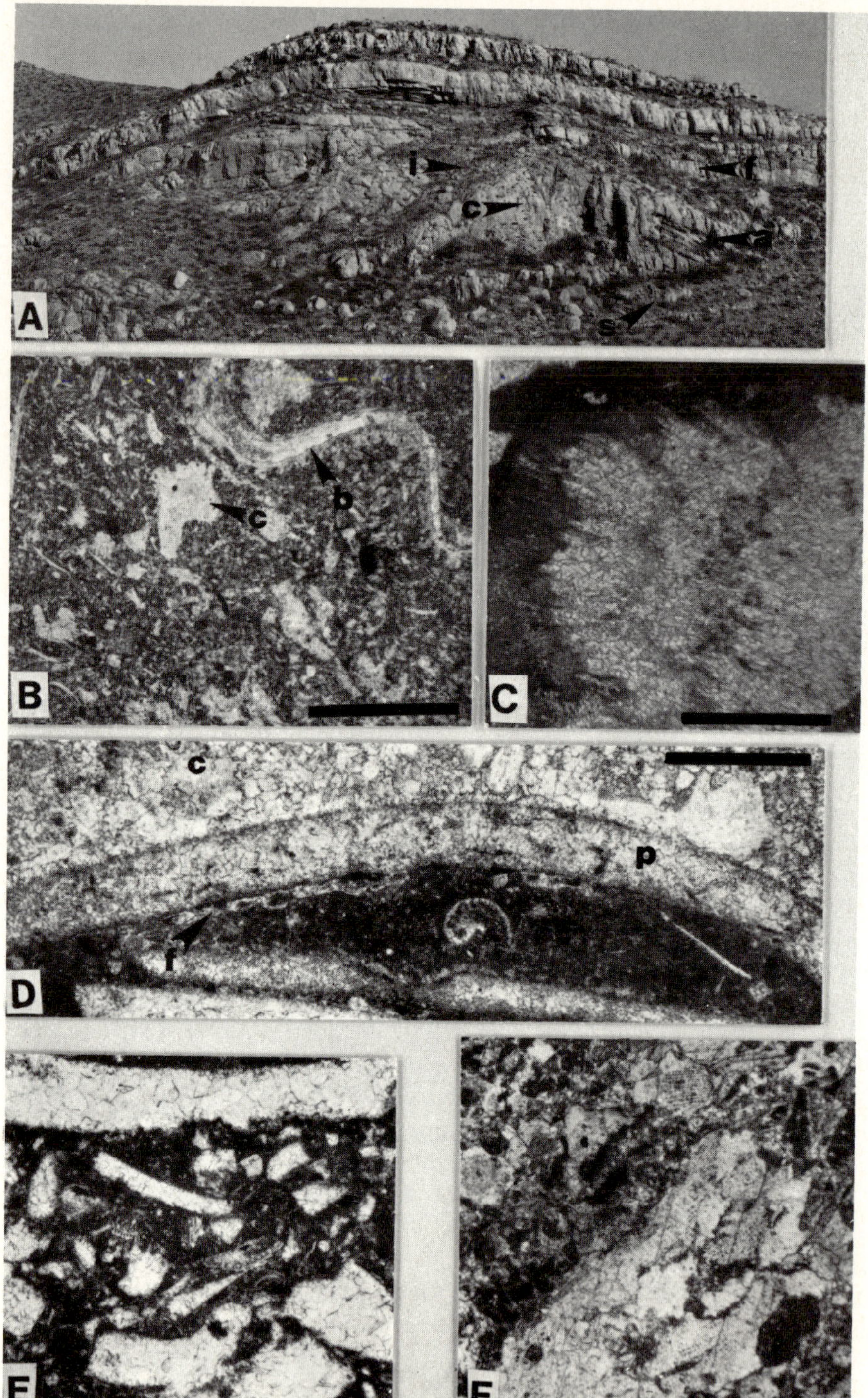

Fig. 10A–F

Initiation of Phylloid Algal Mounds. Directly above the algal flat facies is an interval of "small" mounds measuring less than 10 feet (3 m) thick and less than 25 feet (8 m) wide (Fig. 10A). The "small" mounds grade laterally into a horizontally bedded packstone of the intermound algal flat facies. The rock in the intermound area is compositionally similar to the mounds, although there are important biotic and textural differences. The major difference in the community structure of the intermound facies is a decrease in the abundance of dasycladacean algae and an increase in the amount of phylloid algae. Texturally, the grains from the intermound facies show a greater degree of abrasion and breakage than do those from the small mound facies. The large number of complete, unbroken phylloid algal blades shows that most grew in situ, whereas a smaller number of broken algal blades appear to have been transported from nearby mounds. Intergranular spaces are filled with micrite, but coarse calcite cement is found in the fossil molds.

The major biotic component of this facies is the phylloid Codiacean alga *Eugonophyllum*, while dasycladacean algae (*Epimastopora*) were found to be common to abundant. Other fossils include brachipods, fusulinids (*Triticites*), pelecypods, gastropods, bryozoans, crinoid columnals, and the ubiquitous tubular and encrusting foraminifers (*Calcivertella*, see Fig. 10B).

The "small mound" zone is interpreted as indicating deepening water conditions, and drowning of the lagoonal dasycladacean-rich environment. At this stage the sea floor consisted of extensive algal flats dotted with widely spaced, small phylloid algal mounds. The crests of these mounds were shallow enough to initiate prolific dasycladacean algal growth, but the algal flats appear to have been dominated by dense "meadows" of phylloid algae which preferentially restricted overall biotic diversity (Toomey 1976).

Massive Phylloid Algal Mounds. Compositionally, the massive phylloid algal mounds are very similar to the "small mounds". The major difference is one of scale. Whereas the "small mounds" are no thicker than 10 feet (3 m), the massive mounds range in thickness from 60 to 85 feet (18–26 m, see Fig. 10A). Another major difference

Fig. 10A–F. Textural relationships of the mound facies of the middle carbonate unit. **A** Oblique aerial photograph of the mound facies, including "small mounds" (*s*), massive mound core (*c*), accretionary bedding (*a*), flank beds (*f*), and intermound areas (*i*). (Photo by D.F. Toomey). **B** Thin-section photomicrograph of the "small mound" zone with abundant brachipods (*b*) and mollusc fragments, in addition to phylloid algal fragments (not shown), crinoids (*c*), and unidentified fossil hash; (sample, RH-1-P-4, plane light, bar = 1 mm). **C** Thin-section photomicrograph of a former aragonite botryoid, identified by the chisel-shaped crystal terminations; (sample, RH-5-L-6, plane light, bar = 1 mm). **D** Thin-section photomicrograph of a phylloid algal blade with sheltered micrite; note encrusting forams on the underside of the blade. Texture is similar to massive mound core, with the exception of the abundant cement (*c*); (sample, RH-1-Q-1b, plane light, bar = 1 mm). **E** Thin-section photomicrograph of a phylloid algal packstone of the flank beds; note the large quantity of transported, broken phylloid algal blades and other fine fossils; (sample, RH-3-T-4, plane light, bar = 1 cm). **F** Thin-section photomicrograph of the intermound area characterized by a crinoid packstone/grainstone; (sample, RH-1-Q-6, plane light, bar = 1 mm) is the overall increase in the abundance of phylloid algae (Fig. 10D) and concurrent decrease in dasycladacean algae, possibly suggesting somewhat deeper water conditions. Conspicuous submarine cementation occurs throughout the mound facies. Syndepositional precipitation of aragonite fans may have produced a structure that was relatively rigid and somewhat resistant to wave and current action (Fig. 10C). Mazzullo and Cys (1977, 1979) describe similar cementation by aragonite in Permian organic buildups from the Guadalupe and Sacramento Mountains of New Mexico.

The massive mounds probably formed as a result of active accretionary growth from the "small mound" stage. The differential sedimentation rates between mounds caused some to grow at faster rates, and these appear to have been able to bury their less active neighbors. Seemingly, the large mounds formed where conditions were most favorable for carbonate production. The thick phylloid algal "meadows" were very productive because of their dense rapid growth (Toomey et al. 1977), as well as the great amount of epibiont growth (Land 1970, Patriquin 1972). The mound crest probably grew in very shallow water, as evidenced by the presence of abundant dasycladacean algae, especially in the intermound areas. Depositional relief of over 40 feet (12 m) would not be unreasonable during the climax phase of mound accretion.

Several sub-environments were defined in this massive mound zone from outcrop and petrographic work (Fig. 3). These include the following: (a) mound core, (b) accretionary beds, (c) intermound area, and (d) flanking beds. Repeated periods of subaerial exposure may have occurred during deposition of the massive mounds facies.

a) Mound Core. The mound core is composed of phylloid algal wackestone and packstone (Fig. 10D). The algal blades are, for the most part, intact and do not show evidence of appreciable breakage or transport. The large amount of micrite may be attributed to the current-baffling effects of the rapid growth rate of broad-leaved (phylloid) algae and the high productivity of the biotic community. Quite common in the core subenvironment are tubular encrusting foraminifers, brachiopods, and minor dasycladacean algae.

b) Accretionary Bedding. The texture of the accretionary beds is similar to the mound core, but is characterized by a decrease in the quantity of intergranular micrite and an increase in calcite pore-filling cement (Fig. 10D). This is interpreted as an indication of higher energy conditions which winnowed out the micrite and other fine particles. In outcrop, accretionary beds occur along the southern edge of a massive mound core (Fig. 10A). As the name implies, these beds record depositional accretion of massive mound rock.

c) Intermound Areas. Intermound areas in the "Reef" Hill mound complex are less well represented than the other facies, but where they occur they are composed of fusulinid (*Triticites*) and crinoidal packstones and grainstones (Fig. 10F). The minor amounts of micrite indicate high energy, possibly due to an increase in current velocity around the obstructing mound. In this respect, the accretionary beds and the intermound areas are related in terms of energy, although biogenetically they are very different.

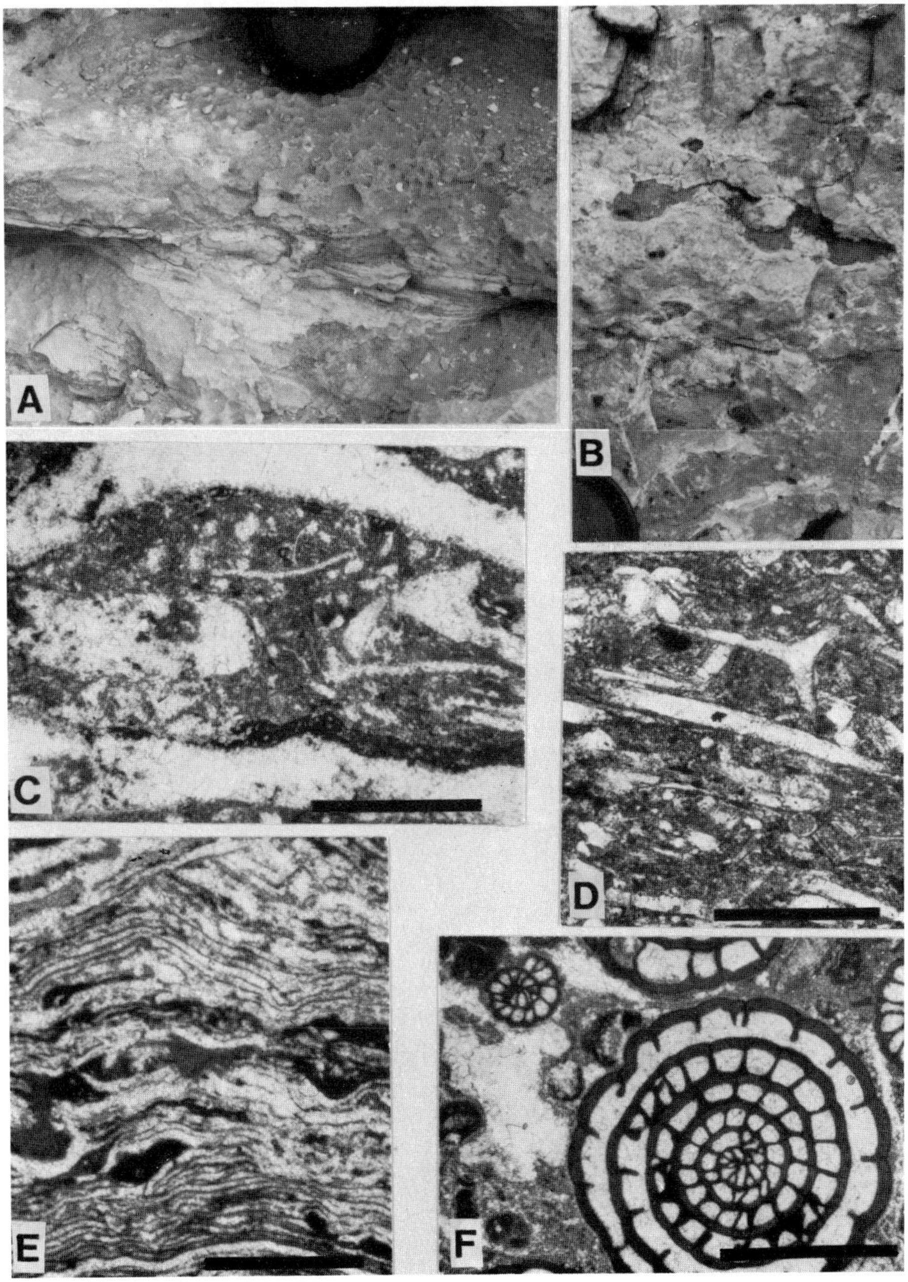

Fig. 11A–F. Petrographic evidence for subaerial exposure of the mound complex between periods of carbonate sedimentation. **A** Solution vug filled with laminated shale [note, lens cap = 2 inches, (5 cm)]. **B** Hematite-stained shale filled vug. **C–F** Thin-section photomicrographs of a vertical sequence on the seaward margin of the mound complex, indicating rapidly changing environmental conditions, interpreted to record a sea level drop (all photos plane light, scale bar = 1 mm). **C** Phylloid algal packstone/wackestone of the mound facies; (sample, RH-6-G-3). **D** Spiculite shaley wedge laterally equivalent to mounds; (sample, RH-6-H-1). **E** Large oncolite from an oncolitic and fossil packstone zone overlying the spiculite; (sample, RH-6-J-2). **F** Fusulinid phylloid algal packstone and wackestone, mound facies; (sample, RH-6-K-1)

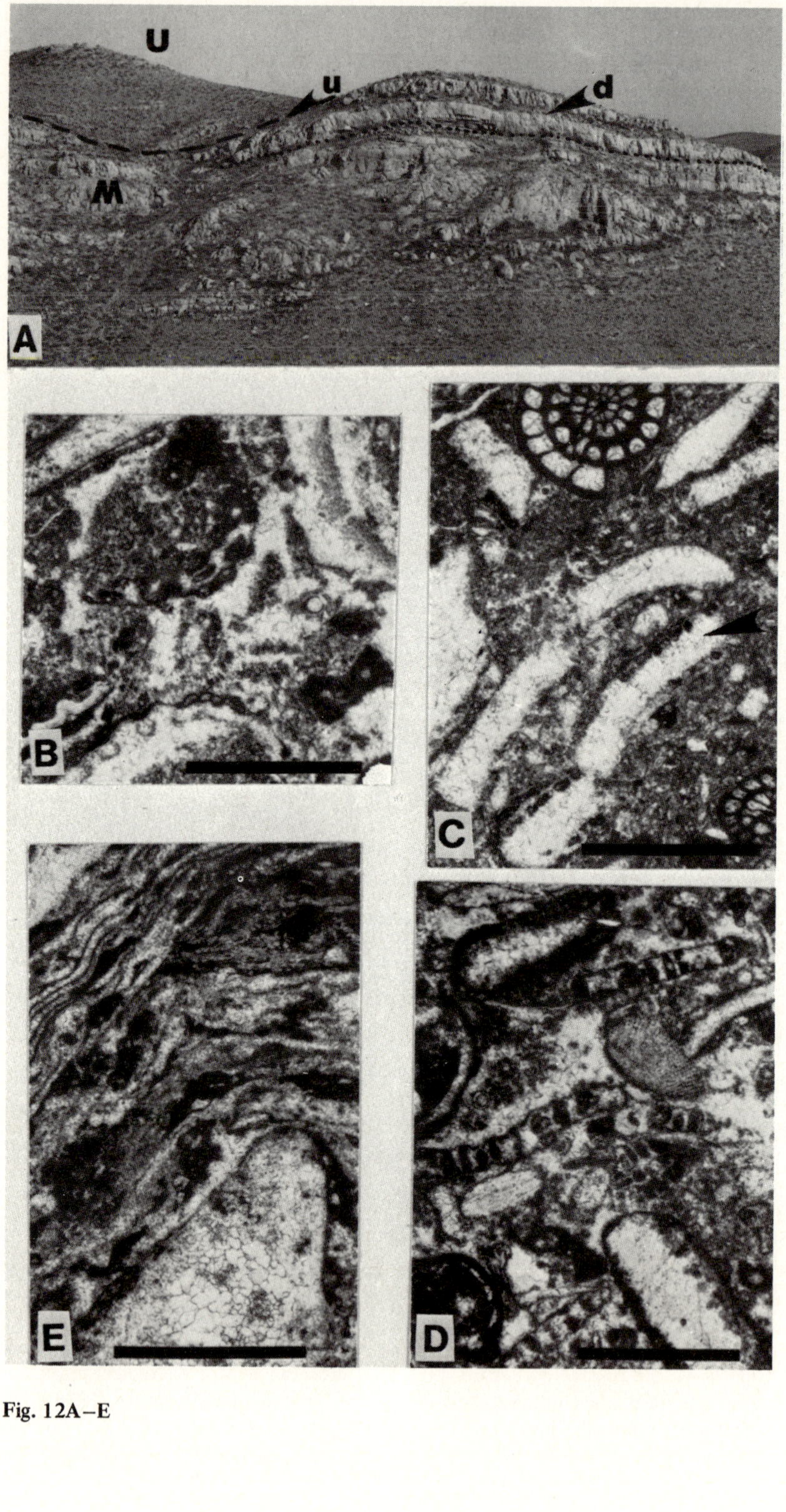

Fig. 12A–E

d) Flank Beds. Flank bed sediments are interpreted to have originated on the crest of the mound complex and then washed off by wave or current energy towards the lee side of the bioherm (Fig. 10A,E). The sediments contain an almost identical faunal and floral community, but the grains are more broken, abraded, and indicative of transport (Fig. 10E). Outcrop geometry indicates that the wave or current energy was towards the southeast (Fig. 3). The deposition of the flank beds filled the low areas between mounds and the lee side of the mound complex. The net effect of this process was a decrease in bathymetric relief of the complex, and increasingly shoaling conditions. Decreasing subsidence rates coupled with high carbonate production are probably the main factors responsible for the shallowing.

Subaerial Exposure. Evidence exists which suggests repeated intervals of subaerial exposure of the mound complex between periods of carbonate sedimentation (Fig. 11). First is the occurrence of large, clay-filled cavernous vugs, some up to nine inches (23 cm) in diameter. Also, hematitic staining is pervasive in and around these solution features. The vugs, however, could have developed during major uplift of the region at the end of Pennsylvanian time, and not necessarily between periods of mound sedimentation.

Further evidence is the occurrence of a cluster of "small mounds" that were deposited after the accretionary beds of one mound but before deposition of the flank beds (Fig. 3). This cluster is interpreted as evidence of renewed carbonate sedimentation during a period of increasing water depth following a minor relative drop in sea level.

On the seaward side of the complex, the sea level drop resulted in deposition of a spiculitic shale several meters thick. The spiculite beds are encapsulated between phylloid algal-rich limestones and are stratigraphically equivalent to the mound and drape-bed facies of the crest area. The vertical facies change are depicted by the series of photomicrographs in Fig. 11.

The interpreted sequence includes the following events: (a) initial relative rise in sea level and termination of clastic deposition in the lower clastic unit, (b) carbonate deposition uninterrupted from the colonization stage through deposition of the massive mound cores, (c) one or more periods of subaerial exposure, and (d) renewed carbonate sedimentation on the crest of the submerged mound complex and deposition of the flank beds.

Fig. 12A–E. Drape bed sediments exposed on "Reef" Hill. **A** Oblique aerial photograph, drape beds (*d*) overlie the mound complex; upper surface marks unconformable boundary (*u*) between middle carbonate unit (*M*) and the upper clastic unit (*U*). **B** Dasycladacean algal packstone; (sample, RH-1-R-6, plane light, bar = 1 mm). **C** Phylloid algal-fusulinid packstone; encrusting forams and brachiopod spines are abundant; phylloid algae (*Eugonophyllum*) identified by the preserved utricles; (sample, RH-1-R-4, plane light, bar = 1 mm). **D** Phylloid algal-fusulinid packstone with abundant dasyclad algal fragments and crinoids; (sample, RH-1-R-5, plane light, bar = 1 mm). **E** Large oncolite with algae, and encrusting foraminifer coating a large, unidentified grain; (sample, RH-4-J-2, plane light, bar = 1 mm)

Drape Bed Deposition. The last sequence present in the Pennslyvanian carbonate beds of the "Reef" Hill outcrop is a final, shallowing-upward cycle. The drape bed zone is characterized by thin- to medium-bedded limestones grading upward to a massive limestone (Fig. 12A). Very thin shale streaks form the distinct bed boundaries. The zone is composed of alternating beds of fusulinid packstone, dasycladacean algal-fusulinid packstone, and phylloid algal wackestone/packstone (Fig. 12B–D, respectively). Once again, tubular and encrusting foraminifers are common to abundant.

The shallow water environment was established after the deposition of the flank beds reduced the overall bathymetric relief of the area. The broad, gentle crest of the mound complex allowed re-establishment of dasyclad algal flats as well as fusulinid shoals.

Oncolites are common to abundant in massive beds near the upper part of the drape bed zone (Fig. 12E). The oncolites consist of blue-green algae coating almost any type of grain available. Broken phylloid algal blades, however, seem to make up the bulk of oncolitic nuclei. The oncolites are interpreted to have formed in a quiet, moderately agitated lagoonal environment. Near the top of the zone the quantity of fossils begins to diminish, until only barren intraclast packstone and pellet wackestone exist in the uppermost portion of the outcrop. Brecciated, black limestone with hematitic staining on the uppermost limestone is thought to indicate the culminating erosional event at the end of the Pennsylvanian Period.

This zone is interpreted as a shallowing upward sequence. Depositional environments grade upward from a shallow subtidal (lagoonal) to intertidal. Shallowing may have resulted from depositional filling due to high rate of carbonate production, or tectonic uplift of the area, or, most probably, from a combination of these two factors. Uplift continued until subaerial exposure and erosion occurred at the end of Late Pennsylvanian (Virgil) time. Renewed clastic deposition occurs above and has produced a conspicuous angular unconformity.

Discussion

Late Pennsylvanian carbonates of the southwestern United States have been extensively studied by many workers. A number of sedimentological and diagenetic studies indicate many similar characteristics in the morphology, depositional environment, growth history, and diagenetic history of phylloid algal mound complexes.

Wilson (1977) found that phylloid algal mounds tend to develop along shelf margins. Within this paleodepositional setting a near-parallelism develops between the trend of the shelf edge and the trend of the mound facies tract. The mounds, although usually isolated, tend to develop a "loaf-shaped" form with the longest axis in a near-parallel orientation with respect to the shoreline or shelf break.

An ideal depositional sequence includes the following microfacies (Wilson 1975, Toomey et al. 1977): (1) "basal bioclastic micrite debris pile", (2) "algal plate micritic core", (3) "crestal boundstone facies", (4) "veneer facies" (developed only in mounds from a protected environmental setting), (5) "flank beds", (6) "marine talus or breccia on the seaward flank of the mound", and (7) "shoal grainstone" capping bed (see also Fig. VI-25 in Wilson 1975).

The depositional facies of the "Reef" Hill carbonate sequence have characteristics in common with the ideal mound facies scheme of Wilson (1975). The "basal bioclastic debris pile" is equivalent to the colonizing rugose coral bed in "Reef" Hill. The "algal plate core" facies and "flank bed" facies are the same as those previously described in this paper. Finally, the "shoal grainstone" facies of Wilson (1975) are the same as the drape beds described above. In the "Reef" Hill facies there is no evidence for th existence of a "crestal boundstone" facies. This may be related to interpreted periods of subaerial exposure, in which a sharp drop in relative sea level may have occurred before the mound crest approached wave base. Subsequent inundation allowed the re-establishment of the phylloid algal community, bypassing the boundstone phase.

The absence of a veneer facies is probably a result of higher energy conditions. According to Wilson (1975) veneer facies form in mounds from a lagoonal or protected environment. The absence of the seaward talus deposit indicates that the slopes on the Late Pennsylvanian sea floor in this area were not steep enough to cause slumping of biohermal sediments.

Summary

A Late Pennsylvanian mixed carbonate-clastic sequence is well exposed on "Reef" Hill in the Hueco Mountains, El Paso County, West Texas. The sequence begins with deposition of a progradational fan-delta system and marine and tidal flat carbonates. This unit is dominated by calclithite and shale, with minor interbeds of shallow-water carbonates (calcareous mudstones and wackestones). Shallow-water spiculites are commonly associated with these limestones. A thick carbonate unit overlies the clastics and is composed principally of limestone thought to be indicative of sedimentation during or just after a major local transgression. The carbonate sediments were deposited on a submerged delta platform with the following sequence of events: (1) colonization of the shallow platform by rugose corals, (2) establishment of shallow-water dasycladacean algal flats, (3) increasing domination of the environment by phylloid algae in response to increasing water depth conditions, and (4) accretion of phylloid algal sediments and formation of mounds. Directly overlying the dasyclad algal flats are a number of "small" mounds formed by accelerated sedimentation rates within the phylloid algal "meadows". High productivity coupled with rapid growth rates of the phylloid algae, in conjunction with their sediment-trapping ability allowed sedimentation to keep up with sea-level rise. Large bioherms resulted, but because of the difference in accretion rates of various mounds, some grew more rapidly than their neighbors. One or more periods of subaerial exposure are thought to have occurred between stages of mound facies deposition. Finally, (5) re-establishment of shallow-water dasyclad algal flats occur as a result of shoaling of mound crests and sedimentary filling of the leeward side of the mound complex (flank beds). Tectonic uplift of the Diablo Platform halted all carbonate sedimentation and produced a major Lower Permian angular unconformity in the Hueco Mountains.

Acknowledgements. Dr. Donald F. Toomey, Conoco Inc., suggested this topic for study, and served as field advisor and as a member of the thesis committee. Dr. Robert Folk of the University of Texas, Austin, served as thesis advisor along with Drs. L.S. Land and A.J. Scott, and all willingly shared their time and experience. This study was in part funded by Cities Service Company through the auspices of Don Toomey, and the necessary thin sections were prepared at their research facility in Tulsa by Bill Batten.

References

Bathurst RGC (1967) Subtidal gelatinous material, sand stabilizer and food, Great Bahama Bank. J Geol 75:736–738

Creaser PH (1977) Lithogenesis and diagnostic features of recent and ancient terrigenous limestones (calclithites). Unpubl M thesis, Aust Nat Univ, Canberra, 392 pp

Dunham RJ (1962) Classification of carbonate rocks according to depositional texture. In: Ham WE (ed) Classification of carbonate rocks. Am Assoc Pet Geol Mem 1:108–121

Flügel E (1977) Environmental models for Upper Paleozoic benthic calcareous algal communities. In: Flügel E (ed) Fossil algae: Recent results and developments. Springer, Berlin Heidelberg New York, pp 314–343

Folk RL (1959) Practical petrographic classification of limestones. Am Assoc Pet Geol Bull 43(1):1–38

Folk RL (1962) Spectral subdivision of limestone types. In: Ham WE (ed) Classification of carbonate rocks. Am Assoc Pet Geol Mem 1:62–84

Folk RL (1974) Petrology of sedimentary rocks. Hemphill, Austin, 82 pp

Ginsburg RN, Rezak R, Wray JL (1971) Geology of calcareous algae (short course notes). Sedimenta I. Comp Sediment Lab, Univ Miami, FL, 64 p

King PB, King RE, Knight JB (1945) Geology of the Hueco Mountains, El Paso and Hudspeth Counties, TX, US Geol Surv Oil Gas Inv Prelim Map No 36, 2 sheets

Land LS (1970) Carbonate mud, production by epibiont growth on *Thalassia testudinum.* J Sediment Petrol 40:1361–1363

Mazzullo SJ, Cys JM (1977) Submarine cements in the Permian boundstones and reef associated rocks, Guadalupe Mountains, west Texas and southeastern New Mexico. In: Hilman MH, Mazzullo SJ (eds) Upper Guadalupian Facies, Permian Reef Facies, Guadalupe Mountains, New Mexico and West Texas. Soc Econ Paleontol Min, Permian Basin Sec Spec Publ 77–16:151–200

Mazzullo SJ, Cys JM (1979) Marine aragonite sea-floor growth and cements in Permian phylloid algal mounds, Sacramento Mountains, New Mexico. J Sediment Petrol 49(3):917–936

McKee ED, Crosby EJ (1975) Paleotectonic investigations of the Pennsylvanian system in the United States. US Geol Surv Prof Pap 853

Nelson LA, Haig BR (1958) Franklin and Hueco Mountains, TX. West Texas Geol Soc Guideb Nov 6, 7 and 8, 1958 Field Trip:18–27

Neumann AC, Gebelein CD, Scoffin TP (1970) The composition, structure, and erodability of subtidal mats, Abaco, Bahamas. J Sediment Petrol 40:274–297

Patriquin DG (1972) Carbonate mud production by epibionts on *Thalassia* : an estimate based on leaf growth data. J Sediment Petrol 42:687–689

Pol JC (1982) Sedimentation and diagenesis of an Upper Pennsylvanian (Virgilian) mixed carbonate-clastic sequence, Hueco Mountains, El Paso County, TX. Unpubl M thesis, Univ Tex, Austin, 212 pp

Pol JC (1983) Sedimentation of an Upper Pennsylvanian (Virgilian) phylloid algal mound complex, Hueco Mountains, El Paso County, TX. Program Abstr 3rd Int Symp Fossil Algae, Golden, CO, August 1983, p 23

Pray LC (1961) Geology of the Sacramento Mountains Escarpment, Otero County, NM. New Mex Bur Mines Min Res Bull 35:144

Seewald KO (1968) Pennsylvanian and Lower Permian stratigraphy, Hueco Mountains, Texas. West Tx Geol Soc Guideb Publ 68–55, Delaware Basin Explor, 45–49

Smith AG (1981) Phanerozoic equal-area maps. Geol Rundsch 70(1):91–127

Stevens CH (1965) Faunal trends in near-shore Pennsylvanian deposits near McCoy, Colorado. Mount Geol 2(2):71–77

Tasch P (1980) Paleobiology of the invertebrates: Data retrieval from the fossil record. Wiley, New York, 975 pp

Toomey DF (1976) Paleosynecology of a Permian plant dominated marine community. Neues Jahrb Geol Palaeontol Abh 152:1–18

Toomey DF, Babcock JA (1983) Precambrian and Paleozoic algal carbonates, West Texas-southern New Mexico. Colo Sch Mines Prof Contrib 11:345

Toomey DF, Cys JM (1979) Community succession in small bioherms of algae and sponges in the Lower Permian of New Mexico. Lethaia 12:65–74

Toomey DF, Wilson JL, Rezak R (1977) Evolution of Yucca Mound Complex, Late Pennsylvanian phylloid-algal buildup, Sacramento Mountains, New Mexico. Am Assoc Pet Geol Bull 61(12):2115–2133

Waters DL, Holland FD (1983) Corals of Madison Group (Mississippian), Williston Basin, ND (abs). Am Assoc Pet Geol Bull 67(8):3160

Wilson JL (1967) Cyclic and reciprocal sedimentation in Virgilian strata of southern New Mexico. Geol Soc Am Bull 78:805–818

Wilson JL (1972) Influence of local structure in sedimentary cycles of Beeman and Holder Formations, Sacramento Mountains, Otero County, New Mexico. In: Elam JG, Chuber S (eds) Cyclic sedimentation in Permian Basin, 2nd edn. West Tex Geol Soc Publ 72–60:100–143

Wilson JL (1975) Carbonate facies in geologic history. Springer, Berlin Heidelberg New York, 471 pp

Wilson JL (1977) Regional distribution of phylloid algal mounds in Late Pennsylvanian and Wolfcamp strata in southern New Mexico. In: Butler JH (ed) Geology of the Sacramento Mountains, Otero County, New Mexico. West Tex Geol Soc Publ 1967–1968:1–7

Wray JL (1968) Late Paleozoic algal limestones in the United States. Proc 23rd Int Geol Congr, Prague, 1968, 8:113–119

Wray JL (1977) Calcareous algae: Developments in paleontology and stratigraphy, no 4. Elsevier, Amsterdam, 186 pp

15 Lower Permian (Wolfcampian) *Archaeolithoporella-Tubiphytes*-Sponge Boundstones from the Subsurface of West Texas

G.P. WAHLMAN[1]

Introduction

Following the extinction of the Siluro-Devonian coral-stromatoporoid reef community during the Late Devonian, there was slow evolution of a new reef community which reached its acme in the Late Paleozoic. Here, the dominant type of organic buildup in the Pennsylvanian and Early Permian (Wolfcampian) of the southwestern United States were phylloid algal mounds. These mounds consist mainly of unbound bafflestones composed of delicate phylloid algal blades and carbonate mud (Wray 1968), and locally, *Archaeolithophyllum*-bound sediments (Wray 1964), and capping *Tubiphytes* boundstones (Malek-Aslani 1970), or a consortium of phylloid algae, encrusting foraminifers and bluegreen algae (Toomey et al. 1977) (Fig. 1).

However, Schatzinger (1983) recently described an early Late Pennsylvanian (Missourian, Canyon Group) boundstone community from the subsurface of West Texas, composed of calcareous sponges, fenestrate and encrusting bryozoans, encrust-

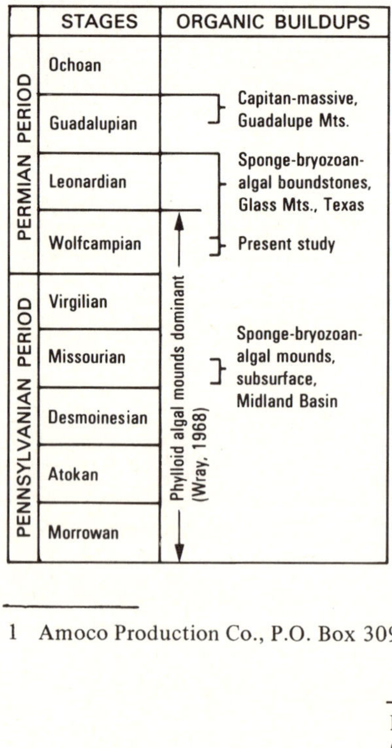

Fig. 1. Time-stratigraphic chart and types of organic carbonate buildups in the Pennsylvanian and Permian Systems of the southwestern United States

1 Amoco Production Co., P.O. Box 3092, Houston, TX 77253, U.S.A.

Paleoalgology: Contemporary Research and Applications
(eds. D.F. Toomey and M.H. Nitecki)

ing red algae, *Tubiphytes*, laminar blue-green algae, and encrusting foraminifers. This boundstone community constructed mounds up to 27 m (89 ft) thick along the margins of the Horseshoe Platform of the Midland Basin. Previously, boundstones of this type were known primarily from the late Wolfcampian and Leonardian-aged strata of the Glass Mountains, West Texas (Bain 1967; Fig. 1).

The best-known example of Late Paleozoic reef community development is the Capitan reef of West Texas and New Mexico. Babcock (1977) and Yurewicz (1977) described the boundstones of the Capitan-massive as consisting predominantly of calcareous sponges, bryozoans, and phylloid algae, encrusted and bound by the problematical organisms *Tubiphytes* (hydrozoan?) and *Archaeolithoporella* (red alga?), and by primary marine radial fibrous cement (Fig. 1).

The middle Wolfcampian-aged boundstones described in this paper contain biotic elements and community relationships that seem to represent an intermediate stratigraphic position between Pennsylvanian (Missourian) boundstones of Schatzinger (1983), and the Permian (Guadalupian) boundstones of Babcock (1977) and Yurewicz (1977). Comparisons can also be made to the slightly younger (late Wolfcampian – Leonardian equivalent) Trogkofel buildups in the southern Alps of Europe (Flügel 1981). The boundstones described here are apparently the earliest known that contain both *Archaeolithoporella* and radial fibrous cement as binding agents.

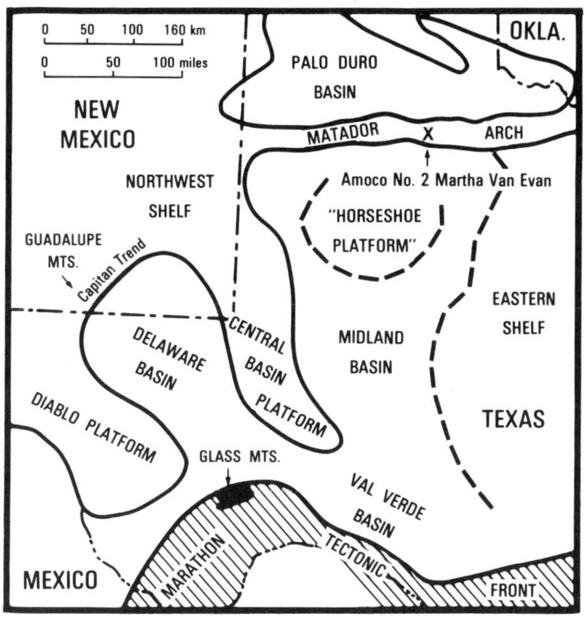

Fig. 2. Index map of Permian Basin of West Texas and New Mexico showing paleogeographic-structural features and location of the Amoco No. 2 Martha Van Eman drill core from Motley Co., Texas. (After Ross 1965; Fig. 1)

Geologic Setting and Biostratigraphy

The material is from a core (Amoco Production Co., No. 2 Martha Van Eman well) in the Roaring Springs Field, Motley County, Texas. This core was taken just south of the crest of the Matador Arch, which formed the northern edge of the Midland Basin during Permian time (Fig. 2).

The core was dated as middle Wolfcampian in age based on the occurrence and identification of sparse fusulinids. The fusulinids *Paraschwagerina kansasensis* occurs near the top of the core at 1–3 ft interval, and *Leptotriticites* cf. *L. tumida* was identified at 41 ft interval.

Intermixed Lithologies

The boundstones in the core are intermixed with skeletal wackestones and peloidal packstones. The skeletal wackestones are composed mainly of crinoidal, gastropodal, and spiculitic grains in dark gray carbonate mudstone matrix.

The peloidal packstones are commonly distinctly layered. Layered peloidal sediment in the form of stacked hemispheroids are thought to be attributable to the activities of stromatolitic blue-green algae (Fig. 3). Other layered sediment appears to be the result of gravitional sedimentation within the interstices of the framestones and bafflestones.

Boundstones and Zonation

The core can be divided into six lithologic zones, five of which contain boundstones.

Zone I (1–3 ft) consists of encrusting foraminiferal-skeletal packstones and boundstones (Plate 1, Fig. 1) that coarsen upward into a lithoclastic packstone. Lithoclasts vary from green shaly mudstone to bafflestone containing well preserved erect blades of *Archaeolithophyllum.*

Zone II (3–8 ft) consists of boundstones having a framework of phylloid algae and calcareous sponges bound by radial fibrous cements, *Tubiphytes,* and laminar red algae (Plate 1, Fig. 2).

Zone III (8–18 ft) is very rich in radial fibrous cement, which forms the groundmass of the boundstone (Plate 1, Fig. 4). Calcareous sponges are the main framebuilders, and these are encrusted by *Tubiphytes* and *Archaeolithoporella,* and sometimes by stromatolitic algal masses (Plate 1, Figs. 3–5; Plate 2, Fig. 6).

Zone IV (18–23 ft) consists mainly of medium-gray mudstones and skeletal wackestones. Near the top is a packstone concentration of problematic small concentric cement-rich spheres, which commonly have irregular micritized outer surfaces (Plate 1, Fig. 6). Such spheres are sparsely scattered throughout the core.

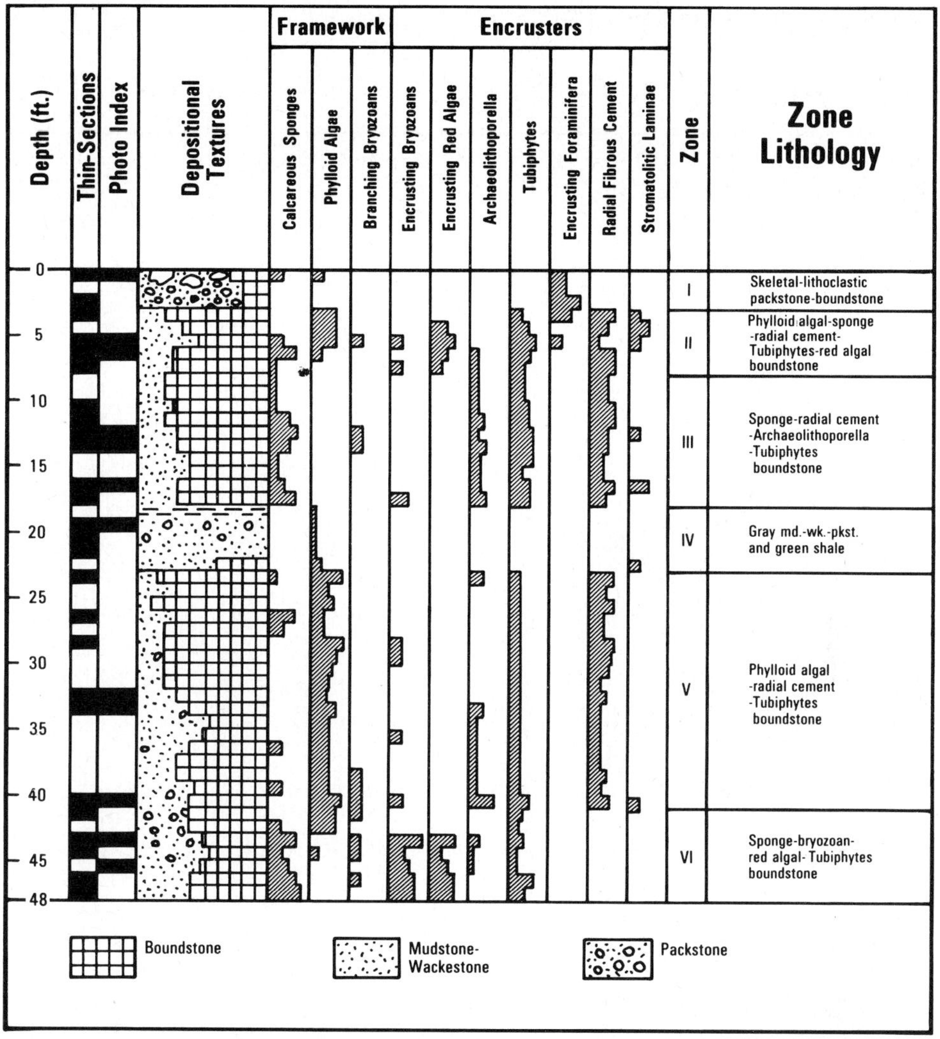

Fig. 3. Chart showing relative distributions of depositional textures and boundstone constituents, and delineation of lithologic zonation. Relative proportions are subjective estimates

The top of the zone is a thin green shaly layer capped by a thin cherty conglomerate band. This discontinuity possibly reflects a diastemic contact.

Zone V (23–41 ft) is characterized by an abundance of phylloid algae, which commonly occur as curved blades having a common orientation throughout the zone (Plate 2, Fig. 1). These curved blades are thought to be codiacean remains, as some have remnants of utricles preserved. One thin subzone contains prostrate overlapping algal blades that have the macroscopic appearance of a twisted length of rope (Plate 2, Fig. 3). These phylloid algal blades sometimes preserve internal cells and are probably red algae. Cavities are common beneath phylloid blades (Plate 2, Fig. 1–3). These cavities may be filled with mosaic or radial fibrous cements, and are

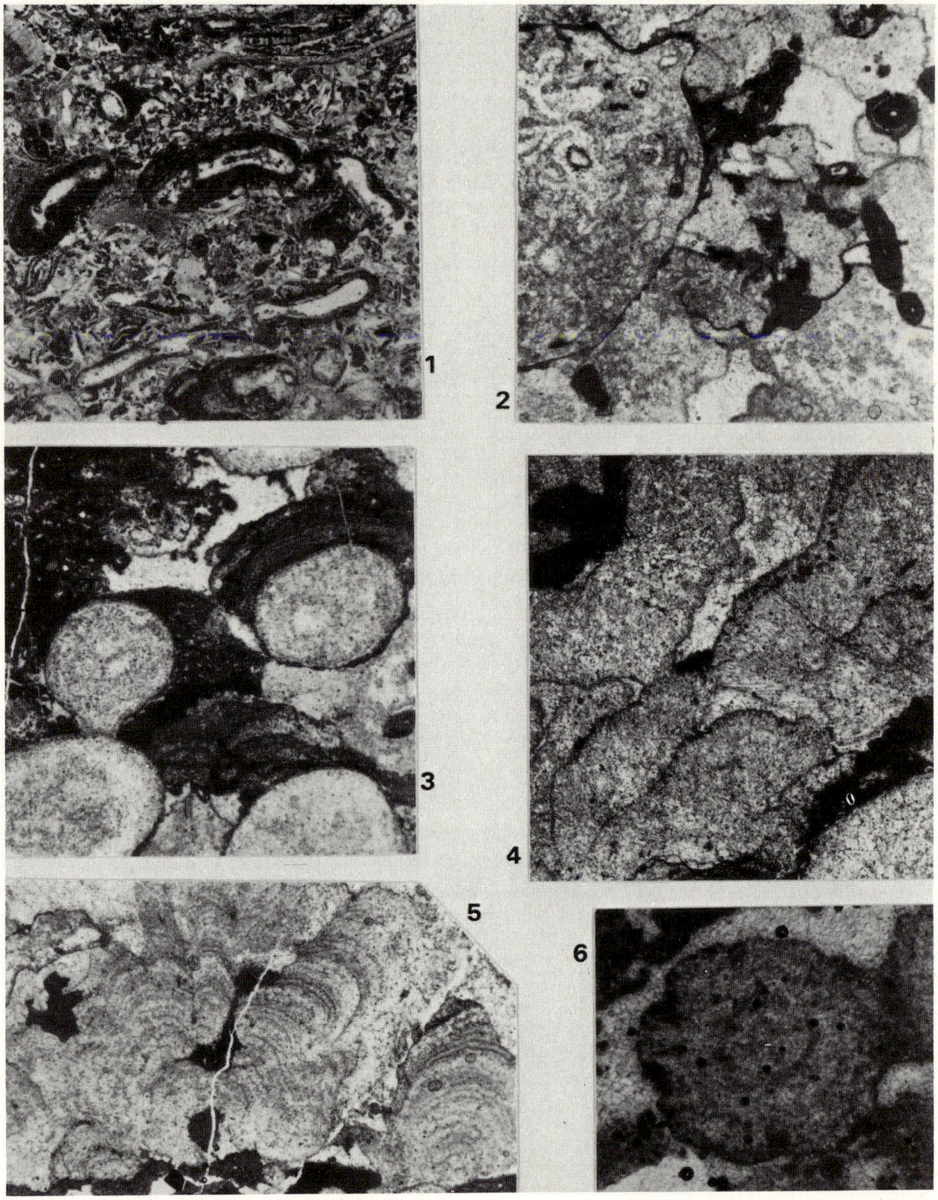

Plate 1. Thin-section photomicrograph of:

Fig. 1. Encrusting foraminifera-skeletal packstone to boundstone locally; (Zone I, 1 ft). (X 4.2)

Fig. 2. Sponge – *Tubiphytes* – radial fibrous cement boundstone; note partial recrystallization of sponge and cavities within radial cements; (Zone II, 6 ft). (X 4.2)

Fig. 3. Recrystallized sponges and radial fibrous cement with perched stromatolitic sediment; dark interstitial wackestone in upper left; (Zone III, 12 ft). (X 4.2)

Fig. 4. Radial fibrous cement groundmass; note cavity; (Zone III, 12 ft). (X 4.2)

Fig. 5. Apparent stromatolitic mass in radial fibrous cement-rich rock; (Zone III, 12 ft). (X 4.2)

Fig. 6. Small concentrically layered cement-rich spheres of uncertain origin, which are scattered throughout the core, but are concentrated at the top of Zone IV, as here at 19 ft. (X 17.5)

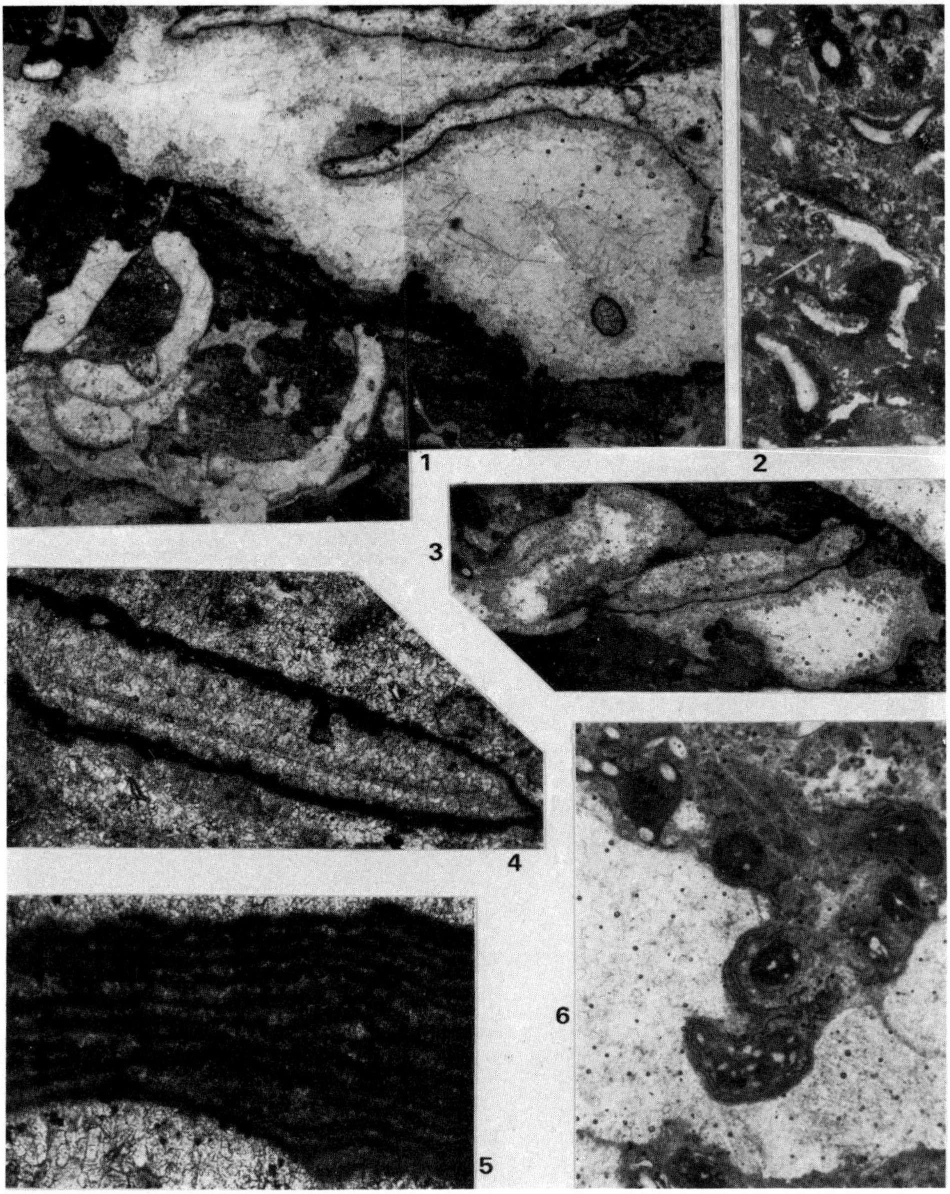

Plate 2. Thin-section photomicrographs of:

Fig. 1. Large shelter cavity in phylloid algal bafflestone; note that the hummocky layered peloidal sediment flooring cavity appears crustose; radial fibrous cement to lower left. Phylloid algal blades are commonly curved and oriented as shown throughout Zone V, (40 ft). (× 4.2)

Fig. 2. Small agglutinated tubes of uncertain origin which occur in abundance locally; (Zone II, 5 ft). (× 4.2)

Fig. 3. Irregularly-shaped phylloid algal blades that occur in horizontal overlapping "chains"; note series of cells in center blade, suggesting a red alga. Chains of phylloid algal blades are often underlain by cement-filled cavities, (Zone V, 33 ft). (× 4.2)

Fig. 4. Phylloid algal blade encrusted by *Archaeolithoporella* in recrystallized radial cement-rich rock, (Zone V, 32 ft). (× 17.5)

Fig. 5. Layers of *Archaeolithoporella* in recrystallized radial cement-rich rock, (Zone VI, 45 ft). (× 70)

Fig. 6. *Tubiphytes* boundstone encrusted by *Archaeolithoporella* and surrounded by radial fibrous cement, (Zone III, 13 ft). (× 10.5)

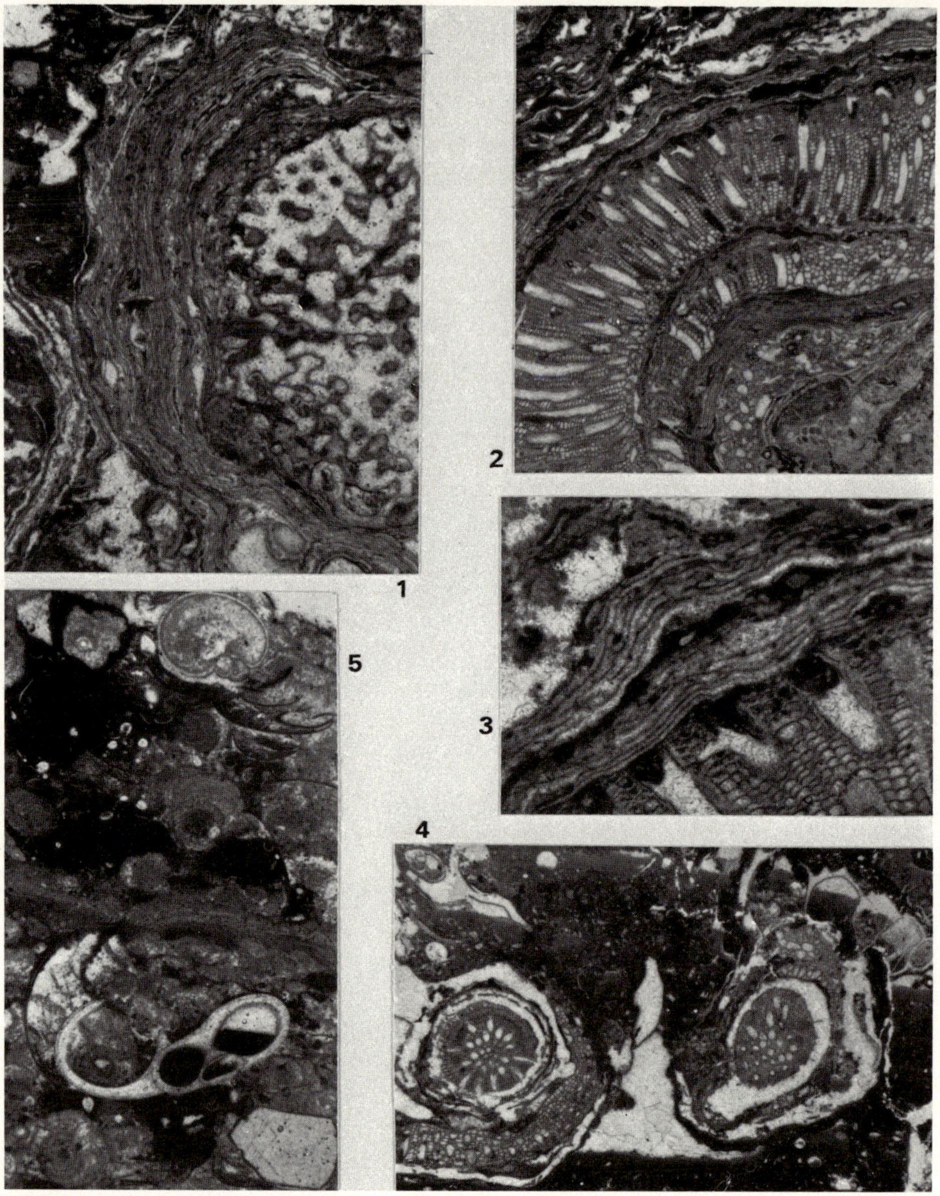

Plate 3. Thin-section photomicrographs of:

Fig. 1. Laminar encrusting red algae binding sponges; note dark interstitial wackestone, (Zone VI, 43 ft). (× 4.2)

Fig. 2. Interlayered encrustations of laminar red algae and fistuliporid bryozoans over calcareous sponge, (Zone VI, 43 ft). (× 4.2)

Fig. 3. Close-up of laminar red alga from Fig. 2 (Zone VI, 43 ft). (× 14)

Fig. 4. Branching bryozoans encrusted by laminar red algae and fistuliporid bryozoans; note geopedal fabric in gastropods at *upper right* and *left,* and the dark wackestone matrix, (Zone VI, 43 ft). (× 4.2)

Fig. 5. Intermixing of radial fibrous cement clouded by fine sediment (*bottom third* and *upper right*), layered peloidal sediment (*across center*), and fine dark skeletal wackestone (*upper left*); note geopedal fabric in gastropod, and small problematical cement-rich spheres (*lower left* and *upper center*), (Zone III, 16 ft). (× 4.2) floored by layered peloidal sediment that often gives the impression of having formed cemented or bound crusts. This framework is bound by radial fibrous cement, *Tubiphytes,* and *Archaeolithoporella.*

Zone VI (41–48 ft) contains large irregular masses and rhodoliths consisting of calcareous sponges and some branching bryozoans encrusted by alternating layers of laminar red algae, encrusting bryozoans, and *Tubiphytes* (Plate 3, Figs. 1–4). Articulated crinoidal columnals are particularly common in this zone, and crinoid roots may have played a minor role in the construction of this lower boundstone.

Summary

The biotic constituents and community relationships of the lower Permian (middle Wolfcampian) boundstones described here suggest that they represent an intermediate position in Late Paleozoic reef-community evolution between Late Pennsylvanian (Missourian) boundstones of the Canyon Group, and the Permian (Guadalupian) boundstones of the Capitan Limestone. It is thought that these Wolfcampian boundstones may have formed below wave-base within an upper slope setting.

Acknowledgements. Thanks are given to Jack Babcock for his advice and encouragement. The writer is grateful to Amoco Production Co. for permission to publish this study, and to Mr. Henry Cass for preparation of the thin-sections.

References

Babcock JA (1977) Calcareous algae, organic boundstones, and the genesis of the upper Capitan Limestone (Permian, Guadalupian), Guadalupe Mts., West Texas and New Mexico. In: Hileman M, Mazzullo SJ (eds) SEPM Permian Basin Sect Publ 77-16, pp 3–44

Bain RJ (1967) Paleoecology of some Leonardian patch reefs in the Glass Mts., Texas. Brigham Young Univ Stud Geol 14:195–236

Flügel E (1981) Lower Permian *Tubiphytes/Archaeolithoporella* buildups in the southern Alps (Austria and Italy). In: Toomey DF (ed) SEPM Spec Publ No 30, Eur Fossil Reef Models, pp 143–160

Malek-Aslani M (1970) Lower Wolfcampian reef in Kemnitz Field, Lea County, New Mexico. Bull Am Assoc Pet Geol 54:2317–2335

Schatzinger RA (1983) Phylloid algal and sponge-bryozoan mound-to-basin transition: a Late Paleozoic facies tract from the Kelly-Snyder Field, West Texas. In: Harris PM (ed) SEPM Core Workship No 4, pp 244–304

Toomey DF, Wilson JL, Rezak R (1977) Evolution of Yucca mound complex, Late Pennsylvanian phylloid algal buildup, Sacramento Mountains, New Mexico. Bull Am Assoc Pet Geol 61:2115–2133

Wray JL (1964) *Archaeolithophyllum,* an abundant calcareous alga in the limestones of the Lansing Group (Pennsylvanian), southeastern Kansas. Kans Geol Surv Bull 170:pt 1, 1–13

Wray JL (1968) Late Paleozoic phylloid algal limestones in the United States. Proc 23rd Int Geol Congr Prague 8:113–119

Yurewicz DA (1977) The origin of the massive facies of the lower and middle Capitan Limestone (Permian), Guadalupe Mts., New Mexico and west Texas. In: Hileman M, Mazzullo SJ (eds) SEPM Permian Basin Sect Publ 77-16, pp 43–93

16 The "Coralligène" of the Mediterranean – a Recent Analog for Tertiary Coralline Algal Limestones

D.W.J. BOSENCE[1]

Introduction

During the 1950's and 1960's, considerable work was undertaken by French biologists and geologists on the nature of Recent coralline algal buildups ("coralligène") of the Mediterranean. Much of this work is summarized in Pérès (1967), Blanc (1968), and Laborel (1961). The best known example of the "coralligène" is the intertidal "trottoir" or "corniche" constructed principally of *Lithophyllum tortuosum* on steep rocky coasts (Blanc and Molinier 1955, Laborel 1961). Similarly, large coralline buildups occur on sublittoral rocky slopes (Laborel 1961). The coralligène de plateau, the subject of this paper, occurs in deeper shelf waters and passes laterally into carbonate sands and gravels or terrigenous sands and muds (Laborel 1961). This coralligène is found widely distributed in the Mediterranean between depths of 20 to 160 m. The shallow occurrences of 30–50 m at Marseille and of 20–40 m at Banyuls (see below) are in more turbid coastal waters. Examples at 100–160 m are found in the southern Mediterranean (Pérès 1967). The corallines involved in construction of the coralligène are *Mesophyllum lichenoides, Pseudolithophyllum expansum, Neogoniolithon mamillosum* and *Lithothamnium philippii.* Three modes of growth have been recognized in these corallines by Laborel (1961): (1) foliaceous (feulléte), bifurcating thalli of *M. lichenoides, L. philippii* and *P. expansum;* (2) branching (fructiculeux), thick, adherent crusts of *N. mamillosum* and (3) thin, closely layered (concentrique) thalli of *P. expansum.* Laborel (1961) also describes the growth cycle for the coralligène de plateau of shelly gravel and rhodolith accumulation, stabilization by encrusting corallines, crustose coralline construction, bioerosion, and collapse. The similarity between the associations of coralline genera and the occurrence of crustose frameworks, rhodoliths, and coralline gravels has led to comparisons between the coralligène de plateau and Miocene coralline algal limestones (Bosence 1983a,b, Bosence and Pedley 1982, Studencki 1979). However, because of the lack of detailed study of the internal structure of the coralligène these comparisons have been only of a general nature. This paper examines the internal structure of dredged and diver-collected slabs of "coralligène" from the Cote de Vermeille of the southern Golfe du Lion (Laubier 1965). This preliminary investigation permits a comparison between Tertiary and Recent buildups which concludes that the coralligène can be considered to be a good Recent analog for some Tertiary coralline limestones of the Tethyan region.

1 Department of Geology, University of London, Goldsmiths' College, Creek Road, Deptford, London, SE8 3BU, Great Britain

Paleoalgology: Contemporary Research and Applications
(eds. D.F. Toomey and M.H. Nitecki)

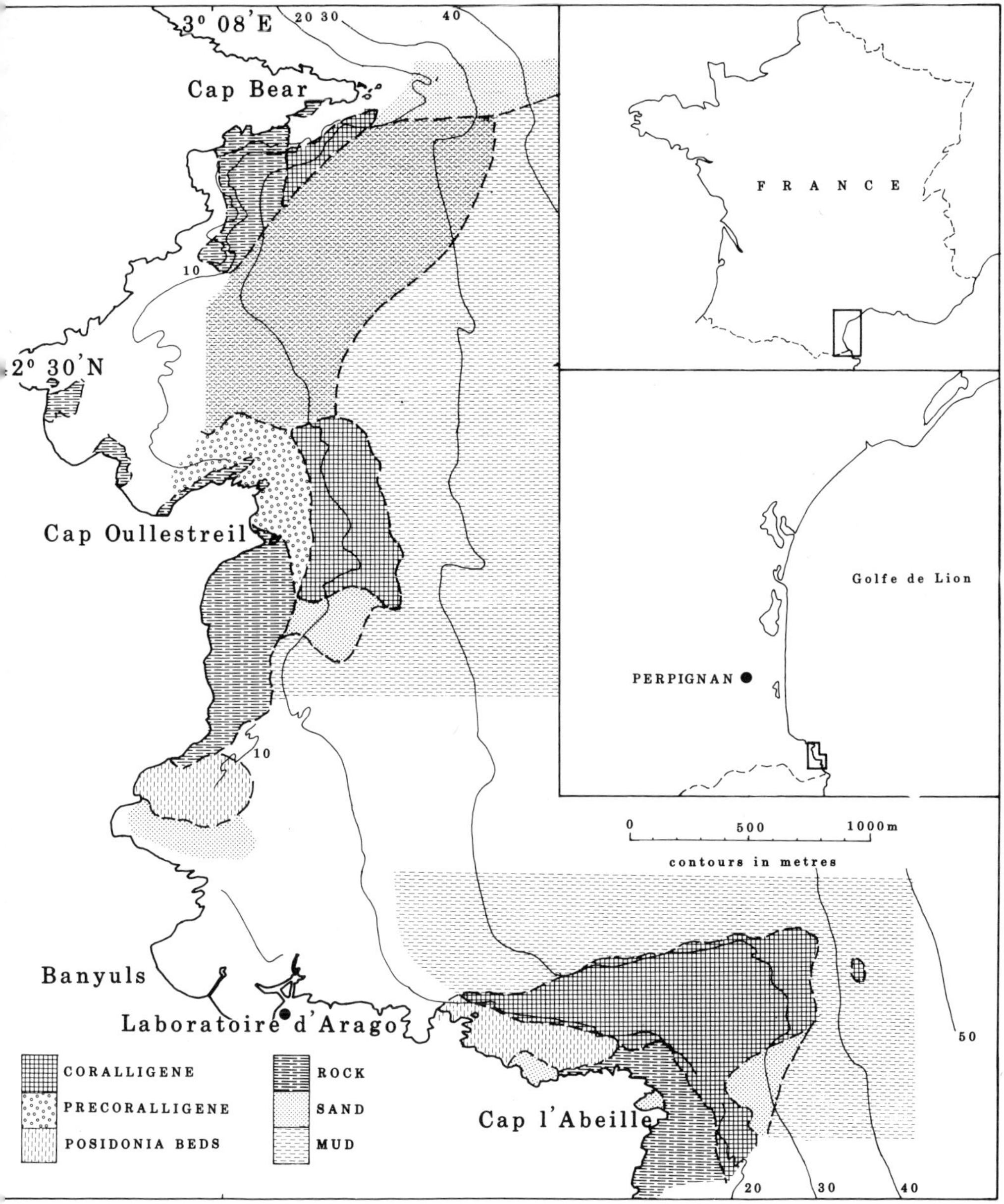

Fig. 1. Location and sedimentary environment of the coralligène of Banyuls. (After Laubier 1965)

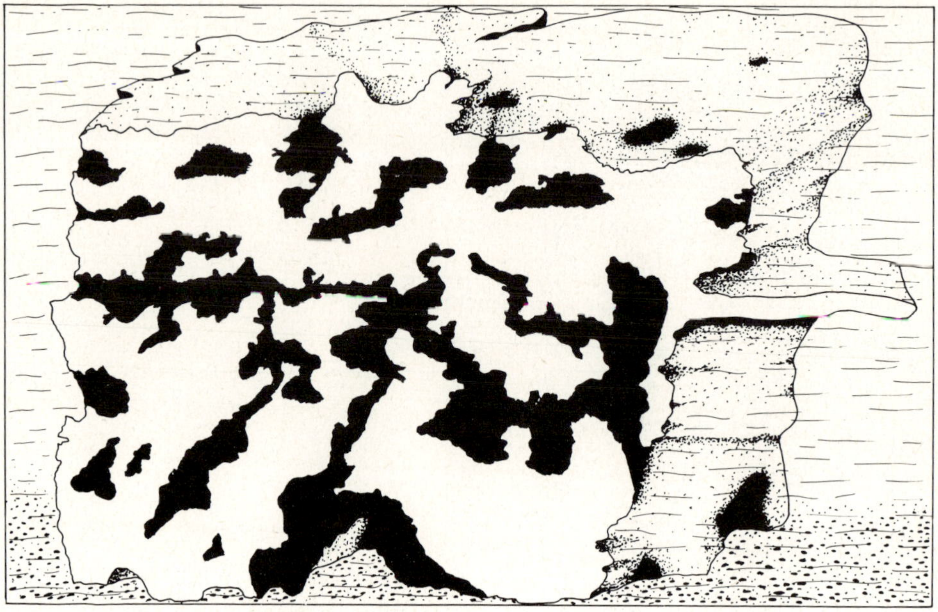

Fig. 2. Section through reef of coralligène illustrating porous texture, ledges, lips and pillars. (Composite sketch after Laubier 1965)

Environment and Occurrence

The coralligène of Banyuls on the Cote de Vermeille occurs off rocky promontaries at depths of 20–40 m (Fig. 1). The coralline algal buildups form 1–4 m raised areas of the sea bed. These areas of coralligène are up to 1 km across and surrounded by carbonate gravels, terrigenous sands and muds, and some rocky areas. Currents around the coralligène are negligible during normal weather conditions, but Laubier (1965) notes an occasional weak southerly flowing current. Following storms Laubier (1965) observed a thin, temporary layer of mud over the reefs. Water temperature varies seasonally from 10° to 23°C and salinity from 37 to 38‰ (Laubier 1965).

Reef Morphology and Biota

The sublittoral reefs of the coralligène form reefs up to 4 m high and 50 m across. Between these buildups are sand-, gravel-, or mudfloored channels. The reefs are of complex morphology and have many ledges, lips, chambers and pillars and some are mushroom shaped (Fig. 2, divers observations 1982; Laubier 1965). The reefs are all cavernous with large and small tunnels and galleays, many filled with sediment.

The coralligène of Banyuls supports a very rich fauna. Some 530 species are listed by Laubier (1965) inhabiting vagile, epifaunal, crevice, and endolithic environments. An epiphytic population of weakly calcified *Peysonnelia polymorpha, Halimeda tuna* and *Udotea petiolata* covers most of the surface between the live corallines.

Fig. 3A–G. Dredged blocks of Coralligène, Banyuls, southern France. **A** Section through block of coralligène illustrating typical porous structure with remnants of coralline framework, live corallines on surface, encrusters on lower surface and internal sediments. (× 0.35). **B** Section through coralligène illustrating well preserved crustose framework of *Mesophyllum lichenoides* with surpuiid tubes and internal sediment. (× 1.05). **C** Surface of coralligène illustrating overgrowth of *Lithothamnium philippii* (*lower left*) by rough surfaced crusts of *M. lichenoides.* (× 0.7). **D** Surface view of leafy crusts of *Mesophyllum lichenoides.* (× 0.7). **E** Surface view of crusts and columns of *Lithothamnium philippii.* (× 0.7). **F** Surface view of thick leafy crusts of *Pseudolithophyllum expansum.* (× 0.7). **G** Surface view of live *Pseudolithophyllum expansum* infested with *Cliona.* Note raised collars around osculae. (× 0.7)

Coralline Frameworks

Live corallines cover an average of 30% of the outer surface of dredged blocks from the Cap d'Abeille coralligène. *Peysonnelia polymorpha* is the next most abundant epiphyte and commonly reaches 20% cover. *Mesophyllum lichenoides* (Figs. 3B,D and 4A,B) is the commonest coralline on the surface and in slabbed section from within the reefs. The next most abundant corallines are *Lithothamnium philippii* (Figs. 3C,E and 4E),*Pseudolithophyllum expansum* (Figs. 3F and 4G,H), and *Tenarea* sp. (Fig. 4B,D,E).

Mesophyllum lichenoides occurs as thin (0.2–0.5 mm), pale pink rugose crusts with white growing edges (July 1982). The crust margins are commonly upturned and free of the substrate imparting a leafy shape to this coralline (Figs. 3D and 4C). Concentric growth lines, paralleling crust margins, are common. Crusts frequently bifurcate and rejoin, which, combined with the leafy growth, results in an open crustose framework (Figs. 3D and 4A). The majority of bifurcations are produced by new crusts growing from previous perithallial tissue. (Figs. 4A). The new crust subsequently develops its own hypothallus and perithallus. Columnar perithallial tissue also occurs within the framework with diameters of 3 mm and heights of 3–4 mm. These frequently bear fertile conceptacles (Figs. 4B). Growth ridges formed by neighboring crusts attempting to overtop each other are common. Intergrown within this framework are thin crusts of *Tenarea* sp. (Fig. 4D), serpulid tubes, encrusting foraminifers, and bryozoans. The framework is commonly bored by sponges.

Lithothamnium philippii forms thick (1–2 mm), smooth, undulose, pink crusts. These crusts overgrow cavities, bifurcate, and overgrow neighboring crusts (Fig. 3D,E). Columnar growths are common (5–6 mm high and 4–8 mm in diameter) and frequently contain the multipored conceptacles (Fig. 4E). Thus a framework of larger scale, but of similar mode of construction to the *M. lichenoides* framework, is produced. Serpulid tubes are again common and the framework is bored by bivalves as well as by sponges.

Pseudolithophyllum expansum occurs as thick (1–2 mm) pink crusts that are smooth and often bifurcate through growth of new crustose perithallial tissue. Crust growth is either adherent or, as with the previous two species, free growing over

Fig. 4A–I. Thin-section photomicrographs of Coralligène, Banyuls, France. **A** *Mesophyllum lichenoides* framework with abundant botryoidal aragonite cements and internal sediment. Crust bifurcation in lower centre. (× 13.5). **B** *M. lichenoides* framework with columnar tissue bearing fertile conceptacles, crustose overgrowth, and *Tenarea* crusts. (× 13.5). **C** Old bored and infilled coralligène overgrown by leafy crust of *M. lichenoides.* (× 13.5). **D** Framework of *M. lichenoides* and *Tenarea* sp. (× 13.5). **E** Columnar growth of *Lithothamnium philippii* with multipored conceptacles; Serpulid tube encrusted by *Tenarea* (*lower right*). (× 13.5). **F** Irregularly eroded, stained and infilled *L. philippii* with infilled sponge borings. (× 13.5). **G** Thick, bifurcating crusts of *Pseudolithophyllum expansum.* (× 9). **H** Detail of *Pseudolithophyllum expansum* crust illustrating coaxial hypothallus and thick fertile perithallial tissue. (× 27). **I** Coralligène reef rock, showing eroded, infilled and cemented internal sediment with relics of original coralline algal framework. (× 13.5)

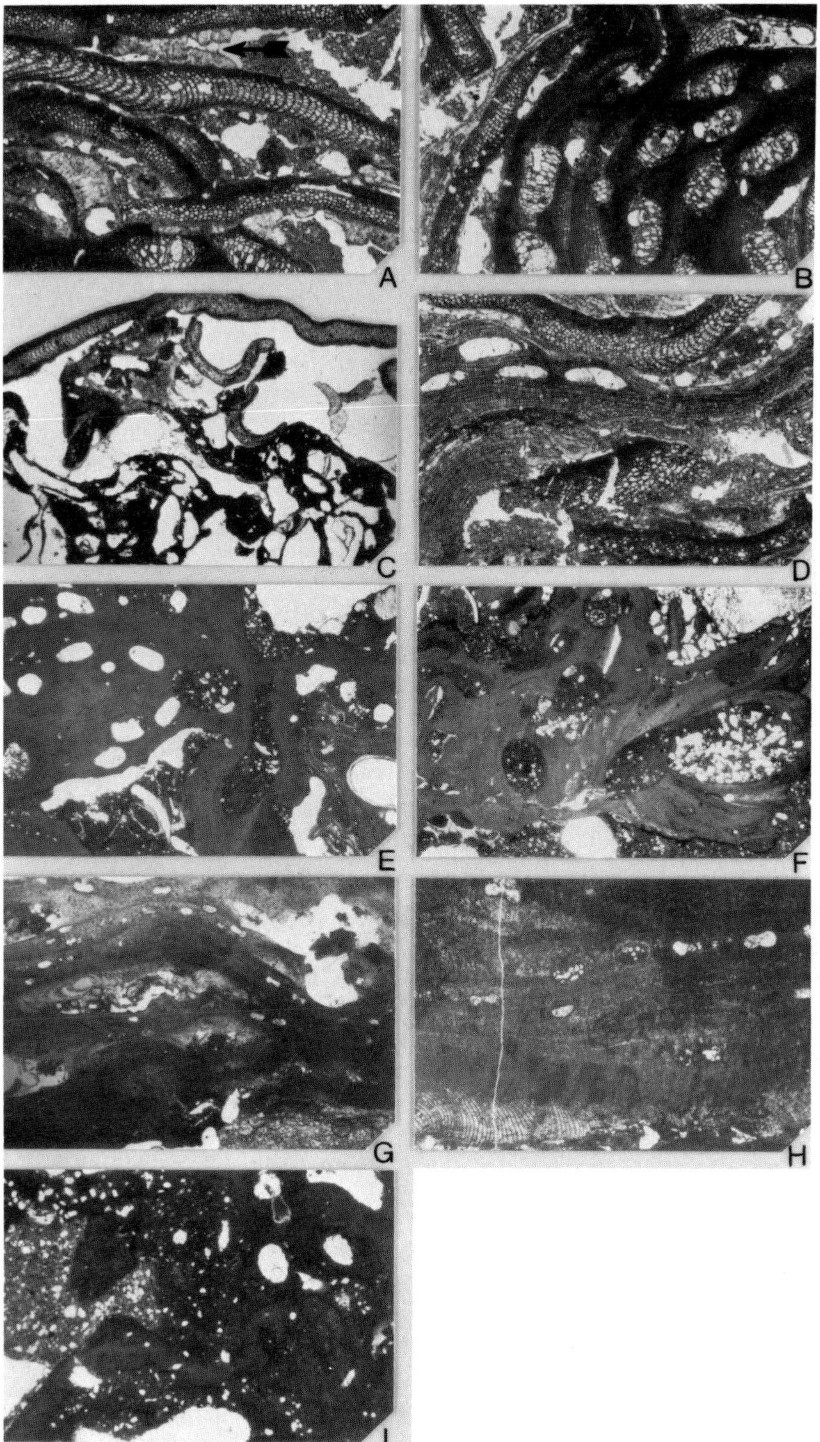

Fig. 4A–I

neighboring corallines (Fig. 3F). Intergrowth with *Tenarea* sp., serpulids, and bryozoans is common, and sponge boring is ubiquitous. There is a close association between this coralline alga and *Cliona viridis* which results in regularly spaced osculae within the crust (Fig. 3G). The alga reacts by building a collar around each osculum (cf. Cotte 1914). Laubier (1965) suggests a larval attraction of *C. viridis* to this coralline.

To summarise, the three major corallines construct broadly similar frameworks. All three corallines have leafy crusts which bifurcate and rejoin to form an open crustose framework. The differences are in the smaller size of the *Mesophyllum* crusts and the more frequent columnar growth in *Pseudolithophyllum* than in *Lithothamnium.* The commonest occurrence on the reef surface is to find individual species growing in separate patches. Within these patches individuals of the same species compete for space by overgrowth or growth ridges. Similarly, in slabbed reef sections the frameworks are monospecific on a centimeter scale. Therefore, the competitive growth between neighboring crusts usually involves the same species or different parts of the complex individual making up the framework (Fig. 3B). Less common is the overgrowth of one species by another (Fig. 3C). I have suggested earlier (Bosence 1983a) that such leafy growths are competitively superior to adherent crusts on hard substrates.

The frameworks described here are most similar to Laborel's (1961) feullete type. The branching *Neogoniolithon mamillosum* was not collected although it is noted present by Laubier (1965).

Occasional multispecific rhodoliths are dredged up with the coralligène. These are of ellipsoidal and spheroidal shape and of concentric laminar to columnar construction. All examples sectioned show alternations from *Lithothamnium philippii* to *Tenarea* together with encrusting serpulids and bryozoans.

Bioerosion, Sedimentation, and Diagenesis

Sections through the reefs indicate that the reef framework is sometimes reduced down to an area of about 5% of the reefrock (Fig. 3A). The main agent of erosion appears to be clionid sponge boring (Figs. 3G and 4F). Bivalve borings are also common but not as ubiquitous as sponge borings. No evidence of physical erosion is preserved and it would not be expected to be present in these quiet waters. Borings are usually filled with muddy sediment (composition 23% calcite, 12 mol% Mg, and the remainder terrigenous and organic matter) with various mixtures of reef debris and sand- to pebble-sized terrigenous material (Fig. 4I). The origins of terrigenous pebbles within reefs, 1–4 m high, is at first puzzling. The simplest explanation would be for the sediment level in the inter-reef channels to fluctuate, possibly during storms. This would allow large grains to be washed into the cavernous reef. This process could also explain the large ledges protruding from the sides of reefs. All coralline growth laminae in these ledges are horizontal and the right way up, suggesting that the reef was built out over a previous substrate, at a higher level, which has since been removed.

Large areas (up to 60%) of the internal sediment are cemented to a thoroughly indurated limestone. The coarsest cements are botryoidal growths of blade-shaped aragonite crystals (Fig. 4A). These occur most commonly on the undersides of coralline crusts and within conceptacles. They are similar to cements from coralline reefs of Bermuda (Ginsburg and Schroeder 1973). More widespread, but less spectacular, is a brown micrite cementing the internal sediment. X-ray diffraction of cemented and uncemented sediment indicates a high magnesium calcite cement (15 mol% Mg). Blanc and Molinier (1955) report a cryptocrystalline high magnesium carbonate (13 mol% Mg) in the intertidal coralligène (trottoir).

The cemented sediment occurs within centimetres of the reef surface and is therefore subjected to further bioerosion, and sections show many generations of the cycle of framework growth, bioerosion, sedimentation, and cementation so characteristic of reef rocks.

Discussion

From this preliminary examination of the coralligène at Banyuls a number of comparisons can be made with Tertiary limestones from the Tethyan region.

Firstly, the generic composition of Miocene coralline algal limestones is very similar to that of the coralligène (Bosence 1983a,b, Studencki 1979). Dominant genera in both buildups are *Mesophyllum, Lithothamnium* and *Lithophyllum. Pseudolithophyllum* is not usually recognized as a distinct fossil genus (Wray 1977). These genera are found over a wide geographic range from cool temperate to tropical seas. Their depth ranges, however, vary from intertidal in Britain (Adey and Adey 1973) to moderate depths in the warm, temperate Mediterranean to depths greater than 50 m in the tropics (Adey et al. 1982). This same association can, however, occur in shallow cryptic environments in tropical reefs (Adey 1975), suggesting that light rather than temperature controls distribution.

The growth forms exhibited by the Recent corallines are very similar to those from the Miocene coralline algal biostrome of Malta (Bosence and Pedley 1982). In particular the Tertiary *Mesophyllum commune* produces an almost identical framework to the Recent *Mesophyllum lichenoides* (cf. Fig. 11, Bosence and Pedley 1982; Fig. 3B above). These open *Mesophyllum* frameworks appear to occur in quiet water settings of moderate depth. They contrast with the more densely constructed *Mesophyllum* frameworks found on undersurfaces of Recent reefs in St. Croix (Adey 1975, Bosence 1983b, 1984). The gross reef morphology of the coralligène has not to my knowledge been recognized in Tertiary sequences. The in situ crustose coralline buildups in Malta are of lower relief and are not characterized by such extensive early cementation and bioerosion. However, the cements and reef-rock fabric of the coralligène are in all aspects identical with the Eocene cor-algal reef described in this volume (Taberner and Bosence, this vol.; cf. Figs. 3G and 4I). In summary, there are many features in common between the Tertiary and Recent coralline buildups, and the coralligène de plateau appears to be the best known Recent analog for some Tertiary coralline limestones. It remains to be seen whether or not this is the same facies which has continued through Tertiary to Recent times.

Summary

Coralline algal buildups (coralligène) occur in the intertidal zone (trottoir) and at depths of 20–40 m (Coralligène de Plateau) off the rocky Albère coastline of southern France. The deep water examples from areas of small reefs with a relief of 1–4 m and diameters of 10 s of meters. The reefs are surrounded by basement rocks, locally derived carbonate gravels, rhodoliths, and terrigenous muds and sands.

Reef frameworks are constructed by foliaceous crusts of the coralline algae *Mesophyllum, Lithothamnium, Pseudolithophyllum, Neogoniolithon,* and *Tenarea.* Lightly calcified *Peyssonnelia* and *Halimeda* also occur as epiphytes. Bioerosion by sponges, bivalves, polychaetes and algae is common. The many cavities are usually infilled with sediment. Patches of the reef interior are cemented by void-filling, blade-shaped crystals of aragonite.

The recent reefs have many features in common with Miocene coralline limestones from Tethyan areas. They are considered to be the best known Recent analogs for these limestones.

Acknowledgments. Field work for this project was undertaken at the Laboratoire d'Arago, Banyuls. The Director and his staff are thanked for their assistance in obtaining samples of the "Coralligène". Funding for this work is acknowledged with thanks from the Goldsmiths' Geology Department. I also thank John Maddicks, Tom Easter, and Doreen Norman for technical and secretarial assistance. Financial support from the Royal Society and Goldsmiths' College to present this paper is gratefully acknowledged.

References

Adey WH (1975) The algal ridges and coral reefs of St. Croix and their Holocene development. Atoll Res Bull 187:67

Adey WH, Adey PJ (1973) Studies on the biosystematics and ecology of the epilithic crustose corallines of the British Isles. Br Physol J 8:343–407

Adey WH, Townsend R, Boykins W (1985) The crustose coralline algae (Rhodophyta: Corallinaceae) of the Hawaian Islands. Smithson Contrib Mar Sci 15:74

Blanc JJ (1968) The sediments of the Mediterranean Sea. Oceanogr Mar Biol Annu Rev 6:373–545

Blanc JJ, Molinier R (1955) Les formations organogenes construites superficielles en Méditerranée occidentale. Bull Inst Oceanogr Monaco 1067:26

Bosence DWJ (1983a) Coralline algae from the Miocene of Malta. Palaeontology 26:147–173

Bosence DWJ (1983b) Coralline algal reef frameworks. J Geol Soc London 140:365–376

Bosence DWJ (1984) Construction and preservation of two Recent coralline algal reefs, St. Croix, Caribbean. Palaeontology 27:549–574

Bosence DWJ, Pedley HM (1982) Sedimentology and palaeoecology of a Miocene coralline algal biostrome from the Maltese Islands. Palaeoclimatol Palaeoecol Palaeogeogr 38:9–43

Cotte J (1914) L'association de *Cliona viridis* et de *Lithophyllum expansum.* C R Soc Biol 76: 739–740

Ginsburg RN, Schroeder H (1973) Growth and submarine fossilization of algal cup reefs, Bermuda. Sedimentology 20:575–614

Laborel J (1961) Le concretionnement algal 'coralligène' et son importance geomorphologique en Méditerranée. Recl Trav Stu Mar Endonme Marseille Fasc Hors Ser 23:37–60
Laubier L (1965) Le coralligène des Albéres; Monographic biocoenotique. Ann Inst Oceanogr 43:137–316
Pérès JM (1967) The Mediterranean benthos. Oceanogr Mar Biol Annu Rev 5:449–533
Studencki W (1979) Sedimentation of algal limestones from Busko-Spa environs (middle Miocene, Central Poland). Palaeogeogr Palaeoclimatol Palaeoecol 27:155–165
Taberner C, Bosence DWJ (this vol) Ecological succession from corals to coralline algae in Eocene patch reefs of Northern Spain
Wray JL (1977) Calcareous algae. Elsevier, Amsterdam, 185p

17 Ecological Succession from Corals to Coralline Algae in Eocene Patch Reefs, Northern Spain

C. TABERNER[1] and D.W.J. BOSENCE[2]

Introduction

Both corals and coralline algae can be dominant reef-building organisms in the Recent and Tertiary. The best-documented co-occurrence is where corallines replace corals in the intertidal zone producing the coralline algal ridges to reefs (Adey 1975, Littler and Doty 1975). In this paper we investigate the repeated overgrowth of corals by thick coralline algal encrustations in an Eocene patch reef. In this case the overgrowing coralline genera are not those typical of exposed intertidal algal ridges, but today occur either at moderate depths in the tropics or as shallow shaded floras. The evidence in these reefs points to a shallow shaded flora, since the coralline crusts occur both over and around the corals and, in this siliciclastic setting, were probably living in turbid waters.

Geological Setting

The coralgal buildups described in this paper occur in the southeast margin of the Palaeogene Vic Basin (Fig. 1) of northeastern Spain. The regional stratigraphy and sedimentology of this area was described by Reguant (1967) and Taberner (1982). The present boundaries of the basin approximate to its Paleogene margins with land areas to the north (Pyrenees) and south (El Montseny massif and Prelittoral range) (Fig. 1A). Early Tertiary times were characterized by alluvial fan deposition (Figs. 1B and 2A). These sediments were later flooded by transgressive (Lutetian-Biarritzian) fan-delta systems with marginal reworked beach sediments (Figs. 1B and 2A). Northwards, and to the center of the basin, the terrigenous deposits pass into a carbonate ramp with nummulitid mounds in the shallower areas (Taberner 1982) (Fig. 1B). These are followed by southerly derived deltas which infill the basin by upper Eocene to Oligocene times.

Carbonate sediments were deposited on, and reflect the morphology of, the underlying siliciclastic beach and fan-delta sediments (Santisteban and Taberner

1 Departamento de Petrologia i Geoquimica, Facultad de Geologia, Universidad de Barcelona, Gran Via 585, 08007, Barcelona, Spain

2 Department of Geology, University of London Goldsmiths' College, Creek Road, London SE8 3BU, Great Britain

Paleoalgology: Contemporary Research and Applications
(eds. D.F. Toomey and M.H. Nitecki)

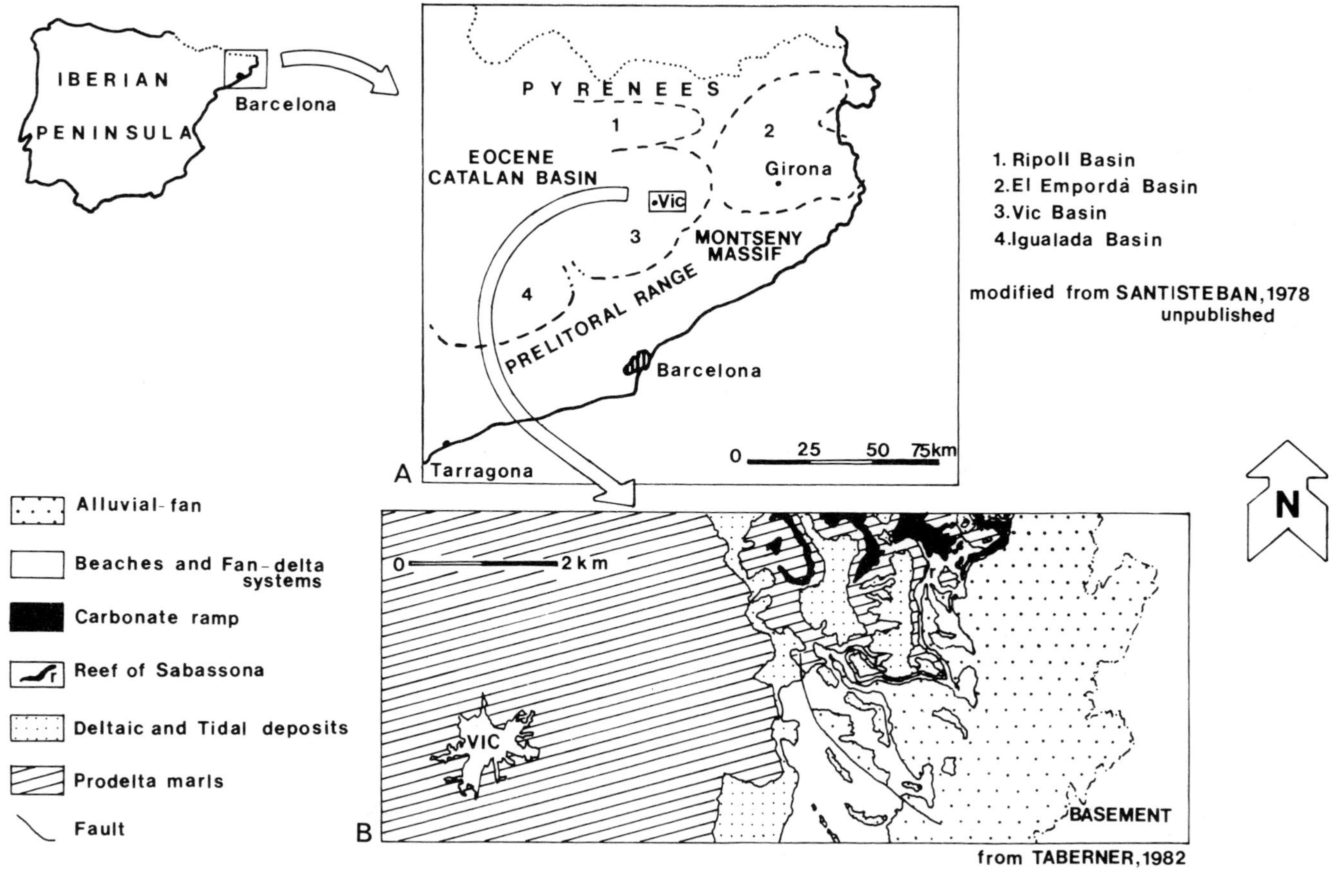

Fig. 1. Geological setting of Eocene coralgal reef at Sabassona

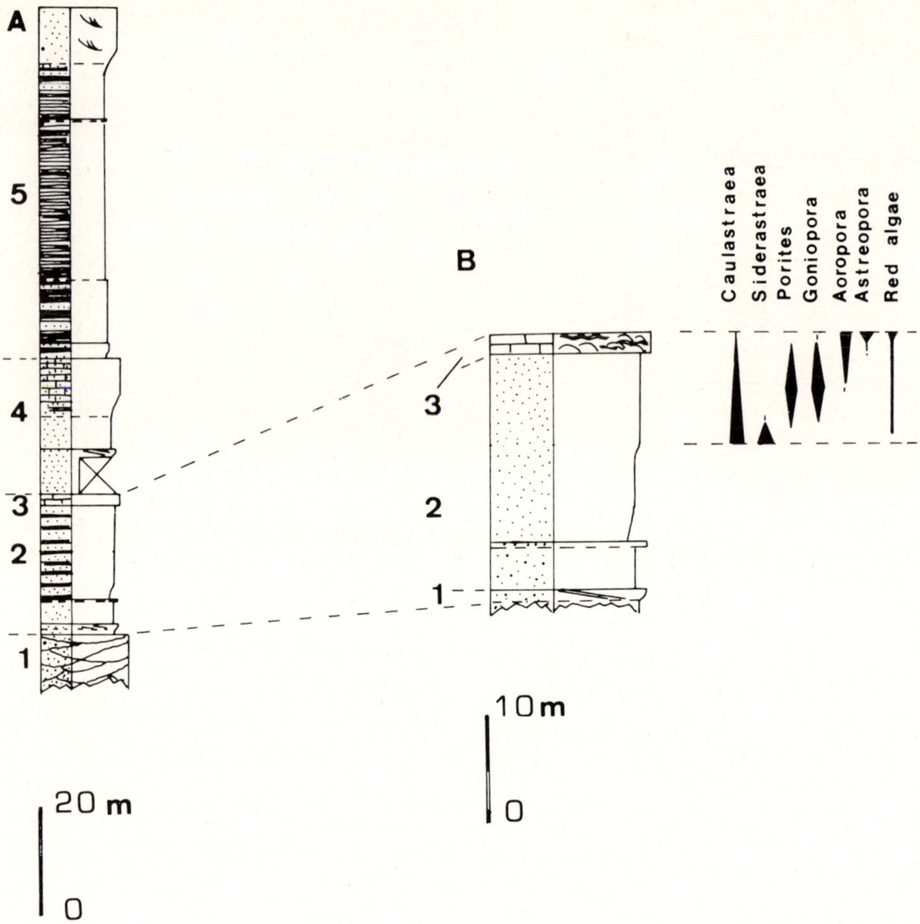

Fig. 2A,B. Stratigraphic section of Eocene sediments of Sabassona. **A** *1* Alluvial fan conglomerates and sands; *2* beach and fan-delta sands; *3* reef and associated calcarenites; *4* beach and foreshore calcarenites; *5* delta front marls. **B** Detail of reef and associated calcarenites from **A** with relative abundances of rock-forming organisms

1979, 1980). The upper Lutetian coralgal patch reefs at Sabassona described in this paper grew in a tabular shape following the basinward edge of a beach and fan-delta system. Similar reef settings have recently been described by Gvirtzman and Buchbinder (1978), Westcott and Ethridge (1980). Santisteban (1981) and Hayward (1982). The Sabassona carbonates are then succeeded by beach and foreshore mixed terrigenous and carbonate sands followed by prodelta marls (Figs. 1B and 2A).

The Sabassona Coralgal Reef

The Sabassona reef is isolated within the siliciclastic deposits (Fig. 2A) and its greatest thickness is found on the break of slope of the pre-existing fan-delta deposits

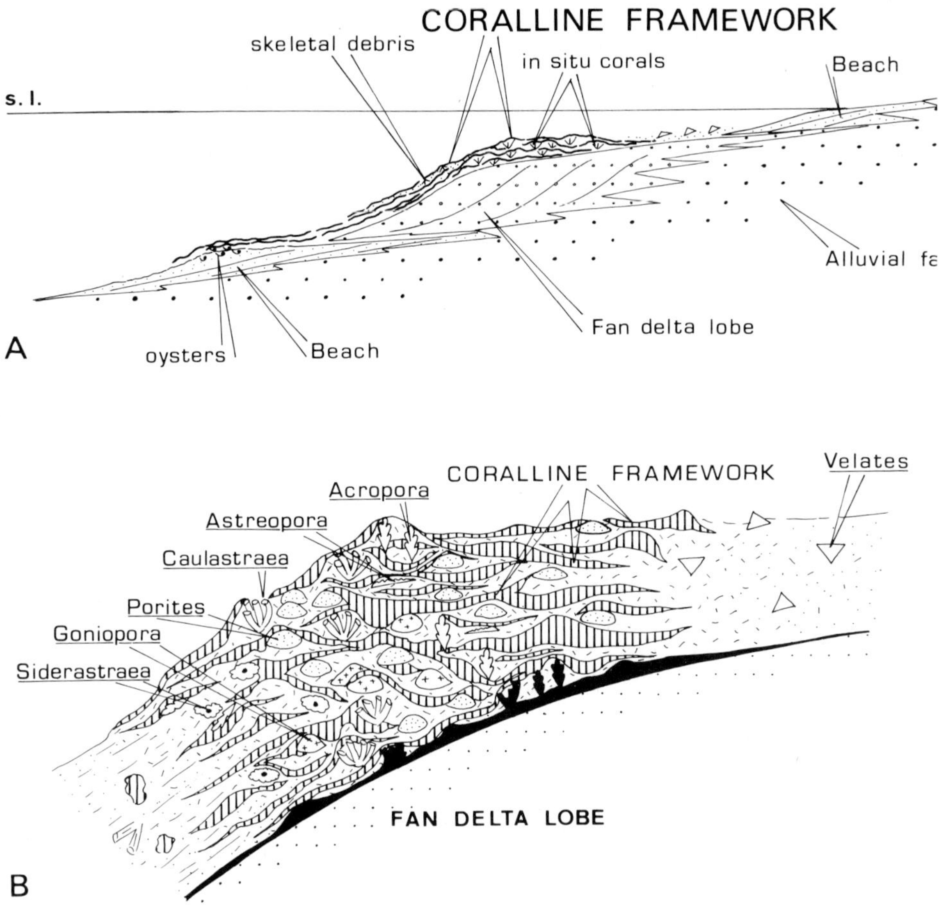

Fig. 3. A Sketch section idicating facies distribution in Sabassona reef. Note reef foundation on fan delta lobe. **B** Sketch reconstruction of occurrence of main coral genera and coralline algal frameworks

(Fig. 3). Carbonates occur in an arcuate buildup constructed by corals and coralline algae. In the reef facies algae alternate with corals preserved in life position. The reef has a tabular shape (1.5 m thick and about 200 m^2 in plan) reflecting the basinward margin of the siliciclastic deposits. Three facies have been differentiated in the rocks making up the reef and surrounding areas.

Inner Facies. The sediments of this facies are a medium to coarse-grained calcarenites with a packstone texture. *Velates* and *Alveolina* are common and *Porites* occurs adjacent to the back of the thickest part of the reef. This facies rests directly on conglomerates of the fan-delta system and is laterally adjacent to the basinward reef facies and the shoreward fan-delta and beach deposits.

Reef Facies. Reef sediments are irregularly bedded nodular limestones with a framestone texture. The nodules contain in situ corals, many of which are encrusted by coralline algae. Colonies of *Caulastraea* and *Siderastrea* are common in the lower part of the reef, whereas *Porites* and *Goniopora* dominate the middle part and *Acropora,* with subsidiary *Astreopora* and *Caulastraea,* occur at the top of the carbonate build-up. *Porites* is also locally common at the top of this facies. Encrusting bryozoans and serpulids also occur.

Coralline algae (*Mesophyllum, Archaeolithothamnium, Lithoporella* and *Lithothamnium*) form crusts up to 30 cm thick. These crusts have been found from the base to the top of this facies, alternating with corals and also encrusting the clastic sediments at the base of the reef. The matrix of the sediments in this facies is mainly micritic and shows many generations of erosion, infilling and cementing. This facies is mainly located on the break of slope of the underlying fan-delta deposits.

Outer Facies. The outer facies is mainly constructed of coralline algal crusts with scarce corals (*Porites*). Towards the upper part of this facies siliciclastic grains become more important and the sediment passes into a calcarenite packstone similar to that of the inner facies. However, elsewhere carbonate grains dominate over terrigenous grains which occur as laminae in the micritic matrix. In this upper part of the facies the carbonate allochems are bryozoans, corals (*Acropora* and *Porites*), coralline algae and mollusc fragments. This facies occurs along the slope of the previous siliciclastic deposits.

Coralline Algal Crusts in Reef

The corals within the reef are overgrown and surrounded by up to 30 cm thick encrustations of coralline algae (Figs. 3–5). Coral calices are encrusted with little visible erosion of the coral (Figs. 5 and 6A). However, later erosion frequently exposes truncated corals and corallines on the reef surface. The most widespread and thickest encrustations occur over corals (Fig. 5) and as sideways accretions in between coral heads (Fig. 6B,C). Commonly the algal crusts are eroded and tabular fragments are redeposited within the reef sediments. Some of these fragments have continued to grow as leafy, concentric structured rhodoliths. Frameworks alternate on a centimeter scale from crustose algae, mainly *Mesophyllum,* to sediment layers deposited over or against the eroded coralline framework. The undersurface of the algal layers may also be eroded indicating undercutting and infilling of the framework.

The framework is constructed of foliaceous crusts of *Mesophyllum vaughani* (Howe) (Fig. 4D), *Archaeolithothamnium affine* (Howe) (Fig. 4H) with thin adhering crusts of *Lithoporella melobesioides* (Foslie) and an occasional *Lithothamnium* sp. The crusts frequently form domed sites of preferential growth. The framework is mainly made up of thin (0.1–3 mm) bifurcating crusts of *Mesophyllum,* but these alternate with crusts of the thicker (0.4–10 mm) *Archaeolithothamnium* (Fig. 6). Where seen, the sequence following erosion is always: erosion – sediment – *Meso-*

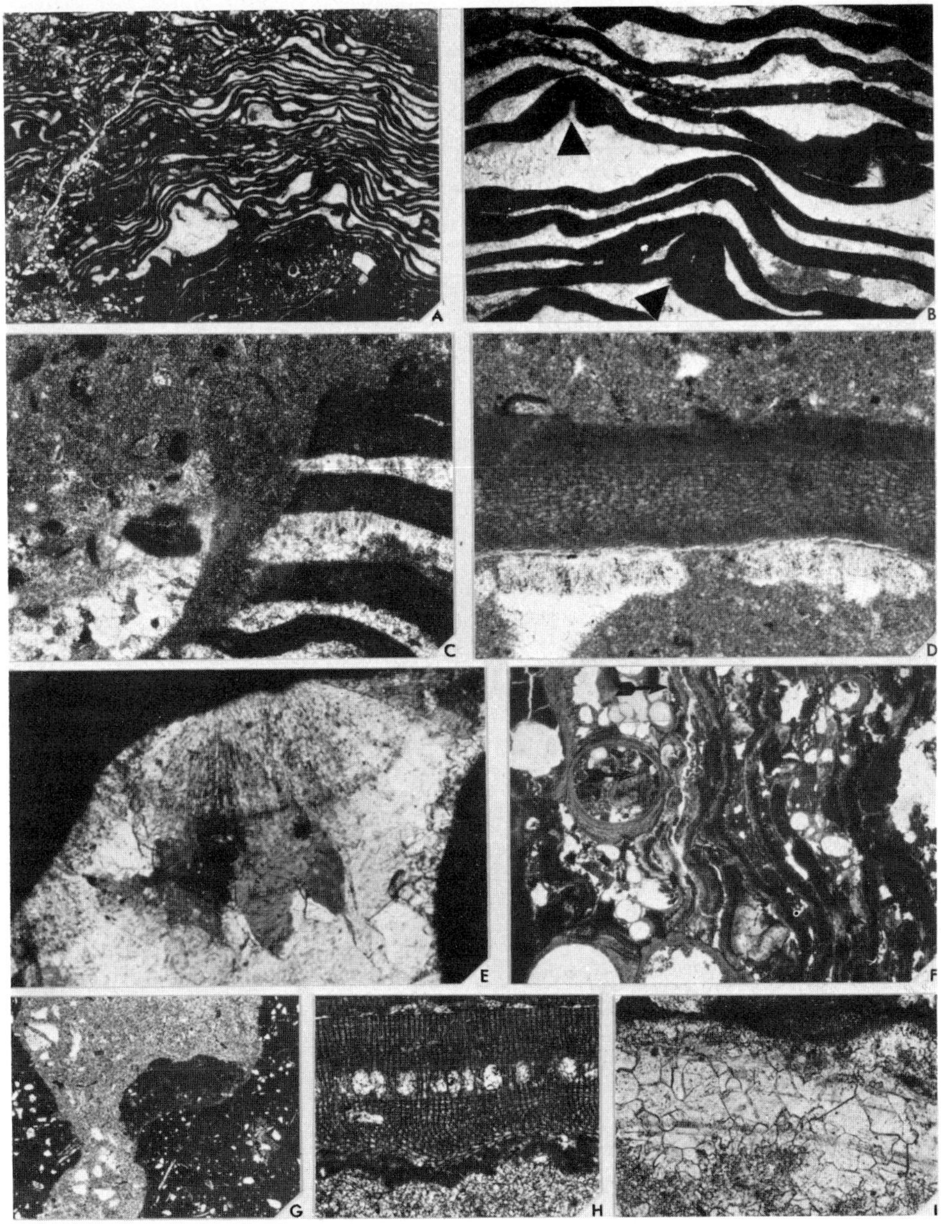

Fig. 4A–I. Thin section photo micrographs from Sabassona Reef. **A** *Mesophyllum* crusts and interlayered sediment. (× 2.8). **B** *Mesophyllum vaughani* crusts; sections of growth ridges (*arrows*) where crusts attempt to overgrow neighbors. (× 9.8). **C** *M. vaughani* crusts with early botryoidal cements cut by three generations of boring sediment infill and cementation. (× 14). **D** Crust of *M. vaughani* with typical thick noncoaxial hypothallus. Botryoidal early cements underlie crust and later micritic sediment. (× 35). **E** Early botryoidal cement of fibrous crystals now recrystallized to low magnesium calcite spar. (× 56). **F** Framework of *Mesophyllum syntrophicum* with vermetid gastropods (V) and early botryoidal aragonite (*arrow*), from overhang on side of Recent reef, St. Croix, U.S. Virgin Is. (× 7). **G** Early cemented reef rock with three generations of erosion, infilling and cementation. (× 7). **H** *Archaeolithothamnium affine* (Howe) with thin non-coaxial hypothallus and thick perithallus with individual sori. (× 28). **I** Neomorphic spar recrystallizing mollusc shell and matrix. (× 28)

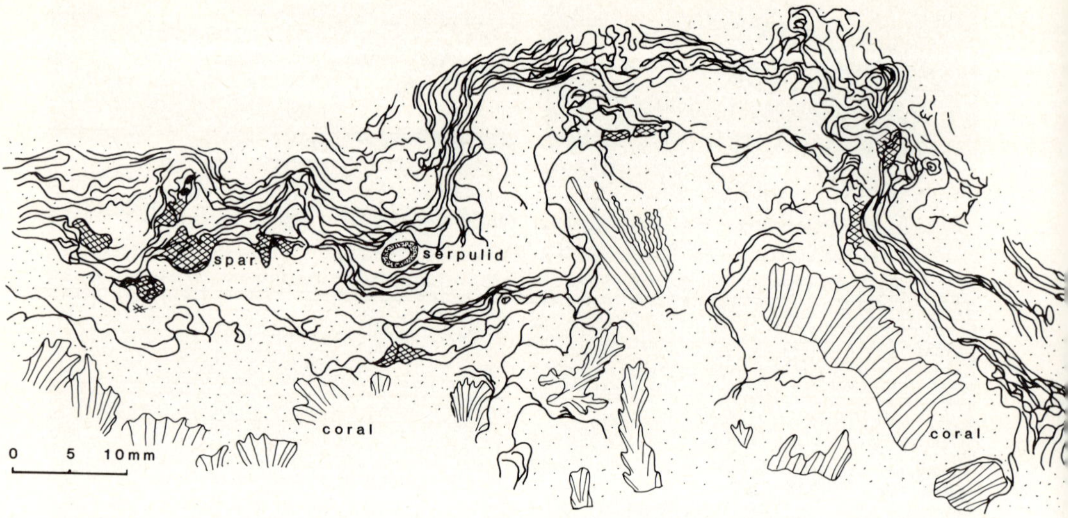

Fig. 5. Camera lucida drawing from peel of reef slab illustrating in situ corals with overgrowing crusts of coralline algae (each *line* represents one crust)

phyllum – *Archaeolithothamnium* – erosion, suggesting ecological succession from *Mesophyllum* to *Archaeolithothamnium* (cf. Bosence 1984).

In detail the framework is constructed by a combination of foliaceous crustose growth, crust bifurcation and some columnar growth. Crust bifurcation normally arises through rejuvenation of the perithallial meristem, resulting in upward or sideways growth of the new crusts which developes a new hypothallus. Similarly, overgrowing crusts may grow down onto older crusts and fusion takes place with additional perithallial growth. Growth ridges (Fig. 4B) are common and indicate competitive growth of neighboring crusts. This mode of framework construction by *Mesophyllum* is very similar to that produced by Miocene *M. commune* in a relatively deep water setting (Bosence 1983a) and also to shallow frameworks of *M. syntrophicum* in shallow shaded areas in Recent coralline algal reefs of St. Croix (Bosence 1984).

The matrix to the framework is mainly a carbonate mud (brown micrite) with various mixtures of terrigenous grains (quartz, mica, lithic fragments and feldspar), reef debris, foraminifera and molluscan debris. Between the crusts many generations of early cemented sediment are present which sharply cross cut earlier sediments or framework (Fig. 4G). Frequently the truncation surfaces are iron stained and thus older sediments are the darkest. These multigenererational internal sediments are common in present day patch reefs; Ginsburg and Schroeder (1973) suggest a solution mechanism for erosion, whilst I (Bosence 1984), have suggested bioerosion. In the Sabassona reef a solution mechanism appears likely from the irregular eroded outlines (Fig. 4G) which are quite distinct from the morphology of borers (Fig. 4C).

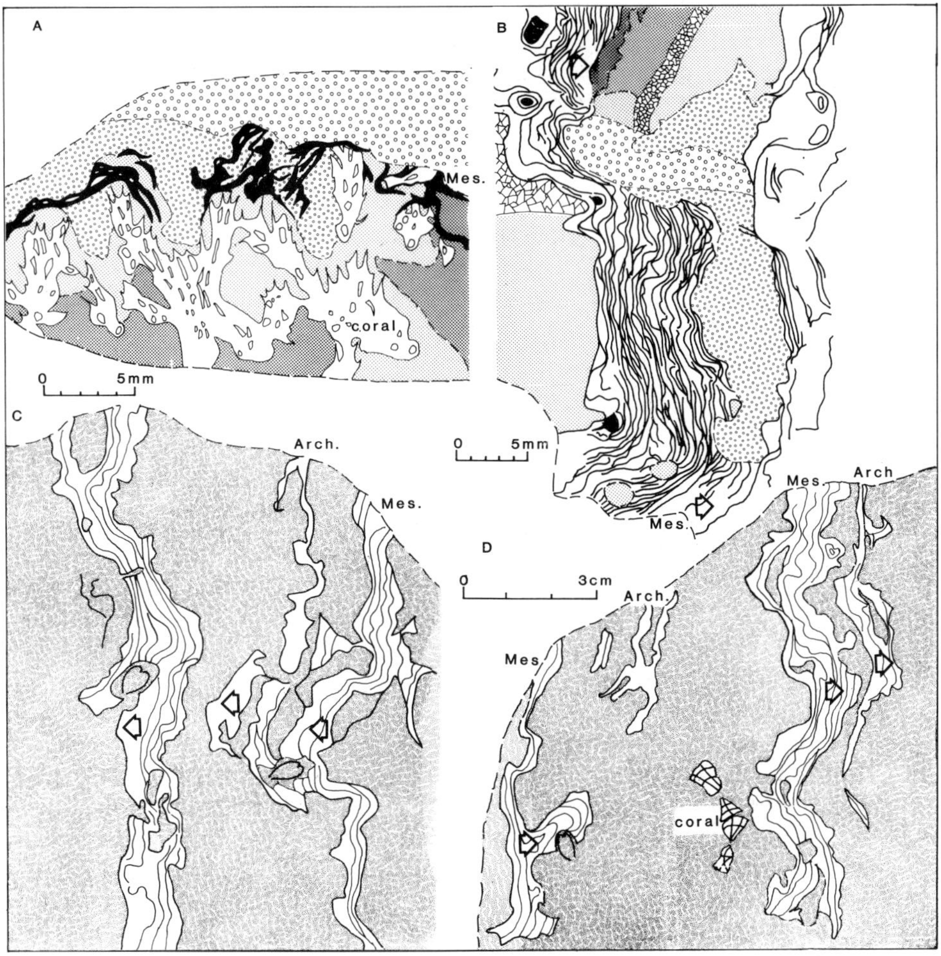

Fig. 6A–D. Camera lucida drawings of sections through coralline algal frameworks. **A** Thin-section illustrating *Mesophyllum* (*Mes.*) overgrowing coral (*Acropora*) and three generations of internal reef sediment (*darkest stippling* is oldest, *size* of *stipples/circles* relative to grain size). Note final erosion of corallines and deposition of calcarenite. **B** Polished slab with *Mesophyllum* framework (each *heavy line* indicates one crust), erosion surfaces and five generations of reef sediment (*stipple size* reflects grain size and the *darkest stippling* indicates the oldest sediment). Geopetal (*left center*) indicates way up and *arrows* indicate direction of crust accretion. **C,D** Vertical sections through frameworks showing lateral crust buildup and alternation with sediment (*stippled*). Note overgrowth of *Mesophyllum* (*Mes.*) by *Archaeolithothamnium* (*Arch.*)

Diagenesis

Early Submarine. The commonest early cement is now present as a brown micrite forming sediment layers around crusts. The evidence for its early nature is that it is sharply cut by marine borers, and, that successive generations of the micrite are sharply cut, iron stained and infilled with fresh sediment (Fig. 4G). This micrite is also encrusted by corallines, but this is not conclusive evidence for a hard substrate on its own (Bosence and Pedley 1982). The other common early cement occurs as botryoidal growths of fibrous crystals below coralline crusts. These cements are surrounded by micrite or late spar and are cut by marine borers (Fig. 4C). The cements are presently calcite (less than 2 mol% Mg) but contain fibrous ghosts suggesting an aragonite precurser (Fig. 4D,E).

Late Stage. Much of the micrite and some adjacent coralline crusts have recrystallized to microspar. Similarly, earlier matrix, cement and aragonitic shells have recrystallized to neomorphic spar (Fig. 4I). Corals are replaced by a void filling spar, as are shelter cavities in the frameworks (Fig. 5).

Discussion

The ecological succession from reef building corals to coralline algae is known from a number of environments today. Similarly, there are several Recent examples of buildups by the crustose coralline genera occurring in the Sabassona reef. The best known ecological replacement of corals by coralline algae occurs in the formation of a coralline algal ridge on the top of a coral reef. The causes for this replacement are still not entirely clear. Adey (1975) has described coralline ridges arising from algal growth on dead *Acropora* stands in the Caribbean. Because of the exclusion of grazers by pounding waves the coralline algae flourish in the intertidal. In contrast, I (Bosence, in press) describe the overgrowth of live corals by coralline algae in the shallow subtidal and intertidal, quieter water back-reef areas of the Florida Keys. Working in the Pacific, Doty (1974) and Littler and Doty (1975) have argued that conditions of high hydraulic and light energy are better tolerated by coralline algae than by corals and uncalcified algae. Therefore, in each of these cases the succession is taking place in the intertidal, where for a number of reasons, it appears the normally faster growing corals are at a competitive disadvantage to the more robust corallines. If this is applied to the Sabassona reef then repeated small changes in water depth could be the cause of the alternating coral and coralline algal frameworks.

When comparing the Sabassona reef with Recent coralline algal ridges there is, in addition to the smaller scale, a major difference in the coralline algal genera present. The primary frameworks of the Pacific and Caribbean ridges are constructed by species of *Porolithon* and *Lithophyllum.* Species of these genera have been involved in reef construction since the Tertiary and have been found to useful palaeoecological indicators (Adey (1979). However, they are not present in the Sabassona reef. The association of corallines present at Sabassona (*Mesophyllum, Archaeolithothamnium, Lithoporella* and *Lithothamnium*) is today form primary frameworks in water depths in excess of 30 to 40 m. Examples of leafy *Mesophyllum* frameworks are from the Miocene (Bosence and Pedley 1982) and the Recent coralligène of the Mediterranean (Bosence, this vol.). Both these occurrences are in water depths greater than 40 m. There are a number of arguments against a deep water setting for the Sabassona reefs. Firstly, the shallow-water corals and corallines alternate and intergrow within the reef. Secondly, the reefs are laterally adjacent to, and are overlain by beach-foreshore sands. Thirdly, the siliciclastic setting of these reefs may well have been associated with relatively turbid waters. All these lines of evidence point to a shallow water origin for the reef. Similar coralline algal frameworks are also found in the shallow water St. Croix algal reefs. The shaded, vertical walls of Shark Reef, for example have a thick crustose coralline algal build-up of *Mesophyllum syntrophicum* which in many respects is identical to the frameworks described above (Fig. 4F) (Adey 1975, Bosence 1984). Therefore, it appears most likely that the Eocene coralline algal crusts were growing in and around the corals in a shaded situation, possibly in turbid, shallow subtidal waters.

Summary

Middle Eocene sedimentation on the southeastern margin of the Catalan Basin (northeast Spain) is characterized by alluvial fan and deltaic sedimentation. Northwards these clastic sediments pass into a carbonate ramp, with nummulitid mounds in shallower areas. Coralgal reefs develope in an arcuate band on preexisting delta lobes. The small patch reefs are constructed by an intergrowth of corals (*Caulastrea, Porites, Astreopora* and *Acropora*) and coralline algae.

The coralline algae overgrow the corals with up to 30 cm densely superposed crusts of *Mesophyllum, Archaeolithothamnium* and *Lithoporella.* Some crusts are built laterally around coral heads and others are eroded and redeposited. Abundant previous submarine cements are now present in the form of calcite micrite and spar botryoids.

The association of corallines within the crusts is typical of relatively deep tropical waters, which contrasts with the shallow marine corals and surrounding fan delta and beach sediments. The coralline crusts are thought to represent a shallow shaded flora growing over and between corals, in relatively turbid waters.

Acknowledgments. We thank Dr. C. Santisteban (Universidad de Valencia) for fruitful discussion and help in the field. C.T. is also grateful to Dr. A. San Miguel for providing laboratory facilities and encouragement whilst undertaking a PhD at the University of Barcelona. D.W.J.B. acknowledges financial support from the Royal Society and Goldsmiths' Research Fund to present the paper.

References

Adey WH (1975) The algal ridges of St. Croix: their structure and Holocene development. Atoll Res Bull 187:1–67

Adey WH (1979) Crustose coralline algae as micro-environmental indicators for the Tertiary. In: Boucot A, Gray J (eds) Historical biogeography, plate tectonics and the changing environment. Oregon State Univ Press, pp 459–464

Bosence D.W.J. (1983a) Coralline algae from the Miocene of Malta. Palaeontology 26:147–173

Bosence DWJ (1983b) Coralline algal reef frameworks. J Geol Soc London 140:365–376

Bosence DWJ (1984) Construction and preservation of two Recent coralline algal reefs, St. Croix, Caribbean. Palaeontology 27:549–574

Bosence DWJ (1984) The morphology and ecology of *Neogoniolithon strictum,* a mound building coralline alga from the Florida Keys. Palaeontology 28:189–206

Bosence DWJ, Pedley HM (1982) Sedimentology and palaeoecology of a Miocene coralline algal biostrome from the Maltese Islands. Palaeoclimatol Palaeoecol Palaeogeogr 38:9–43

Doty MS (1974) Coral reef roles played by free living algae. Proc 2nd Coral Reef Symp 1:27–33

Ginsburg RN, Schroeder JH (1973) Growth and submarine fossilization of algal cup reefs, Bermuda. Sedimentology 20:575–614

Gvirtzman G, Buchbinder B (1978) Recent and Pliestocene coral reefs and coastal sediments of the Gulf of Elat. In: 10th Int Congr Sedimentol, Jerusalem. Guidb PT II, Excursion Y4, pp 163–191

Hayward AB (1982) Coral reefs in a clastic sedimentary environment: Fossil (Miocene, S.W. Turkey) and modern (Recent, Red Sea) analogues. Coral Reefs 1:109–111

Littler MM, Doty M (1975) Ecological componants structuring the seaward edges of tropical Pacific reefs: The distribution, communities and productivity-ecology of *Porolithon.* J Ecol 63:117–129

Reguant S (1967) El Eoceno marino de Vic. Mem IGME LXVII:330

Santisteban C (1981) Petrologia y sedimentologia de los materiales del Mioceno superior de la cuenca de Fortuna (Murcia), a luz de la "Teoria de la Crisis de Salinidad". Dep Petrol Geoquim Univ Barcelona XII-VI:722

Santisteban C, Taberner C (1979) Relacion entre sedimentos terrigenos costeros y facies arrecifales. El modelo de Centelles y su aplicacion regional. Acta Geol Hisp 14:229–236

Santisteban C, Taberner C (1980) The siliciclastic environments as a dynamic control in the establishment and evolution of reefs. Sedimentary Models. IAS Abstr 1st Reg Meet, Bochum, Germany, pp 208–211

Taberner C (1982) Evolucion ambiental y diagenetica de los depositos del Terciario inferior (Palaeoceno y Eoceno) de la cuenca de Vic. Tesis Doctoral. Dep Petrol Geoquim Univ Barcelona, 1400 pp

Westcott WA, Ethridge FG (1980) Fan Delta sedimentology and tectonic setting. Yallahs fan delta, southeast Jamaica. AAPG Bull 64:374–399

18 Depth Zonation and Growth Form of Crustose Coralline Algae: Flower Garden Banks, Northwestern Gulf of Mexico

G.A. MINNERY[1], R. REZAK[2], and T.J. BRIGHT[2]

Introduction

The East and West Flower Garden Banks are located on the outer continental shelf in the northwestern Gulf of Mexico, approximately 190 km south-southeast of Galveston, Texas, at 27°55'N, 93°36'W and 27°52'N, 93°49'W, respectively (Fig. 1). These bathymetric highs are surface expressions of salt domes, with the seafloor elevated to approximately 20 m of the surface by the upward movement of the Jurassic Louann Salt. These two banks are similar in origin and each is presently capped by living coral reefs at its crest.

The biology and geology of the East Flower Gardens have been studied intensively for the past decade by Texas A & M Oceanographers. These studies were initiated in respose to a concern about possible environmental impact of oil and gas drilling and production operations on the Texas-Louisiana continental shelf in the vicinity of coral reefs and fishing banks. The Flower Gardens represent the northernmost, tropical, shallow-water coral reefs in the Gulf of Mexico. Eighteen species of hermatypic corals have presently been identified from the crests of each bank. The upper reef (15–36 m) is very similar to the *Diploria-Montastrea-Porites* community described by Logan et al. (1969) for typical southern Gulf of Mexico coral reefs. Crustose coralline algae cover approximately 10–15% of the hard substrate on the coral reefs, making them important contributors to reef growth and sedimentation. Although corallines are subordinate to corals in terms of abundance on the crests, they are by far the dominant frame builders and sediment contributors on the banks between 50–100 m depths (Fig. 2). At present, 11 genera of crustose coralline algae (Melobesioideae) and one squamariacean have been identified from the East and West Flower Gardens.

Hydrography and Sedimentology

The Flower Gardens are bathed in clear, oceanic waters with surface salinities seldom falling outside 34–36 ppt. Water temperatures on the reefs are subtropical, ranging from a winter minimum of about 18°C (February-March) to late summer highs of

1 Present Address: Chevron U.S.A., Inc., P.O. Box 599, Denver, CO 80201, U.S.A.
2 Department of Oceanography, Texas A & M University, College Station, TX 77843, U.S.A.

Paleoalgology: Contemporary Research and Applications
(eds. D.F. Toomey and M.H. Nitecki)

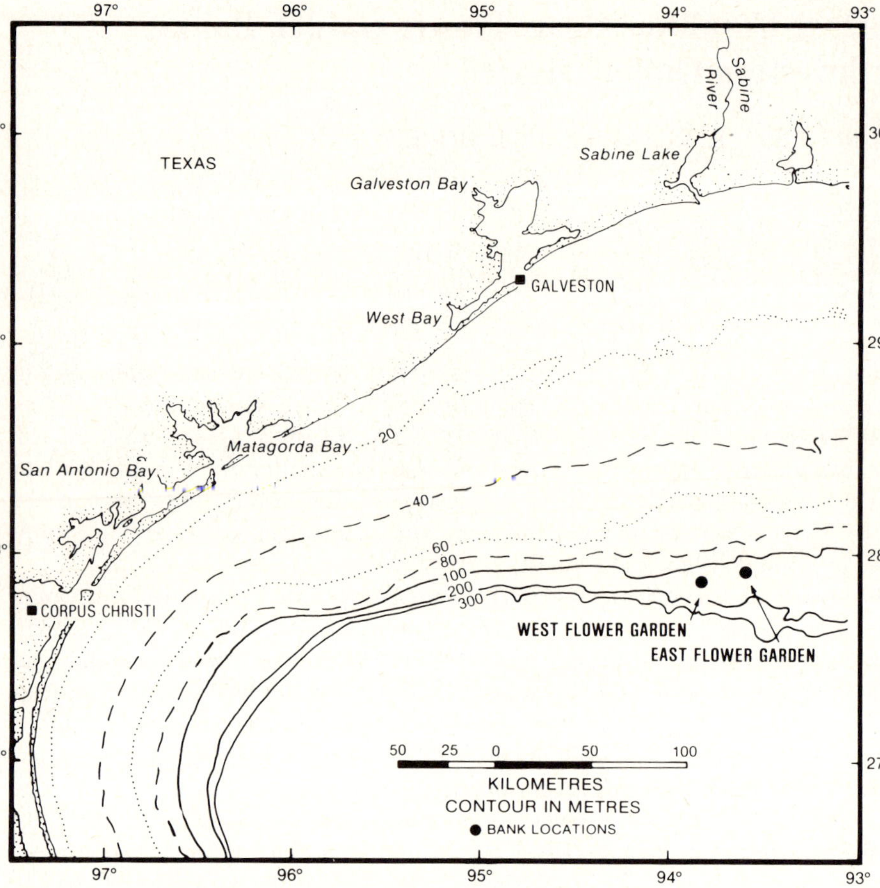

Fig. 1. Map showing location of East and West Flower Garden Banks

30°C (August-September). Water currents on the shelf flow persistently toward the east, occasionally attaining velocities approaching 100 cm/s in the upper portions of the water column. The bottom waters move around the banks rather than flowing up and over the reef crests, forming large-scale sand waves oriented normal to the isobaths (McGrail and Horne 1981).

The composition and bathymetric distribution of carbonate sediments (all autochthonous) on both banks is identical (Edwards 1971, Rezak 1982). Carbonate sediments on the Flower Gardens have been divided into five depth-related zones roughly equivalent to the biotic zonation. A comparison of the biotic zones and the sediment facies is given in Table 1.

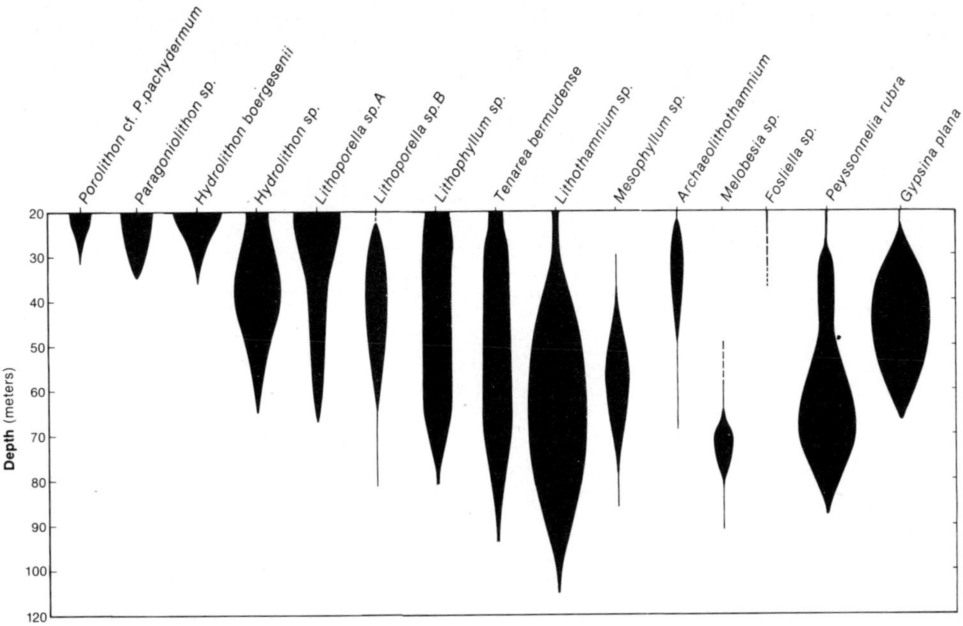

Fig. 2. Depth distribution of coralline algae, *Peyssonnelia* and *Gypsina plana* at the Flower Garden Banks

Table 1. Relationship between sediment facies and biological zones at the East Flower Garden Bank

Sediment facies	Depth (m)	Biological zone	Depth (m)
1. Coral reef	15–50	1. *Montastrea-Diplora-Porites*	15–36
a) Living reef (massive limestone)	15–45		
b) Coral debris (coarse sand and gravel)	25–50	2. *Madracis*	28–46
		3. *Stephanocoenia*	36–52
2. *Gypsina-Lithothamnium* (coarse gravel and massive limestone)	50–75	4. Algal-sponge	46–88
3. *Amphistegina* sand (medium to coarse sand, muddy at depths greater than 85 m)	75–90		
		5. Transition	88–89
4. Quartz-planktonic Foraminifers (sandy mud)	90+	6. Nepheloid	89
5. Molluscan hash (muddy sand)	90+		

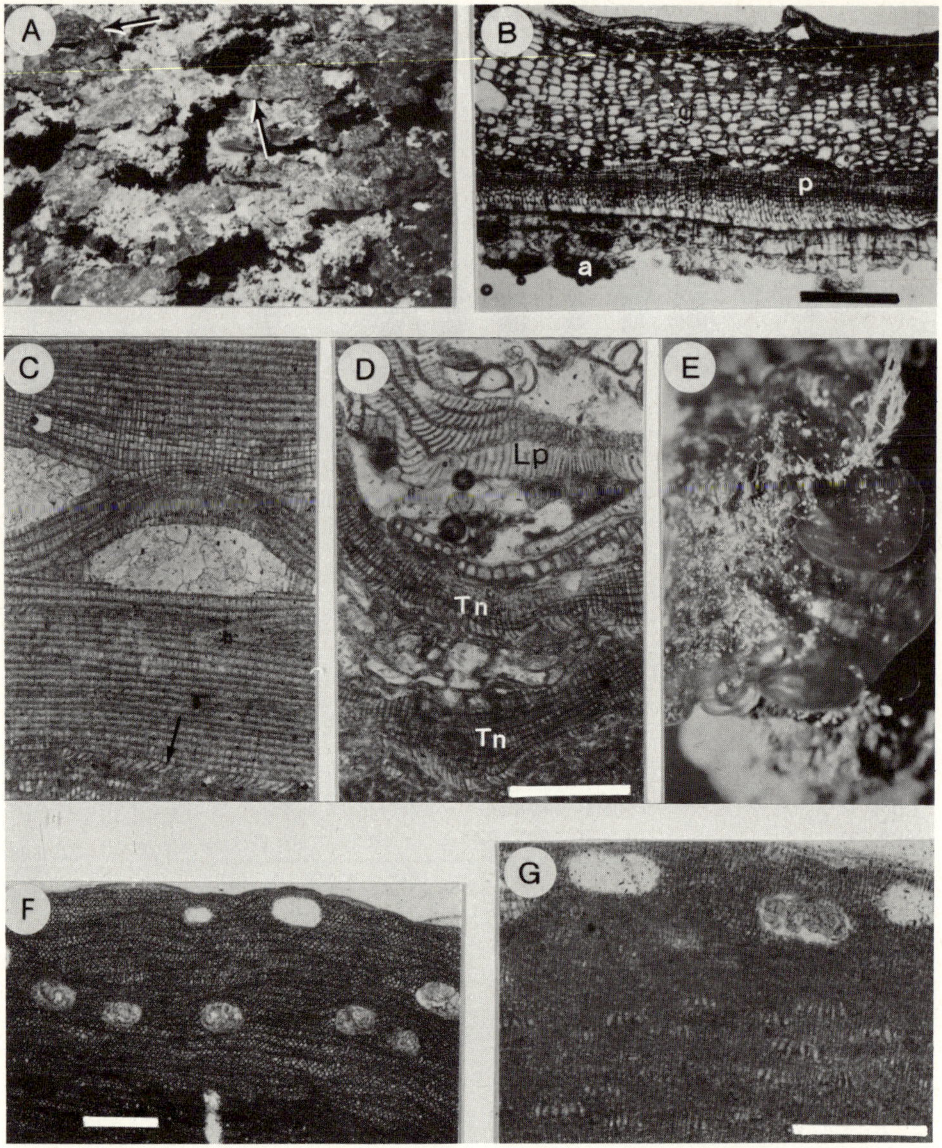

Fig. 3. A Surface of one of the numerous drowned reefs encrusted by coralline algae. *Arrows* point to areas covered by *Peyssonnelia rubra. Lighter areas* are *Lithothamnium* and *Lithophyllum.* **B** Thin-section through the outer 1.5 mm of an algal nodule showing *Gypsina plana* (*g*) overlying a crust of *Peyssonnelia rubra* (*p*). A living, single layer of *Lithoporella* is encrusting the outer surface. Note the botryoidal aragonite cement (*a*) growing from the underside of *Peyssonnelia rubra;* scale: 500 μm. **C** Thin section through the thallus of *Tenarea bermudense.* Note the very regular rectangular cells in the perithallus and the single layer of oblique cells in the hypothallus (*arrow*); conceptacle is 400 μm wide. **D** Multiple crusts of coralline algae covering coral rock at the crest of the West Flower Garden reef (27 m). The algae are: *Tenarea bermudense* (*Tn*) and *Lithoporella* sp. (*Lp*), scale: 200 μm. **E** Free crusts of *Peyssonellia rubra* growing from an algal nodule. Small circular crust is 1.5 cm in diameter. **F** Outer perithallic crust of a thick (15 mm) layer of *Hydrolithon boergesenii* encrusting dead coral at the crest of the East Flower Garden reef; depth: 23 m, scale bar: 300 μm. **G** Perithallus of *Prolithon* cf. *P. pachydermum* from the crest of the East Flower Garden reef; depth: 23 m, scale bar: 200 μm

Methods

Samples of coralline algae encrusting dead coral on the reef crests have been collected during SCUBA operations on the Flower Garden Banks during the past several years. Corallines inhabiting deeper water on the bank slopes, either in the form of crusts, nodules, or pieces of reef rock, were collected by submersible, grabs, and dredges. The coralline taxa comprising these samples were identified by standard plastic impregnation and thin section techniques. The taxonomy used in this study is that of Adey and MacIntyre (1973) and Adey et al. (1982).

Results and Discussion

Crustose coralline algae collected from the Flower Gardens generally fall into four overlapping zones: the main coral reefs, the *Madracis*-algal ridges, the algal nodule zone (*Gypsina-Lithothamnium* facies), and the deep-water algal reefs (Minnery 1984). Each zone will be described separately.

The Coral Reefs

The coralline community inhabiting the coral reefs at the Flower Gardens is dominated by *Hydrolithon boergesenii* (Fig. 3F) and *Tenarea bermudense.* This approximates the deep-reef coralline community on Curacao examined by Van den Hoek et al. (1975). In light intensity experiments conducted on the shallow water (1–2 m) algal ridges in St. Croix (Adey and Vassar 1975), *Hydrolithon boergesenii* typically grows under cryptic, low-light conditions. Under what are probably similar light levels on the Flower Gardens, these same species encrust open, light-exposed reef rock. Surprisingly, *Porolithon* cf. *P. pachydermum* (Fig. 3G) is a fairly common coralline on the Flower Garden reefs. The presence of *Porolithon,* generally considered to be a shallow water genus, is attributed to the exceptional water clarity at the banks (visibility usually 30 + m).

Crustose corallines cover approximately 10–15% of the hard substrate on the coral reefs at the Flower Gardens (measured from numerous photographic transects).

The Madracis-Algal Ridges

On the periphery of the main reef at the East Flower Garden between 30–40 m depths, large linear ridges of *Madracis mirabilis* rubble have been developed. Living thickets of the finger coral, *Madracis mirabilis,* are the only significant hermatypic coral presently growing on the ridges. The crests of these ridges have been entirely covered by a thin, 5–10 mm thick crust of coralline algae. Figure 4A shows a hole that has been punched through the coralline crust, exposing the underlying, unconsolidated *Madracis mirabilis* debris. A cross section through a portion of this crust is illustrated in Fig. 4B.

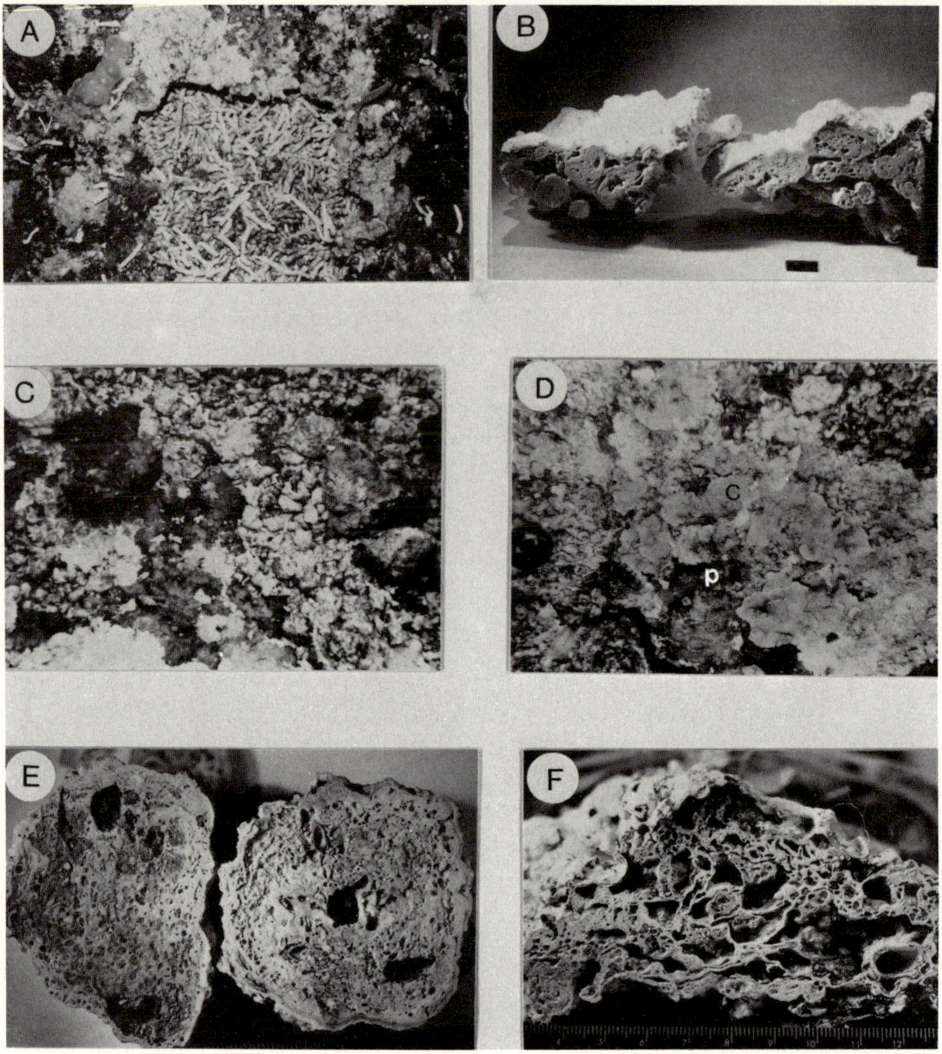

Fig. 4. A Crest of a *Madracis*-algal ridge at the East Flower Garden Bank with a section of the crustose coralline veneer removed exposing unconsolidated *Madracis mirabilis* rubble beneath. Width of hole in crust about 30 cm; water depth: 33 m. **B** Cross section of portion of the coralline algae veneer shown in **A**. The circular bodies are cross sections of the coral *Madracis mirabilis* that have been incorporated into the crust; scale: 2 cm. **C** Concentration of algal nodules at the West Flower Garden Bank at a depth of 72 m. **D** Pavement of coralline algae (*c*) and *Peyssonnelia rubra* (*p*) at the West Flower Garden Bank; water depth: 70 m. **E** Slabs of two small algal nodules from 50 m depth at the West Flower Garden Bank showing the typical dense concentric algal laminations. The large secondary voids and numerous smaller pores (1 mm or less) are borings; scale in millimeters. **F** Slab of a large, flattened nodule collected from 80 m depth at the West Flower Garden Bank. Note the numerous, large constructional voids caused by irregular crustal growth of coralline algae; scale in millimeters

The diversity of corallines forming the ridge crusts is high with 8 genera identified. From most common to least common these genera are *Lithophyllum, Tenarea, Paragoniolithon, Lithothamnium, Lithoporella, Hydrolithon, Archaeolithothamnium,* and *Porolithon. Lithophyllum, Peyssonnelia rubra* (Squamariaceae), and the encrusting foraminifer *Gypsina plana* are the dominant crust-formers on the ridges. Lush assemblages of green and brown algae (mostly *Dictyota* and *Lobophora*) overgrow the coralline crust during spring and fall blooms, but these die and slough off to the ridge troughs in the summer and winter months.

Similar ridges have been found on the West Flower Garden, but these have not been examined in detail.

Gypsina-Lithothamnium Facies

Aprons of coral sand and gravel surrounding the main reefs at the Flower Gardens provide nuclei for the next commonly described algal structure, the algal nodule (or rhodolith). Algal nodules, ranging in size from 1–15 cm, are by far the dominant sediment type in this facies (50–75 m depths) (Fig. 4C). The nodules cover anywhere from 50–90% of the bottom in this zone, and are underlain by coarse carbonate sands. The calcareous green algae, *Halimeda* and *Udotea,* are also important sediment producers in this zone.

An overall increase in nodule size with depth has been observed during submersible transects down the slopes of the Flower Garden Banks. This is probably the result of downslope migration of the nodules as they increase in age and, hence, size. In the upper nodule zone (50–65 m depths), 8 genera of coralline algae have been identified. From most to least abundant, these are *Lithothamnium, Lithophyllum, Lithoporella, Hydrolithon* (?), *Mesophyllum, Archaeolithothamnium,* and *Melobesia.* At greater depths in this zone (65–80 m), *Lithothamnium* and *Tenarea bermudense* (Fig. 3C,D) are overwhelming dominants. *Peyssonnelia rubra* and *Gypsina plana* are also major contributors to algal nodule formation, sometimes being as abundant as all corallines combined (Fig. 3B). Observations on successional changes in coralline taxa or dominance from the interior to the exterior of algal nodules (indicating downward transport) were usually confounded by intense bioerosion of nodule interiors. Interestingly, *Gypsina plana* has been observed only in the outer layers of the nodules, making it an apparent late-comer to the Flower Gardens. The same observation has been made on the Yucatan shelf by Logan et al. (1969). Other significant biotic components of the nodules include bryozoa, sabellid worm tubes, and the encrusting foraminifer *Homotrema.* The primary borers in nodules are bivalves, clionid sponges, and polychaetes (Fig. 4E).

Preliminary examinations of sectioned nodules made it apparent that specimens collected from the upper slopes were more dense (with tight, concentric algal laminae) than the nodules collected from deeper waters (with irregularly oriented crusts and numerous, large constructional voids). Compare photos of nodule sections in Fig. 4E,F. Thin section point count data quantified this observation by demonstrating an increase in constructional voids with water depth (Fig. 5). This change in the internal structure of nodules with depth may be the result of a shift in the coral-

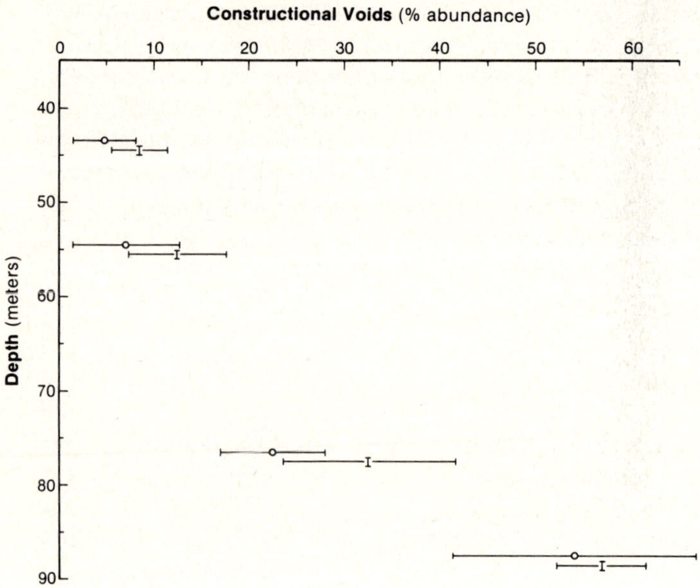

Fig. 5. Percent abundance of coralline algae constructional voids plotted against water depth. —I— nodule intern, —o— nodule extern

line taxa, from adherent crust formers on the upper slopes to nonadherent crust, knob, and column formers in deeper waters.

In the lower limits of the algal nodule zone (75–85 m), the nodules decline in abundance, and pavement-like crusts of coralline algae and *Peyssonnelia* develop between or over the tops of existing nodules (Fig. 4D). Although samples of these crusts were not collected, it is likely that the algae forming the pavements are the same ones observed growing on the deep-water nodules, mainly *Lithothamnium, Lithophyllum,* and *Peyssonnelia. Peyssonnelia* is easily recognized from the submarine by its dark red color.

Deep-Water Algal Reefs

The deep-water algal reefs are low-lying, 1–2 m high flattopped structures, most of which are 5–10 m in diameter. The majority of the reefs occur below the algal nodule zone and are scattered in an apparent haphazard fashion around the banks at depths of from 70 to 90 m. The reefs represent relict (Pleistocene) patch reefs that have since been drowned by sea level rise and covered by a veneer of coralline algae. Approximately 60–90% of the reef surfaces are covered by living crusts of corallines and *Peyssonnelia* (Fig. 3A,B,E), with thick laminar crusts often projecting horizontally from the sides of some reefs in shelf-like fashion.

The dominant algal reef formers at the Flower Gardens are *Lithothamnium, Peyssonnelia, Lithophyllum,* worm tubes, and bryozoans. The biotic composition and internal morphology (thin, irregularly oriented crusts with large constructional voids) distinguish these deep-water reefs from the shallow subtidal and intertidal algal ridges that have been widely reported from the Pacific (Munk and Sargent 1954, Littler and Doty 1974) and the Atlantic (Boyd et al. 1963, Adey and MacIntyre 1973, Ginsburg and Schroeder 1973). The shallow-water algal reefs examined by Ginsburg and Schroeder (1973; Figs. 5, 7 and 11) appear to have thicker, more regular algal laminations. The common shallow-water ridge builders are *Porolithon, Lithophyllum,* and *Neogoniolithon,* often containing small head corals of *Diploria, Montastrea,* and crusts of *Millepora* on their subtidal flanks.

The 90-m isobath at the Flower Gardens is generally the upper limit of the bottom nepheloid layer, composed primarily of continually resuspended clays and silts. Reefs below 90 m are largely void of living encrusters, their surfaces now covered by thin layers of fine sediment deposited out of the nepheloid layer.

Although coralline algae are subordinate to corals in shallow tropical reef environments, they can be the primary sediment contributors and reef builders at depths greater than normal hermatypic coral growth (approx. 50 m). The growth form of crustose coralline algae and taxonomic zonation appears to be depth related. With increasing water depth, the internal morphologies of algal nodules and algal reefs change from relatively dense, laminar, adherent crusts to non-adherent irregularly oriented crusts and columns that form numerous constructional voids.

Summary

The East and West Flower Garden Banks are located in the northwestern Gulf of Mexico, approximately 190 km SSE of Galveston, Texas. These bathymetric highs are surface expressions of salt domes, and are two of the many offshore banks that have been extensively examined during the past 10 years by the Department of Oceanography, Texas A & M University. Although the crests of the East and West Flower Garden Banks are capped by a living coral reef with 18 species of hermatypic corals (15–50 m depths), crustose coralline algae are the dominant sediment contributors and framework builders below 50 m. Depth zones spanning 20–200 m have been established on the banks and slopes by using eleven genera of corallines (*Lithothamnium, Mesophyllum, Melobesia, Archaeolithothamnium, Lithophyllum, Tenarea, Hydrolithon, Porolithon, Paragoniolithon, Lithoporella,* and *Fosliella*), one squamariacean (*Peyssonnelia rubra*), and an encrusting foraminifer (*Gypsina plana*).

The growth form and abundance of corallines can be correlated with depth; however, in this apparent relationship, currents, available substrate, and grazing must also be considered. The general growth forms of corallines consist of small crusts and knobs on bare reef rock in the upper reef zone (15–45 m), algal nodules on the bank slopes and terraces (50–75 m), pavement-like crusts in the deeper nodule zone (70–80 m), and living, 1–2 m high algal reefs scattered between 75–100 m. With increasing depth, the internal structure of algal nodules has reduced concentric growth, favoring a more knobby and irregular platy design with numerous and sizeable constructional voids. The deep-water algal reefs may consist of a living coralline (and *Peyssonnelia*) veneer over drowned algal reefs that began their development in shallower water during the Pleistocene.

References

Adey WH, MacIntyre IG (1973) Crustose coralline algae: a re-evaluation in the geological sciences. Geol Soc Am Bull 84:883–904

Adey WH, Vassar JM (1975) Colonization, succession, and growth rates of tropical crustose coralline algae (Rhodophyta, Cryptonemiales). Phycologia 14:55–69

Adey WH, Townsend RA, Boykins WT (1982) The crustose coralline algae (Rhodophyta: Corallinaceae) of the Hawaiian Islands. Smithson Contrib Mar Sci No 15:1–74

Boyd DW, Kornicker LS, Rezak R (1963) Coralline algae microatolls near Cozumel Island, Mexico. Wyo Univ Contrib Geol 2:105–108

Edwards GS (1971) Geology of the West Flower Garden Bank. Tex A & M Univ Sea Grant Publ 199:71–215

Ginsburg RN, Schroeder JH (1973) Growth and fossilization of algal cup reefs, Bermuda. Sedimentology 20:574–614

Littler M, Doty MS (1974) Ecological components structuring the seaward edges of tropical Pacific reefs: the distribution, communities, and productivity-ecology of Porolithon. J Ecol 63:117–129

Logan BW, Harding JL, Ahr WM, Williams JD, Snead RG (1969– Carbonate sediments and reefs, Yucatan Shelf, Mexico. Am Assoc Petrol Geol Mem 11:1–98

McGrail DW, Horne D (1981) Water and sediment dynamics (Flower Garden Banks). In: Rezak R, Bright T (eds) Northern Gulf of Mexico features study. Tex A & M Univ Techn Rep No 81-2T 3: part B

Minnery GA (1984) Distribution, growth rates, and diagenesis of coralline algal structures on the Flower Garden Banks, northwestern Gulf of Mexico. Dissertation, Tex Abm Univ, College Station, TX 177 pp

Munk H, Sargent M (1954) Adjustment of Bikini Atoll to ocean waves. USGS Prof Pap 260C: 275–280

Rezak R (1982) Geology of the Flower Garden Banks. In: Environmental studies at the flower gardens and selected banks. Fin Rep Minerals Manage Serv, Contr AA851-CT0-25, Chap II, NTIS Order No PB83-101303, pp 19–37

Van den Hoek C, Cortel-Breeman AM, Wanders JB (1975) Algal zonation in the fringing coral reefs of Curacao, Netherlands Antilles, in relation to the zonation of corals and gorgonians. Aquat Bot 1:269–308

Calcification, Microstructure and Growth

19 Occurrence and Preservation of Eocene Squamariacean and Coralline Rhodoliths: Eua, Tonga

B. BUCHBINDER[1] and R.B. HALLEY[2]

Introduction

A widespread rhodolith facies occurs within middle Eocene limestones of Eua, Tonga (Fig. 1). These limestones, first described by Hoffmeister (1932), represent a portion of a broad, early Tertiary platform that developed in the Tonga area prior to disruption and uplift by later Tertiary plate movements (Kroenke and Tongilava 1975). Algal rhodoliths form beds several meters thick within Eocene limestones and occur at localities several kilometers apart along the length of Eua.

The rhodoliths are of two types that occur together at individual outcrops. One type is composed exclusively of squamariacean algae, and the second type is composed of coralline algae together with encrusting foraminifers. This study documents some ecological aspects and the diagenetic evolution of Eocene squamariacean algae and coralline algae from Eua, Tonga.

Previous Studies

Squamariaceae, a family of red algae that resemble coralline algae in its thallus structure, has attracted interest through the years because of its problematic taxonomy and aragonitic mineralogy (e.g., Elliott 1963, Massieux and Denizot 1964, Denizot 1968, Wray et al. 1975). Despite its subtle but widespread occurrence in modern tropical-to-subtropical shallow, marine environments, and its occurrence in early Cenozoic limestones (Wray 1977), there are few recent studies using modern techniques to document squamariacean microstructure and diagenesis (e.g., Flajs 1977a,b).

The two extant genera *Peyssonnelia* and *Ethelia* (redefined as *Polystrata,* see below) are also known as fossils (Wray 1977). Fossil *Ethelia* was first described by Pfender (1936) as *Pseudolithothamnium.* Denizot (1968), however, redefined the genus *Ethelia* Weber Van Bosse as a previously designated genus *Polystrata* Hydrich. *Polystrata* has been reported in algal rhodoliths (e.g., *Pseudolithothamnium album,* Orszag-Sperber et al. 1977; *Ethelia alba,* Racz 1979).

1 Geological Survey of Israel, 30 Malkhei Yisrael Street, Jerusalem 95501, Israel
2 U.S. Geological Survey, Box 25046, Denver Federal Center, Denver, CO 80225, U.S.A.

Paleoalgology: Contemporary Research and Applications
(eds. D.F. Toomey and M.H. Nitecki)

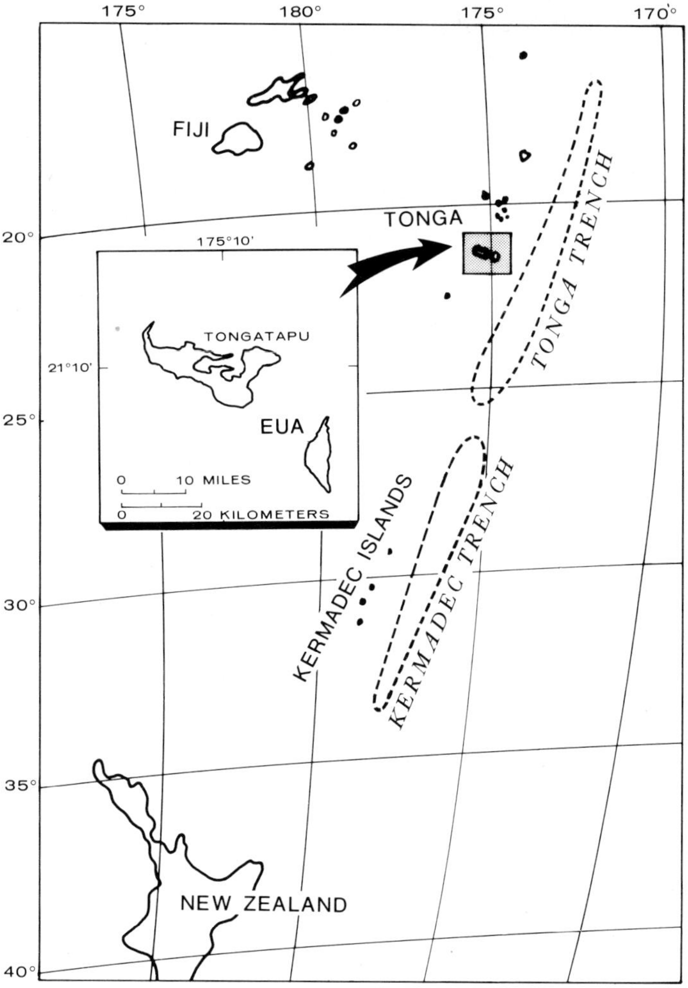

Fig. 1. Location map

Morphology and Microstructure

Euan rhodoliths consist of low mg-calcite and are divided into two distinct types: (1) alternating crusts of coralline algae (predominantly *Archaeolithothamnium* and *Lithoporella*) and encrusting foraminifers, but with a conspicuous absence of squamariacean algae (Fig. 2); and (2) almost exclusively squamariacean algae forming superimposed crusts (0.2–0.6 mm thick) of *Polystrata alba* with very rare encrusting foraminifers and coralline algae (Fig. 3). Externally, however, the two types appear very similar; they are generally spherical to oblong, lack a distinguished non-algal core, typcially range from 4 cm to 9 cm in diameter with a maximum diameter of 15 cm, and have no prominent irregularities on their outer surface (Figs. 4 and 5).

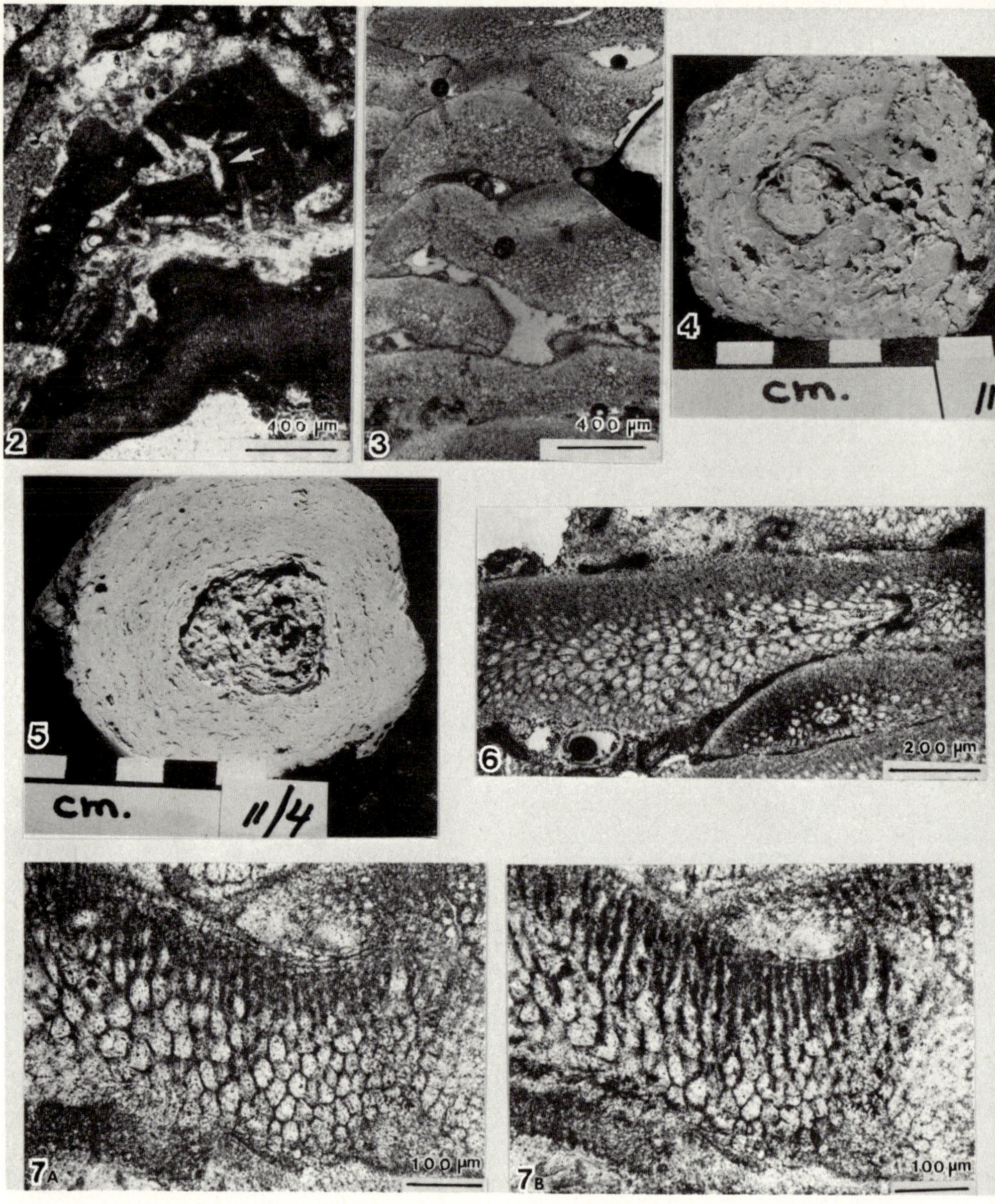

Figs. 2–7

Fig. 2. Photomicrograph (plane-polarized light) showing alternating crusts of coralline algae (*Archaeolithothamnium*) (*dark*) and encrusting foraminifers (*light*) in a rhodolith. Note borings in the upper crust (*arrow*)

Fig. 3. Photomicrograph (plane-polarized light) showing superimposed crusts of *Polystrata alba* in a rhodolith. Note absence of other encrusting organisms

Fig. 4. A rhodolith composed of coralline algae and encrusting foraminifers. Note the irregular encrustations and borings

Both types of rhodoliths are found together at individual outcrops. Although the coralline algae crusts are extensively bored, the squamariacean crusts are intact.

Under the microscope, the *Polystrata* crusts show the typical central mesothallium (= medullary hypothallium) with its horizontally arranged cells (Figs. 6 and 9) and the two upper and lower perithallii fanning out of the mesothallium (see Massieux and Denizot 1964, p. 32, Denizot 1968, p. 157–169). The lower perithallium, however, is often poorly developed.

The crusts also show the transparent golden tint typical of fossil squamariaceans (Elliott 1963, Wray 1977). Under crossed polarizers, calcite crystals along the filament walls (= the elongated cell walls) show a uniform parallel extinction (Figs. 7a,b), and under plane polarized light, the same crystals show pseudo-pleochroism with the maximum absorption (crystals appear darker) when filament walls are vertical (Fig. 8). Some of the cell-filling crystals adjacent to the cell wall show the same optical orientation as the wall crystals.

The cell walls of the *Archaeolithothamnium* and *Lithoporella,* however, have a uniform dark appearance typical of coralline algae in general (Fig. 2); and under crossed polarizers, the crystals do not show any preferred extinction pattern, indicating that their optical axes are randomly arranged.

SEM study of the *Polystrata* thallus reveals that the perithallium consists of tubular-shaped filaments about 12–18 μm in diameter (Figs. 9 and 10). Horizontal partitions into separate cells are very faint or absent. Filament walls consist of flat calcite crystals up to 2.5 μm in size and less than 0.5 μm in thickness (Figs. 10 and 11). Larger crystals appear to be formed from composites grown together with their long dimension preferentially oriented parallel to the filament walls. The filament tubes (or cell cavities) may be filled with equant micron-size crystals (micrite) randomly arranged within the filaments, or they may be filled by microspar crystals. In the mesothallium, however, individual cells are clearly distinguished (Fig. 6). The cells are filled with either randomly oriented, micritic crystals, or with microspar (Fig. 12). The cell walls, however, may be leached, leaving empty molds (Fig. 12).

SEM study of *Archaeolithothamnium* shows poor preservation of cell configurations and random orientation of crystals in the cell wall (Fig. 13). These features appear to be characteristic of coralline algae that have been altered to low-magnesian calcite (Buchbinder 1979, Fig. 11G).

Fig. 5. A rhodolith composed exclusively of squamariaceans (*Polystrata alba*). Note regular encrustations and absence of borings

Fig. 6. Photomicrograph (plane light) showing a nearly vertical section of *Polystrata alba* crust. Note that the medullary hypothallium (= mesothallium) in the middle of the crust fans upward into a well developed perithallium and downward into a poorly developed perithallium

Fig. 7. a Photomicrograph (plane-polarized light) of an oblique section through *Polystrata alba* showing large hypothallium cells and vertically arranged filaments of the perithallium. Note micritic microstructure of cell walls. **b** As above, under cross-polarized light, showing uniform parallel extinction of vertical cell walls (= filament walls). Note branching habit of perithallium filaments

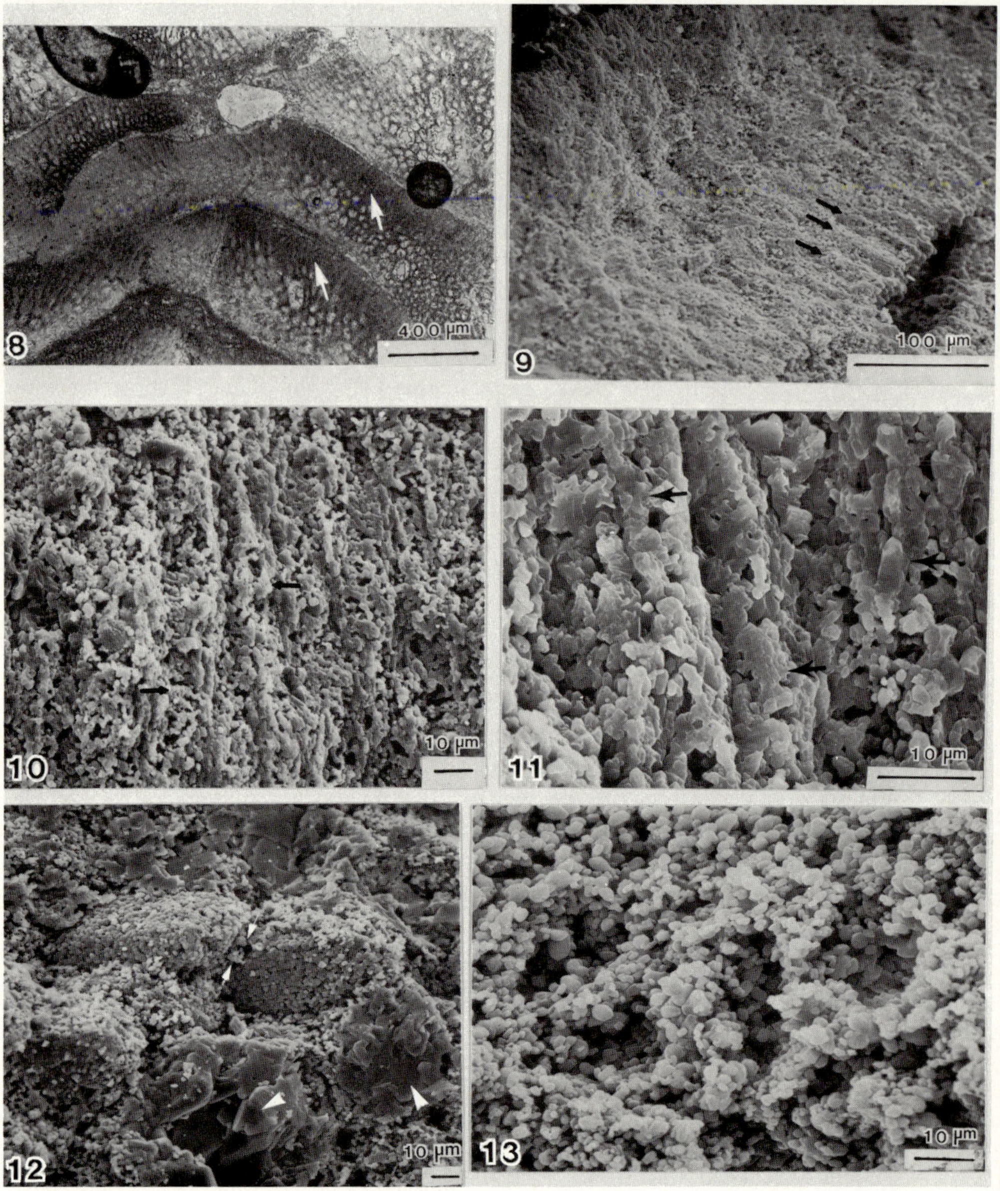

Figs. 8–13

Fig. 8. Photomicrograph under plane-polarized light showing pseudo-pleochroism expressed by maximum light absorption (crystals appear *darker*) of vertically oriented filaments (*arrow*)

Fig. 9. SEM micrograph showing broken surface of *Polystrata* thallus. Note the medullary hypothallium and the two symmetrical (upper and lower) perithallii showing tubular filaments (*arrows*)

Fig. 10. SEM micrograph showing perithallium filaments in *Polystrata*. Note faint horizontal partition into separate cells (*arrows*), and parallel arrangement of crystals along the filament walls

Discussion

Ecologic Equivalence of Squamariacean and Coralline Rhodoliths. Of the two fossil squamariacean genera (*Polystrata* and *Peyssonnelia*), *Polystrata* is more commonly described in rhodolith facies (Massieux and Denizot 1964, Pl. 2, Fig. 4; Orszag-Sperber et al. 1977, Racz 1979). The exclusiveness of *Polystrata alba* to certain rhodoliths in Eua without intermixing with either coralline algae or encrusting foraminifers is very striking. Racz (1979) reported exclusiveness of *Ethelia alba* (= *Polystrata alba*) in the entire rhodolith population of a Paleocene reef facies in Oman. In contrast, Orszag-Sperber et al. (1977) reported Miocene *Pseudolithothamnium album* (= *Polystrata alba*) sharing the same rhodoliths with *Lithothamnium.*

It seems, therefore, both in the case of the Paleocene rhodoliths from Oman and the Eocene rhodoliths from Eua, *Polystrata alba* is a dominant species which successfully competes with other encrusting organisms for space, and, is apparently capable of withstanding boring activity. In the case of samples from Eua, it is obvious that the dominance of *Polystrata* in certain rhodoliths is not due to environmental conditions favorable to squamariacean, since coralline algae and encrusting foraminifers in the other rhodoliths equally share the same environment of deposition.

Preservation of Skeletal Detail in Polystrata. The uniform parallel extinction of the calcitic filament walls in the fossil squamariacean indicates that the crystals in the walls are uniformly arranged with their C-axes parallel to filament walls. This structure resembles the microstructure of the primary layer in living corallines (Flajs 1977a,b, Giraud and Cabioch 1979), where crystals of high mg calcite (up to 1 μm long and 0.15 μm thick) are arranged with their long axis parallel to the longitudinal cell wall. Similar crystal elongations in *Polystrata* are occasionally observed in the Eua material (Fig. 11). Flajs (1977a,b) claims, however, that in the living squamariacean genus *Peyssonnelia,* this primary layer is not calcified but occupied by organic membranes, and that the skeletal aragonite (in *Peyssonnelia*) is precipitated in concentric layers in the secondary layer with the crystals oriented radially from the cell walls, in the same fashion described for the secondary layer of coralline algae (Flajs 1977b, Fig. 10). In the absence of published material about the microstructure of living *Polystrata,* it is impossible to compare the parallel extinction of the crystals in the filament walls of fossil *Polystrata* with microstructure of living *Polystrata.* Still, it is very likely that the preferred orientation of the crystals in the fossil squamariacean was inherited from a precursor aragonitic microstructure. By analogy, this precursor microstructure may be similar to the primary calcite layer of living coralline algae.

Fig. 11. As above, showing composite crystals preferentially oriented elongate parallel to the filament walls (*arrows*)

Fig. 12. SEM micrograph showing hypothallial cells of *Polystrata.* Note microspar filling of cells (*large arrows*) and partially leached cell walls (*small arrows*)

Fig. 13. SEM micrograph of *Archaeolithothamnium.* Cellular microstructure is poorly preserved, and cell walls crystals show random orientation

Significance to the Formation of Neomorphic Calcite. Presumbably, the alteration of the rhodoliths took place in fresh water during uplift and exposure of Eua. In fresh water, aragonite is commonly dissolved and replaced by calcite cement but some aragonite may be neomorphically altered to calcite, the latter process preserving skeletal textures (Bathurst 1975). During the neomorphism of Eua squamariacean rhodoliths, not only have filament walls and cell configurations been preserved, but internal cell wall microstructure is preserved as evidenced by uniform optical properties and SEM micrographs.

In general, high-magnesian calcite shells (i.e., foraminifers, coralline algae) are better preserved than aragonite skeletons (Gavish and Friedman 1969). This is because magnesian calcites, as generally found in coralline algae, are more soluble than aragonite and recrystallize more quickly to stable, magnesian-poor calcite (Mackenzie and Bricker 1971, Walter 1983). The high solubility of magnesian calcites may cause pore waters to become highly supersaturated with respect to low-mg calcite, and thus produce immediate and local (and random) nucleation and precipitation. In contrast, aragonite dissolution may be more gradual because of its lower solubility, and result in the development of moldic porosity.

In both squamariaceans and corallines, solution and reprecipitation processes take place on a micron size, unlike the process of wholescale skeletal leaching observed in other aragonitic shells like molluscs and *Halimeda*. However, unlike coralline algae, the reprecipitated squamariacean calcite nucleates in optical continuity with the crystals being replaced. This indicates a degree of substrate control during the replacement of aragonite by calcite. Substrate control, in turn, suggests a lower level of supersaturation during alteration, a level that does not allow random nucleation but one which requires a solid (in this case, aragonite) which acts as a template for precipitation (Wollast 1971). Even so, it is not clear why squamariacean algae undergo faithful neomorphic replacement rather than wholesale leaching. Such variations in the rate of aragonite dissolution and calcite precipitation might be caused by a variety of factors including grain size, surface area, microarchitecture, organic matrices, pore-fluid composition, and other variables.

Hudson (1962) found similarly preserved Jurassic pelecypods exhibiting traces of the original aragonitic shell structure preserved in brown-colored, pseudo-pleochroic calcite. Although high-magnesian calcite is commonly preserved in great detail, the examples described by Hudson (1962) and Eua squamariacean rhodoliths illustrate that replaced aragonite may also faithfully record minute skeletal details. These examples point to the uncertainty involved in reducing the explanation of the preservation of detailed microstructure to nothing more than original mineralogy.

Conclusions

1. Middle Eocene squamariacean and coralline rhodoliths grew simultaneously and intermixed on a relatively shallow platform partially exposed on Eua, Tonga.
2. Unlike coralline rhodoliths, squamariacean rhodoliths are composed of a single algal species (*Polystrata alba*), with very rare encrusting foraminifers, and appear to be more resistant to boring activity.

3. Although both types of algae altered to low magnesian calcite, *Polystrata alba* exhibits optical and ultrastructural characteristics indicating a higher degree of preservation of original skeletal microstructure than do the coralline algae.
4. Despite the common observation that magnesian-calcite skeletal material is better preserved from skeletal aragonite, Eua squamariaceans illustrate that aragonite skeletons may be extremely well preserved. Thus, preservational detail should not be solely attributed to original mineralogy.

Summary

Outcrops of Middle Eocene platform limestones on Eua, Tonga, contain mixtures of squamariacean and coralline rhodoliths. Squamariacean rhodoliths from these outcrops are composed of a single algal species and are less frequently encrusted or bored than coralline rhodoliths from the same outcrops. Both coralline algae (originally composed of magnesian calcite) and squamariacean algae (formerly aragonite) are now low-magnesian calcite and both were presumably altered during exposure to fresh water. Surprisingly, microstructural detail is better preserved in the squamariacean rhodoliths than in coralline rhodoliths.

References

Bathurst RGC (1975) Carbonate sediments and their diagenesis. Elsevier Sci Publ Co, New York, 658 p

Buchbinder B (1979) Facies and environments of Miocene reef limestones in Israel. J Sediment Petrol 49:1323–1344

Denizot M (1968) Les algues floridees encroutantes (à l'exclusion des corallincées). Lab Cryptogamie, Mus Nat Hist Nat, Paris, 280 p

Elliott GF (1963) Problematical microfossils from the Cretaceous and Palaeocene of the Middle East. Palaeontology 6:293–300

Flajs G (1977a) Skeletal structure of some calcifying alga. In: Flügel E (ed) Fossil algae. Springer, Berlin Heidelberg New York, pp 225–231

Flajs G (1977b) Die Ultrastrukturen des Kalkalgenskeletts. Palaeontographica 160:69–128

Gavish E, Friedman GM (1969) Progressive diagnesis in Quaternary to Late Tertiary carbonate sediments: sequence and time scale. J Sediment Petrol 39:980–1006

Giraud G, Caboich J (1979) The example of Recent corallinaceae. Bull Cent Rech Explor-Prod Elf-Aquitaine 3:601–610

Hoffmeister JE (1932) Geology of Eua, Tonga. Bernice P Bishop Mus Bull 96:93

Hudson JD (1962) Pseudo-pleochroic calcite in recrystallized shell-limestone. Geol Mag 99: 492–501

Kroenke L, Tongilava SL (1975) A structural interpretation of two reflection profiles across the Tonga Arc. S Pac Mar Geol Not 1:9–15

Mackenzie FT, Bricker OP (1971) Cementation of sediments by carbonate minerals. In: Bricker OP (ed) Carbonate cements. John Hopkins Univ Press, Baltimore, pp 239–246

Massieux M, Denizot M (1964) Rapprochement du genre *Pseudolithothamnium* Pfender avec le genre actuel *Ethelia* Van Bosse (algues Florideae, Squamariaceae). Rev Micropaleontol 7: 31–42

Orszag-Sperber F, Poignant AF, Poisson A (1977) Paleogeographic significance of rhodolites: some examples from the Miocene of France and Turkey. In: Flügel E (ed) Fossil algae. Springer, Berlin Heidelberg New York, pp 288–294

Pfender J (1936) Sur un organisme constructeur de calaire crétacés et nummulitique: *Pseudolithothamnium album, nov. gen. nov. sp.* Bull Soc Geol Fr Ser 5: t 6, 303–308

Racz L (1979) Paleocene carbonate development of Ras Al Hamra, Oman. Bull Cent Rech Explor-Prod Elf-Aquitaine 3:767–779

Walter LM (1983) New data on relative stability of carbonate minerals: implications for diagenesis and cementation (abs). Bull Am Assoc Pet Geol 67:566

Wollast R (1971) Kinetic aspects of the nucleation and growth of calcite from aqueous solutions. In: Bricker OP (ed) Carbonate cements. John Hopkins Univ Pres, Baltimore, pp 265–273

Wray JL (1977) Calcareous algae: Developments in palaeontology and stratigraphy, vol IV. Elsevier, Amsterdam, 185 p

Wray JL, James NP, Ginsburg RN (1975) The puzzling Paleozoic phylloid algae – Holocene answer in squamariacean calcareous red algae (abs). Am Assoc Pet Geol Annu Meet 2:82–83

20 Growth Rate and Carbonate Production in *Halimeda opuntia:* Marquesas Keys, Florida

J. HAROLD HUDSON[1]

Introduction

The importance of the calcareous green alga *Halimeda* as a major sediment producer was first cited by Finckh (1904) and his colleagues, who reported on the core and dredged material recovered during the historic deep borings at Funafuti Atoll. Although these remarkable plants have been recognized for over three quarters of a century as significant contributors to the geologic record, their growth rate has, until very recently, remained virtually unknown. Ironically, Finckh (1904) was also the first to attempt field observations on growth in this species. His singular and entirely accidental success was an unknown species of *Halimeda* that in 6 weeks grew up through a hole in a submerged board to form a hemispherical clump 3 inches (7.6 cm) in diameter. Dry weight of the calcareous matter thus produced was 14.38 g. No detailed studies on growth in this species were attempted until Colinvaux et al. (1965) maintained numerous green algae, including six species of Caribbean *Halimeda,* in laboratory aquaria for 19 months. This work was followed by that of Merton (1971), who conducted a 10-month ecological study of *H. macroloba* on Guam. A similar investigation was carried out by Bach (1979), who estimated standing crop, growth, and production of calcareous algae (which included three species of *Halimeda*) over a period of 12 months in a south Florida lagoon. A third, and to the author's knowledge, most recent study is that of Wefer (1980), who used the Alizarin Red-S staining technique to calculate growth of *H. incrassata* and two other calcareous algal species at various intervals over a period of 27 days in Harrington Sound, Bermuda. In addition to the above publications, the treatise on *Halimeda* by Hillis-Colinvaux (1980) includes growth and environmental data on this species previously unpublished.

The present study is part of an ongoing geologic investigation of the large accumulation of *Halimeda* sands in the Marquesas Keys-Quicksands area and was initiated because of the lack of quantitative data on rates of calcium carbonate production in *H. opuntia,* a major contributor to sediments there.

1 U.S. Geological Survey, Fisher Island Station, Miami Beach, FL 33139, U.S.A.

Paleoalgology: Contemporary Research and Applications
(eds. D.F. Toomey and M.H. Nitecki)

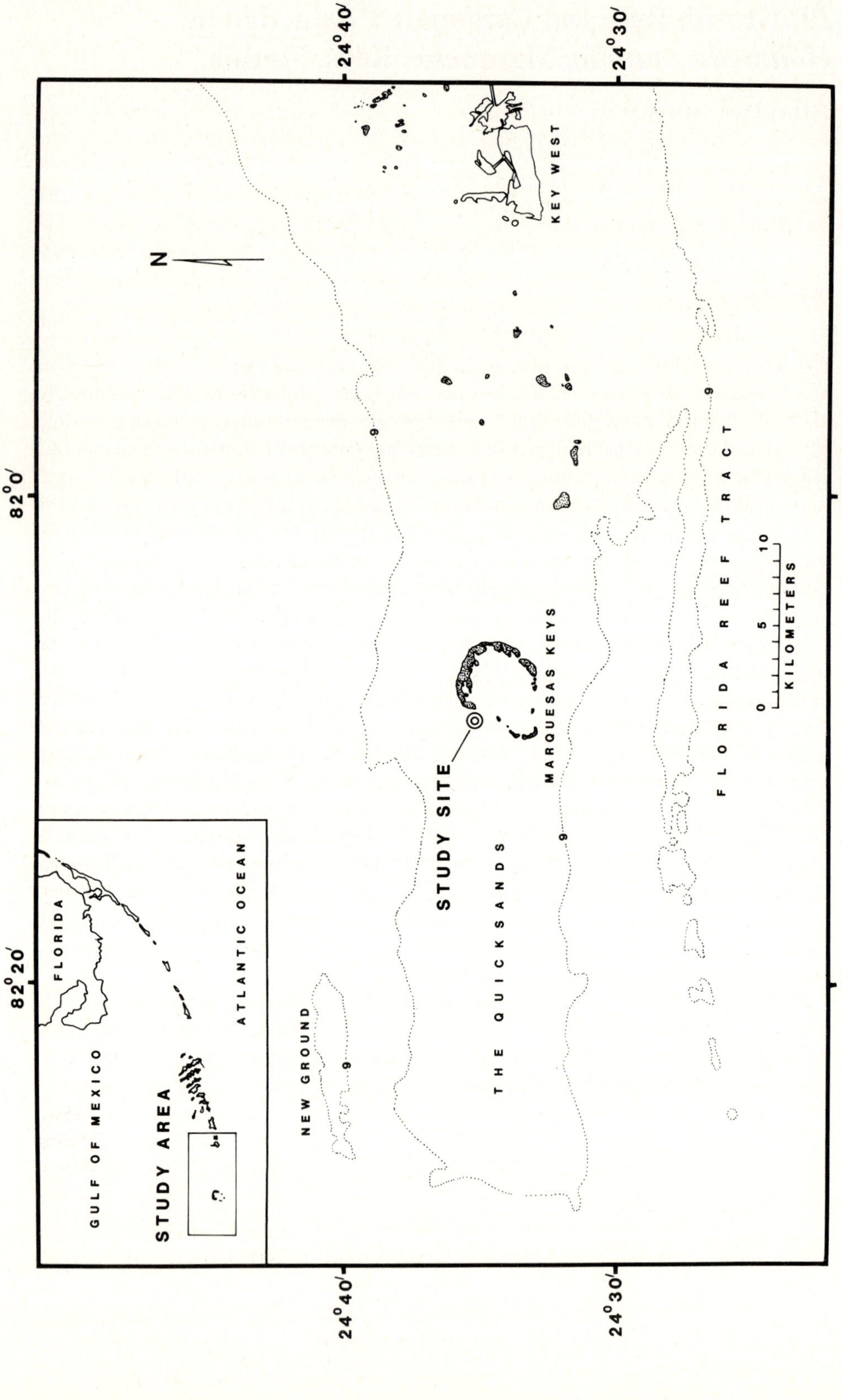

Fig. 1. Map of Marquesas Keys area showing location of study site. Inset indicates location of study area in the Florida Keys

Description of Area and Study Site

The Marquesas Keys, located 29 km west of Key West, Florida, are a circular group of low (average elevation less than 0.5 m) mangrove-covered islands whose atoll-like configuration forms and partially encloses a shallow (average depth less than 0.5 m) sediment-filled lagoon (Fig. 1). Water flow in and out of the lagoon is controlled by shallow seagrass-covered mudbanks and 2- to 3-m-deep tidal channels that penetrate the atoll along its south and west rim. A series of piston cores and rock borings taken on the islands and across the lagoon (unpublished data, U.S. Geological Survey, Fisher Island Station) indicate that the entire atoll down to Pleistocene bedrock is composed almost entirely of modern unconsolidated *Halimeda* sediments. These sand deposits extend westward from the Marquesas Keys to form the Quicksands (Fig. 1), a 13 × 29 km area of high energy *Halimeda* sand accumulation that is up to 12 m thick (Shinn et al. 1982). A shallow (less than 2 m) seagrass-covered shoal on the atoll's west edge was selected as a study site because the biota there appeared to be representative of the *Thalassia-Halimeda* community that dominates large portions of the study area.

Methods

Techniques for measuring growth in *Halimeda* have included sequential photographs of developing plants (Colinvaux et al. 1965), diver observation of individual specimens identified by coded tags secured around plant thalli with electrical wire (Merton 1971), cropping of plant material growing up through monofilament netting stretched over the substrate (Bach 1979), and in situ staining of algae with Alizarin Red-S to produce a time marker for subsequent measurement of new plant growth (Wefer 1980). Of these, in-situ staining was selected as the only practical method of determining growth rates in *H. opuntia,* since its major growth form in the study area consists of a loosely interlocking series of connected plates that coalesce to form compact clumps and large cushion-like mats, some of which exceed 3 m in diameter. Conventional techniques, such as photography, observation, and cropping, I felt would be useful but only as accessory tools in a long-term field study of an alga that grows both rapidly and in three dimensions.

Healthy plants with a diameter between 20 and 50 cm were selected for staining, since observations at the study site indicated that this size range was typical of *H. opuntia* colonies growing there. Plants selected for study were snugly enclosed in a thin sheet of clear plastic held in place by an iron chain sufficiently long enough to encircle the entire colony. Staining was accomplished by injecting a concentrated solution of Alizarin and seawater underneath the plastic covering with a syringe made from polyvinyl chloride pipe and plastic tubing. Stain was added in this manner until the entire area inside the plastic cover was opaque. After two hours, the plastic was removed and the location marked with a labelled plastic pipe driven into the sediment alongside the colony. Stain solutions were mixed in the field as needed in 30 × 30 cm self-sealing plastic bags in the following proportions: 3.7 g of Alizarin Red-S to 1250 ml of seawater.

Because of the remoteness of the study area and the often unpredictable nature of the weather there, frequent monitoring was not possible. To insure replicate samples, therefore, and in the event a colony died or was destroyed between visits, several plants (as many as three) were stained each time the site was visited. In addition, as part of the monitoring schedule, condition and size of colonies were recorded with underwater photographs. Salinity was measured with an optical densitometer, and water temperature data were collected by recording thermographs placed at the study site and nearby New Ground Shoal (Fig. 1). Monitoring of growth, sample collecting, and sample processing were as follows.

Previously stained colonies were subsampled during each visit (a small cluster of plates was removed down to the stain level) and the collected pieces immersed in full strength household bleach (20% sodium hypoclorite) for 3 min, then thoroughly rinsed. This process removed plant chlorophyll, revealing newly formed white plates attached to older, easily identifiable segments that were stained. Whole colonies were harvested by gently removing them from the substrate and floating the entire mass into a plastic collecting bag. When the bag was raised to the surface, fine sediment and debris along with excess water were drained from the bag and its contents labelled, sealed, and placed on ice to prevent breakdown of plant-connective fibers. In the laboratory samples were thoroughly rinsed in fresh water to remove adhering mud and other extraneous material. Any particles recognizable as *Halimeda* plates were retained with the sample and assumed to be residue shed from the colony. All cleaned material was bleached as described earlier, soaked and rinsed in fresh water, and dried in the sun for 8 h. A dissecting scalpel and scissors were used to remove new plant growth from the older stained area of the colony. After new plate growth was separated from the stained portion of the colony, both were dried in an electric oven at 200°F for 30 min to remove residual moisture, then sealed in plastic bags. All samples were treated in the manner described and weighed on a Mettler electronic weighing device to the nearest thousandth gram.

Results

Growth rates of four *Halimeda opuntia* colonies were monitored from July 1982 to May 1983 and their rate of carbonate sediment production calculated by determining total dry weight of new plates produced by each colony at various times during the year (Fig. 2). Production of new plates was found to be strongly seasonal, and all *H. opuntia* observed in the vicinity of the study area, regardless of size, appear to follow a common pattern of growth (Fig. 2). Formation of new plates begins in the spring between March and May and continues unabated until fall (November). During November, growth drops sharply, and with the advent of winter, plants revert to a near dormant state, generating few new plates until spring, whereupon the growth cycle is repeated.

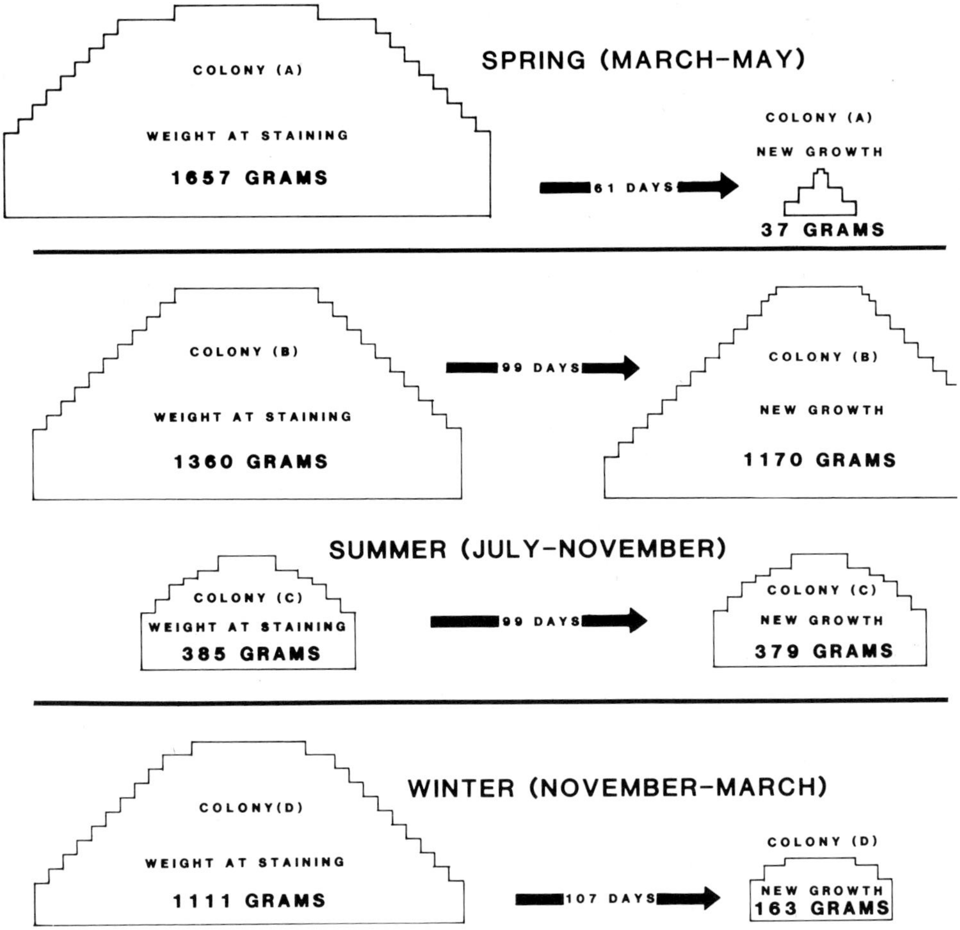

Fig. 2. Seasonal growth rate in *Halimeda opuntia.* (Data shown do not include months of May and June). Graphic representation of algal colonies are drawn to scale for size comparison. Colony (*B*) dimensions at staining: height = 24 cm, width = 25 cm, length = 48 cm

Assuming a negligible loss of plant material from shedding during the year and barring subsequent loss of plates during harvest, the data presented in Fig. 2 indicate that the *H. opuntia* studied were capable of doubling their weight during a year's growth cycle. Observation of growth patterns in Marquesas Keys *H. opuntia* revealed that as many as seven new plates can develop along the upper edge of a single mature plate, and since growth in this species is a function of plate segments proliferating from existing ones, it follows that, barring interference from outside sources, plate production could (in theory) proceed at a near exponential rate. Obviously, this does not happen, because, in great part, of the restriction of lateral growth by crowding from adjacent plates that confines major growth to the periphery of the colony. In addition, according to Goreau (1963) and Colinvaux et al. (1965), growth in *Halimeda* is not continuous but episodic, even over relatively brief time periods. Determining

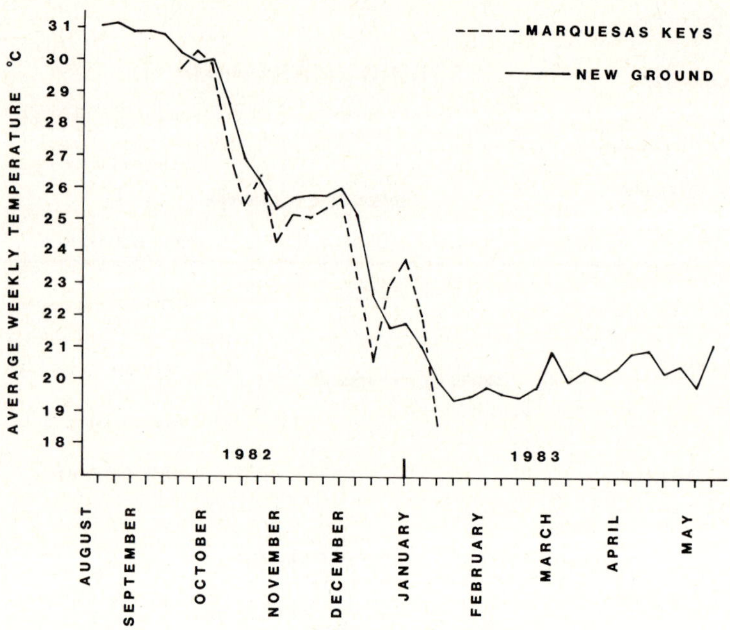

Fig. 3. Weekly plot of bottom water temperature at study site and New Ground. Deeper water (5 m) at New Ground apparently moderates both summer and winter temperature fluctuations seen at shallow (1.5 m) study site. Missing study-site data due to instrument malfunction

fluctuations in growth over time intervals less than one month was not possible in this study because of the infrequency of site monitoring.

For the purposes of this study, extrapolation of growth rate data was confined to a 100 m^2 area surrounding the study site where *H. opuntia* was estimated by observation and underwater photography to cover a minimum of 12.5% of the substrate. Thus, a colony with a height of 25 cm and a basal area 25 × 50 cm would occupy each square meter of seabed. Assuming a production rate comparable to that of colony B (Fig. 2) (since colony sizes are nearly equal), a plant of these dimensions should produce a minimum of 1250 g dry weight of plates per square meter per year. Yearly plate production within the 100 m^2 area would be 12,500 kg. Production of calcium carbonate was calculated to be 1088 g per square meter per year and 10,880 kg for the entire 100 m^2 area which calculation was based on the work of Böhm (1973), who found that Jamaican *H. opuntia* collected from a depth of 2 m contained an average of 87% (dry weight) $CaCO_3$.

Field observations and available data indicate that water temperature (Fig. 3), light availability, and overgrowth by algal epiphytes are the major factors controlling growth in *H. opuntia* in the Marquesas Keys area. Salinity ranged only from 34 to 39 ppt during the entire year and is presumed not to have greatly influenced observed growth rates.

Summary

Seasonal growth rates and carbonate production in the codiacean alga *Halimeda opuntia,* a major contributor to modern marine sediments, were measured at the Marquesas Keys 29 km west of Key West, Florida, from July 1982 to May 1983. Growth rates were determined by harvesting and analyzing entire colonies of algae that had been previously stained in situ with Alizarin Red-S. Increase in weight of Marquesas *H. opuntia* plates over time was calculated by weighing over 10,000 plates collected during the year at the study site. Plate formation was most rapid during summer months from June through September and slowest from late fall until early spring (November through March). One colony, 24 m high by 48 cm in diameter, produced 1.2 kg (dry weight) of algal plates in 99 days, increasing its original weight by 85%.

Acknowledgments. The author thanks Daniel M. Robbin, Charles Sherrer, Van Mount, and Lee Kump for assistance in field monitoring. Special appreciation is due my wife, Elizabeth, who assisted me on numerous field trips. Thanks arc also due Eugene A. Shinn for providing ideas and discussion during the study. Barbara Lidz provided many helpful suggestions during manuscript preparation and deserves speciel credit for her critical editing and typing of the final manuscript.

References

Bach SD (1979) Standing crop, growth and production of calcareous Siphonales (Chlorophyta) in a south Florida lagoon. Bull Mar Sci 29(2):191–201

Böhm EL (1973) Studies on the mineral content of calcareous algae. Bull Mar Sci 23(2):177–190

Colinvaux LH, Wilber KM, Watabe N (1965) Tropical marine algae: growth in laboratory culture. J Phycol 1(2):69–78

Finckh AE (1904) Biology of the reef-forming organisms at Funafuti Atoll. In: The Atoll of Funafuti. Borings into a coral reef and the results. R Soc London, pp 125–150

Goreau TF (1963) Calcium carbonate deposition by coralline algae and corals in relation to their roles as reef-builders. Ann N Y Acad Sci 109:127–167

Hillis-Colinvaux LH (1980) Ecology and taxonomy of *Halimeda:* primary producer of coral reefs. In: Blaxter JHS, Russel Sir Frederick S, Yonge Sir Maurice (eds) Advances in marine biology, vol 17. Academic Press, London New York, pp 1–327

Merton Sister MJ (1971) Ecological observations of *Halimeda macroloba* Decaisne (Chlorophyta) on Guam. Micronesica 7(1-2):27–44

Shinn EA, Holmes CW, Hudson JH, Robbin DM, Lidz BH (1982) Non-oolitic, high-energy carbonate sand accumulation – the Quicksands, southwest Florida Keys. Am Assoc Pet Geol Annu Meet Calgary, Book Abstr, pp 110–111

Wefer G (1980) Carbonate production by algae *Halimeda, Penicillus* and *Padina.* Nature (London) 285:323–324

21 Calcification in Fossil Neomereae (Dasycladales)

P. GÉNOT[1]

Introduction

Among the Dasycladales, aragonitic calcification covers the outer layer of the thallus in many species. The accuracy of the reconstruction of fossils depends on the extent of this calcification during the life of the alga and on the state of post-mortem preservation.

Although the literature on the systematic and morphology of fossil and recent Dasycladales is relatively abundant, studies of calcification are rare and generally limited to a number of selected species of calcareous algae. This paper deals with one group of Dasycladales, the Neomereae, specifically the genera *Neomeris* and *Cymopolia.* These living genera are also known as fossils, which permits comparative studies between living and extinct taxa. In addition, many Cenozoic species of *Neomeris* and *Cymopolia* have retained their original aragonite while other calcareous fossil Dasycladales are more frequently calcitic.

The morphology of the 18 Cenozoic species of Neomereae has been previously studied (Génot 1980). In *Neomeris,* each primary branch (R1) bears two secondary branches (R2) positioned on both sides of the fertile ampulla. In *Neomeris* (*Larvaria*), the secondary branches are below the plane containing the fertile ampullae. In *Cymopolia,* each R1 terminates with more than two R2 around the fertile ampulla.

Variability of the Degree of Calcification (Figs. 1–3)

The morphology of fossils may be extremely variable, depending upon the extent of the initial calcification, and will produce the following results:

1. Heavy calcification around the fertile ampullae and the branches produces cylindrical specimens (Fig. 1a): e.g., *Neomeris herouvalensis.*
2. Light or no calcification around the anatomical elements, notably the rows of R2 in circles, produces ringed calcification enveloping the fertile organs (Fig. 1b): e.g., *Neomeris limbata.*
3. Fossilization frequently leads to breakup of the cylindrical or ringed specimens (Fig. 1c,d): e.g., *Neomeris pseudo-eruca.*

---

1 Département des Sciences de la Terre, Université de Nantes, 44072 Nantes Cédex, France

Paleoalgology: Contemporary Research and Applications
(eds. D.F. Toomey and M.H. Nitecki)

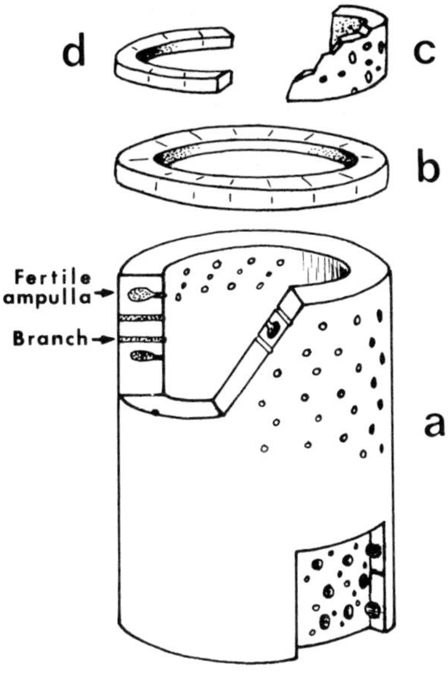

Fig. 1. Calcification and preservation of fossil *Neomeris*

The variability in the degree of calcification within each species of *Neomeris* is presented in Figs. 2 and 3. With the exception of the outermost surface of the R2, almost complete calcification of the stem, ampullae, and branches has been observed in only two species of *Neomeris*, *N. larvarioides* and *N. scrobiculata* (Fig. 3c,h). In the 14 other species, calcification is always around the reproductive organs. It is rare on the R1, but when present is limited to the distal part of these branches (Fig. 3i). In turn, axis calcification is visible on certain specimens of *N. auversiensis*, *N. fragilis*, *N. filiformis*, and *N. encrinula* (Figs. 2d,f,g and 3d). But in all species and in all individuals, calcification is certainly most irregular along the R2. The central part of the organs (around the greatest diameter of the fertile ampullae) is frequently preserved. Bases and tips of branches are often missing; however, calcification in these regions has been observed in *N. auversiensis*, *N. craniphora*, *N. arenularia*, and *N. scrobiculata* (Figs. 2a,g and 3c,i). The outermost facet of R2 is not preserved.

Irrespective of the extent of calcification and the conditions of fossilization, the shape of the calcareous envelope surrounding each of the fertile ampullae determines the appearance of the fossils. These calcareous envelopes with their undulated surfaces impart great rigidity and a tubular shape to certain individuals of *N. arenularia* (Génot 1980).

The calcification of fossil and living *Cymopolia* is more extensive than the calcification of *Neomeris*, since, with the exception of the upper facets of the distal extremities of the R2, it covers almost the entire organs.

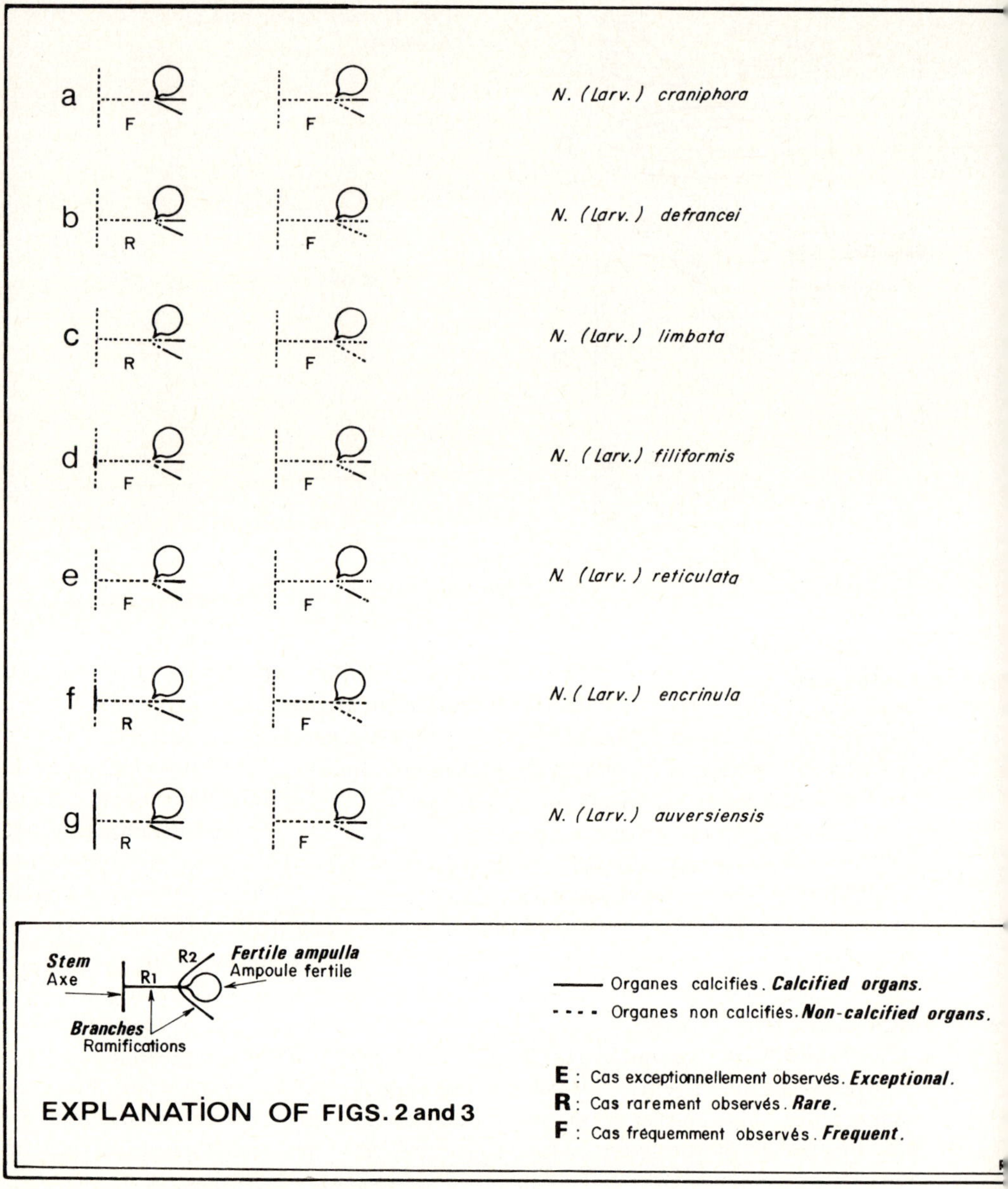

Fig. 2. Calcification of *Neomeris* (*Larvaria*) species

Ultrastructure of the Calcareous Envelope (Plates 1–3)

In only the following five fossil and five extant species of Neomereae has the ultrastructure of the envelope been analyzed: fossils – *Neomeris limbata* (Bignot 1977), *N. arenularia, N. scrobiculata* (Flajs 1977a,b), *N. craniphora* (= *Larvaria fragilis*

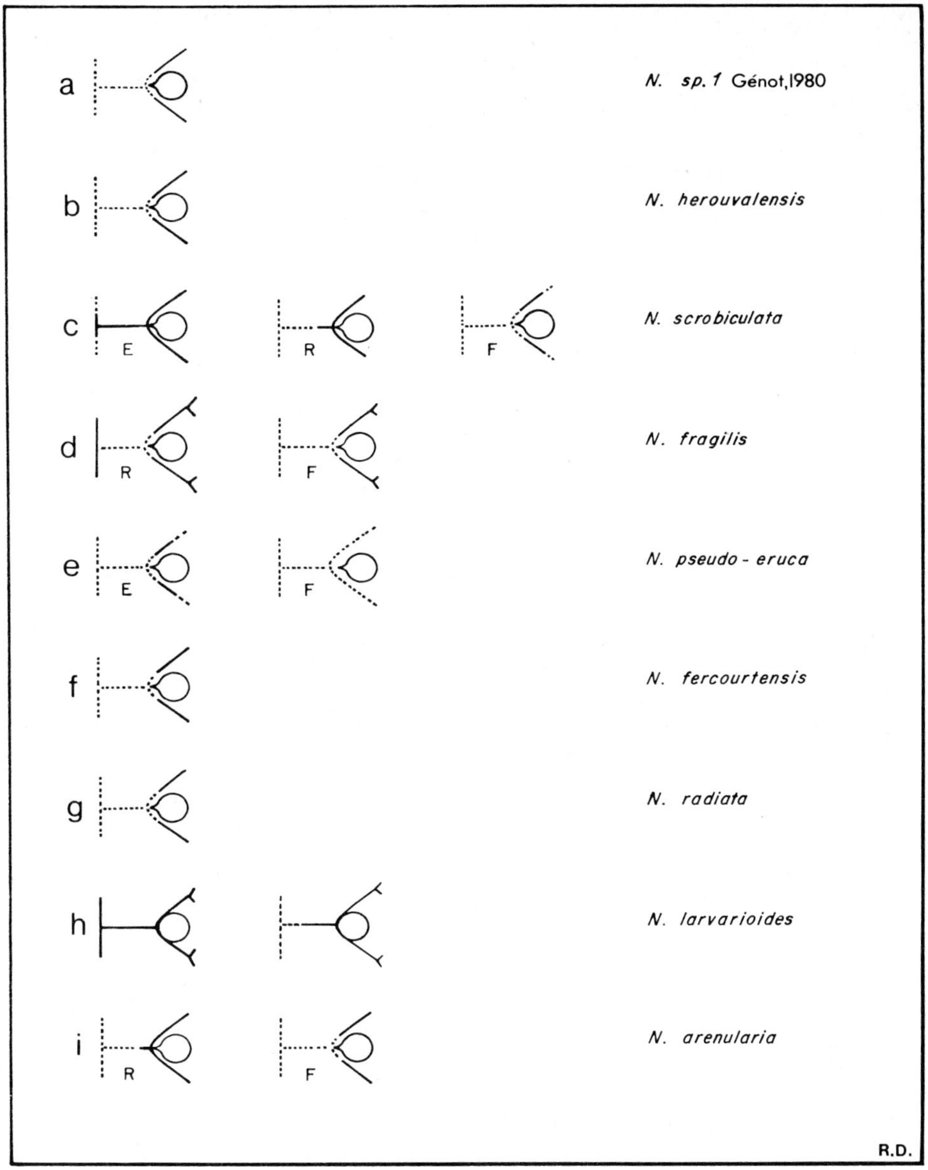

Fig. 3. Calcification of *Neomeris* (*Neomeris*) species

Defr.), and *Cymopolia elongata* (Flajs 1977a); living – *N. annulata* (Bignot 1977, Flajs 1977a), *N. eruca, N. dumetosa* (Flajs 1977a), *Cymopolia barbata* (Flajs 1977a, Bassoullet et al. 1979), and *C. mexicana* (Flajs 1977a,b; synonymous with *C. barbata,* Valet 1969).

The studies of the 13 other species (shown in Figs. 2 and 3) allowed us to enlarge further upon the descriptions presented by the above authors.

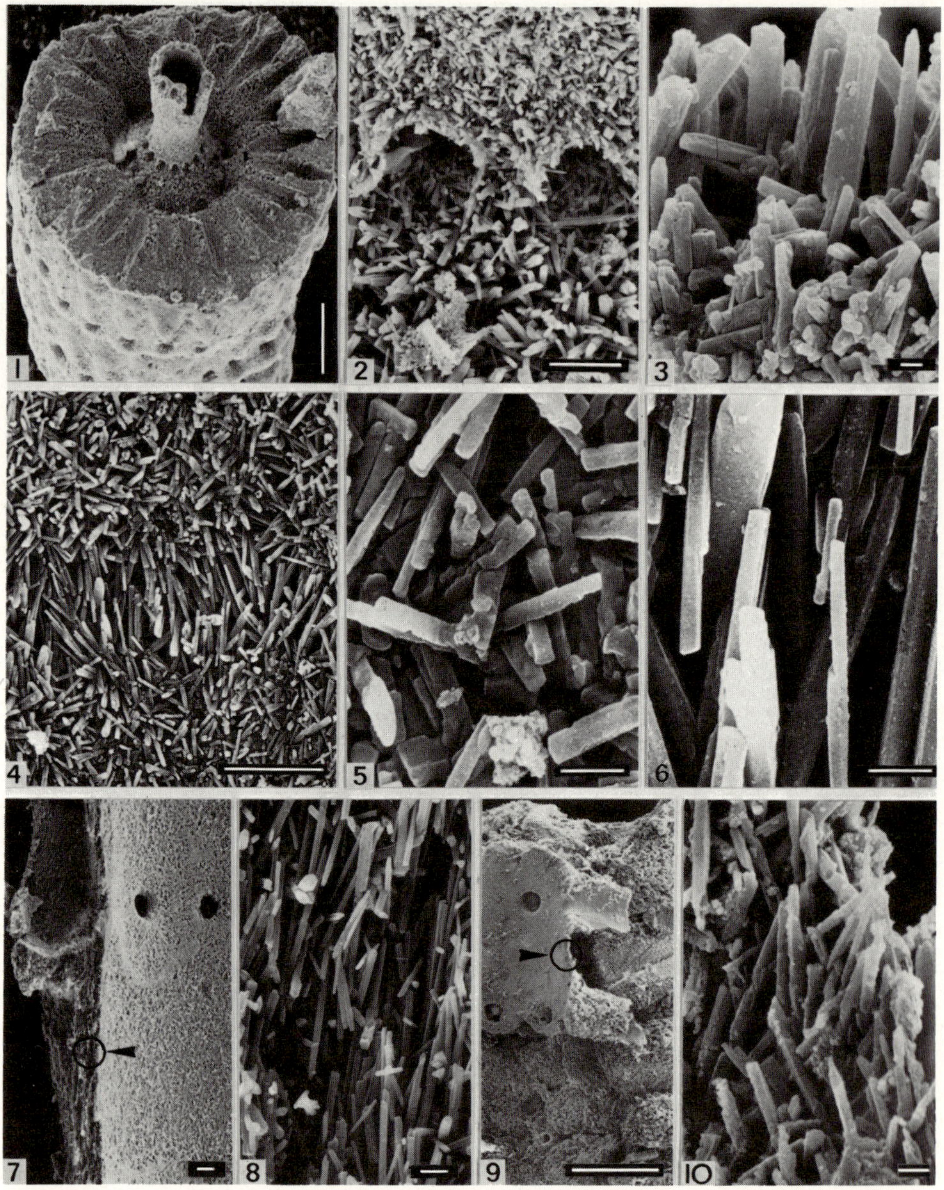

Plate 1. Scale: 1 μ for Figs. 3, 5, 6, 8, 10; 10 μ for Figs. 2, 4, 7; 100 μ for Figs. 1, 9.
1–6 *Neomeris auversiensis* (Morellet), Auversian, Auvers-sur-Oise, France. **1** Specimen with axial calcification. **2** Calcification around proximal extremities of R1. **3** Detail of outer surface of axial calcification. **4** Inner surface of axial calcification. **5** Details of upper part of calcification shown in **4**. **6** Details of central part shown in **4**.
7, 8 *Neomeris fragilis* (Defrance), Lutetian, Grignon, France. **7** Longitudinal section through axial calcification (*arrow:* details shown in **8**). **8** Detail of **7**.
9, 10 *Neomeris scrobiculata* (Gumbel), Cuisian, Pierrefonds, France. **9** Fragment of axial calcification (*arrow:* details shown in **10**). **10** Details of **9**

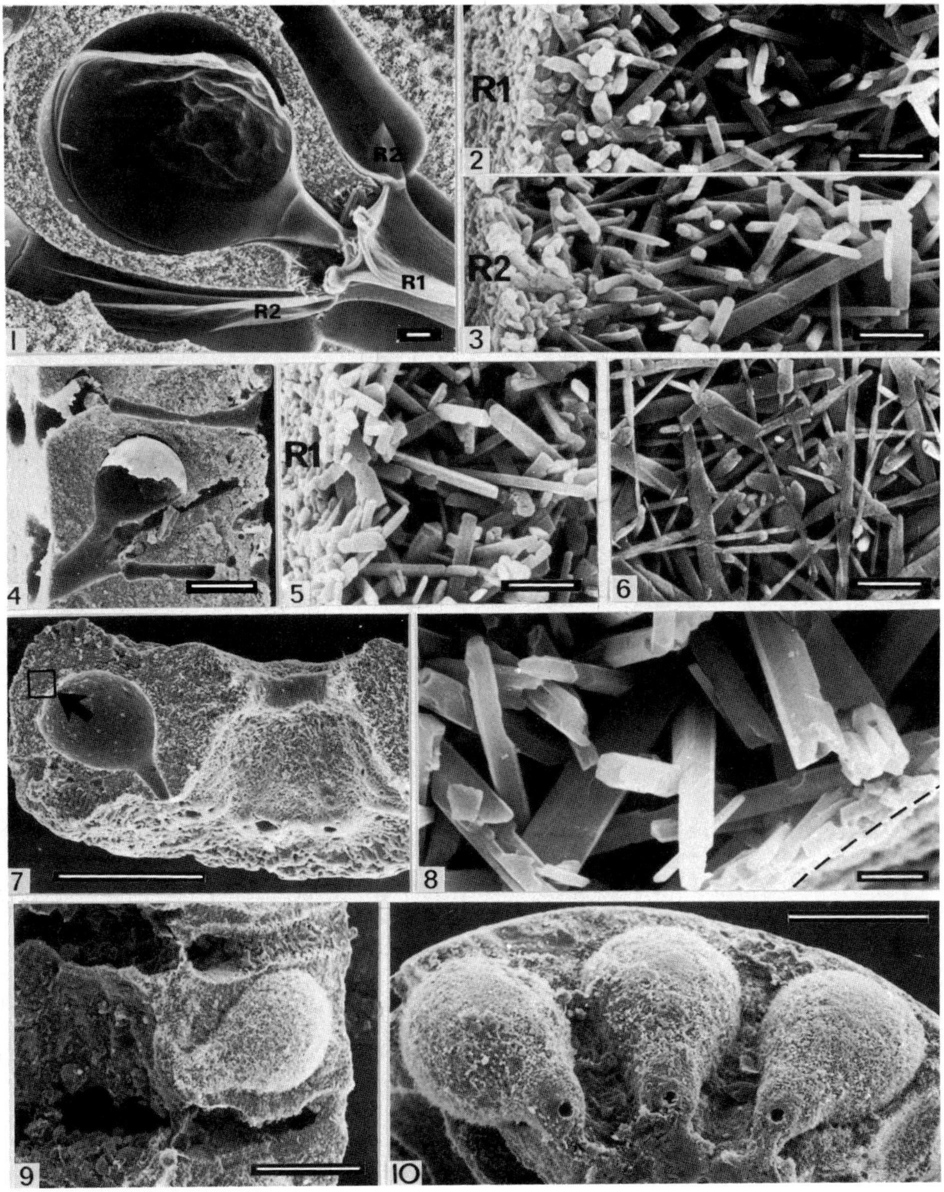

Plate 2. Scale: 1 μ for Figs. 2, 3, 5, 6, 8; 10 μ for Fig. 1; 100 μ for Figs. 4, 7, 9, 10.

1, 2, 3, 6 *Cymopolia barbata* (Linné), Recent, Andros Island, Bahamas. **1** Branches and fertile ampulla (organic wall present inside cavities). **2** Calcification along a R1. **3** Calcification along a R2. **6** Calcification around a fertile ampulla.

4, 5 *Cymopolia elongata* (Defrance), Lutetian, Grignon, France. **4** Longitudinal section of branches and fertile ampulla. **5** Calcification along R1.

7, 8 *Neomeris encrinula* (Defrance), Auversian, Attainville, France. **7** Longitudinal section of a fertile ampulla (*arrow:* details shown in 8). **8** Details of 7. Calcification along a fertile ampulla (*right corner*), layer with crystals tangential to the surface of the ampulla (*dotted line*).

9, 10 *Neomeris filiformis* (Morellet), Lutetian, Damery, France. Longitudinal and transversal sections of the wall showing a thin calcareous envelope around fertile ampullae

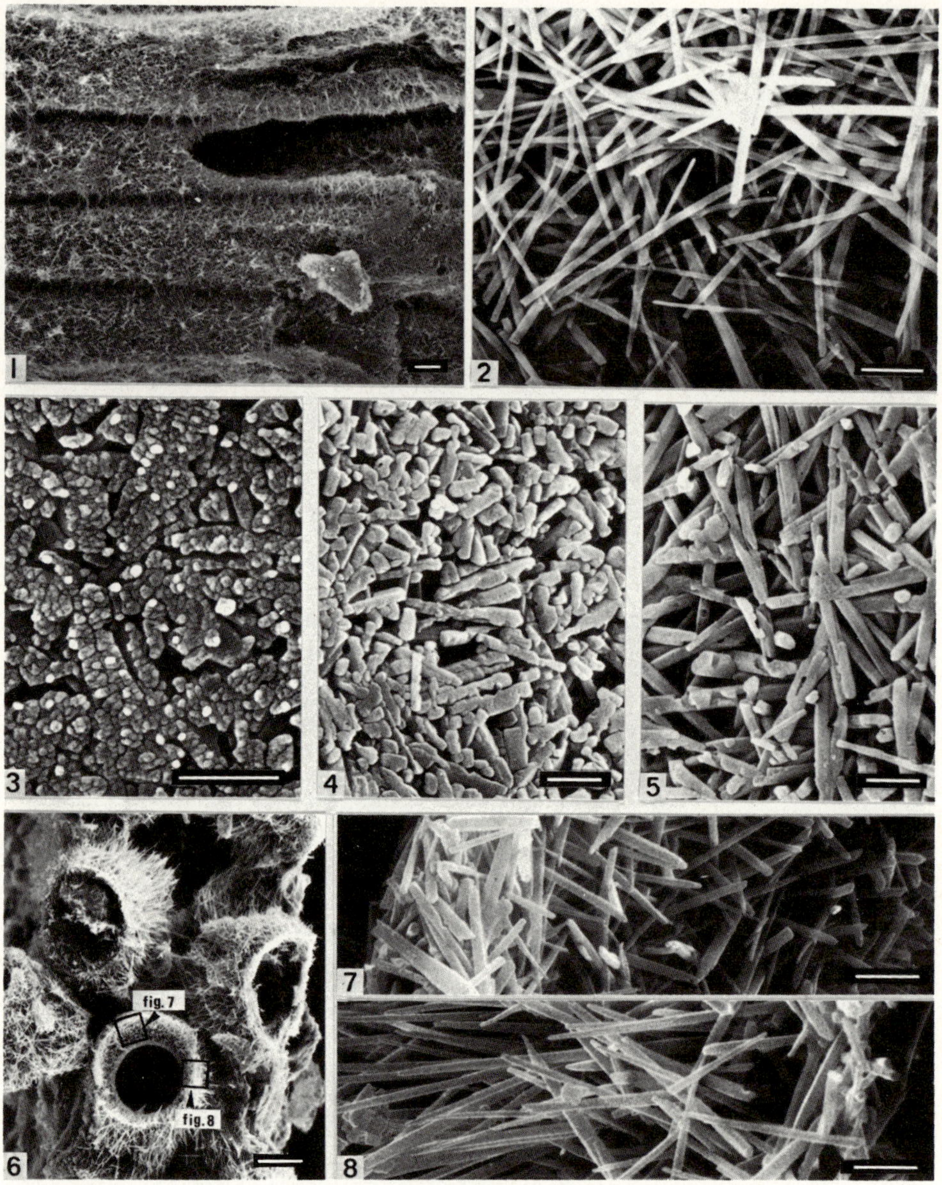

Plate 3. Scale: 1 μ for Figs. 2, 3, 4, 5, 7, 8; 10 μ for Figs. 1, 6.

1, 2 *Neomeris annulata* Dickie, Recent, Rose Island, Bahamas. 1 Calcification around 4 R1. 2 Needle-shaped crystals around R1.

3 *Cymopolia elongata* (Defrance), Lutetian, Grignon, France; granular crystals around fertile ampulla.

4 *Neomeris herouvalensis* Mun.-Ch., Cuisian, Hérouval, France; calcification around fertile ampulla.

5 *Neomeris arenularia* Mun.-Ch., Auversian, Auvers-sur-Oise, France; calcification around fertile ampulla.

6–8 *Neomeris larvarioides* (Morellet), Lutetian, Croix-Blanche, France. 6 Calcification around R1 (*arrows:* details shown in 7, 8). 7 Details of 6; crisscrossed crystals. 8 Detail of 6; crystals perpendicular to the surface of R1

Aragonitic crystallization appears in two main forms: (1) in isolated grains or aggregates, most often situated along the anatomical structures (Plate 3, Fig. 3), and (2) in prisms, sometimes needle shaped with more or less clear-cut facets (Plate 2, Fig. 8; Plate 3, Fig. 2). These are between the branches and ampullae, and are generally randomly oriented; however, in certain regions of the wall they are parallel or perpendicular to the main axis of organs.

Calcification Around the Stem. In rare specimens in which the axis is calcified (Plate 1, Fig. 1), the inner surface of the axis is mostly composed of aciculary and nonoriented crystals (Plate 1, Fig. 5). In certain regions on these surfaces, however, the orientation of crystals is more or less parallel to the axis (Plate 1, Fig. 6). The transition between these two types is abrupt (Plate 1, Fig. 4) and does not seem to be related to the insertion of whorls of primary branches along the axis. This arrangement of prisms, more or less parallel to the main axis of the alga, has also been observed in *N. fragilis* (Plate 1, Figs. 7, 8) and *N. scrobiculata* (Plate 1, Figs. 9, 10). The outer surface of axial calcification consists most often of a criss-cross of crystals with occasional prisms which are arranged perpendicularly to the axis of the alga (Plate 1, Fig. 3).

Calcification Around Branches and Fertile Ampullae. In all species studied, the spaces between the different organs of the alga are filled with irregularly arranged prismatic crystals (Plate 2, Figs. 2, 3, 5, 8). In the immediate neighborhood of the organs (branches and ampullae), the direction of the needles or lathlike aragonite is more or less tangential to the surface of the laterals, particularly around the fertile ampulla (Plate 2, Fig. 6; Plate 3, Figs. 4, 5). In this region, the crystals form a compact layer which is often clearly delineated and which breaks off from the wall when the specimen is fractured. The shape of the organs can then be reconstructed (Plate 2, Figs. 4, 9, 10).

Grains and slightly elongated prisms of aragonite sometimes form a layer of variable thickness which occurs between the outer surface of organs and the zone of elongated crisscrossed crystals (Plate 3, Fig. 3). The granular zone is not always present, but the internal surface of cavities (indicating the location of the organs), shows that the region of crisscrossed crystals exists in most specimens studied (Plate 3, Fig. 5). The distribution of crystals along the organs is not uniform, and the network of aragonitic needles and lathlike crystals is more or less dense and varies from species to species, depending on the region of the thallus, and even within individual alga. Sometimes, there is also a change in the direction of the crystals around an organ. Thus, in *N. larvarioides,* certain crystals along a primary branch are more or less perpendicular to the surface, while others form a crisscrossed layer (Plate 3, Figs. 6–8).

The three phases of calcification – (1) granular layer, (2) needles perpendicular to the surface of the organs, (3) irregularly arranged aragonite needles, described by Flajs (1977a,b) – have not been observed on any of the 18 species of Neomereae nor on *Cymopolia barbata* (Plate 2, Figs. 1–3, 6). This type of crystallization cannot therefore be considered typical of all Dasycladales.

Summary

All Neomereae have aragonitic skeletons whose degree of calcification varies from species to species and from individual to individual. A study of this variability leads to a more accurate definition of numerous species, particularly Cenozoic species. Aragonite crystals, especially lathlike and prismatic needles, are most often unoriented inside the wall. However, they can be parallel or perpendicular to the surface of the organs of the alga.

This study of the extent of calcification and the ultrastructure of skeletons will be developed in a future analysis of other species of Dasycladales, particularly of Bornetelleae and Parkerelleae.

Acknowledgments. I am grateful to A. Barreau (University of Nantes, France) for scanning electron micrographs; R. Dufour (Museum of Natural History, Bordeaux, France) for drawings, and G. Riegl for the English translation.

References

Bassoullet JP, Blanc P, Jaffrezo M (1979) La structure du squelette calcaire de quelques Dasycladales actuelles et fossiles. Trav Lab Micropaléontol Univ Paris 6:175–184

Bignot G (1977) L'observation des tissus organiques minéralisés au M.E.B.; 4: Les thalles calcifiés de quelques Algues marines, vivantes et fossiles. Trav Lab Micropaléontol Univ Paris 6:121–125

Flajs G (1977a) Die Ultrastrukturen des Kalkalgenskeletts. Palaeontographica Abt B 160 (4–6): 62–128

Flajs G (1977b) Skeletal structures of some calcifying algae. In: Fossil algae. Springer, Berlin Heidelberg New York, pp 225–231

Génot P (1980) Les Dasycladacées du Paléocène supérieur et de l'Eocène du Bassin de Paris. Mém Soc Géol Fr 138:40

Valet G (1969) Contribution à l'étude des Dasycladales. Nova Hedwigia 17:551–644

Algae and Sediments

22 Origin of Stromatactis – a Replacement of Colonial Microbial Accretions

H.H. TSIEN[1]

Introduction

Since 1881, and especially in recent decades, *Stromatactis* has been the subject of numerous studies. Despite wide interest in its nature, there is, however, still no agreement concerning its genesis (Bathurst 1959, 1981, Lees 1964, H. Dupont 1969, Wilson 1975, Burchette 1981; see also PREVIOUS WORK below, and Table 1. Most specialists regard *Stromatactis* as a cavity system, others believe that it is recrystallized algal colonies (Tsien 1971), or tufa-like microbial accretions (Monty 1982).

This paper presents a description of *Stromatactis* in Belgium, based on field analyses available from wire-sand cut quarry walls and blocks and from thin sections, and suggests that *Stromatactis* is a replacement of colonial microbial accretions by algae, cyanobacteria, and/or bacteria.

Table 1. Origin of *Stromatactis* – Current Interpretations

- I. As a cavity system.
  - 1. Organic origin: through the decay of a soft-bodied organism (Prentice 1950, Belliére 1953, Bathurst 1959, Philcox 1963, Lees 1964).
  - 2. Organism-controlled origin.
    - a) Unknown organic process (Bathurst 1959).
    - b) Formed in a sediment made patchily coherent by algal mats (Pratt 1982).
    - c) Related to sponge networks (Bourque and Gignac 1983).
  - 3. Inorganic origin – as a sedimentary structure.
    - a) As water escape structures (Heckel 1974).
    - b) Related to submarine-cemented crusts (Bathurst 1980, 1982, Burchette 1981).
    - c) Cavity opening during creep, slumping, or compaction of the mud (Schwarzacher 1961).
- II. As recrystallized Stromatoporoids (E. Dupont 1881).
- III. As precipitated calcite possibly resulting from decomposition of algae (Lecompte 1937).
- IV. As recrystallized calcite mudstone (Black 1952, Ross et al. 1975).
- V. As recrystallized algal colonies (Tsien 1971, 1977).
- VI. As tufalike microbial accretions (Monty 1982).
- VII. As recrystallized colonies of microorganic (algae, cyano-bacteria, bacteria) accretions (this paper).

1 Laboratoire de Paleontologie, Universite Catholique de Louvain, 3, Place Louis Pasteur, B–1348 Louvain-la-Neuve, Belgium

Paleoalgology: Contemporary Research and Applications
(eds. D.F. Toomey and M.H. Nitecki)

Previous Work

The term *Stromatactis* was first applied by Edouard Dupont (1881, p. 266) to irregular crystalline masses "de longues bandes ou de petites masses cristallines irrégulières" that he considered to be organisms, classifying them among stromatoporoids. In 1882, he believed the "calcaire de Waulsort" to be an agglomeration of organisms related to Devonian *Stromatactis* and he named the radiated sparry calcite *Stromatocus bulbaceus* and the "veines bleues" *Ptylostroma fibrosa.* He presumed that some fine structures inside *Stromatactis* are "D'un tissue extrêmement délicat, souvent voilé par une recristallisation, son squellette présente une disposition en bandes parallèles; l'intérieur de celles-ci se divise en chambres irrégulières et perpendiculaires dont les parois ont une structure rayonnée penniforme." It was only in 1885 that E. Dupont defined *Stromatactis* more precisely as "une série de bandes concentriques, à surface rugueuse, dont le tissu blanchâtre et chatoyant au polissage est très finement rayonné dans un sens perpendiculaire," and considered it to be an organism with only the external characters preserved. Delhaye (1908) regarded *Stromatactis* as a calcite concretion. Maillieux (1913a, b) considered *Stromatactis* as organic. In 1937, Lecompte studied giant *Stromatactis* (± 70 cm high) from Bergnonry Quarry, which is only one special type of *Stromatactis.* This type is associated with colonies and debris of *Striatopora gracilis* and shows the aureoles of pigment around the organism (*Striatopora gracilis*). He noted that "La position des organisms est en harmonie avec les particularités structurales des '*Stromatactis*' ... les terrasses paraissent aussi en conformité d'allure avec les digitations cristallines." He proposed that *Stromatactis* is precipitated calcite, possibly resulting from decomposition of algae, "On est assez naturellement porté a penser à des algues enveloppantes autour de ces colonies délicates. Ceux qui ont eu l'occasion d'observer la decomposition rapide des algues d'eau douce à l'intervention des bactéries seront assez portés à admettre cette hypothèse. Celli-ci pourrait, dans une certaine mesure, expliquer la form particulière des '*Stromatactis*' ainsi que leur régularité et résoudre le problème du mode de suspension des petits organisms enclavés." Dumon (1957) regarded *Stromatactis* as organic and believed that the preservation of the external form was due to rapid lithification of the lime mud. H. Dupont (1969), who studied Waulsortien facies in Waulsort, concluded that the appearance of the "calcaire à veines bleues", (= *Ptylostroma fibrosa*) is due to a differential recrystallization and thus frequently different from the so-called *Stromatactis.*

Stromatactis has been observed and studied from Silurian reefs (Lecompte 1938, Lowenstam 1950, Bourque and Gignac 1983), in Carboniferous (Black 1952, Bathurst 1959, Schwarzacher 1961, Orme and Brown 1963, Philcox 1963, Lees 1964), in Ordovician (Ross et al. 1975, Bathurst 1977), and in Devonian (Playford and Lowry 1966, Bellière 1953, Heckel 1974, Logan and Semeniuk 1976, Desbordes and Maurin 1974).

Lowenstam (1950) believed that the *Stromatactis* is an organic framebuilder that exhibits systematically varied forms in different parts of the Silurian reef. This systematic change in morphology across a reef bears a tantalizing similarity to the belts of frame-builders observed in modern reefs. Black (1952) considered that this pattern is a result of recrystallization of calcite mudstone in situ. Bathurst (1959) proposed that *Stromatactis* is a cavity in a sediment preserved by a sediment-binder, and later altered by compaction and internal erosion and solution. Earlier Bathurst (1959) believed that a sediment-binder is responsible for the steep slopes and cavity roofs. He interpreted *Stromatactis* as a cavity fill, consisting of variable quantities of sparry calcite cement and internal sediment. This interpretation has been largely accepted by most specialists, although some have allowed for a slight recrystallization of the host carbonate mud. Since them, the research on *Stromatactis* has been focused on the genesis of the cavity system.

Prentice (1950), Bellière (1953), Bathurst (1959), Philcox (1963), and Lees (1964) proposed that the cavity system was formed by the decay of a soft-bodied organism, yet unequivocal evidence to support this has not been produced (Bathurst 1982). Schwarzacher (1961) postulated cavity opening during creep, slumping, or compaction of the mud. He also proposed that cavities were supported by bryozoan skeletal framework. Heckel (1972) suggested that *Stromatactis* is a sedimentary structure related to a waterescape structure, and therefore *Stromatactis* should not be italicized (Heckel 1974). Since then, specialists have concurred with Heckel, writing "stromatactis" rather than "*Stromatactis.*" Bathurst (1980, 1982) has suggested further that the cavities are related to a submarine-cemented crust and that the masses of radiaxial fibrous calcite are an altered marine cement. This interpretation was, however, questioned by Kendall and Tucker (1973), Ross et al. (1975) and Logan and Semeniuk (1976). Kendall and Tucker (1973, p. 377) noted that some radiaxial fibrous mosaics are possibly a replacement of acicular carbonates of noncement origin. Orme and Brown (1963) and Ross et al. (1975) believed *Stromatactis* to be a product of recrystallization of the host carbonate mud. Logan and Semeniuk (1976) further suggested that *Stromatactis* is a product of recrystallization of carbonate mud and dissolution during low-grade metamorphism associated with pressure-solution.

Monty (1982) regarded radiaxial fibrous calcite (hence *Stromatactis*) as tufalike microbial accretions. Pratt (1982) suggested that *Stromatactis* is a cavity system formed in a sediment that has been made patchily coherent by algal mats. He stated that (1) patchy distribution of algal mats resulted in areas of both bound and unbound sediment, (2) sporadic floodlike episodes of rapid sedimentation temporarily smothered the living algal mat and left layers of unbound sediment, (3) periods of winnowing washed out unconsolidated, unbound sediment and promoted synsedimentary cementation of resulting stromatactoid cavities and algal bound framework. For Bourque and Gignac (1983), *Stromatactis* is a cavity system formed by early marine cementation of growth cavities in the sponge network or created by decay of local uncemented sponge tissue.

Stromatactis of Belgium

Stromatactis is a mass of fibrous calcite enclosed by irregular micrite masses. Its morphology is characterized by a smooth, though often undulose or stepped lower

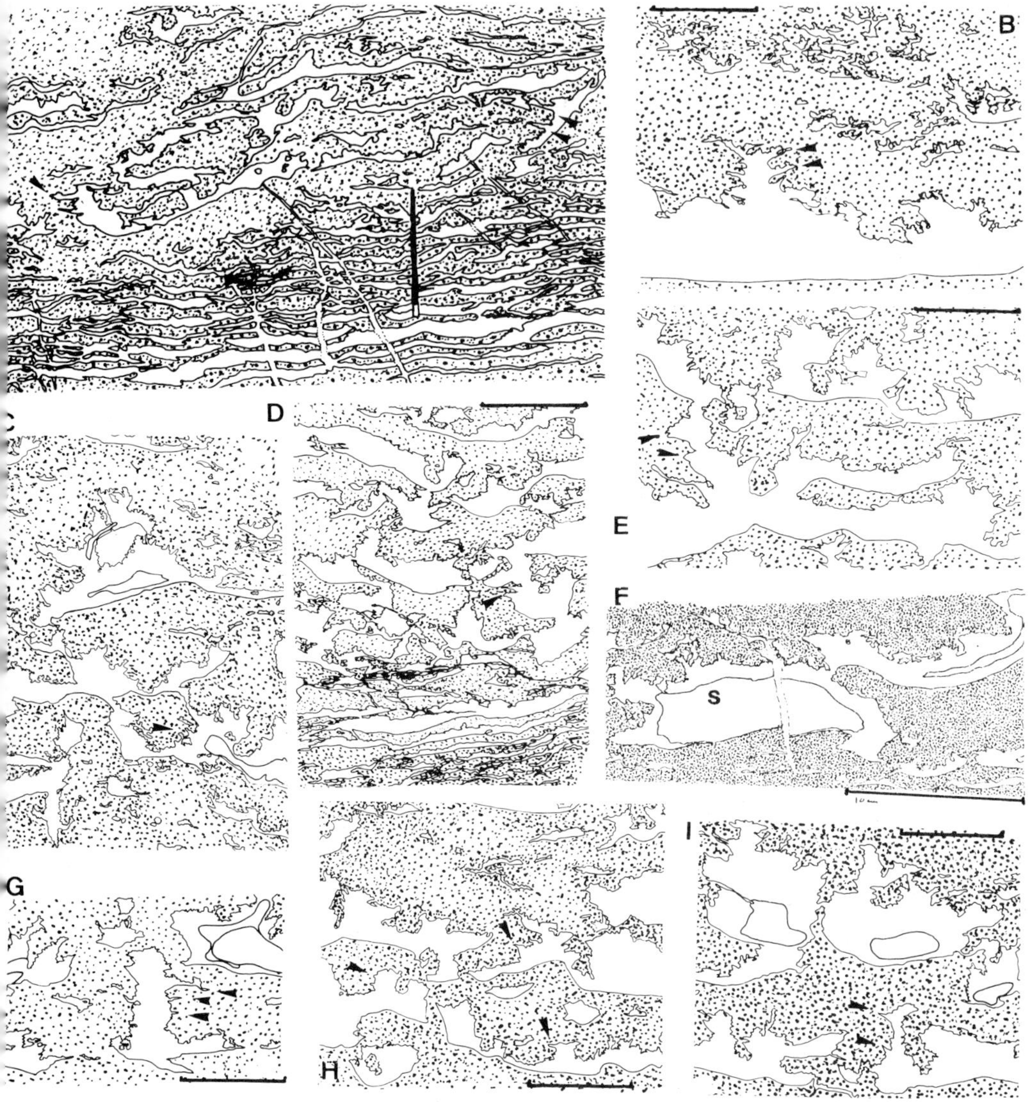

Fig. 1A–I. Diagrammatic representation of morphology of *Stromatactis.* The form varies from zebra to columnar. *Arrows* indicate roughly symmetrical lateral expansions characteristic of *Stromatactis.* All from Carriere des Croisettes, Vodecee. **A** Bar = 15 cm, **B–I** Bar = 10 cm

surface; and by irregular, embayed and digitated flanks on upper surface. Frequently these digitations have roughly symmetrical horizontal lateral expansions and contractions that are often aligned with similar digitations in adjacent *Stromatactis.* The shape varies from a thin few millimeters to two centimeter horizontal layers or bands, often bifurcating, and sometimes termed "zebra structures" to domal and columnar forms with symmetrical lateral digitations. Zebra forms generally occur at the base and reticulate *Stromatactis* occur in the central part of the *Stromatactis*

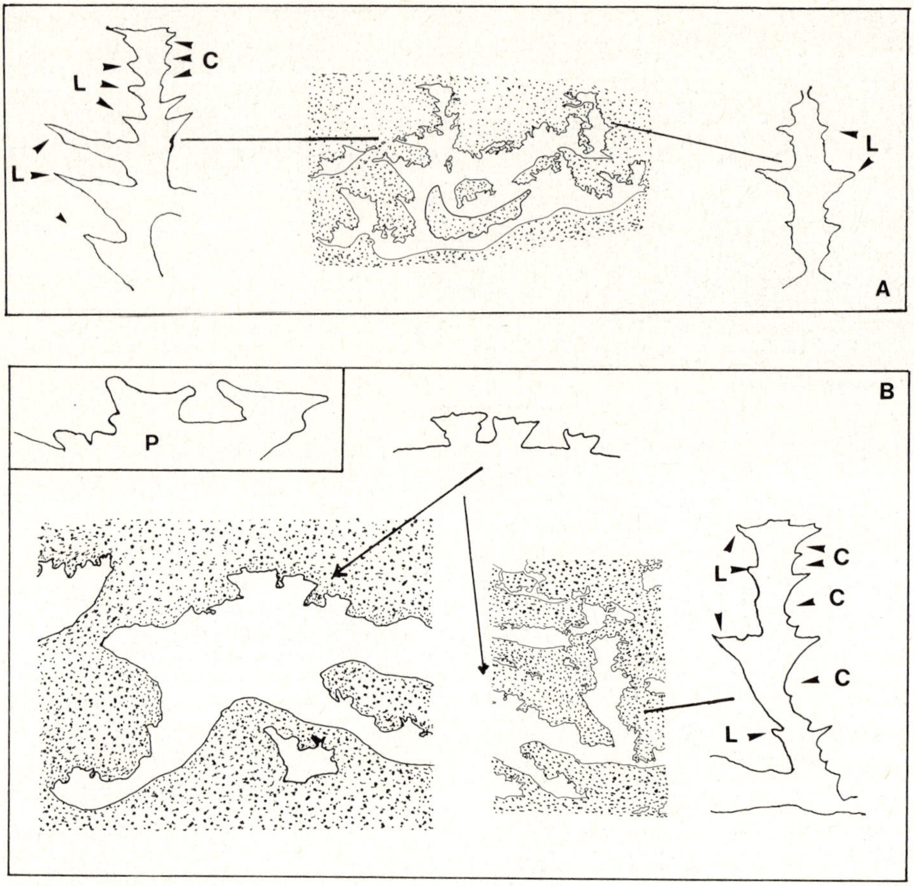

Fig. 2A,B. Diagrammatic representation of *Stromatactis* showing roughly symmetrical lateral expansions (*L*) and contractions (*C*). *Stromatactis* buddings are comparable with those of *Phillipsastrea* (*P*)

facies. The size varies from less than 2 cm across in isolated cases to large irregular shapes ranging up to 70 cm high.

Morphology. Morphology of *Stromatactis* varies from zebra, isolated lamellar, reticulated lamellar, domal, globular, to columnar form (Fig. 1 and Plate 1). *Stromatactis* has a smooth base usually flat, undulose, or stepped. Its upper surface and lateral flanks are generally digitate or dentate, very often with symmetrical lateral fingers extending into the surrounding micrite masses. Small structures of symmetrical forms comparable to the buds of *Phillipsastrea* may be found on the upper surface (Fig. 2).

The type described by E. Dupont (1881) typically has an anastomosing and reticulate structure, while the material described by Lecompte (1937) is a giant *Stromatactis*.

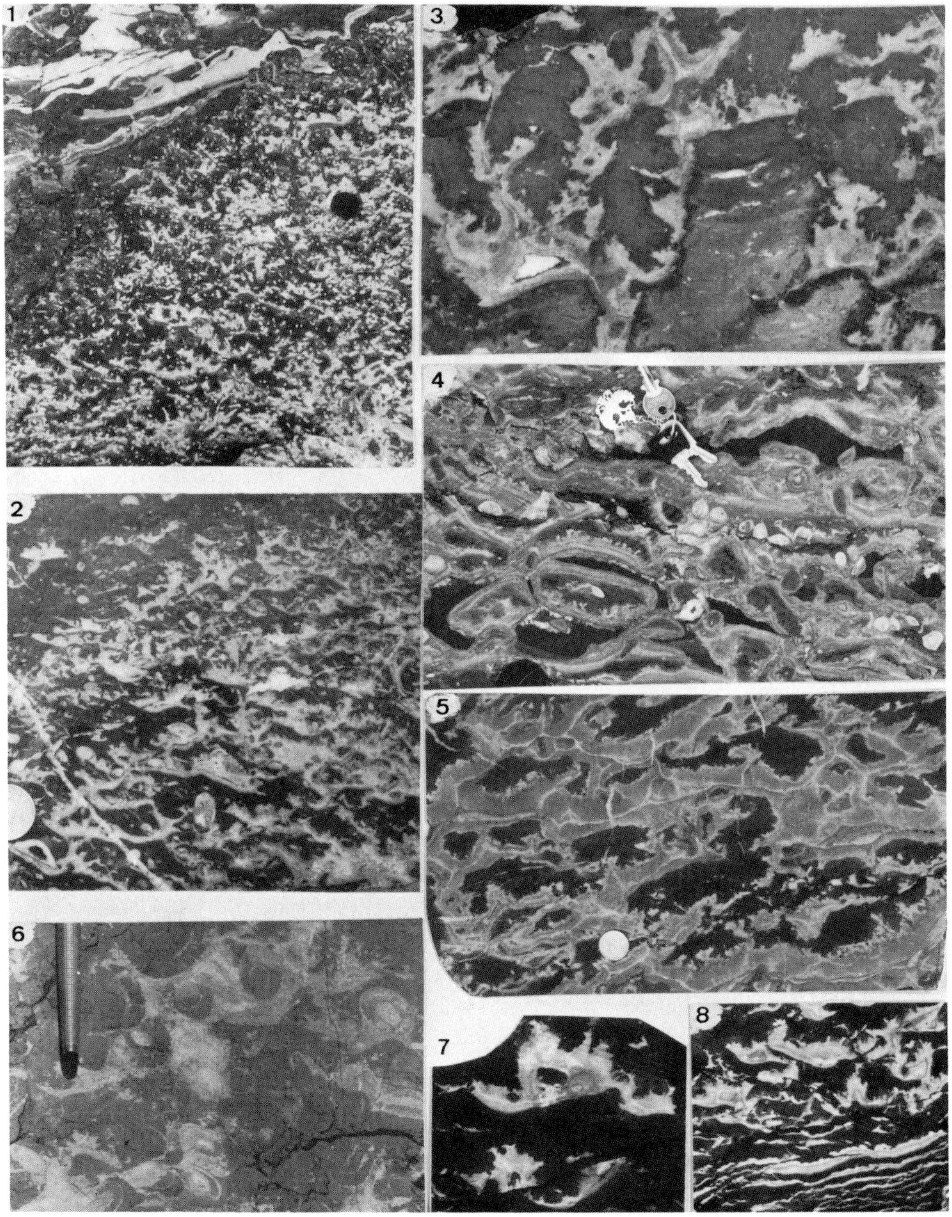

Plate 1. Photograph of *Stromatactis* from wire-sand cut quarry walls and polished slabs. 1 *Stromatactis* with *Phacellophyllum* sp. forms a small mound in the mud mound. Black hole about 8 cm in diameter. 2 *Stromatactis* forms a small mound in the mud mound. 3 Encrusting *Stromatactis*. 4 *Stromatactis* with sponges and brachiopods. 5 Reticulated lamellar *Stromatactis* from St. Remi Quarry. 6 *Stromatactis* with *Receptaculites* sp. 7 Isolated *Stromatactis*. 8 Ecological control of the morphology of *Stromatactis*

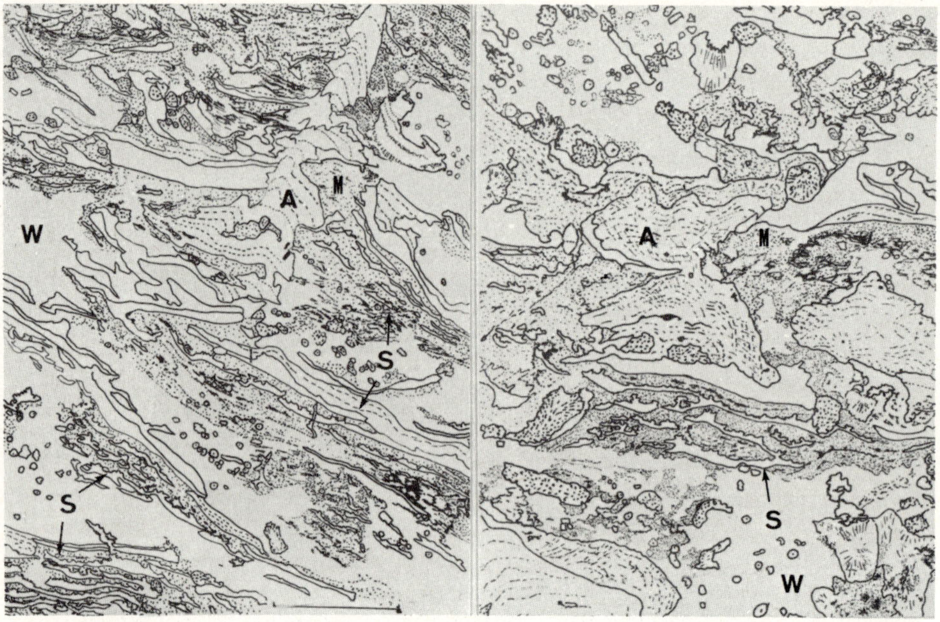

Fig. 3. Diagrammatic representation of Zebra *Stromatactis* (*S*) associated with *Alveolites* (*A*). *M* micrite; *W* wackestone. Arche Quarry, Belgium

Distribution. *Stromatactis* occurs only in the particular facies of red mud mounds (Tsien 1983). It generally forms layers or beds continuous across the mud mounds but thinning and disappearing into the surrounding shales of the mud mounds, in which very often, mini mounds form (Plate 1, Figs. 1, 2). The flanks of the *Stromatactic* mound and the growth direction of *Stromatactis* on the flanks may be inclined up to 45°. This suggests that *Stromatactis* is an organic colony. This angle occurs also in small micrite lenses in or at the upper and lateral margins of the mud mounds. In such case, *Stromatactis* is partially in contact with noncemented argilaceous shale. Therefore, *Stromatactis* is self-supporting, because neither sediment-binders nor cemented micrite frames for the maintenance of steep slopes and cavity roofs are present for *Stromatactis* to be a cavity system.

Stromatactis occurs frequently with *Alveolites*, *Phillipsastrea*, and sponges, and in some instances with *Receptaculites*, algae or with branching corals (Fig. 4, Plate 1, Figs. 1, 4, 6). However, these fossils are not necessarily associated with *Stromatactis*.

The distribution of shape of *Stromatactis* in *Stromatactis* facies shows considerable ecological variation which bears striking similarity to the morphologic variation of *Alveolites*, stromatoporoids, and *Phillipsastrea*. In basal part of facies *Stromatactis* is generally of lamellar or layered zebra shape; in the lower parts or on the flanks it is of isolated lamellar or domal form, whereas the central part is characterized generally by reticulate *Stromatactis* (Tsien 1977, 1980). In some cases, zebra *Stromatactis* are closely associated with *Alveolites* (Fig. 3).

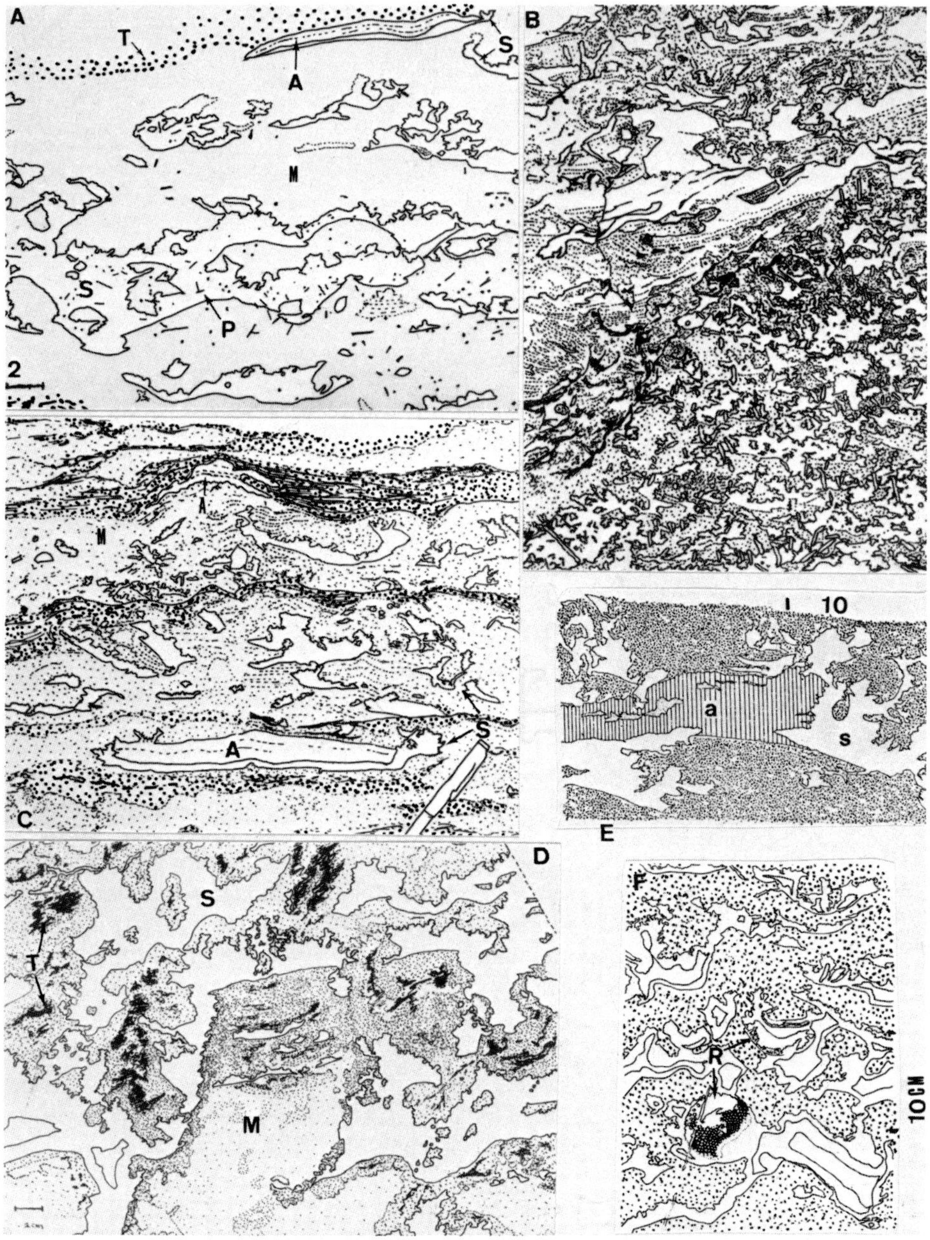

Fig. 4A–E. Diagrammatic representation: **A, C, E** *Stromatactis* occurring with *Alveolites* (**A**,a); **A, B** with branching corals; **E** with *Receptaculites* (*R*); **D** with biogenic micrite (*M*)

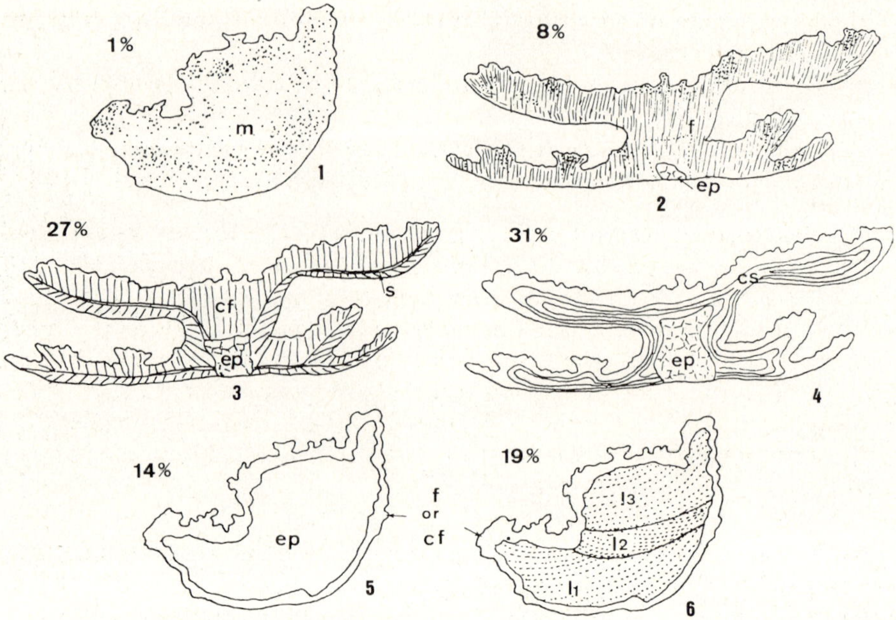

Fig. 5. Diagrammatic representation of types of *Stromatactis. 1–6* Types I–VI: *m* micrite; *f* fine fibrous calcite; *cf* coarse fibrous calcite; *ep* clear equant spar; *I1, I2, I3* three generations of internal laminated micrite

Type of Stromatactis and Sparry Calcite Other Than Cement. Based on the nature of the sparry calcite, *Stromatactis* can be sub-divided into six types (Fig. 5, Plate 2).

I: *Stromatactis* with micrite or microspar patches which are partly altered to fine fibrous calcite. However, vague internal bandings and some delicate remains of algal filaments are found within the blocky micrite or microspar (Plate 2, Fig. 1; Plate 3, Fig. 1). There is approximately 1% of this type of *Stromatactis.*

II: Mainly fine, fibrous sparry calcite with or without suture line which when present is not situated in the middle but which is called median seam, or median interface by some authors (Plate 2, Fig. 2). The longer crystal axes of the fine fibrous calcite are parallel to the growth direction of *Stromatactis.* The fine fibrous calcite is neither normal to the surface of *Stromatactis* nor is in contact with it (Plate 3, Fig. 2). Small equant paraxial calcite may occur at the basal central part (Plate 2, Fig. 2). There is approximately 8% of this type.

III: Coarsely crystallized fibrous calcite with suture line and with large area of equant paraxial calcite. Zoning of the fibrous calcite parallel to the surface of the equant paraxial calcite mass may be present. However, it is not distributed isopachously over the surface of *Stromatactis* (Plate 2, Fig. 3). Approximately 27% present.

IV: Coarsely crystallized fibrous calcite with areas of equant paraxial calcite. Zoning of the fibrous calcite well developed (Plate 2, Fig. 4), and generally parallel to the lower surface of *Stromatactis,* but not parallel to the upper or lateral surfaces (Fig. 5.4; Plate 2, Fig. 4).

V: Radiaxial fibrous spar and large equant paraxial spar (Fig. 5.5; Plate 2, Fig. 5). Approximately 14%.

VI: Radiaxial fibrous sparry calcite and laminated pelletoidal micrite with occasional *Frutexites* or internal sediments (Fig. 5.6; Plate 2, Fig. 6). Approximately 19%.

Intermediate structures exist between all six types of *Stromatactis.* Type I represents, undoubtedly, the primary structure which although rare, is however important. *Stromatactis* of this type has irregular patchy micrite or microspar masses distributed in vague zones. Much of this micrite or microspar is recrystallized into a fine fibrous calcite mosaic. The fibrous calcite is parallel to the lateral and vertical growth direction of *Stromatactis.* Type VI indicates that the cavity stages exist in some, but not all, *Stromatactis.* Different types of *Stromatactis* and sparry calcite suggest that sparry calcite results from replacement of different stages of microbial accretions (Fig. 6).

Organic Debris in the Sparry Calcite. Small fossil debris exists within *Stromatactis,* much of it occurring in the upper part of the suture line, and all supported by fibrous calcite: (1) carapaces of ostracode ± 1.5 mm in length found above the suture line, (2) unoriented small branching debris 1–3 mm in length that are very different from long stems, and (3) small debris of *Striatopora gracilis* a few millimeters in size embedded in giant *Stromatactis* 70 cm high.

The presence of organic debris suggests that *Stromatactis* is not a cavity system.

Host Sediment. The host sediment surrounding *Stromatactis* may be: (1) sponge-like micrite or microspar irregular masses, (2) porcelaneous milky micrite masses, (3) peletoidal micrite or microspar masses, (4) laminated or cross-laminated micrite masses, and (5) uncemented shale rich in terrigenous material generally occurring at the top of *Stromatactis.*

Stromatactis found in contact with uncemented shales also suggests that it is not a cavity but a self-supporting system.

Source of $CaCO_3$. Carbonate muds have a low permeability hence the exchange between cavity water and sea water was almost impossible. Consequently, the source of $CaCO_3$ in *Stromatactis* becomes incompatible with the "cavity system" origin of *Stromatactis.*

Microorganisms must have fluorished in mud mounds. Most of them could use simple carbon compounds as nutrients, the source of carbon being CO_2 in the water. Thus, the carbonate-saturated water may have lost CO_2 and precipitated carbonates. Microorganisms could also precipitate, trap, stabilize, and support lime mud to form carbonate colonial accretions which may have been totally or partly converted to fibrous sparry calcite and/or may have been dissolved to form cavities (Fig. 6B,3). These subsequently could have been filled by cements, internal sediments, and other biogenic micrite (Fig. 7).

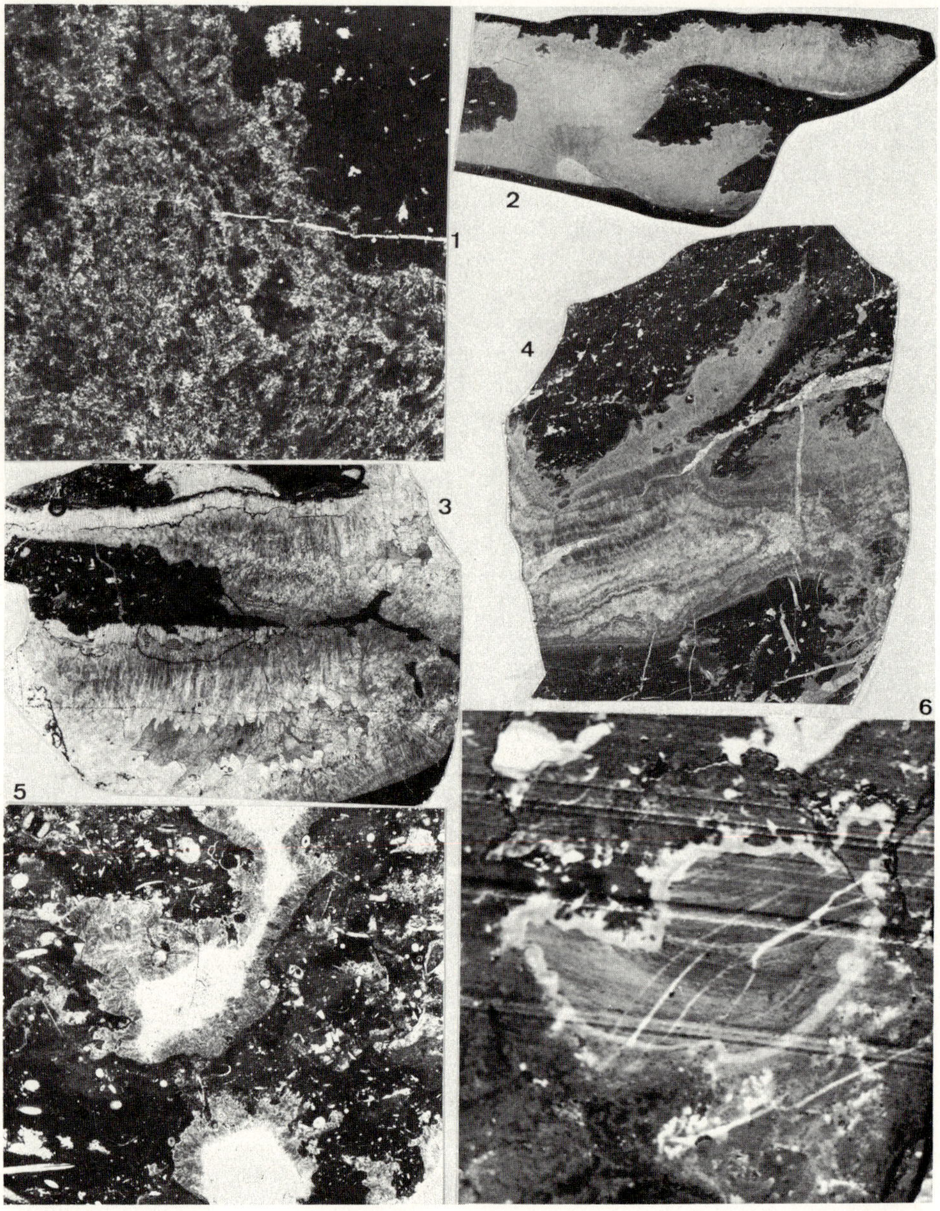

Plate 2. Thin section photomicrograph of types of *Stromatactis.* **1** Type I: *Stromatactis* with micrite or microspar patches which are partly converted to fine fibrous calcite. Note vague zonation and some delicate remains of algal filaments within the blocky micrite or microspar. **2** Type II: *Stromatactis* with fine fibrous calcite with or without suture line. **3** Type III: *Stromatactis* with coarse fibrous calcite. **4** Type IV: *Stromatactis* with zoning of fibrous calcite well developed. **5** Type V: *Stromatactis* with radiaxial fibrous calcite and equant paraxial spar. **6** Type VI: *Stromatactis* with three generations of biogenic laminated micrite fillings

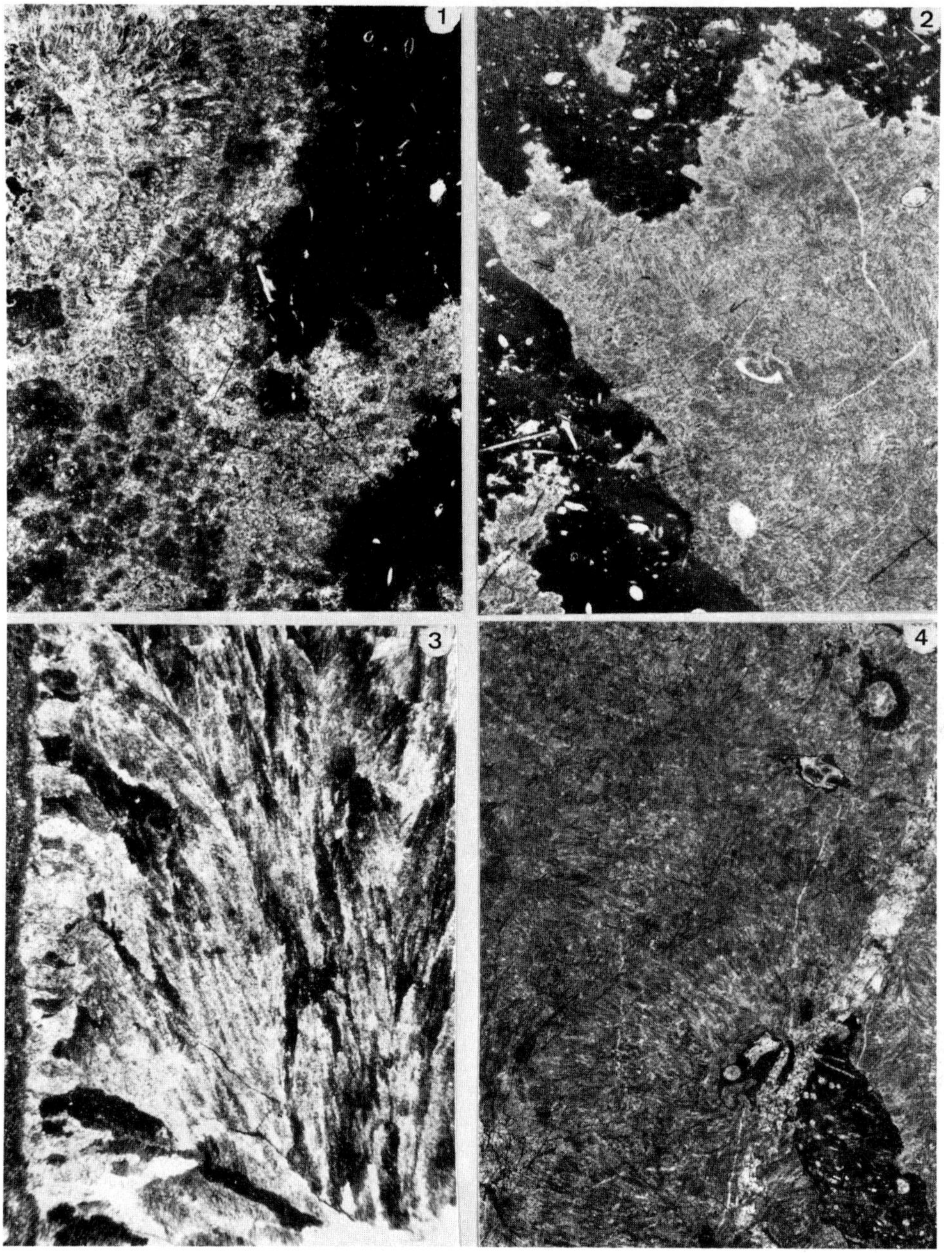

Plate 3. Thin section photomicrographs of nature of the sparry calcite. **1** Sparry calcite of micrite or microspar patches partly altered to fine fibrous calcite. **2** Three ostracode carapaces within a *Stromatactis* are supported by fibrous calcite. **3** The fibrous calcite not normal to *Stromatactis* surface but parallel to the lateral and vertical growth direction of *Stromatactis*. **4** Small branching coral debris floating in three dimensions within a *Stromatactis*

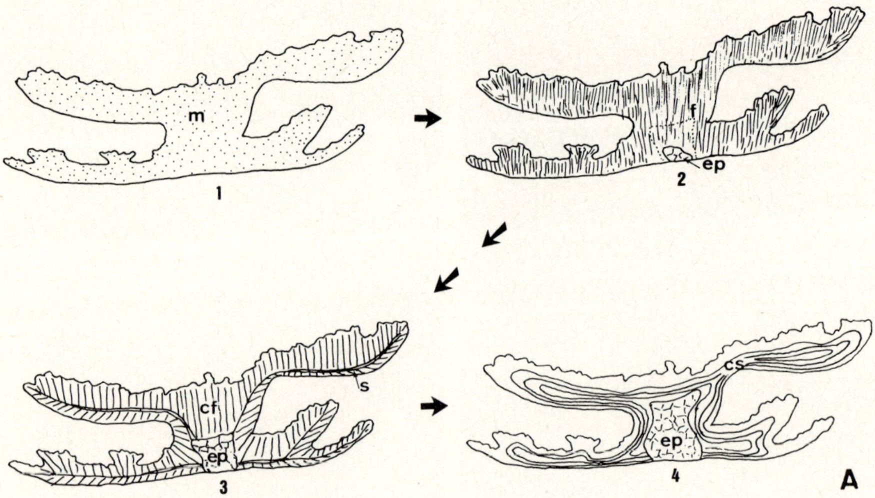

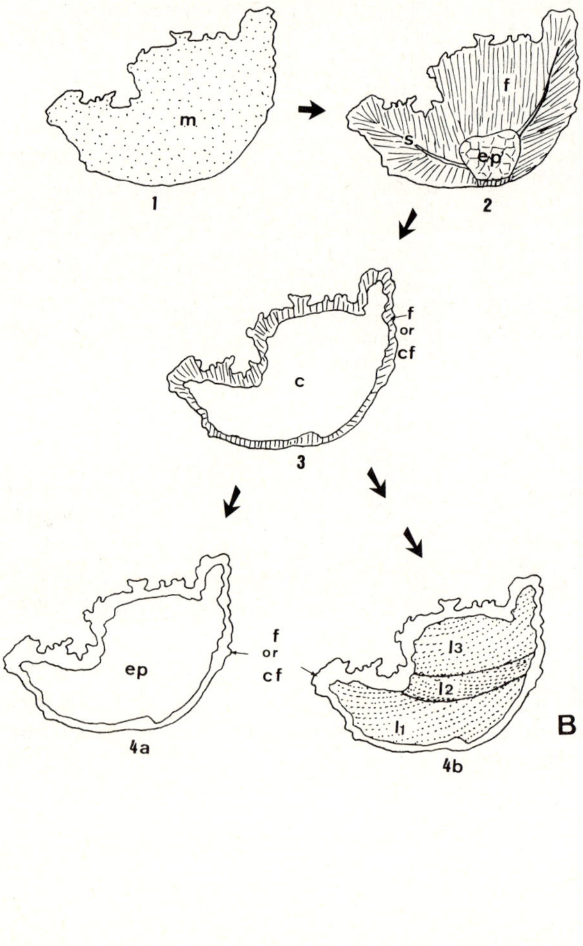

Fig. 6A,B. Diagrammatic representation of different types of *Stromatactis* suggesting that sparry calcite resulted from differential replacement of microbial accretions. Sketches summarize a series of events leading to different types of *Stromatactis*.
A Without cavity: *1* Algae associated with bacteria may produce carbonate to form carbonate colonial accretions. *2* Micrite or microspar may convert to fibrous sparry calcite; areas of clear equant spar may or may not occur. *3* Fine fibrous calcite may convert to coarse fibrous sparry calcite, areas of clear equant spar enlarged. *4* Formation of zoning in the fibrous calcite. **B** With cavity: (*1, 2*). Same as in **A**. *3* Fibrous calcite may dissolve to form cavity; *4a* Cavity filled with equant paraxial calcite; *4b*. Cavity filled with three generations of biogenic, laminated micrite

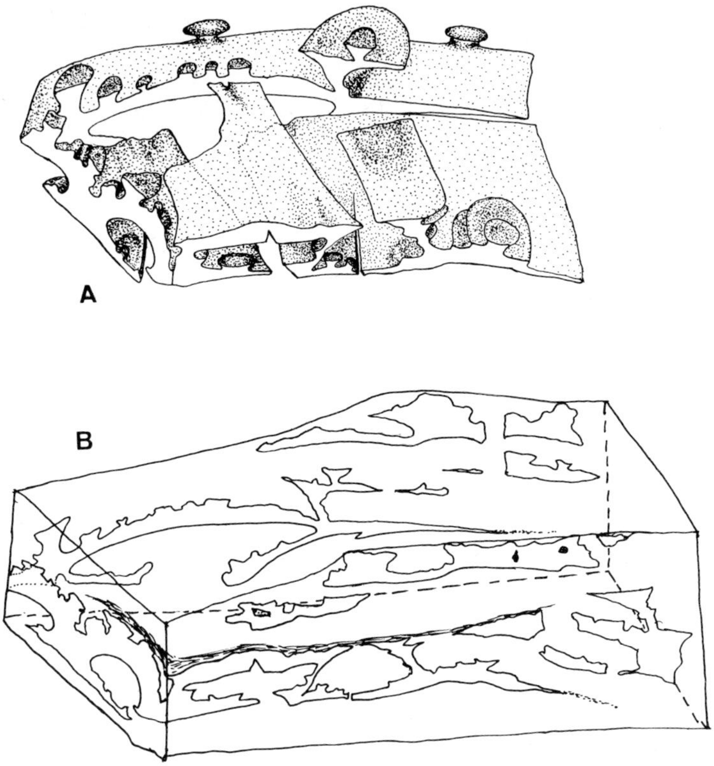

Fig. 7A,B. Diagrammatic representation of *Stromatactis* colonies. **A** Reconstruction from **B. B** A rock section with *Stromatactis*

Recent *Revularia biasolettiana* and *R. haematites* associated with bacteria and forming important carbonate concretions may be considered the modern counterpart of *Stromatactis,* although their colonial form and their distribution are very different.

Spelling of Stromatactis

Stromatactis is not a sedimentary structure. Therefore, Heckel's (1974) proposal to write *Stromatactis* as "stromatactis" is not valied, E. Dupont did not define *Stromatactis* taxonomically, therefore it cannot be considered to represent a generic name. However, Dupont defined it as an unknown group of stromatoporoids and wrote *Stromatactis* in italics. The term has been used for more than one hundred years and hence there is no reason to modify it. Therefore I would suggest that *Stromatactis* be written in italics as proposed in 1881.

Summary

The explanation of the origin of *Stromatactis* must include (1) the formation of its morphology and size (discussed in Tsien 1977, 1980, 1983), (2) the nature of fibrous calcite, (3) the formation of its top, (4) the mod of support of organic debris in the fibrous calcite, and (5) the understanding of depositional environment. These five briefly described aspects demonstrate that the attempts to relate *Stromatactis* to a cavity system of organic or inorganic origin are not satisfactory.

The analyses of *Stromatactis* from the type locality at St. Remi Quarry and from other localities have demonstrated that (1) *Stromatactis* does not occur in all facies of red mud mounds, (2) it occupies an important part of the mounds and shows ecological zonations (Tsien 1977, 1980), (3) its morphology is characterized by a wide variety of ecological forms and sizes that are comparable with other organisms such as *Alveolites,* stromatoporoids, and *Phillipsastrea,* (4) its upper digitations often show symmetric lateral expansions which are also ecological, and (5) small structures of symmetrical shape on the upper surface of *Stromatactis* are comparable with the buds of *Phillipsastrea* and *Alveolites.* These suggest that *Stromatactis* originated from organic colonies.

References

Bathurst RGC (1959) The cavernous structure of some Mississippian *Stromatactis* reefs in Lancashire, England. J Geol 67:506–521

Bathurst RGC (1977) Ordovician Meiklejohn bioherm, Nevada. Geol Mag 114:308–311

Bathurst RGC (1980) Stromatactis – origin related to submarine-cemented crusts in Paleozoic mud mounds. Geology 8:131–134

Bathurst RGC (1982) Genesis of stromatactis cavities between submarine crusts in Palaeozoic carbonate mud buildups. J Geol 139:165–181

Bellière J (1953) Note sur le calcaire famennien de Baelen et ses *Stromatactis.* Ann Soc Geol Belg 76B:115–128

Black WW (1952) The origin of the supposed tufa bands in Carboniferous reef limestones. Geol Mag 89:195–200

Bourque P-A, Gignac H (1983) Sponge-constructed stromatactis mud mounds, Silurian of Gaspé, Québec. J Sediment Petrol 53:521–532

Burchette TP (1981) European Devonian reefs: a review of current concepts and models. In: Toomey DF (ed) Eur Fossil Reef Models. Spec Publ Soc Econ Paleontol Mineral 30:85–142

Delhaye F (1908) Etude de la formation des récifs de calcaire rouge à *Rhynchonella cuboides* (note préliminaire). Ann Soc Geol Belg 35:243–253

Desbordes B, Maurin AF (1974) Trois exemples d'étude du Frasnien d'Alberta, Canada. Compagnie Francaise de petroles. Not Mem 11:293–336

Dumon P (1957) Note sur les marbres rouges de Belgique. Publ Assoc Ing Fac Pol De Mons 3:1–41

Dupont E (1881) Sur l'origine des calcaires dévoniens de la Belgique. Bull Acad R Sci Belg 3rd Ser 2:264–280

Dupont E (1882) Le terrain devonien de l'Entre-Sambre-et-Meuse. Les îles coralliennes de Roly et de Philippeville. Bull Mus R Sci Nat Belg 1:89–160

Dupont E (1885) Sur les calcaires frasniens d'origine corallienne et sur leur distribution dans le Massif Paleozoique de Belgique. Bull Acad R Sci Belg 3rd Ser 10:21–38

Dupont H (1969) Contribution à l'étude des Faciès du Waulsortien de Waulsort. Mem Inst Geol Univ Louvain 24: fs 2

Heckel PH (1972) Possible inorganic origin for stromatactis in calcilutite mounds in the Tully Limestone, Devonian of New York. J Sediment Petrol 42:7–18

Heckel PH (1974) Carbonate buildups in the geologic record: a review. In: Laporte LF (ed) Reefs in time and space. Spec Publ Soc Econ Paleontol Mineral 18:90–154

Kendall AC, Tucker ME (1973) Radiaxial fibrous calcite: a replacement after acicular carbonate. Sedimentology 20:365–389

Lecompte M (1937) Contribution à la connaissance des récifs du Dévonien de l'Ardenne. Sur la presence de structures conservees dans des efflorescences crystallines du type *"Stromatactis."* Bull Mus R Hist Nat Belg 13:1–14

Lecompte M (1938) Quelques types de "récifs" Siluriens et Dévoniens de l'Amerique du Nord. Essai de comparaison avec les récifs actuels. Bull Mus R Hist Nat Belg 14:1–54

Lees A (1964) The structure and origin of the Waulsortian (Lower Carboniferous) "reefs" of west-central Eire. Philos Trans R Soc London Ser B 247:483–531

Logan BW, Semeniuk V (1976) Dynamic metamorphism: processes and products in Devonian carbonate rocks, Canning Basin, Western Australia. Spec Publ Geol Soc Aust 6:1–138

Lowenstam HA (1950) Niagaran reefs of the Great Lakes area. J Geol 58:430–487

Maillieux E (1913a) Nouvelles observations sur le Frasnien et en particulier sur les Paléorécifs de la plaine des Fagnes. Bull Soc Belg Geol 27:67–104

Maillieux E (1913b) Quelques mots sur les récifs frasniens de la plaine des Fagnes, entre Nismes et Chimay. Bull Soc Belg Geol 27:114–125

Monty CL (1982) Cavity or fissure dwelling stromatolites (Endostromatolites) from Belgian Devonian mud mounds. (Extended abstract). Ann Soc Geol Belg 105:343–344

Orme GR, Brown WWM (1963) Diagenetic fabrics in the Avonian limestones of Derbyshire and North Wales. Proc Yorks Geol Soc 34:51–66

Philcox ME (1963) Banded calcite mudstone in the Lower Carboniferous "reef" knolls of the Dublin basin, Ireland. J Sediment Petrol 33:904–913

Playford PE, Lowry DC (1966) Devonian reef complexes of the Canning Basin, western Australia. Bull Geol Surv West Aust 118:1–150

Pratt BR (1982) Stromatolitic framework of carbonate mud-mounds. J Sediment Petrol 53: 1203–1227

Pentice JE (1950) The Carboniferous Limestone of the Manifold Valley region, North Staffordshire. Q J Geol Soc London 106:171–209

Ross RJ, Jaanusson V, Friedman I (1975) Lithology and origin of Middle Ordovician calcareous mudmound at Meiklejohn Peak, southern Nevada. U S Geol Surv Prof Pap 871:48

Schwarzacher W (1961) Petrology and structure of some Lower Carboniferous reefs in northwestern Ireland. Bull Am Assoc Petrol Geol 45:1481–1503

Tsien HH (1971) The Middle and Upper Devonian reef-complexes of Belgium. Petrol Geol Taiwan 8:119–173

Tsien HH (1977) Morphology and development of Devonian reefs and reef complexes in Belgium. Proc 3rd Int Coral Reef Symp, Miami, pp 191–200

Tsien HH (1980) Les regimes recifaux devoniens en Ardenne. Bull Soc Belge Geol 89:71–102

Tsien HH (1983) Paleoecology of Devonian reefs and reef builders of Ardennes (Belgium). In: Mouravieff AN, Tsienn HH (eds) 1st Int Congr Paleoecol Excurs 8B:32

Wilson JL (1975) Carbonate facies in geologic history. Springer, Berlin Heidelberg New York, 471 pp

23 Algal-Bacterial Origin of Micrites in Mud Mounds

H.H. TSIEN[1]

Introduction

Devonian red mud mounds, exposed in Belgium, consist essentially of diverse micrites and carbonate mud. They occur in nodular shales of the Neuville Formation and are confined to a narrow belt along the deeper southern part of the Dinant basin and in the Philippeville massif (Tsien 1983). The Neuville Formation constitutes a facies transition between deeper pelagic shales and limestones of the Matagne Shale, and shallower argillaceous limestones and shales of the Aisemont Formation. Both interfingering and onlapping relationships are present. At their upper and lateral margins, the mounds show several stages of geopetal fillings which are at different angles in sponges and cavities. This suggests that the flanks of the mounds were tilted during deposition. Mounds are flanked at some levels by intertonguing crinoid rudstones and branching coral rudstones, which indicates that current winnowing was important at that time. At other levels mounds are flanked by sponges, *Phillipsastrea* and *Alveolites* colonies, and other organisms embedded in the argillaceous shales. The morphology of these organisms differs only slightly from those in the mounds. This indicates that the ecological conditions in the mound and around the mound are very similar. Depositional dips, both in the mounds and in the flanking sediments, at the latest stage of the mound development, are generally more than 10 degrees and range up to 45 degrees.

There are four possible sources of the micrites and carbonate mud present in the mounds: (1) fine carbonate fractions ($< 60\ \mu$) formed primarily by the disintegration of delicate organisms and/or abrasion in an organism rich environment, with most lime mud of this type remaining in the source area; (2) lime mud transported from the source area that was washed in or settled out from suspension; (3) physicochemical precipitation; and (4) in situ biologically produced mud.

It appears that most of the micrites were, however, produced in situ biologically. Microorganisms appear to have flourished in the mud mounds, and most of these were able to use simple carbon compounds as nutrients, with the source of carbon being carbon dioxide in the water. Thus, the carbonate-saturated water lost CO_2 and precipitated carbonate micrites. Microorganisms could also have trapped, stabilized, and supported lime mud to form biocementstone (Tsien 1982).

1 Institut de Géologie, Laboratoire de Paléontologie, Université Catholique de Louvain, Pl Louis Pasteur 3, B–1348 Louvain-la-Neuve, Belgium

Paleoalgology: Contemporary Research and Applications
(eds. D.F. Toomey and M.H. Nitecki)

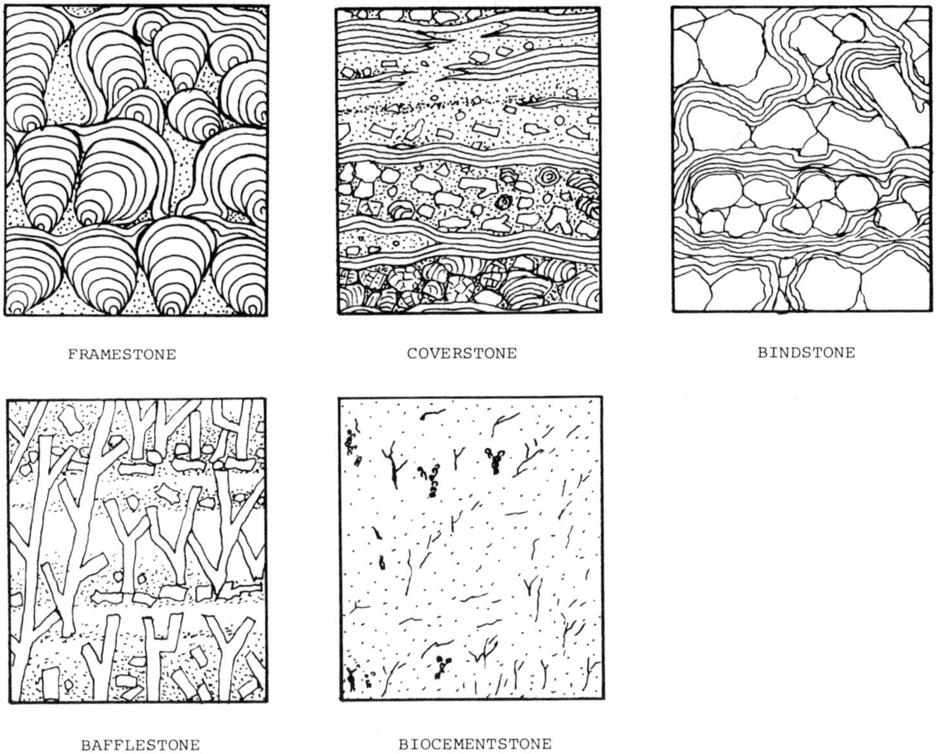

Fig. 1. Various reef-building organisms that constructed the five different types of reef carbonates seen in the Devonian of Belgium: framestones, coverstones, bindstones, bafflestones, biocementstones

Reef-Building Organisms and Their Role in the Construction of Reef Carbonates

Reef-building organisms, including massive, lamellar, branching, encrusting forms and various microorganisms, can build reef carbonates in five dinstinct ways: (1) in situ sedentary massive organisms forming a supporting framework to construct framestones; (2) in situ sedentary tabular or lamellar organisms covering skeletal debris, stabilizing them and forming coverstones (Tsien 1982); (3) in situ sedentary branching organisms retarding water movement, allowing sediment to settle, thus forming bafflestones; (4) organisms encrusting or binding skeletal debris to form bindstones; (5) microorganisms (algae, cyanobacteria and bacteria) producing micrites, and trapping, stabilizing and supporting lime mud to form biocementstones (Fig. 1).

The microbial population of the sea consists of bacteria, algae, protozoans, and fungi. Microorganisms inhabit various regions, extending from the ocean surface to the bottom mud. Most microorganisms are freeliving; some flourish within decaying organisms, and a few form symbiotic or parasitic associations. These organisms are responsible for the transformation of organic and inorganic compounds that serve as nutrients for marine life. In this situation they consume CO_2 and produce carbonate micrites as by-products. They may also grow in protective areas, resulting in the accumulation and deposition of micrite, which may either be washed in or is produced in situ, and which is then trapped, stabilized, and supported by microorganisms. In both cases, the micrite can be cemented very rapidly to form reef carbonates, the biocementstones. These biocementstones could have served as potential reef frames, if sufficiently abundant and if they occurred in association with other reef limestones. Certain other microorganisms (algae, cyanobacteria, bacteria, and fungi) are specialized for living in cryptic habitats and occupy reef cracks and fissures, and interspaces among reef debris. Many of these secrete skeletal material or precipitate carbonate micrite, thus filling the spaces and cementing the reef debris and organic framework into a solid rigid mass, thereby forming secondary reef frameworks (Tsien 1982).

Types of Biogenic Micrite

Micrite or microspar in Devonian red mud mounds of Belgium generally show these features: (1) variation in color, skeletal debris content, and distribution; (2) remains or traces of algal filaments; (3) patched or clotted texture; (4) vague internal banding or zonation; (5) spongy structure; (6) pelleted structure; (7) cryptalgal structure; (8) curved laminae; and (9) a porcelaneous, milky aspect. On the basis of composition of these features, the following five types of biogenic micrites or microspars are recognized (Fig. 2):

1. Laminated or Curved Microbial Structures (Fig. 3.1). In most cases, these occur in or at the bottom of a *Stromatactis,* and either form lenses or mounds in the mud mounds, or form peloidal laminites.

2. Spongy or Sponge-Like Structures (Fig. 3.2). These are irregular in shape and their microstructure consists of a tubular network rather similar to the spicular network of sponges. In rare cases these structures may grade into a sponge spicular network, which suggests that spongy structures may have easily formed in decaying sponge bodies. The tubular networks in the micrite masses may have been occupied by uncalcified algal filaments or microorganic tubes.

3. Cryptalgal Structures (Fig. 3.3). These are irregular in shape, generally encrusting, and their microstructure is similar to those described from shallow-water cryptalgal structures. They are characterized by alternating light and dark micrite (Fig. 2.3B) or by alternating sparry calcite and micrite (Fig. 2.3A), which may have resulted from the periodical precipitation and binding, or the differentiation of microbial communities.

4. Pelleted or Peloidal Structures (Fig. 3.4). These are patches or clotted (Fig. 2.4A) or irregularly shaped masses which may have been associated with decaying organic bodies, forming cryptalgal mat-like structures (Fig. 2.4B).

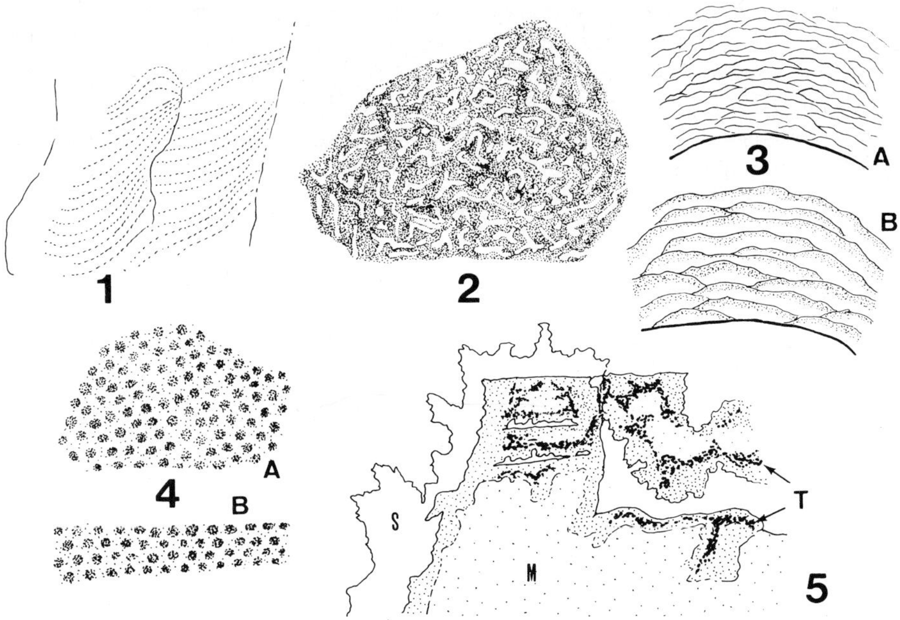

Fig. 2. Types of biogenic micrite seen in Belgian Devonian mud mounds. **1** Curved laminated micrite; **2** spongy or sponge-like micrite; **3** micrite with cryptalgal structures: **3A** alternating sparry calcite and micrite; **3B** alternating light micrite and dark micrite; **4** pelleted micrite: **4A** clotted masses; **4B** cryptalgal mat-like micrite; **5** porcelaneous, milky micrite (*M*), terrigenous sediments (*T*), and *Stromatactis* (*S*)

5. Diversely Clotted Porcelaneous, Milky Micrite Masses (Fig. 2.5). These primary sediments of structureless porcelaneous, milky micrite generally form knobs, nodules, or irregularly shaped structures. Very often they coat organisms or *Stromatactis,* and generally connect together, forming an apparent framework in the mud mounds.

Based on my understanding of Devonian mud mounds (their growth, development, and construction) in Belgium, attempts to attribute the presence of large amounts of micrite to algal binding and submarine cementation of bioclastic lime mud, deposited from suspension (and derived from the shallow shelf and dispersed basinward by currents), or to the baffling out of suspension by bryozoans, pelmatozoans, soft-bodied plants or sponges, have not been satisfactory. The following data strongly suggest an in situ microorganic origin of micrites and microspars:

1. Micrite masses of lenticular form or small mounds. Their size varies from a few centimeters to 5 m.
2. The contrast between distribution of terrigenous mudstones and micrite masses suggests that the micrites are of local origin.
3. Encrusting *Alveolites* on the tops of some micrite lenses suggest very early lithification (Fig. 4B).
4. Microbial structures (Monty 1982, Tsien 1983) and the poorly sorted and disoriented nature of bioclasts in the micrite masses suggest that micrites were not

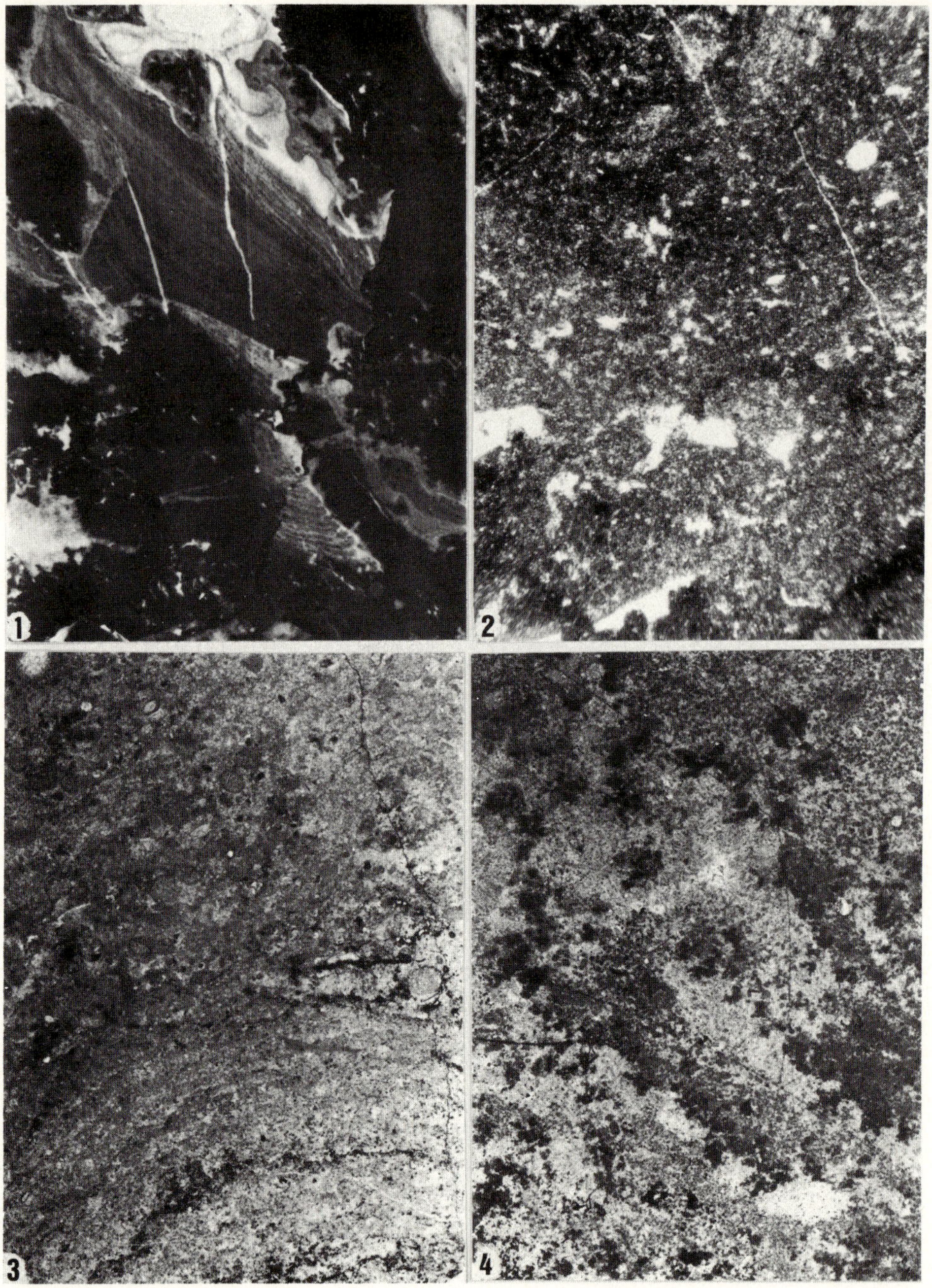

Fig. 3. Thin section photomicrograph of various micrite structures (× 5). **1** Curved laminated micrite; **2** spongy or sponge-like micrite; **3** micrite with cryptalgal structures; **4** pelleted or peloidal micrite

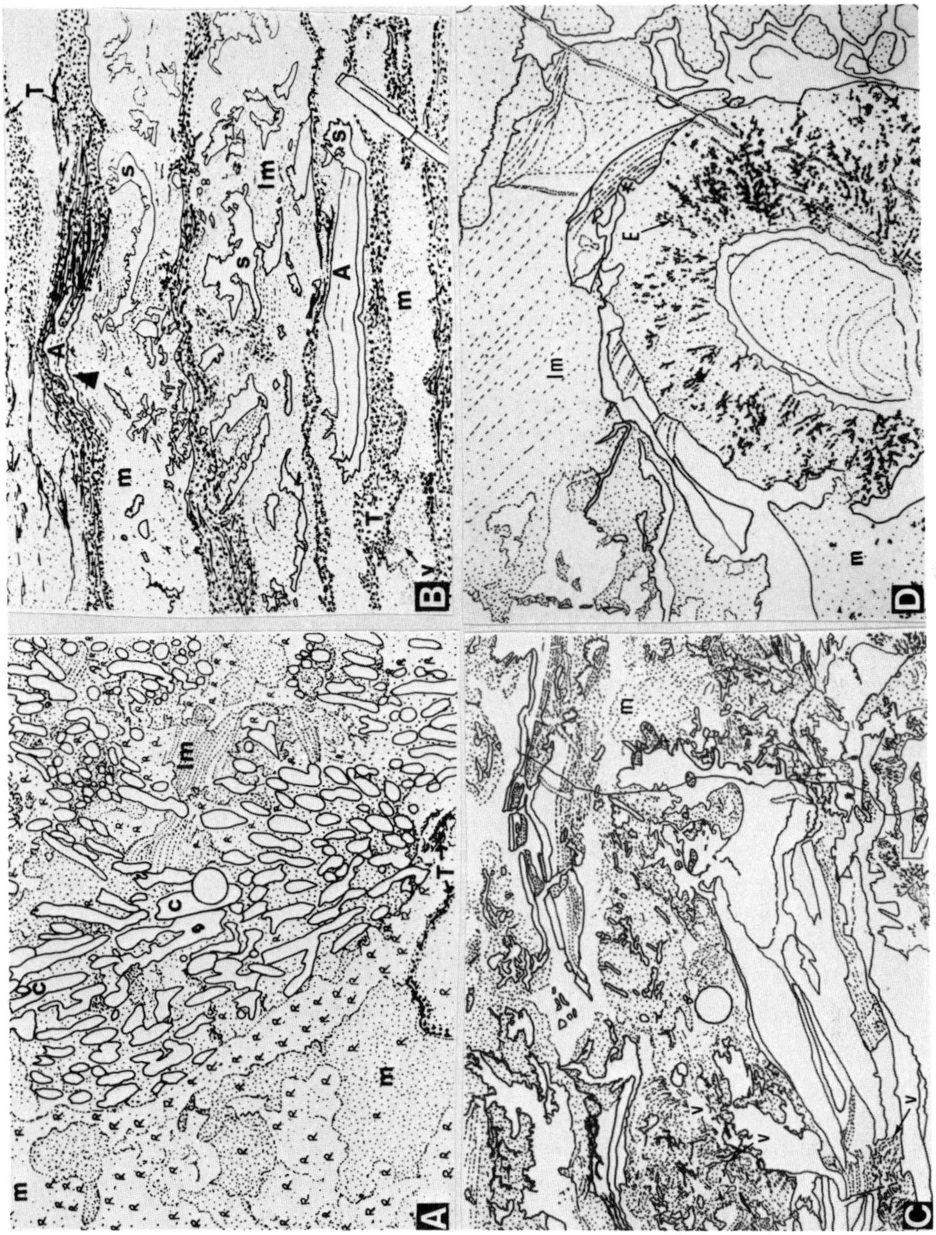

Fig. 4A–D. Examples of biocementstones formed by biogenic micrites. **A** Biocementstone formed by microorganisms which produced micrite, trapped and supported lime mud, and cemented these to form reef limestone. **B** Biocementstone lenses formed mainly by micrite and *Stromatactis.* The presence of the encrusting *Alveolites* on the top of this micrite lens (*black triangle*) suggests very early lithification. (T = terrigenous sediments). **C** Biocementstone formed mainly by cryptalgal micrite (*V*), porcelaneous, milky micrite (*m*), and *Stromatactis* (*S*). **D** Biocementstone formed by curved laminated micrite (1 m), porcelaneous, milky micrite (*m*), and *Epiphyton* (*E*)

physicochemically precipitated in situ (Pratt 1982), but were biologically produced in situ.
5. The presence of vague internal bandings or zonations (V on Fig. 4C) within the micrite masses suggests that particle accretions were probably episodic due to seasonal algal-bacterial activities.
6. Different generations of micrites developed around the in situ branching corals, indicating that they originated locally (Fig. 4A).

Summary

Devonian mud mounds consist largely of clotted porcelaneous, milky micrite. It is thought that most of these micrites were produced biologically in situ. They generally show these features: (1) variation in color, skeletal debris content, and distribution; (2) remains or traces of algal filaments; (3) non-uniform packing of particles resulting in a clotted texture; (4) irregular patches of coarse and fine bioclasts; (5) spongy or vermicular structures; (6) vague internal banding and zonation; and (7) algal-bacterial laminite. The contrast in distribution of terrigeneous mudstones and porcelaneous milky micrite suggests that the micrite is of local origin. Encrusting *Alveolites* on the top of some micrite lenses suggests very early lithification. Microbial structures and poorly sorted, disoriented bioclasts in micrite masses suggest that the micrite was not physico-chemically precipitated in situ, but biologically produced in situ. The presence of a vague internal banding or zonation in the micrite masses suggests that particle or mass accretions were episodic due to algal-bacterial activities that occurred during different seasons. It is proposed that microorganisms (algae, cyanobacteria, bacteria) flourished in these mud mounds, and most of these used simple carbon compounds as nutrients. These microorganisms probably trap, stabilize, and support lime mud to form microbial accretions or biocementstone (Tsien 1982, 1983). The source of carbon was carbon dioxide in the water. Thus carbonates precipitated from the carbonate-saturated water which lost its CO_2.

References

Monty C (1982) Cavity or fissure dwelling stromatolites (Endostromatolites) from Belgian Devonian mud mounds (extended abstract). Ann Soc Geol Belg 105:343–344

Pratt BR (1982) Stromatolotic framework of carbonate mud-mounds. J Sediment Petrol 52: 1203–1227

Tsien HH (1982) Ancient reefs and reef carbonates. Proc 4th Int Coral Symp Manila 1981 1:601–609

Tsien HH (1983) Paleoecology of Devonian reefs and reef builders of Ardenne (Belgium). In: Mouravieff AN, Tsien HH (eds) 1st Int Congr Paleoecol, Excursion 8B, 32 pp

24 Paleodepositional Setting of Rhodoliths from the Upper Pennsylvanian (Virgil) Salem School Limestone of Northcentral Texas

D.F. TOOMEY[1]

Introduction

During Late Pennsylvanian time (Upper Carboniferous-Stephanian), much of the midcontinent United States comprised a slowly subsiding open marine carbonate shelf in which limestone deposition was prominent, although shales and sandstones were also deposited. This region is further exemplified during this time interval by the excellent development of sedimentary cyclic deposition (Moore 1950). Recently, Toomey (1983a) presented the results of a detailed field and laboratory study of the "middle limestone member" of the Oread Limestone (Shawnee Group): the Leavenworth Limestone. This thin (less than 3 feet thick) limestone has long been regarded as a classic example of lateral homogeneity along outcrop for a distance of more than 300 miles. The persistent homogeneity is in part the result of deposition occurring essentially parallel to a Late Carboniferous epicontinental sea. Leavenworth microfacies distribution reflect this overall rock homogeneity in that most of the outcrop belt can be designated as a mud-supported skeletal wackestone. It is only at those localities, present in northern Oklahoma, where the Leavenworth enters and merges into sediments laid down within a clastic trough, that a facies change occurs. Here, the Leavenworth, as identified by detailed tracing of beds, by rock character, faunal associations, and stratigraphic position, changes microfacies to an oncolithic wackestone. This particular microfacies is characterized by the abundance of relatively large algally coated-grains composed of laminar thalli of the red alga *Archaeolithophyllum lamellosum* Wray. In northern Oklahoma, the Leavenworth Limestone, and all other thin Upper Carboniferous limestones, lens-out and merge into a clastic wedge of sediments derived from a paleo-highland source to the south (see Fig. 1). This source area, herein termed the Oklahoma Mountains, had initially been recognized by Powers (1928).

South of the "Oklahoma Mountains," in Young County, northcentral Texas, in what I believe is a stratigraphic position similar to the Leavenworth Limestone to the north (see Fig. 2), is a thin limestone outcrop along the north side of the Brazos River that has been identified as the Salem School Limestone (Lee et al. 1938). The Salem School Limestone, like the Leavenworth Limestone, occurs as a single-prominent bed less than two feet thick (Fig. 3A), is hard, dense, and breaks with a

1 Conoco Inc., Research Center, Ponca City, OK 74603, U.S.A.

Paleoalgology: Contemporary Research and Applications
(eds. D.F. Toomey and M.H. Nitecki)

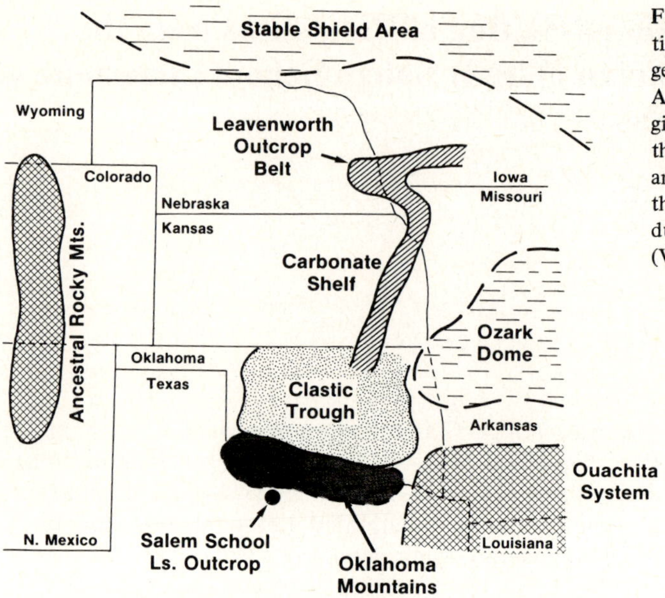

Fig. 1. Schematic representation of the postulated paleogeographic setting of the American midcontinent region during deposition of the Salem School Limestone, and its correlative formation, the Leavenworth Limestone, during Late Pennsylvanian (Virgil) time

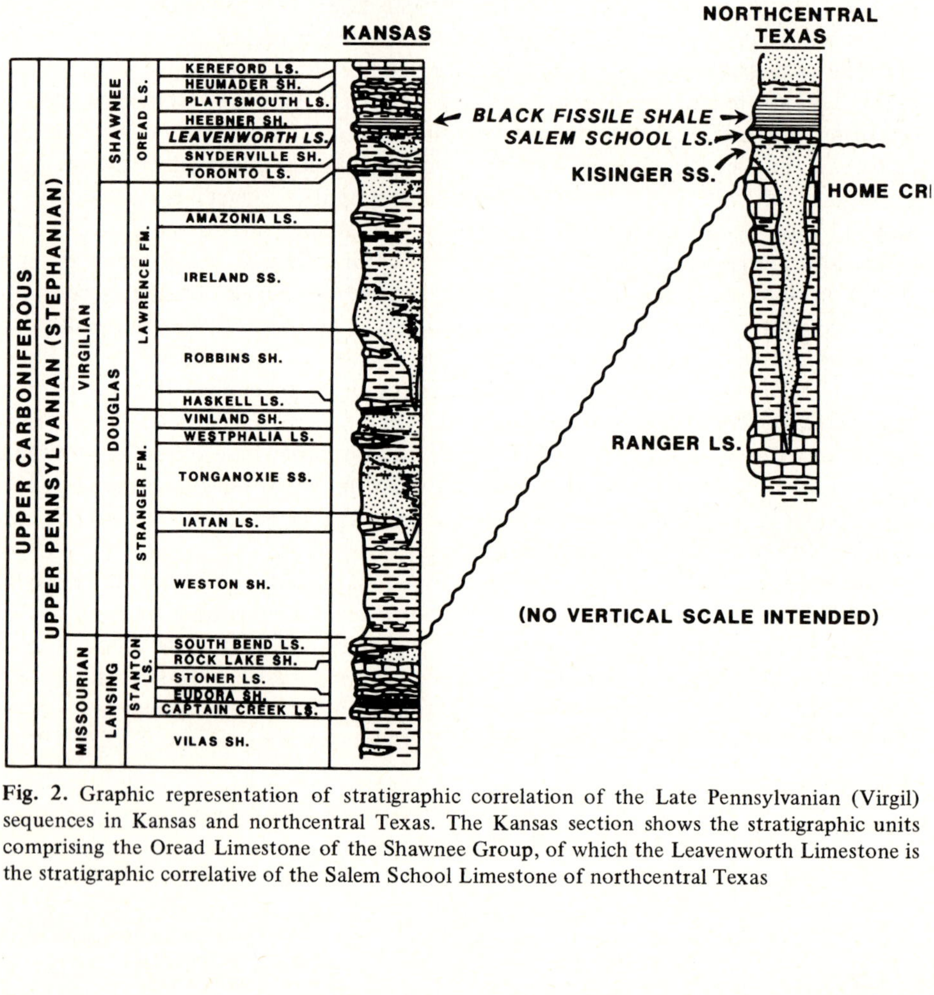

Fig. 2. Graphic representation of stratigraphic correlation of the Late Pennsylvanian (Virgil) sequences in Kansas and northcentral Texas. The Kansas section shows the stratigraphic units comprising the Oread Limestone of the Shawnee Group, of which the Leavenworth Limestone is the stratigraphic correlative of the Salem School Limestone of northcentral Texas

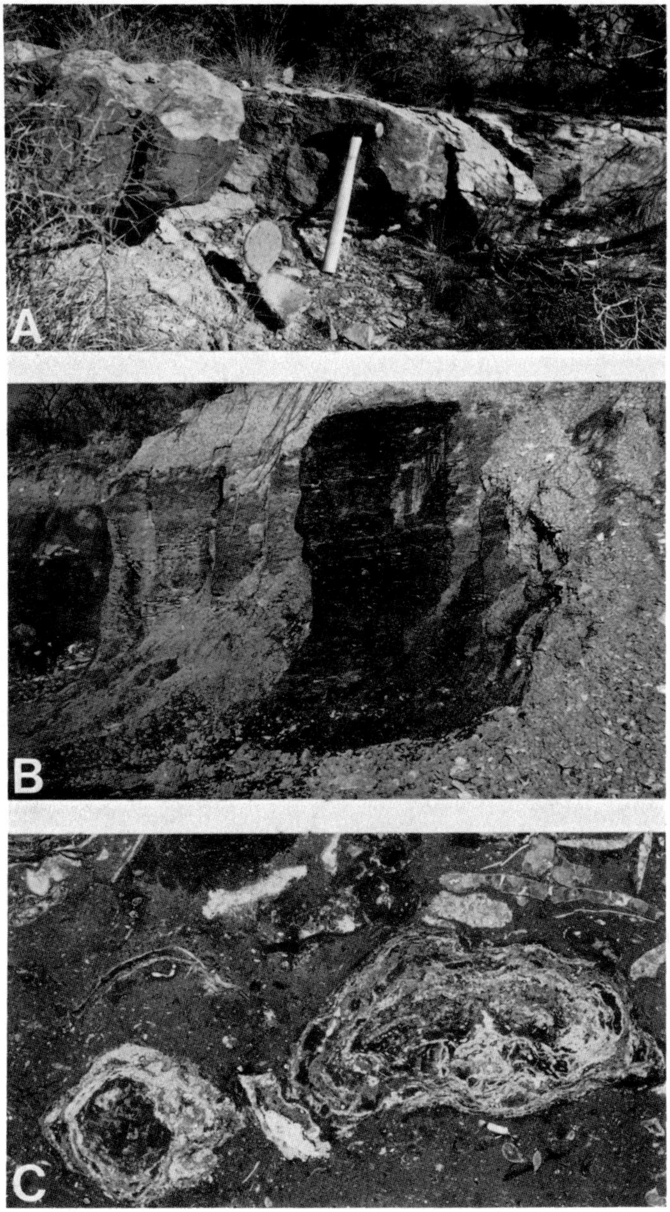

Fig. 3A–C. Photographs of the Salem School Limestone outcrop location in northcentral Texas, and a photomicrograph of typical rhodoliths present in this unit. **A** Outcrop exposure of the Late Pennsylvanian Salem School Limestone as a single limestone bed on the north bank of the Brazos River, Young County, northcentral Texas; height of hammer 17 inches. **B** Black fissile shale overlying the Salem School Limestone at the Herron Bend locality; this distinctive type of lithology is not common in the Pennsylvanian sequence in northcentral Texas; thickness of this shale interval is approximately 15 feet. **C** Thinsection photomicrograph (× 4.8) of characteristic rhodoliths present in the Salem School Limestone and occurring in skeletal wackestones. These rhodoliths show fairly concentric laminations of algal thalli

Table 1. Algal, foraminiferal, and conodont assemblages present in the Leavenworth and Salem School limestones (x = present, o = absent)

Species	Leavenworth	Salem School
ALGAE		
Archaeolithophyllum lamellosum	x	x
Cuneiphycus johnsonii	x	x
C. sp. cf. *C. texana*	x	o
Anthracoporella? sp.	o	x
Epimastopora kansasensis	x	x
Eugenophyllum johnsonii	x	o
Girvanella sp.	x	x
Tubiphytes obscurus	x	x
Rhodoliths	x	x
"Osagia" coated-grains	x	x
Diplosphaerina sp.	x	x
FORAMINIFERS		
Agglutinated smaller forams		
Hyperammina glabra	x	o
H. elegantissima	x	x
H. clavacoidea	x	x
Psammosphaera gracilis	x	o
Saccammina aspera	x	o
Thurammina elliptica	x	o
T. diforamens	x	x
T. papillata	x	o
Ammodiscus annularis	o	x
Tolypammina polyverta	x	x
Ammovertella inclusa	x	x
Reophax sp.	x	x
Textularia eximia	x	o
Bigenerina virgilensis	x	x
B. elongata	x	x
Calcareous smaller forams		
Tuberitina bulbacea	x	x
Endothyra sp.	x	x
Endothyranella gracilis	x	o
Palaeospiroplectammina conspecta	x	x
Bradyina magna	x	x
Pseudobradyina pulchra	x	o
Climacammina moelleri	x	x
Tetrataxis conica	x	x
T. planispiralis	x	x
Globivalvulina bulloides	x	x
Hemigordius harltoni	x	x
Calcivertella sp.	x	x
Apterrinella sp.	x	o
Syzarania bella	x	x

Table 1 (continued)

Species	Leavenworth	Salem School
Fusulinids		
Millerella sp.	x	x
Nankinella sp.	x	x
Waeringella spiveyi	x	x
Triticites sp.	x	x
Kansanella sp.	x	o
CONDONTS		
Hindeodus minutus (A-2 element)	x	x
H. sp. (0-1 element)	x	x
H. sp. (P-element)	x	x
Streptognathodus elegantulus (A-1 and P-elements)	x	x
S. simulator (P-element)	x	o
Adethognathodus lautus (P-element)	x	x
A. gigantus (P-element)	x	o

conchoidal fracture. The surface of this unit is iron-stained and contains relatively numerous megafossils consisting mainly of algally coated-grains, brachiopods (*Composita, Derbyia,* and *Antiquatonia*), small gastropods, echinoid spines, and crinoid columnals. In polished slabs, the limestone is heavily burrowed, with the burrows filled with pyrite or fine-grained mud and pyrite. The base of the unit is irregular and underlain by grey-green shale containing productid brachiopod fragments and crinoid columnals. This shale unit grades downward into a massive channel sandstone containing fossil wood fragments. Overlying the Salem School Limestone is approximately 15 feet of hard, black fissile shale (see Fig. 3B), a lithologic entity common in the cyclic sediments of the midcontinent, but rare in the Late Carboniferous rocks of northcentral Texas.

Time-stratigraphic correlation of the Leavenworth and Salem School Limestones is further strengthened by the presence in both limestones of the distinctive fusulinid *Waeringella spiveyi,* as described by Thompson (1942). This particular fusulinid has a narrow, vertical stratigraphic range and does not occur in other Late Pennsylvanian units below or above the Leavenworth Limestone in the midcontinent, or above or below the Salem School Limestone in northcentral Texas. The similarity of other biotic components with the Leavenworth Limestone, especially the algae, smaller agglutinated and calcareous smaller foraminifers, and conodonts is outstanding and is documented in Table 1 (Ireland 1956, Toomey 1983a).

Salem School Rhodoliths

The name rhodolith, as modified by Toomey (1975), is a descriptive term for nodules and detached branched growths composed principally of laminar growths of coralline algae. Since *Archaeolithophyllum* is regarded as an ancestral coralline alga (Wray 1964), and is the principal component of the Salem School nodules, it does seem proper to identify them as rhodoliths. In essence, this usage aids in delineating various types of algal nodules that very often appear superficially similar, though occurring in quite different environments.

Salem School rhodoliths (Figs. 3C and 4A–E) are somewhat smaller than those in the Leavenworth Limestone (Toomey 1974), attaining a maximum diameter of only 1 1/2 inches; these are appreciably smaller than those well developed Lower Permian rhodoliths from the Laborcita Formation in the Sacramento Mountains of New Mexico (Toomey 1983b). These algally coated-grains consist of a nucleus circumscribed by roughly concentric laminations (thalli) of the red alga *Archaeolithophyllum lamellosum* Wray, and are in turn encrusted by other epizoic organisms, principally encrusting calcareous smaller foraminifers (*Calcivertella, Tuberitina, Tetrataxis*), polychaete worms (*Spirorbis*), and fistuliporoid bryozoa. Some rhodoliths are rather tightly encrusted with thalli of *Archaeolithophyllum,* whereas the majority of rhodoliths are quite loosely encrusted with algal thalli separated by distinct layers of mud. Width of algal thalli vary considerably, some rhodoliths averaging up to 12 laminae per 1/2 inch. Algal origin is demonstrated by the occurrence of hypothallic cells in an arcuate position within the thallus. However, this cellular preservation is erratic, and in most thinsections prepared from the algal nodules, distinctive microstructure is absent and the algal thalli are recrystallized to layers of sparry calcite mosaic. A few Salem School rhodoliths show preferential growth of algal thalli (Fig. 4A) on what appears to be the top, suggesting that some of these accretionary grains were not continually moved about on the sea floor. Other rhodoliths show tight, concentric algal encrustations (Fig. 3C). A distinctive feature of the algal thalli, noted by Toomey (1974, 1975), is the presence, usually at the terminal ends of the thalli, of digitate projections of the thalli (Fig. 4B–D). These "probing fingers" usually extend outward from the rhodolith and appear to be the leading edge of algal thalli seeking out new areas to envelop with an algal mantle; in particular, see Toomey (1975, Fig. 2).

Skeletal grains that can serve as nuclei for rhodolith accretion can vary considerably and appear to be simply a function of availability of skeletal grain type in the immediate depositional environment. Within the Salem School Limestone, calcareous sponges commonly serve as rhodolith nuclei (Fig. 4B). Other skeletal grains serving as nuclei for Salem School rhodoliths are fenestrate bryozoans (Fig. 4D), whole molluscs and mollusc fragments (in particular, gastropods), and brachiopods. None of the skeletal grains serving as nuclei for the rhodoliths have been observed to have been bored, corroded, or altered in a manner attributable to the action of the coating coralline thalli.

It should be mentioned that within a few of the Salem School rhodoliths, usually occurring within mud layers separating some algal thalli, are radial fans of calcite cements which may have originally been aragonitic botryoids. The presence of radial fans of calcite cements is thought to be indicative of early submarine cementation.

Associated with the Salem School rhodoliths, albeit only rarely, are small (up to 1/2 inch in diameter), bean-shaped coated-grains that are commonly referred to the form genus *Osagia* (Twenhofel 1919). These coated-grains are much smaller than the

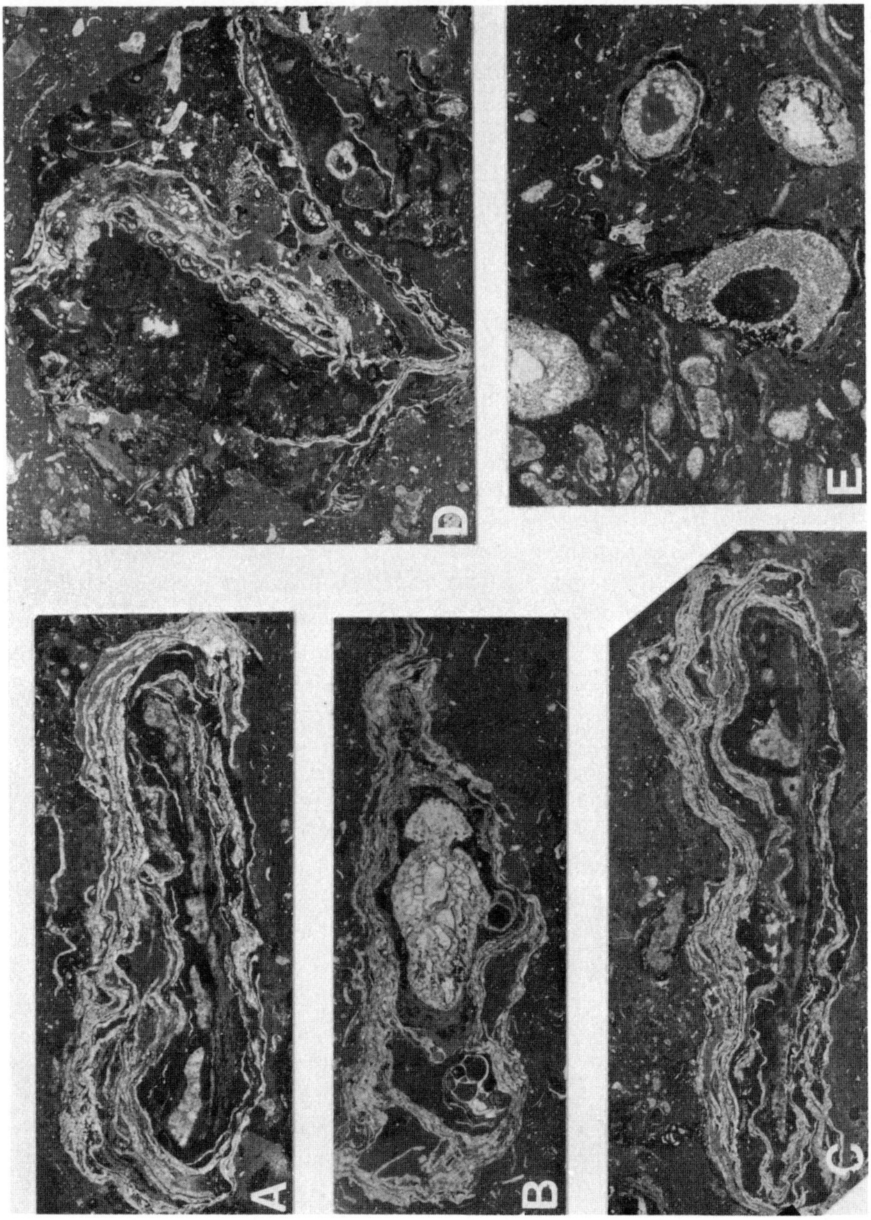

Fig. 4A–E. Thinsection photomicrographs of various algally coated-grains (rhodoliths) present in the Late Pennsylvanian (Virgil) Salem School Limestone of northcentral Texas. **A** Rhodolith (× 4.8) showing concentrations of algal coatings (thalli of the ancestral coralline alga *Archaeolithophyllum lamellosum* Wray) on the upper surface of this coated-grains; nucleus is a mud-encrusted fenestrate bryozoan (much recrystallized). **B** Rhodolith (× 3.2) in which a mud layer surrounds what appears to be the nucleus (calcareous sponge) of this algally coated-grain; note the presence of a small trochoid-shaped gastropod within the muddy portion of the rhodolith. **C** Rhodolith (× 4.8); thinsection cut through another portion of the same rhodolith shown in (**A**). **D** Rhodolith (× 4); this photomicrograph shows a rather loosely organized coated-grain in which the nucleus is a fenestrate bryozoan. Probing "fingers" of algal thalli (upper central portion of photograph) appear to have been in the process of capturing a relatively large portion of muddy substrate in an attempt to evelop it with an algal mantle. **E** Typical Salem School microfacies (× 4.8), composed of skeletal wackestone with rather common calcareous sponges, partially coated with algal laminae rhodoliths and seem composed of an association of algae and encrusting foraminifers. The coated-grain itself and its nucleus are generally riddled or bored throughout by tubelike threads thought to be attributed to the boring and perforating blue-green alga *Girvanella* (Henbest 1963). This type of coated-grain is quite common in the Leavenworth Limestone (Toomey 1983a).

Paleodepositional Setting

It is believed that the paleodepositional setting for the Salem School rhodoliths was essentially similar to that of the coeval Leavenworth Limestone rhodoliths that characterized Leavenworth outcrop locations in northern Osage County, Oklahoma. Here the Leavenworth Limestone changes microfacies from a skeletal wackestone to an oncolithic wackestone reflecting the paleogeographical shift from deposition on an open marine carbonate shelf, passing southward into a somewhat more restricted subsiding clastic trough. Within this clastic trough, the Leavenworth per se loses its identity as it lenses out into a predominantly clastic section termed the Vamoosa Formation. The Leavenworth Limestone can be identified further southward, within the clastic trough, than can any of the other Oread Limestone units.

The source of the clastics was the paleo-highlands ("Oklahoma Mountains"), as recognized by Tanner (1956, 1959). This basin proximity to an actively eroded highland source area is shown in the Leavenworth clay suite. A clay minerals suite composed of chlorite/mixed-layer/illite clay minerals characterizes the Leavenworth from southern Kansas into Nebraska. In the vicinity of southernmost Kansas and northern Oklahoma, the clay mineral kaolinite enters the asemblage in amounts up to 30% of the total clay suite (Toomey 1964, Fig. 17). The occurrence and abundance of kaolinite entering the Leavenworth Limestone is coincident with the change in Leavenworth microfacies. Increase of kaolinite, within clay mineral suites, appears to be a good indicator of proximity to granitic source areas, and Weaver (1958) has noted that kaolinite is most common in continental and nearshore sediments.

Location of the Salem School Limestone outcrop, with its contained rhodoliths, is shown in Fig. 1. This diagram emphasizes the close proximity of the outcrop to the "Oklahoma Mountains" source area. The geographic closeness to a granitic source area also appears to be reflected in the Salem School Limestone clay minerals suite, in which kaolinite comprises 45% of the total clay minerals. The fact that this unit, as a Leavenworth Limestone correlative, can be recognized so far south and in rather close proximity to the source area, and, in addition, contains only minor amounts of quartz silt, suggests to me that the so-called "Oklahoma Mountains" probably did not form a continuous barrier separating northcentral Texas and the region to the north. Most probably, this source area consisted of isolated granitic islands within a relatively shallow sea.

Ideal conditions for rhodolith formation, at least in the Late Paleozoic, appears to have been in relatively shallow waters well within the photic zone, usually close to some land mass, and in waters that carry a rather diverse flora and fauna. The rock type most associated with Late Paleozoic rhodoliths is generally described as an intensively burrowed skeletal wackestone.

Summary

Rhodoliths from the Late Pennsylvanian (Virgil) Salem Limestone, exposed in north-central Texas, consist of accretionary algally coated-grains composed of roughly concentric laminar thalli of the ancestral coralline alga *Archaeolithophyllum lamellosum*. The Salem School rhodoliths are associated with a diverse flora and fauna and occur within a shallow water paleodepositional setting, whose rock lithology can be described as an intensively burrowed skeletal wackestone. Correlation of the Salem School Limestone with its coeval equivalent, the Leavenworth Limestone of the midcontinent region, is based upon unit stratigraphic position, occurrence in both limestones of the distinctive fusulinid *Waeringella spiveyi*, and the occurrence of similar rhodoliths in both units in seemingly identical paleogeographical settings.

References

Henbest LG (1963) Biology, mineralogy, and diagenesis of some typical Late Paleozoic sedentary foraminifers and algal-foraminiferal colonies. Cushman Found Foram Res Spec Publ No 6:44

Ireland HA (1956) Upper Pennsylvanian arenaceous Foraminifera from Kansas. J Paleontol 30: 831–864

Lee W, Nickell CO, williams JS, Henbest LG (1938) Stratigraphic and paleontologic studies of the Pennsylvanian and Permian rocks in northcentral Texas. Univ Tex Publ No 3801:252

Moore RC (1950) Late Paleozoic cyclic sedimentation in central United States. In: Rhythm in sedimentation. 18th Int Geol Congr, Great Britain, 1948, vol IV, Sect C, pp 5–16

Powers S (1928) Age of the folding of the Oklahoma Mountains. Bull Geol Soc Am 39:1031–1071

Tanner WF (1956) Geology of northeastern Osage County, Oklahoma. Okla Geol Surv Circ 40:76

Tanner WF (1959) Permo-Pennsylvanian paleogeography of a part of Oklahoma. J Sediment Petrol 29:326–335

Thompson ML (1942) New genera of Pennsylvanian fusulinids. Am J Sci 240:403–420

Toomey DF (1964) Lateral homogeneity in a "middle limestone member" (Leavenworth) of a Kansas Pennsylvanian megacyclothem. Unpubl PhD Dissertation, Rice Univ Houston, 211 p

Toomey DF (1974) Algally coated-grains from the Leavenworth Limestone (Upper Pennsylvanian, Midcontinent Region, USA). Neues Jahrb Geol Palaontol Monatsh 3:175–191

Toomey DF (1975) Rhodoliths from the Upper Paleozoic of Kansas and the Recent – A comparison. Neues Jahrb Palaentol Monatsh 4:242–255

Toomey DF (1983a) The paleoecology of a "middle limestone member" (Leavenworth) of an Upper Carboniferous (Stephanian) cyclothem, Midcontinent, USA. Facies 8:113–190

Toomey DD (1983b) Early Permian coated-grains from a lagoonal environment, Laborcita Formation, Sacramento Mountains, southcentral New Mexico. In: Peryt TM (ed) Coated grains. Springer, Berlin Heidelberg New York, pp 259–269

Twenhofel WH (1919) Precambrian and Carboniferous algal deposits. Am J Sci 48(4):339–352

Weaver CE (1958) Origin and significance of clay minerals in sedimentary rocks. Bull Am Assoc Petrol Geol 42:254–271

Wray JL (1964) *Archaeolithophyllum*, an abundant calcareous alga in limestones of the Lansing Group (Pennsylvanian), southeastern Kansas. Geol Surv Kans Bull 170: pt 1, 1–13

25 Taxonomic Survey of Lower Permian Algae from the Southern Tatum Basin, Southeastern New Mexico, U.S.A.

J.M. CYS[1]

Introduction

The Tatum Basin (Fig. 1) is located in southeastern New Mexico on the Northwest Shelf of the Permian Basin, a major hydrocarbon-producing region of North America. It is a local subsurface basin that appeared in Early Pennsylvanian time, with maximum development during the Middle Pennsylvanian. As a result of depositional infilling, the area ceased to be a basin sensu stricto by Early Permian (Wolfcamp) time and became a shallow, open carbonate shelf.

Lower Permian sediments of the Tatum Basin are cyclic sequences of interbedded limestones and shales that are assigned to the Hueco Formation. At its base are four shoaling upward cyclic carbonate sequences that have been informally designated the Bough member. Fusulinids date the Bough A, B, and C as Early Permian (Wolfcamp) and the Bough D as Late Pennsylvanian (Virgil) in the southern Tatum Basin. Each Bough interval consists of regressive carbonate cycles separated by interbedded shale and argillaceous limestone (Cys, this vol.). The three most common carbonate lithologies are bioclastic grainstone, phylloid algal bafflestone, and bioclastic wackestone. Algae are an important biotic constituent of the Bough cyclic carbonates and form the subject of this paper.

This paper is based on a core study of the Lower Permian Bough A–C members in the southern Tatum Basin. The fields from which cores were studied are shown in Fig. 1, and the cored wells studied are listed in Table 1. The distribution of the algae in the cored wells is given in Table 2. Very special thanks and appreciation are extended to D.F. Toomey, Midland, Texas, for his initial recognition and identification of *Dvinella* in the Baum Field grainstone, generic identification of some of the tubular foraminifers encountered in the study, preparation of some of the thin sections, and taking the photomicrographs of the Baum Field grainstone biota (Figs. 2B and 3), and reviewing an earlier draft of this paper. Appreciation is expressed to Union Oil Company of California, Dorchester Exploration, Inc., Coastal Oil and Gas Corporation, Mapco Production Company, and Shell Oil Company, all of Midland, Texas, for permission to examine their respective cores. Thanks are also due Mapco Production Company for support of this study, which was done while it employed the author.

---

1 602 West Kansas, Midland, TX 79701, U.S.A.

Paleoalgology: Contemporary Research and Applications
(eds. D.F. Toomey and M.H. Nitecki)

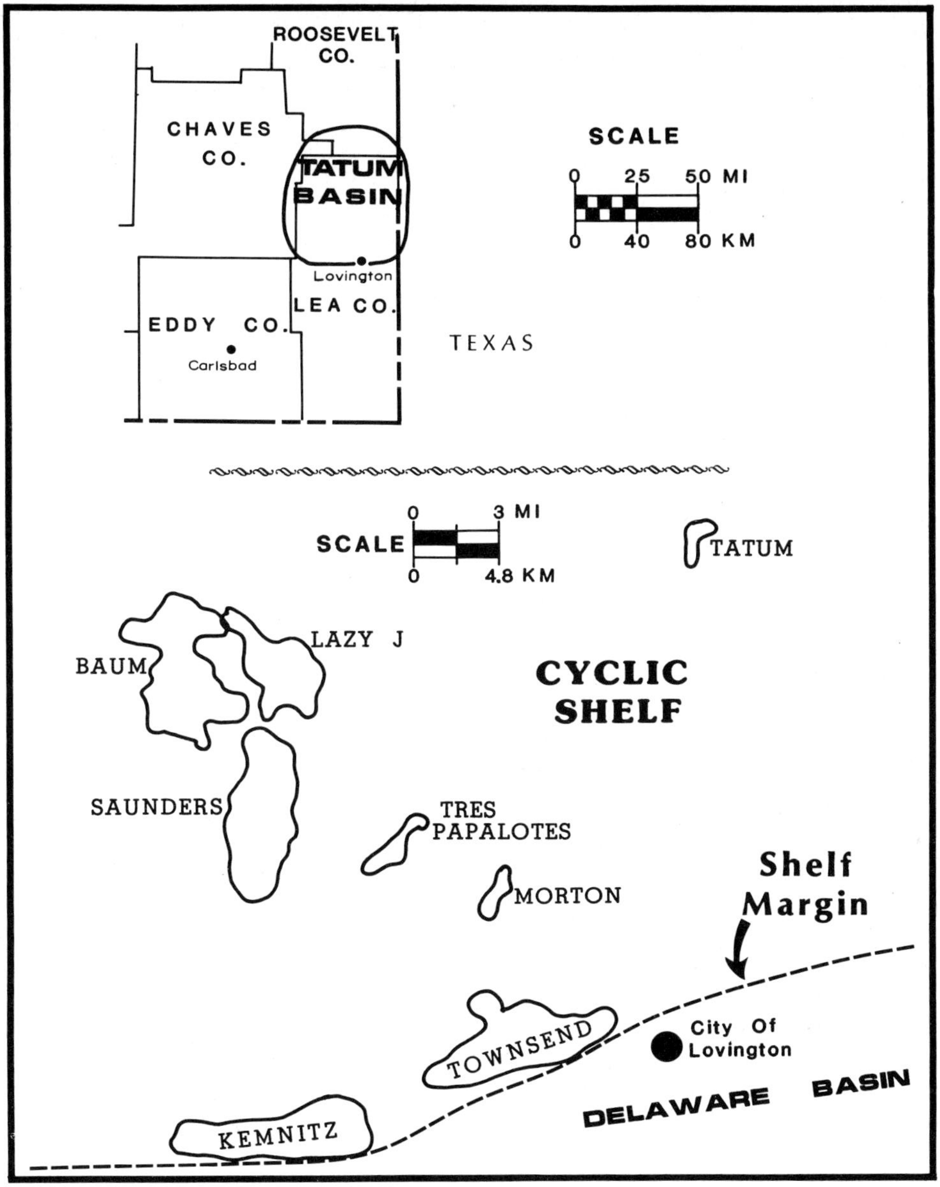

Fig. 1. Index maps showing the general location of the Tatum Basin in southeastern New Mexico and the southern Tatum Basin study area for this paper. Cores were studied from Baum, Lazy J, Saunders, Tatum, Tres Papalotes, and Morton Fields. Also shown for reference are Kemnitz and Townsend Fields, located on the shelf-edge between the Tatum Basin shelf and deep Delaware Basin

Table 1. Wells from which cores were studied for this paper

Well name	Field	Location
Great Western No. 1 Town	Morton	Section 18, T-15-S, R-35-E
Union of California No. 1 State "7"	Morton	Section 07, T-15-S, R-35-E
Union of California No. 1–12 Gulf-Federal	Morton	Section 12, T-15-S, R-34-E
Dorchester No. 1 Morton Solid State Unit	Tres Papalotes	Section 04, T-15-S, R-34-E
Magnolia No. 1 Read-State	Lazy J	Section 34, T-13-S, R-33-E
Shell No. 1-A State	Saunders	Section 34, T-14-S, R-33-E
Shell No. 2 State "A"	Saunders	Section 34, T-14-S, R-33-E
Coastal Oil and Gas Corp. No. 3 State "8"	Baum	Section 08, T-14-S, R-33-E
Union of California No. 1–6 Duncan	Tatum	Section 06, T-13-S, R-36-E
Union of California No. 1–6 Duncan "B"	Tatum	Section 06, T-13-S, R-36-E
Union of California No. 1–5 Atwood Bradley	Tatum	Section 05, T-13-S, R-36-E

Note: All wells are located in Lea County, New Mexico

Table 2. Distribution of algae in cored wells studied for this paper

Taxon	Wells	Abundance
Cyanophyta	All	Uncommon to abundant in all three major facies
Eugonophyllum	All	Abundant in the phylloid algal bafflestone facies. Rare to uncommon (minor) constituent in bioclastic wackestone facies
Dvinella	Coastal Oil and Gas Corp. No. 3 State 8	Abundant in basal part of bioclastic grainstone from 9842 to 9846 feet (2999.8–3001.1 m)
Epimastopora	Union of California No. 1 State 7	Rare constituent in phylloid algal bafflestone facies
	Union of California No. 1 Duncan 6	Uncommon (minor) constituent of bioclastic wackestone (and packstone) facies
	Coastal Oil and Gas Corp. No. 3 State 8	Uncommon (minor) constituent of bioclastic wackestone facies
Tubiphytes	All	Rare to abundant in all three major facies
Diplosphaerina	All	Rare to uncommon in all three major facies

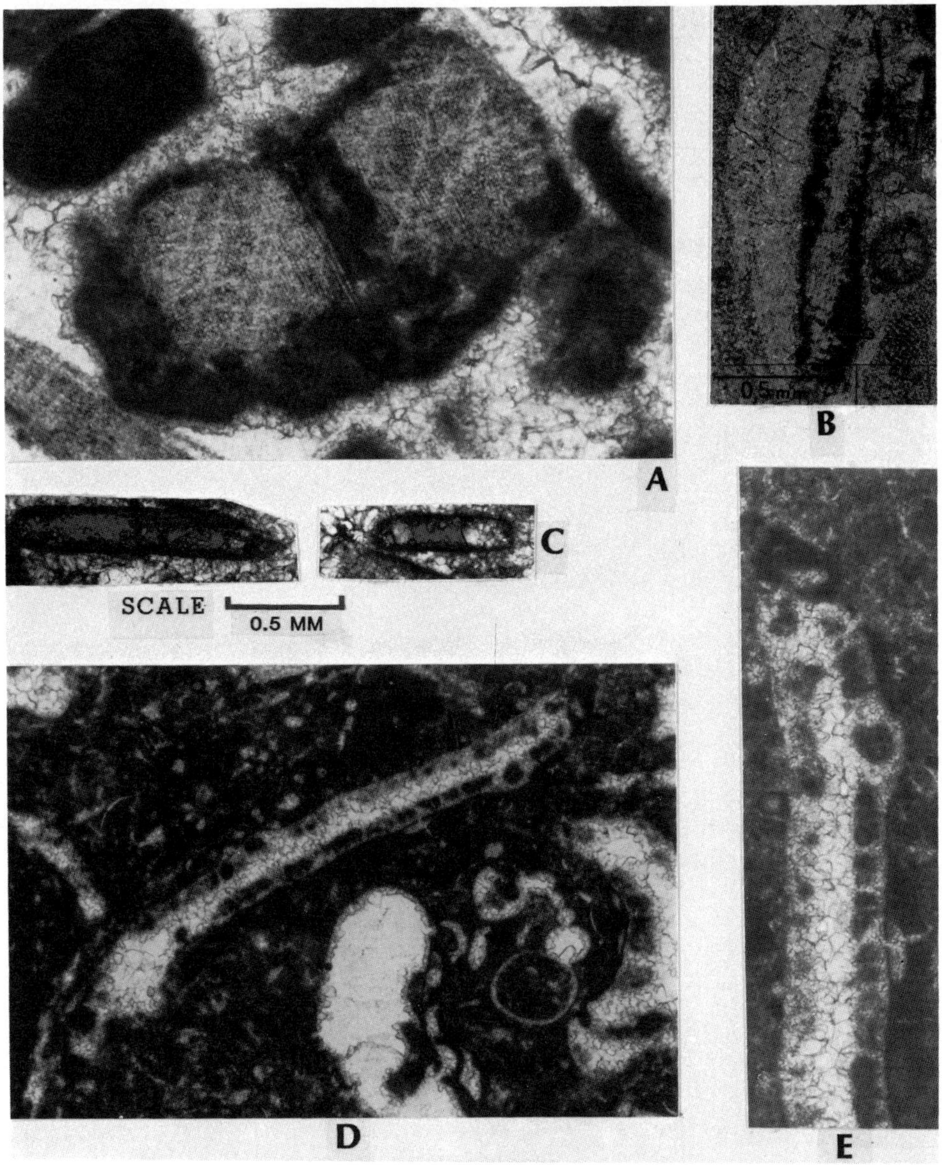

Fig. 2A–E. Thin-section photomicrographs of blue-green and codiacean algae. **A** Bioclastic grainstone in which blue-green algae have encrusted and bored a crinoid fragment. Depth 10,484 feet (3195.5 m), Union of California No. 1 State 7 well, Morton Field. Width of photographic field is 3.2 mm; plain light. **B** Unidentified skeletal grain heavily bored by blue-green alga and/or fungus. Depth 9,845 feet (3000.8 m), Coastal Oil and Gas Corp. No. 3 State 8 well, Baum Field. Crossed-nicols (photo taken by D.F. Toomey). **C** Two examples of leached skeletal fragments (dark gray is blue epoxy-filled porosity) bordered by micrite envelopes. Depth 10,471 feet (3191.6 m), Dorchester No. 1 Morton Solid State Unit, Tres Papalotes Field. Plain light (photos taken by D.F. Toomey). **D** *Eugonophyllum* in phylloid algal bafflestone. Depth 10,459 feet (3187.9 m), Union of California No. 1 State 7 well, Morton Field. Width of photographic field is 4.25 mm; plain light. **E** Closeup of part of the *Eugonophyllum* blade in **D** showing details of the cortical layer with its floating utricles. Note encrusting calcareous foraminifers. Height of photographic field is 3.2 mm; plain light

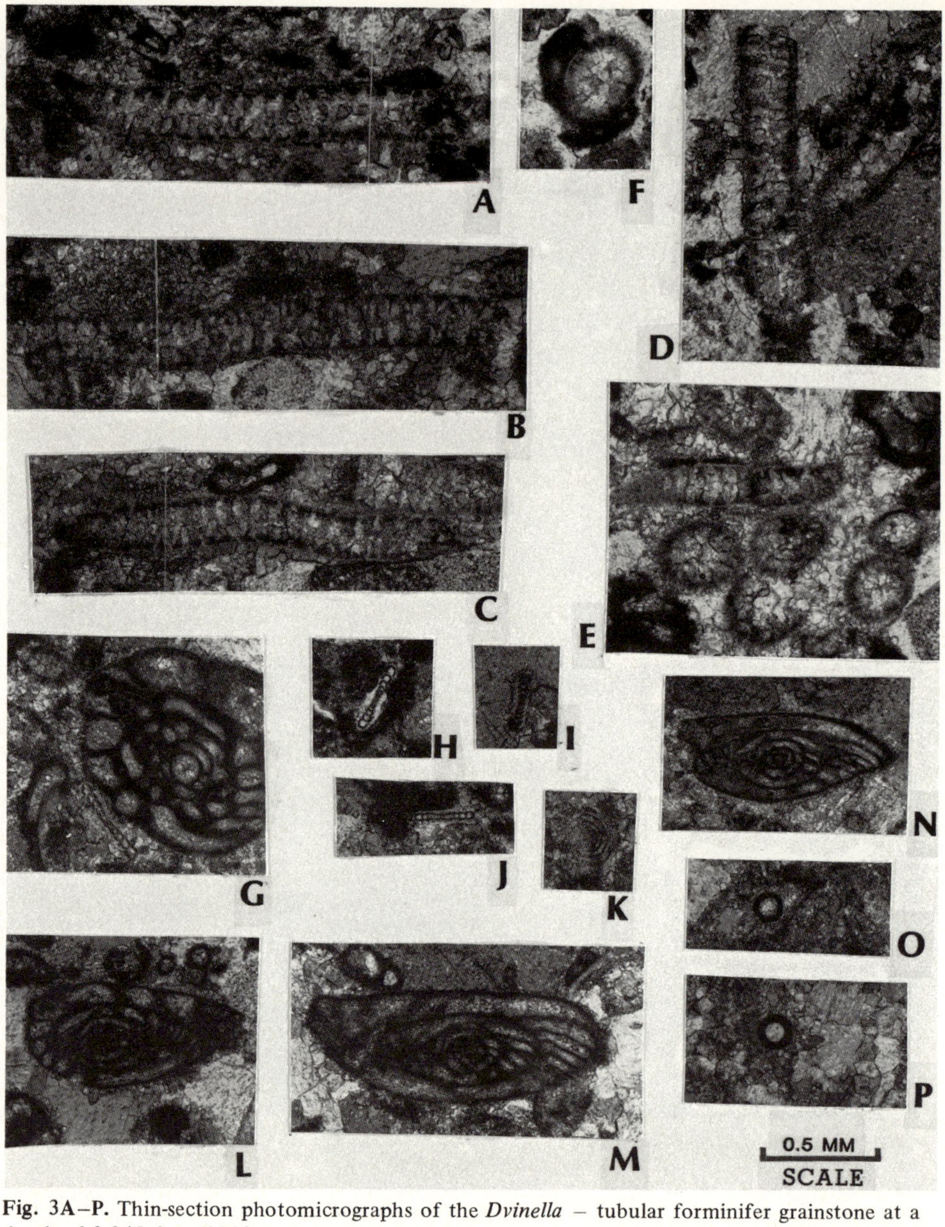

Fig. 3A–P. Thin-section photomicrographs of the *Dvinella* – tubular forminifer grainstone at a depth of 9,845 feet (3000.8 m) in the Coastal Oil and Gas Corp. No. 3 State 8 well, Baum Field. All photographs were taken by D.F. Toomey under crossed nicols. **A–C** Approximate longitudinal sections of *Dvinella.* **D** *Dvinella* fragment showing probable branching. **E** Cross-sections and an approximate longitudinal section of *Dvinella.* **F** Cross-section of *Dvinella* encrusted with calcareous foraminifers. **G–J** Random thin-section cuts of the foraminifer "Hobsonites-Z". Note fusulinid in **G** for size comparison. **K** Approximate longitudinal section of "Hobsonites-Z". **L–N** Unoriented oblique sections of the fusulinid *Schubertella kingi.* **O, P** The algal (?) spore (?) *Diplosphaerina*

Algae

Cyanophyta. Both endolithic and nonendolithic blue-green algae are common constituents of the Bough member in the southern Tatum Basin. Nonendolithic blue-green algae encrust biotic constituents in all three basic carbonate lithologies. Crinoids (Fig. 2A) in the bioclastic grainstone facies are commonly coated with blue-green algae. In the phylloid algal bafflestone facies they occasionally encrust phylloid algae and other skeletal particles. In the bioclastic wackestone facies they encrust shell fragments and sometimes form oncolites with both shell fragments and intraclasts as nucleii.

Endolithic blue-green algae (and possible fungi) are very abundant. They have bored a variety of grains (Fig. 2B) and formed micrite envelopes on various skeletal fragments (Fig. 2C), especially pelecypods (Fig. 4A).

Codiaceae. Phylloid algae are the most important and abundant algal constituent of the Lower Permian rocks of the southern Tatum Basin. They form numerous mound complexes that are major hydrocarbon reservoirs in the area. As is often the case, almost all the blades have been extensively altered during diagenesis. In a few instances some blades preserve the cortical layer with floating utricles (Fig. 20D, E). This double-layered cortical structure is diagnostic of the genus *Eugonophyllum* (Konishi and Wray 1961). In erecting the genus Konishi and Wray (1961) described two new species *Eugonophyllum johnsonii*, characterized by blades 200–700 microns thick, and *E. huecoense*, characterized by blades 500–1250 microns thick. The blade thickness of the southern Tatum Basin specimens examined are within the thickness range of *Eugonophyllum johnsonii* and, on this basis, are assigned to this species.

As noted above, *Eugonophyllum* commonly forms low-relief biohermal and biostromal buildups in the study area. Toward the central part of the Tatum Basin in the area of Tatum and Baum Fields, phylloid algae are minor constituents of the off-mound bioclastic wackestone facies.

As shown in Fig. 2E, phylloid algae are commonly encrusted by a variety of smaller calcareous foraminifers, especially of the calcivertellid type, and they may also be encrusted by blue-green algae and *Tubiphytes*.

Dasycladaceae. In the basal 12 inches (30.5 cm) of a 4 feet thick (1.2 m) grainstone in a core of the Bough A in the Coastal Oil and Gas Corporation No. 3 State 8 well (Table 1) in the southern part of the Baum Field (Fig. 1), there is an abundance of rather poorly preserved tubiform fossils (Fig. 3A–F). As shown by Fig. 3E, F, the fragments are tubular in shape. Their external morphology, perforated wall, and the ladder-like arrangement of partial cross-partitions evident in longitudinal sections (Fig. 3A–C) are all features similar to those of the beresellid dasycladacean (?) *Dvinella*. Note the strong similarity between Fig. 3A and the example of *Dvinella* illustrated by Wray (1977, Fig. 104). On the basis of the above comparisons, these specimens are assigned to *Dvinella*. Figure 3D shows a probable bifurcation of a thallus which Wray (1977) states is also characteristic of the genus.

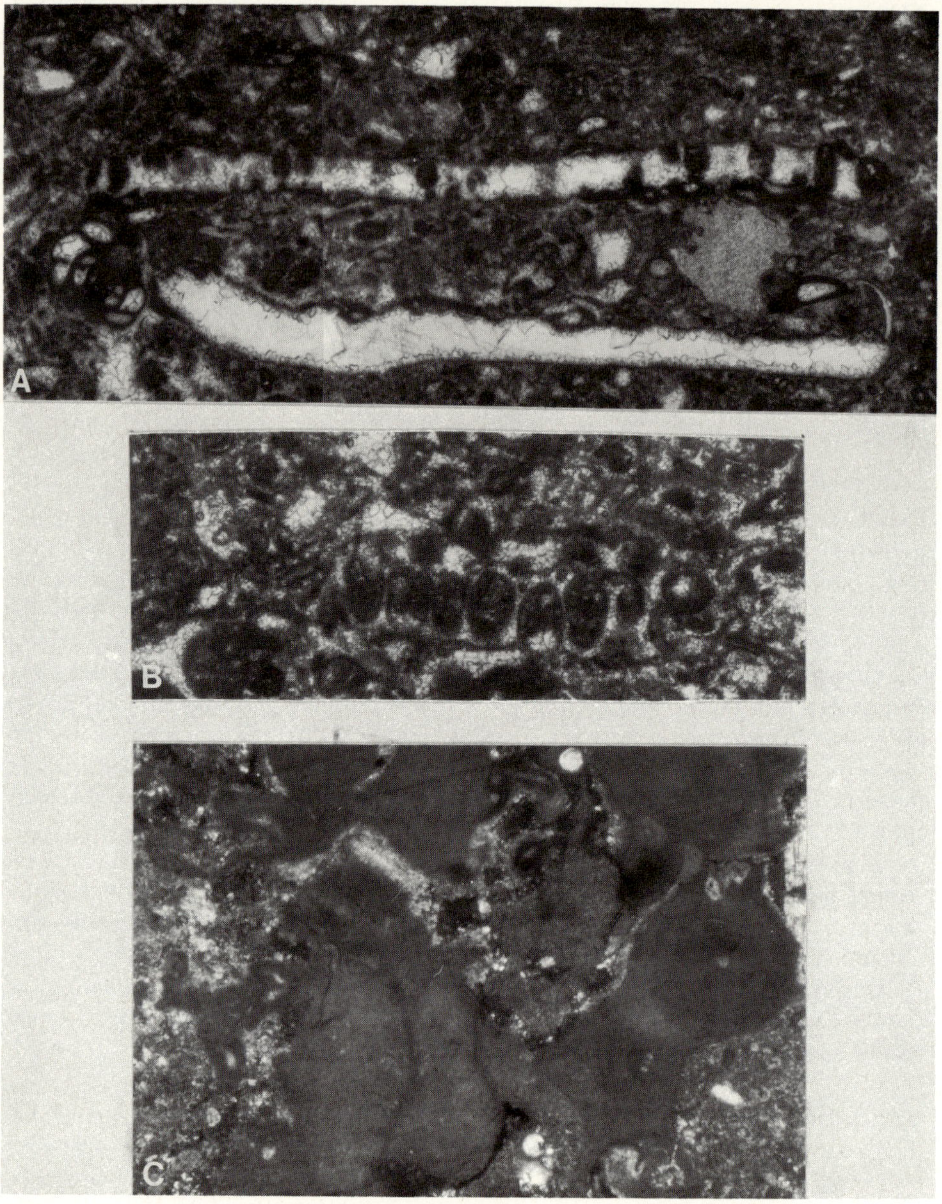

Fig. 4A–C. Thin-section photomicrographs in plain light of dasyclad algae and problematical organisms. **A** The dasyclad alga *Epimastopora* in phylloid algal bafflestone. Note endothyrid foraminifer immediately to the left of large molluscan fragment with micrite envelope. Also note embayed crinoid between the right ends of the mollusc and *Epimastopora.* Depth 10,459 feet (3187.9 m), Union of California No. 1 State 7 well, Morton Field. Width of photographic field is 5.95 mm. **B** *Epimastopora* fragment in an off-mound bioclastic packstone. Depth 10,314 feet (3143.7 m), Union of California No. 1–6 Duncan well, Tatum Field. Width of photographic field is 3.2 mm. **C** Local pocket of *Tubiphytes* boundstone. Depth 10,439 feet (3181.8 m), Union of California No. 1–6 Duncan B well, Tatum Field. Width of photographic field is 4.25 mm

Johnson (1963) states that *Dvinella* is common in Pennsylvanian rocks, but this form has been recorded only once from Permian rocks. This occurrence is in the Lower Permian Akiyoshi Limestone Group of Japan (Endo 1961). Preservation is too poor in the Baum Field specimens and Endo's (1961) description too general to permit specific comparisons. A second Permian occurrence has recently been reported by Mamet et al. (1979) from Lower Permian carbonate rocks of the Canadian Arctic. Specific comparison is not possible because no detailed description, illustrations or species assignment were given. This New Mexico occurrence represents only the third one in Permian rocks of the world and the second in North America. The closest geographic occurrence of *Dvinella* is in Upper Pennsylvanian (Virgil) rocks of the Magdalena Group in the Hueco Mountains of West Texas (Toomey and Babcock 1983). Another occurrence is in Nevada and Utah, where Rich (1967) reports *Dvinella* to be widespread in Lower and Middle Pennsylvanian rocks. The evidence of the Early Permian age of the Baum Field *Dvinella* is the presence in the same grainstone of the diagnostic Early Permian fusulinid *Schubertella kingi* (Fig. 3L–N), along with the flat planispirally-coiled calcareous foraminifer "Hobsonites-Z" (Fig. 3G–K), which, according to D.F. Toomey (pers. comm., 1983), is confined to the Early Permian. In other parts of the core the Early Permian fusulinid *Leptotriticites* (identification by A.M. Reid III) has also been found.

Because of its occurrence as transported fragments in a high-energy grainstone nothing can be said of its paleoecology except that it lived on a shallow, open carbonate shelf.

The dasyclad *Epimastopora* (Fig. 4A, B) occurs as a rare constituent of phylloid algal mounds and a minor constituent of the off-mound bioclastic wackestones, especially toward the central part of the basin in the area of Baum and Tatum Fields (Fig. 1). Of the North American species of *Epimastopora* illustrated by Johnson (1963), these specimens are most comparable to *Epimastopora kansasensis.* However, additional specimens and detailed study are needed before definite specific assignment can be made. Accordingly, they are herein designated only as *Epimastopora* sp. cf. *E. kansasensis.*

Problematica. The problematical organism *Tubiphytes* is an important and moderately abundant biotic constituent of the Bough member in the southern Tatum Basin. It constructs local boundstone patches (Fig. 4C) in phylloid algal mounds and also encrusts mound organisms. In the local boundstone patches it is commonly associated with encrusting tubular foraminifers. *Tubiphytes* also occurs in the bioclastic grainstones that are associated with the mounds and the tubular foraminifer wackestones that usually occur adjacent to the mounds. *Tubiphytes* is a rare constituent of the off-mound normal marine bioclastic wackestone.

Of sporadic occurrence in the study area are small hollow spherical (circular in thin section) objects with thick, dark walls (Fig. 3O, P) that are identical to the algal (?) spore (?) *Diplosphaerina* illustrated by Toomey (1983). Based on this similarity, these spherical forms are also assigned to *Diplosphaerina.* As Toomey (1983) does, the algal spore affinity of *Diplosphaerina* is made on a provisional basis because of morphological differences (namely, the lack of an aperature) from Recent algal spores.

Summary

Lower Permian (Wolfcamp-Assel) rocks of the subsurface southern Tatum Basin of southeastern New Mexico are characterized by shoaling-upward carbonate cycles in which algae are an important biotic constituent. The phylloid alga *Eugonophyllum* is the most common and important algal component and commonly forms low-relief biostromal and biohermal buildups that are major hydrocarbon reservoirs in the study area.

Endolithic algae are very common and appear to have bored a variety of grains, forming micrite envelopes on various skeletal fragments. Nonendolithic blue-green algae encrust phylloid algae and other skeletal fragments and occasionally form oncolites, with both intraclasts and shell fragments as nucleii.

Tubiphytes commonly constructs local boundstone patches in phylloid algal mounds and also encrusts mound organisms. It is also found in grainstones associated with the mounds. *Epimastopora* is a minor constituent of both non-mound and mound rocks.

The tubiform dasycladacean (?) *Dvinella* occurs in a thin grainstone in the Baum Field. This represents only the second reported Permian occurrence of this genus in North America.

The algal (?) spore (?) *Diplosphaerina* occurs sporadically in the area.

References

Endo R (1961) Stratigraphical and paleontological studies of the later Paleozoic calcareous algae in Japan, XVII: Fossil algae from the Akiyoshi Limestone Group. Sci Rep Saitama Univ Ser B Endo Commem Vol, pp 119–142

Johnson JH (1963) Pennsylvanian and Permian algae. Colo Sch Mines Q 58: no 3, 211

Konishi K, Wray JL (1961) *Eugonophyllum,* a new Pennsylvanian and Permian algal genus. J Paleontol 35:659–666

Mamet B, Nassichuk W, Roux A (1979) Algues et stratigraphie du Paléozoique supérieur de l'Arctique Canadien. Bull Cent Rech Explor-Prod Elf-Aquitaine 3: no 2, 669–683

Rich M (1969) *Donezella* and *Dvinella,* widespread algae in Lower and Middle Pennsylvanian rocks in east-central Nevada and west-central Utah. J Paleontol 41:973–980

Toomey DF (1983) The paleoecology of a "middle limestone member" (Leavenworth) of an Upper Carboniferous (Stephanian) cyclothem, mid-continent, U.S.A. Facies 8:113–190

Toomey DF, Babcock JA (1983) Precambrian and Paleozoic algal carbonates, West Texas–Southern New Mexico. Colo Sch Mines Prof Contrib No 11:345

Wray JL (1977) Calcareous algae. Elsevier, New York, 185 p

26 Dasyclad Algae Within Permian (Leonard) Cyclic Shelf Carbonates ("Abo"), Northern Midland Basin, West Texas

D.F. TOOMEY[1]

Introduction

Over the past two decades, a series of papers have appeared relating to the subsurface Permian "Abo" sequence of West Texas and southeastern New Mexico. These include works by Grills (1981), Jeary (1978), LeMay (1960a,b, 1972), McCaslin (1961), Nottingham (1966), Pitts (1977), and Runyan (1965). Most of these papers suggest a number of common features for this rock sequence: (1) that the "Abo" interval comprises basinal, fore-reef, reef, and back-reef sediments, (2) that this sequence rims the Delaware basin and a good portion of the northern Midland basin, and (3) that this represents a continuous barrier-reef complex of similar age, suggesting rather identical paleodepositional settings along this so-called barrier-reef.

More recently, Mazzullo (1982) has suggested that the "Abo" can be subdivided into two portions – a lower carbonate sequence of slope to basinal rocks and an upper sequence of shelf carbonates and evaporites. These rocks reflect upward shoaling and display a progradational relationship between shelf facies and slope to basinal facies. Mazzullo's contention is that favorable petroleum reservoirs can occur in discontinuous oolite shoals situated up-dip along a shelf margin, rather than in a barrier-reef complex, a contention that has been supported by most workers. In fact, my study of three cores (almost 1200 feet) from wells in the Kingdom Abo field, Terry County, Texas (called an "Abo" reef by Grills 1981), has disclosed nothing in the way of so-called reef fabric. The lower "Abo" portion can be described instead as an oolite shoal complex, with some intervals even displaying excellent cross-bedding features within these nonskeletal sands. Study of four additional cores from Lamb and Hockley Counties, West Texas (Fig. 1), has suggested to me that the entire "Abo" interval comprises a progradational sedimentary series ranging from basin to inner shelf. The overall distribution of "Abo" facies represents upward-shoaling sequences with facies belts arranged along what I consider bathymetric contours. This suggests a ramp model (Ahr 1973) as the paleodepositional setting for the "Abo" interval. Perhaps even more accurately, it suggests a distally steepened ramp sensu Read (1982).

While studying a number of "Abo" cores, I was particularly struck by the occurrence of rather common dasyclad algal remains in some of the thin shelf carbonates of the upper "Abo". Accordingly, four cores were studied in detail, and special efforts were

1 Conoco Inc., Research Center, Ponca City, OK 74604, U.S.A.

Paleoalgology: Contemporary Research and Applications
(eds. D.F. Toomey and M.H. Nitecki)

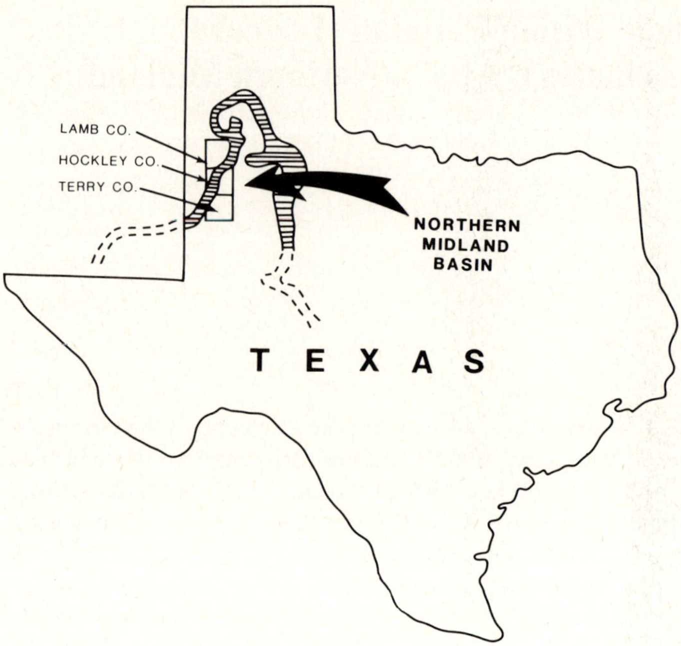

Fig. 1. Map of the northern Midland basin with outline of the "Abo" trend (*ruled area*) delineating the periphery of the basin and the three West Texas counties from which "Abo" cores were studied

Table 1. Thin section control and core footage studied

Core	Thin sections	Footage
Threshold-Jones	134	210
Threshold-Browning	216	178
Citco-Roberts "E"	134	368
Gulf-Sadler	125	415
Total	609	1171

made to sample these thin, dasyclad-rich carbonates. To accomplish this, 609 thin sections were prepared from four cores, with a total footage of somewhat less than 1200 feet (see Table 1). The thickest core (415 feet), the Gulf-Sadler, was used as the "type core" for the overall electric log correlations and for descriptive comparisons. The location of these cores along the "Abo" trend is shown in Fig. 2.

General Stratigraphy

Stratigraphic relationships of the "Abo" interval have been somewhat confused over the last few decades. Mazzullo's (1982) discussion and attempt to place the "Abo"

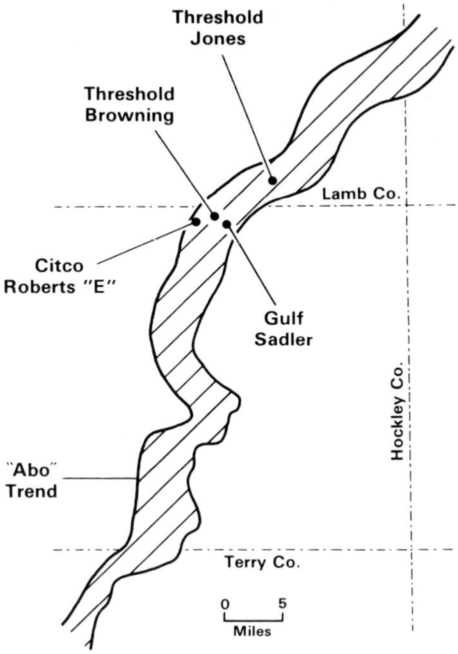

Fig. 2. Index map of Lamb, Hockley, and Terry Conties, West Texas, with the "Abo" trend delineated and showing the locations of four cores used in this study

interval with a workable stratigraphic framework has been partially successful, in the sense that the overall subsurface Permian sequence has now been placed at least within proper perspective. The difficulty in obtaining a usable stratigraphic framework suitable to all workers is the general absence of definitive biostratigraphic subdivision, even though this is an area of high density well control. Until more cores are available and definitive fusulinid biostratigraphy is applied to critical cores, stratigraphic control and designation as to just what constitutes the "Abo" interval will remain a problem.

A graphic representation of the cored "Abo" interval present in the Citco-Roberts "E" well (Fig. 3), Hockley County, shows a typical two-part "Abo" subdivision. This subdivision is well-illustrated on the electric log gamma ray tracing shown to the left of the log and supplemented by the plotted postulated sequence of depositional environments shown to the right of the log. The gamma ray tracing of the upper "Abo" delineates a "dirty" signature typical of thin-bedded, "impure" cyclic carbonates. These carbonates run the gamut from supratidal to shallow subtidal sediments. The lower "Abo" is characterized on the gamma ray tracing as a massive clean porous carbonate (CPC) thought to have been deposited within a shallow subtidal setting.

It should be emphasized that this contact between the upper and lower "Abo" is easily identified or "picked" on electric logs and probably represents a major regional unconformity. On one "Abo" core that I have described, I observed an irregular karsted surface at this contact in which dark gray-green shale derived from the overlying "dirty" interval was present in fractures for a depth of 14 feet into the lower "Abo" clean porous carbonate. Hence, this erosional surface is of significance

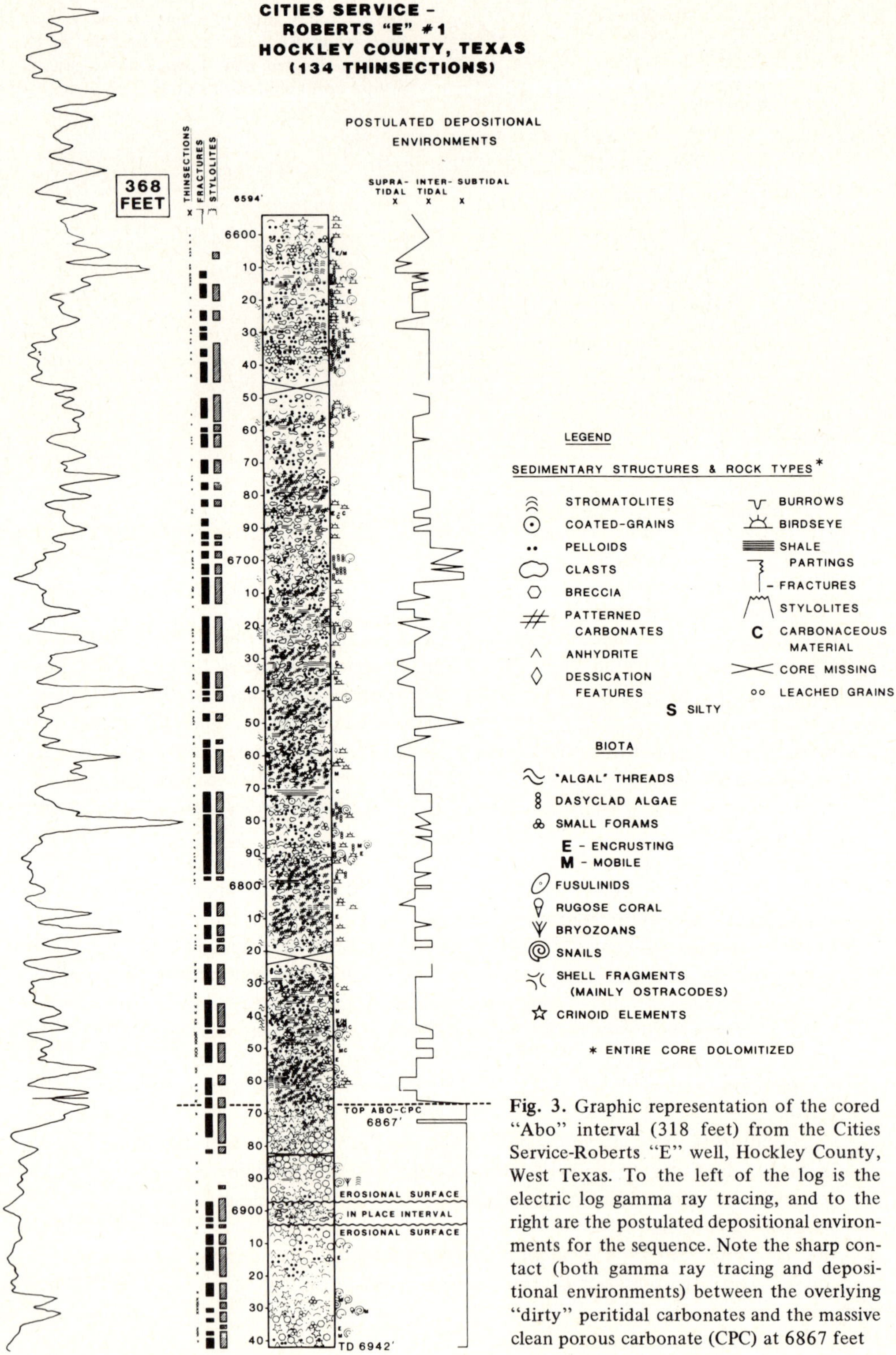

Fig. 3. Graphic representation of the cored "Abo" interval (318 feet) from the Cities Service-Roberts "E" well, Hockley County, West Texas. To the left of the log is the electric log gamma ray tracing, and to the right are the postulated depositional environments for the sequence. Note the sharp contact (both gamma ray tracing and depositional environments) between the overlying "dirty" peritidal carbonates and the massive clean porous carbonate (CPC) at 6867 feet

and may have formed at a time when primary fracturing was initiated within the massive "Abo" dolomites and which later has proved to be of such importance to hydrocarbon entrapment.

"Abo" Lithologies

As noted above, dolomitized shelf sediments are generally thin-bedded, cyclic, of varied lithology and biota, and span a spectrum of rock types usually accepted to have been laid down in supratidal, intertidal, and shallow subtidal depositional realms. Some typical "Abo" examples are shown in Fig. 4, and the overall suite of "Abo" lithologies can be described and subdivided as follows:

A. Supratidal Sediments

1. Stromatolitic boundstones
   a) Laminar algal mats with clasts and peloids.
   b) Small, digitate stromatolite colonies with associated clasts in the trough areas.
2. Patterned carbonates as described by Dixon (1976) with much pyrite, convoluted, and with a sparse restricted biota; thought to represent sedimentation within ponds situated on the low supratidal portion of the tidal flat; generally associated with gray-green shales (muds) which are commonly laminar and pyrite-bearing.

B. Intertidal Sediments

1. Fenestral fabric with birdseye structures along with coated-grains and clasts as described by Shinn (1968), see Fig. 4D.
2. Dasyclad algal-rich skeletal wackestones, usually peloidal, and associated with clasts; algally coated-grains; mollusc fragments; and, more rarely, fusulinids (see Figs. 4C, E, F, 5 and 6).

C. Coastal Subtidal Sediments

1. Oolitic and/or peloidal grainstones, sometimes with abraded fusulinids and crinoidal debris (Fig. 4B).
2. Fusulinid-rich (*Parafusulina*) skeletal wackestones with shell debris (mainly molluscs) and disarticulated crinoidal components (Fig. 4A).
3. Skeletal wackestones with a relatively abundant and diverse biota (fusulinids and smaller foraminifers, bryozoans, brachiopods, molluscs, ostracodes, and crinoidal debris).

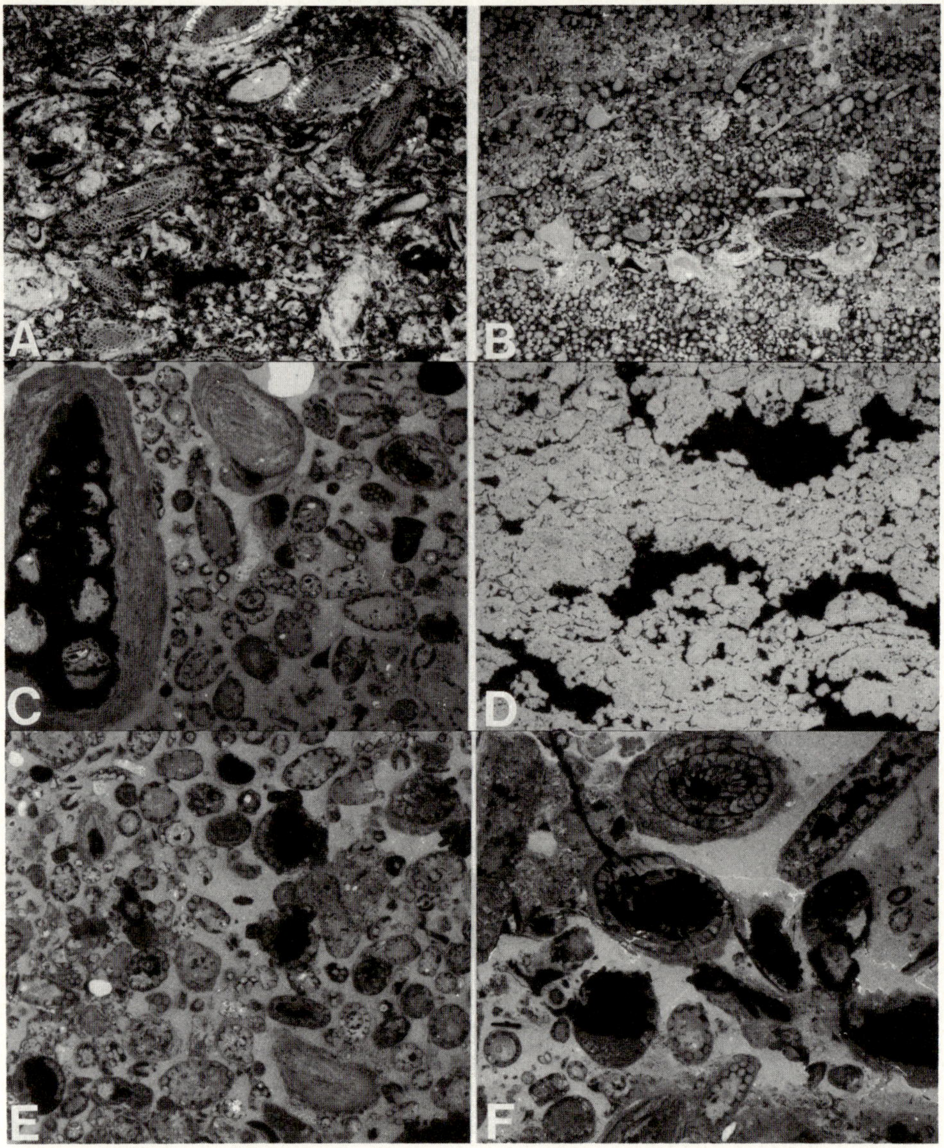

Fig. 4A–F. Thin section photomicrographs (negative prints) of various microfacies encountered in cores from the Permian upper "Abo" interval in the northern Midland basin, West Texas. **A** Skeletal wackestone to packstone with common fusulinids (*Parafusulina*), some of which are partially silicified, and peloids. Dark irregular-shaped areas are plugged with anhydrite; Threshold-Jones core, 6635.10 feet (X 2.5) Lamb County. **B** Oolitic/peloidal grainstone with clasts, coated-grains, fusulinids, and other skeletal debris; Threshold-Jones core, 6630.8 feet (X 2.5) Lamb County. **C** Dasyclad algal wackestone with conspicuous algally coated gastropod. Dark areas in photomicrograph are various skeletal grains replaced and plugged with anhydrite; CITCO-Roberts "E" core, 6703.10 1/2 feet (X 4.1) Hockley County. **D** Fenestral fabric with prominent birdseyes plugged with anhydrite; Threshold-Browning core, 6939.10 feet (X 4.1) Hockley County. **E** Dasyclad algal wackestone with algally coated-grains and anhydrite-plugged fusulinids; CITCO-Roberts "E" core, 6703.9 1/2 feet (X 4.1) Hockley County. **F** Skeletal wackestone with dasyclad algae (note long cylinder, cf. *Macroporella*, in upper right-hand corner, along with relatively large anhydrite-plugged fusulinids (*Parafusulina*); CITCO-Roberts "E" core, 6704 1/4 feet (X 4.1) Hockley County

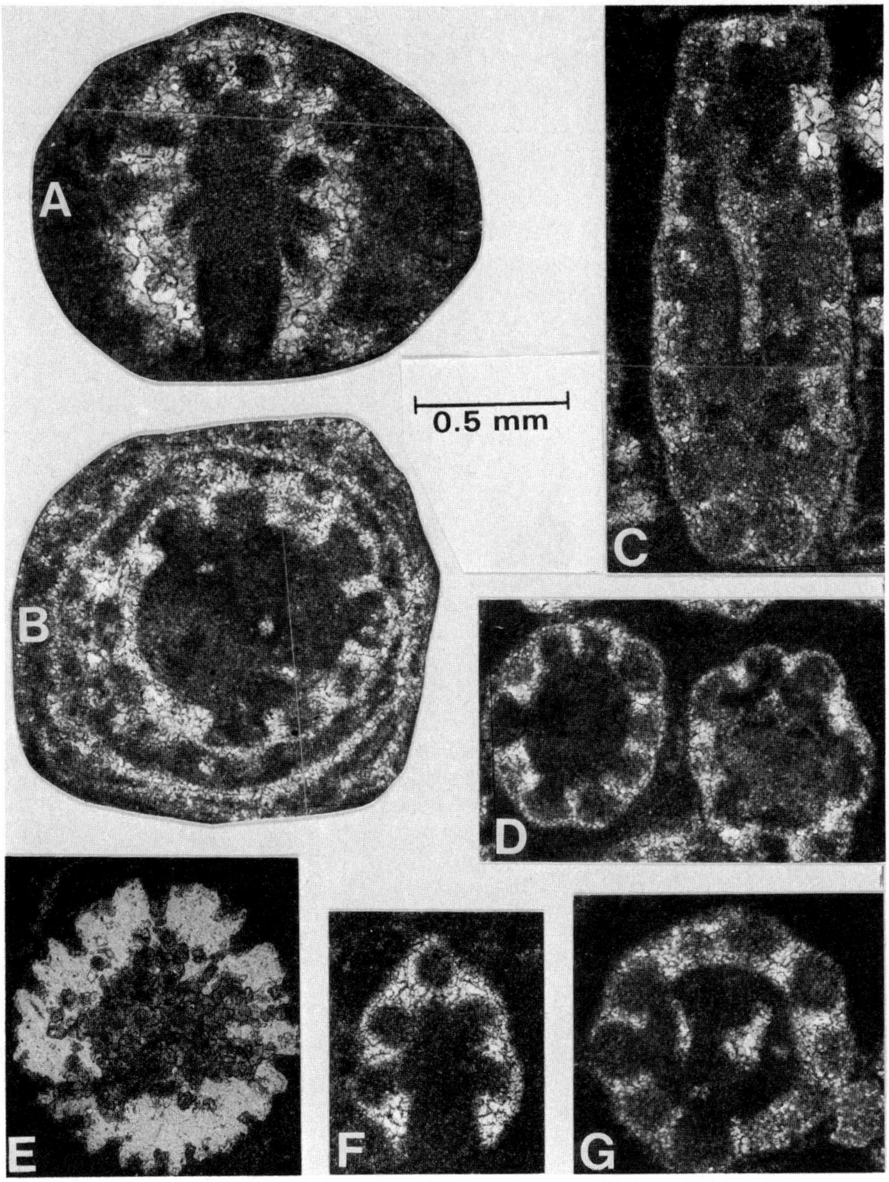

Fig. 5A–G. Thin section photomicrographs of various *Mizzia*-like dasyclad algae encountered in skeletal wackestones of the Permian upper "Abo" dirty interval in the northern Midland basin, West Texas. **A** Tip of a dasyclad thallus coated with recrystallized algal material; CITCO-Roberts "E" core, 6703.9 1/2 feet, Hockley County. **B** Dasyclad alga circumscribed with recrystallized algal coatings; CITCO-Roberts "E" core, 6702.2 feet, Hockley County. **C** Longitudinal thin section cut of a dasyclad "cylinder;" CITCO-Roberts "E" core, 6703.8 1/4 feet, Hockley County. **D, G** Typical dasyclad algal fragments in muddy sediment; CITCO-Roberts "E" core, 6703.10 1/2 feet, Hockley County. **E** Dasyclad algal fragment totally replaced with anhydrite; Threshold-Jones core, 6652.9 1/2 feet, Lamb County. **F** Tip of dasyclad thallus; Threshold-Jones core, 6653.5 feet, Lamb County

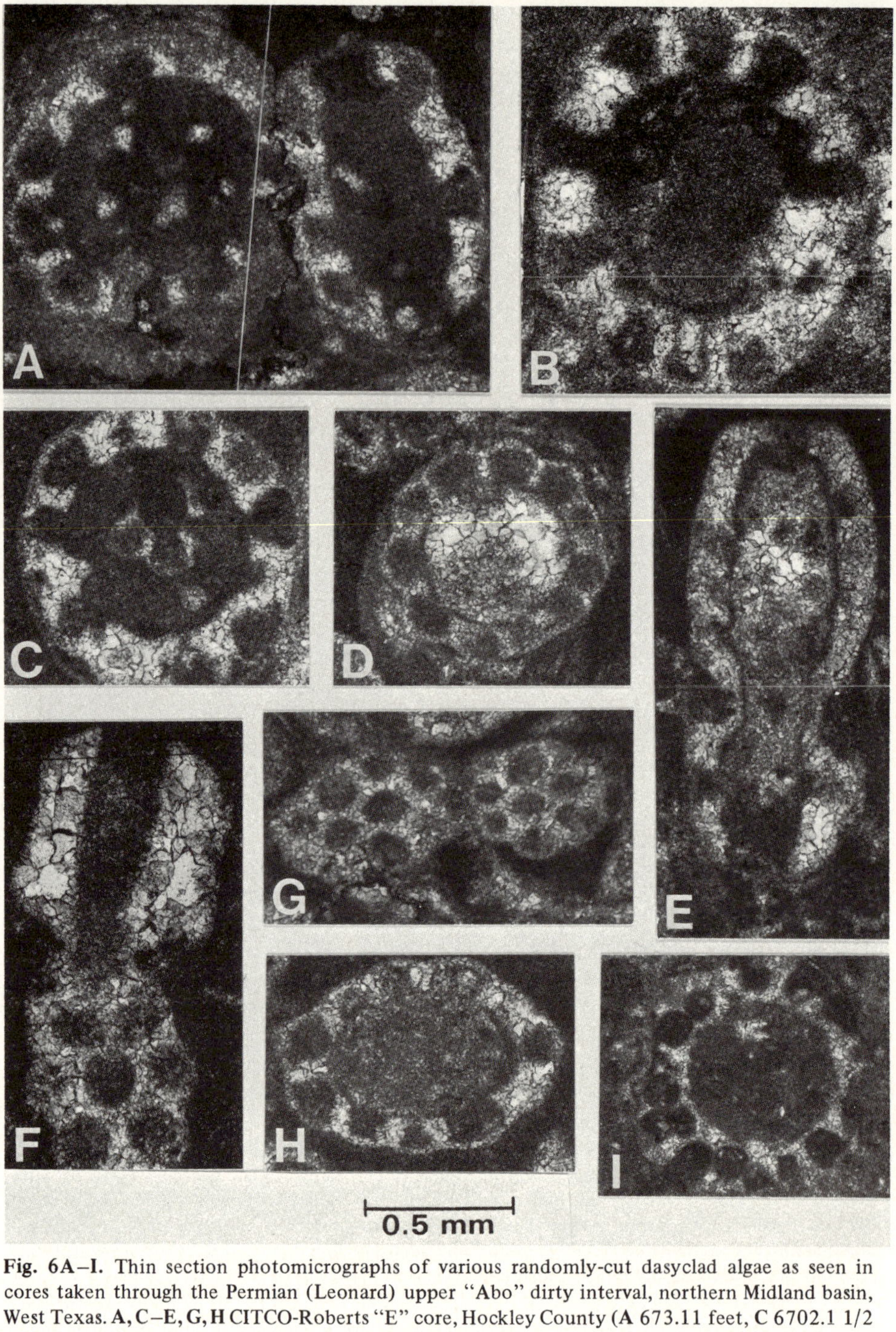

Fig. 6A–I. Thin section photomicrographs of various randomly-cut dasyclad algae as seen in cores taken through the Permian (Leonard) upper "Abo" dirty interval, northern Midland basin, West Texas. **A, C–E, G, H** CITCO-Roberts "E" core, Hockley County (**A** 673.11 feet, **C** 6702.1 1/2 feet, **D** 6703.10 1/2 feet, **E** 6703.8 feet, **G** 6702.7 1/4 feet, **H** 6703.10 1/2 feet). **B, F** Threshold-Jones core, Lamb County (**B** 6753.2 feet, **F** 6652.9 feet). **I** Threshold-Browning core, Hockley County, 6642.2 feet

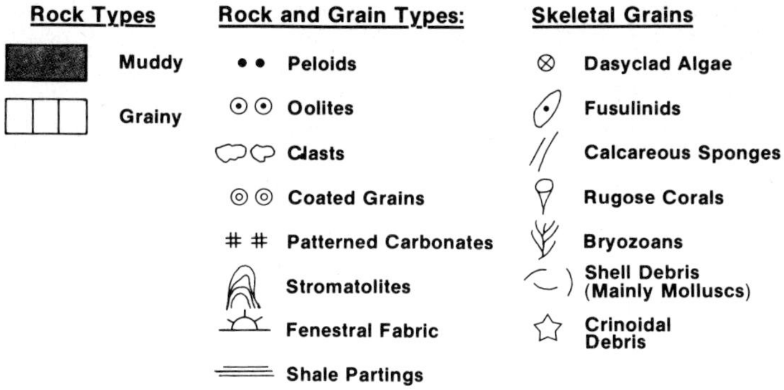

Fig. 7. Legend of graphic symbols of rock types and various skeletal and nonskeletal grains as shown in Figs. 8–11

Cyclicity and Dasyclad Occurrence

Cyclicity within typical Permian shelf sediments of the West Texas subsurface, especially cycles occurring in the younger San Andres interval, have been noted and discussed by Chuber and Pusey (1969), Silver and Todd (1969), and Todd (1976), and in a more general way, as casebook examples of Recent and fossil counterparts, in 1975 by Ginsburg (ed.) and by James (1979). Most recently, Mazzullo (1982) has noted the cyclicity of "Abo" shelf sediments.

In an attempt to demonstrate various cycle types present within "Abo" shelf carbonates, I have prepared stratigraphic section (Figs. 7–10) of relatively thin rock sequences representing various upper "Abo" intervals from three cores taken along the "Abo" trend which, I believe, show a distinctive range of difference within various cycles.

Figure 8 is a graphic representation of a 20-foot interval taken from within the Threshold-Jones core in Lamb County (see Fig. 2). It should be noted that this well was drilled on the more "seaward" side of the "Abo" trend; accordingly, the cycles are consistently "more marine" in aspect, with sediments characteristic of the intertidal flat to shallow subtidal zone. In essence, the interval indicates more open marine conditions, in contrast to other more restrictive cycles shown in Figs. 9 and 10. The biota reflects this more open marine aspect in that appreciably more fossils (fusulinids, e.g., staffellids and *Parafusulina,* smaller foraminifers (*Globivalvulina* and palaeotextularids), calcareous sponges, bryozoans, brachiopods, molluscs, echinoid spines, crinoid components, and ostracodes) are present in most of these somewhat thicker stacked carbonates that comprise the upper "Abo" interval in this core. The contact between the upper and lower "Abo" in this well was "picked" on a change in electric log characteristics at 6705 feet, coincident with a change in lithology from muddy skeletal wackestone to oolitic grainstone/packstones.

The cycles shown in Fig. 8, located some footage above the contact, reflect an oscillating shift in depositional environments from what I consider high-to-low intertidal to shallow subtidal. Dasyclad algae occur within three muddy horizons

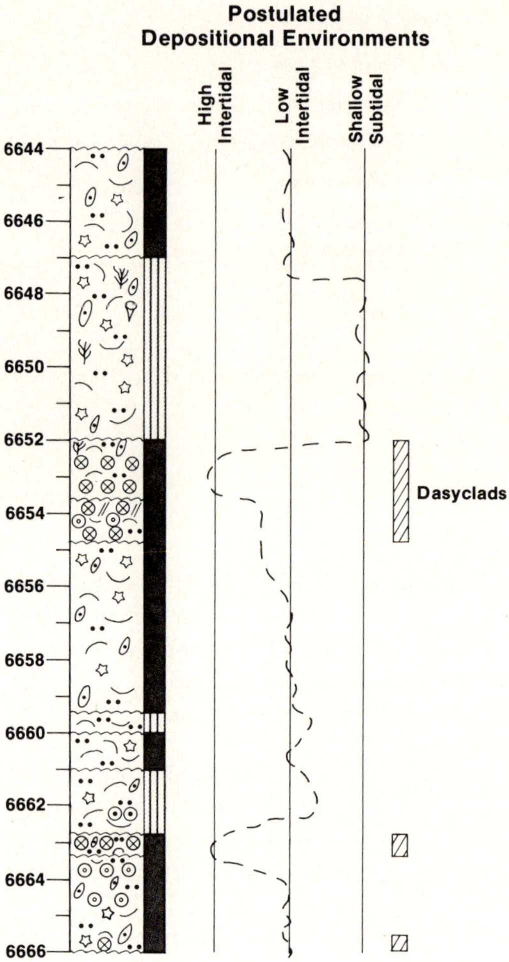

Fig. 8. Graphic representation of the Permian "Abo" dirty interval of peritidal carbonates encountered in the Threshold-Jones core from Lamb County, northern Midland basin, West Texas. This diagram shows postulated depositional environments and the location within the interval of dasyclad-rich beds (refer to Fig. 7 for explanation of symbols)

(Fig. 8). The upper horizon (6652–6653.6 feet) contains, along with common dasyclad algal remains, peloids, rare fusulinids, and some mollusc fragments thought to have been deposited in a somewhat high intertidal setting. Directly below this horizon (6653.7–6654.7 feet) is an interval which contains reduced numbers of dasyclad algae, broken fragments of calcareous sponges, algally coated-grains, mollusc fragments, and peloids. This subtle difference in overall biotic change appears to reflect a slight shifting in depositional setting more seaward into a low intertidal position. This major dasyclad horizon is overlain by a skeletal grainy rock containing a diverse marine biota including rugose corals, bryozoans, echinoid spines, and believed to have been deposited in a more seaward shallow subtidal setting. The dasyclad horizon is underlain by a muddy, peloidal, skeletal wackestone with fusulinids, mollusc fragments, and scattered crinoidal components thought to be indicative of deposition within the low intertidal tidal flat area.

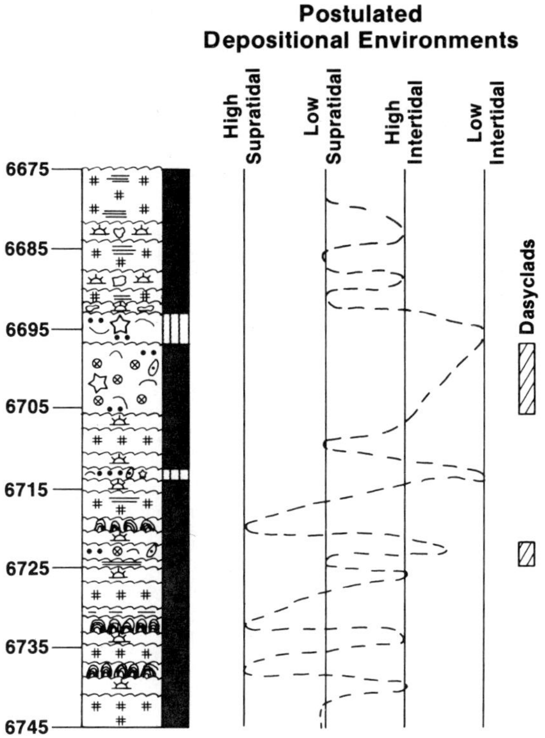

Fig. 9

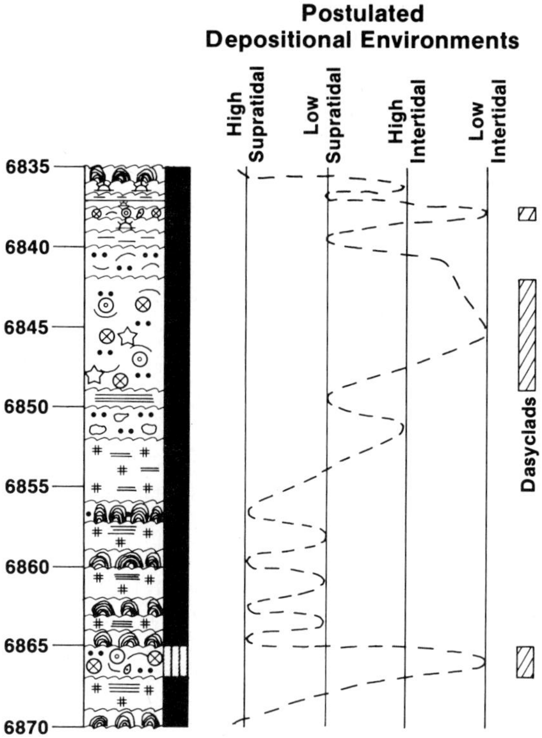

Fig. 10

Fig. 9. Graphic representation of cyclical Permian ("Abo" dirty interval) peritidal carbonates encountered in the CITCO-Roberts "E" core (70-foot interval), Hockley County, northern Midland basin, West Texas. The diagram shows postulated depositional environments and the position of dasyclad algal occurrences within the interval

Fig. 10. Graphic representation of cyclical Permian ("Abo" dirty interval) peritidal carbonates (35-foot interval) encountered in the Threshold-Browning core, Hockley County, northern Midland basin, West Texas. The diagram shows sediment cyclicity, postulated depositional environments, and the position of dasyclad-rich units within the sequence

The second occurrence of only minor dasyclads is in a thin (less than one foot) horizon at 6663 feet. This thin unit is overlain with skeletal grainstone with scattered oolites and is underlain by muddy wackestone with algally coated-grains. The third occurrence of dasyclads, again, in only limited abundance, is in a thin interval at 6665–6666 feet. This horizon is overlain with skeletal wackestone containing algally coated-grains, fusulinids, mollusc fragments, and peloids in sediments thought to have been laid down within the low intertidal portion of the tidal flat.

Cycles shown in Fig. 9 represent a 70-foot interval (6675–6745 feet) within the "Abo" shelf sediments encountered in the Citco-Roberts "E" well in Hockley County. The location of this core (Fig. 2) appears to be in a more up-dip (restricted) position, and as a consequence, cyclic units are thinner, more numerous, better-developed, and appear to go to completion. These cycles also range through a broader depositional spectrum with sediments shifting back and forth across the supratidal-intertidal regime.

Dasyclad algae occur within this cyclic sequence in two intervals. A thin, lower interval from 6723–6724.5 feet containing only minor amounts of dasyclads is underlain by gray-green shale partings thought to have been deposited within the low supratidal area and overlain by a very thin horizon of fenestral fabric sediment with birdseye structures characteristic of the high intertidal regime. The thickest, major dasyclad interval occurs at 6696 to 6706 feet, is underlain by a thin bed of fenestral fabric sediment of the high intertidal regime, and overlain by four feet of grainy, skeletal wackestones with peloids, mollusc fragments, and crinoidal components thought to have been deposited within the low intertidal regime.

Cycles shown in Fig. 10 comprise a 35-foot interval (6835–6870 feet) within the "Abo" shelf sediments encountered in the Threshold-Browning core, Hockley County. The location of this core along the "Abo" trend is basically similar to that of the Citco-Roberts "E" core and shows a comparable cyclic sequence. Two intervals of dasyclad-rich carbonates are shown within this sequence (Fig. 10).

The lowest occurrence of dasyclads is in a thin (2-foot thick), grainy bed at 6865–6867 feet in association with algally coated-grains, fusulinids, mollusc fragments, and peloids. This unit is underlain by two feet of interbedded, patterned carbonates and gray-green shales of the lower supratidal regime and is overlain by a thin (less than one foot) layer of stromatolitic boundstone of the upper supratidal zone. The thickest (7 feet) dasyclad-rich interval occurs from 6842–6849 feet and contains algally coated-grains, mollusc fragments, crinoid components, and peloids within a muddy matrix and is thought to have been deposited along the low intertidal portion of the tidal flat. This horizon is underlain by laminar gray-green shales of the upper supratidal realm and overlain by peloidal, skeletal wackestone with mollusc fragments and thought to have been laid down in an intermediary position between the low and high intertidal portion of the tidal flat. The thinnest dasyclad occurrence is a one-foot interval at 6837.5 to 6838.5 feet within a muddy rock containing algally coated-grains, fusulinids, and mollusc fragments. This algal horizon is both overlain and underlain by fenestral fabric sediment with birdseye structures characteristic of the high intertidal depositional realm.

The dasyclad algae represented in the "Abo" dolomitized shelf sediments discussed above consist primarily of *Mizzia*-like forms (Figs. 5 and 6), although in the Citco-Roberts "E" core dasyclad algae referred to *Macroporella* (Fig. 4F) also occur. Most dasyclad algal specimens are completely dolomitzed, and some have been secondarily replaced by anhydrite (Fig. 5E).

Examples of eight occurrences of dasyclad-rich intervals within upper "Abo" cyclic shelf sediments from three cores in the northern Midland basin demonstrate that primary dasyclad-rich intervals are most common in sediments thought to have been laid down within the low intertidal portion of the tidal flat. Abundant dasyclad algae are generally associated with algally coated-grains, calcareous sponges, molluscs, and rare fusulinids and crinoid components, along with occasional bryozoans and rugose corals. The rugose corals and bryozoans are thought to have been derived from more seaward shallow subtidal settings and intermixed on the low tidal flat.

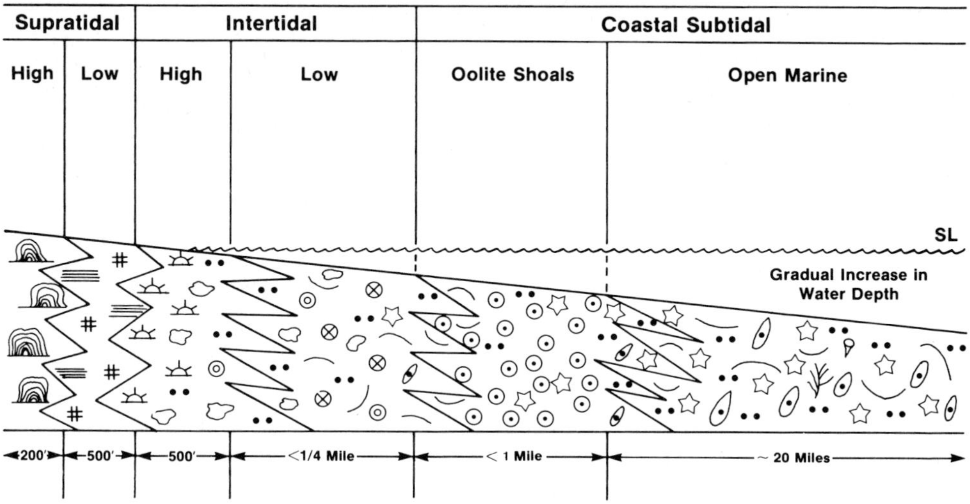

Fig. 11. Schematic representation showing the postulated paleoenvironmental setting for deposition of the Permian (Leonard) "Abo" carbonates of the northern Midland basin, West Texas. This diagram emphasizes the overall ramp aspect of this model, with a low regional dip, and facies belts essentially occurring along bathymetric lines; note postulated width of facies belts (refer to Fig. 7 for explanation of symbols)

When dasyclad algae occur within thin carbonates regarded as having been deposited on the high intertidal tidal flat, they are much reduced in numbers, generally abraded, and only associated with sparse broken mollusc shells. Sediments underlying and overlying dasyclad-rich intervals vary, and can be represented by distinctive rock types characteristic of the high intertidal regime (e.g., fenestral rock with birdseyes) or both low and high supratidal sediments (e.g., laminar shales, patterned carbonates, and stromatolites). Hence, dasyclads present in Permian "Abo" cyclic shelf deposits form an intermediary cyclical wedge separating high intertidal-supratidal sediments up-dip from shallow subtidal sediments down-dip.

Paleoenvironmental Interpretation

Regarding both the upper "Abo" cyclical, "dirty" interval, and the lower "Abo" clean porous carbonate as an entity, one can envision a suitable paleodepositional model encompassing upward shoaling sedimentation occurring on a carbonate ramp of low initial dip (Fig. 11). Facies belts would in essence reflect bathymetric contours of distinctive shallow-water sediments and biota. By projecting this model further seaward, one would encounter over a relatively appreciable distance, typical dark-colored basinal carbonates and shales. In essence, the overall paleoenvironmental reconstruction would be more akin to that proposed by Read (1982) – the distally steepened ramp.

Summary

Study of a number of cores taken along the Permian (Leonard) "Abo" trend of the northern Midland basin suggests that this interval represents an upward shoaling sequence that comprises two distinctive suites of rocks. A lower "Abo" interval is composed principally of massive coastal oolite shoals and shallow subtidal dolomites, separated perhaps by a major regional unconformity, manifested by some karsting, and overlain by a sequence of relatively thin-bedded, impure shelf carbonates whose electric log imprint causes them to be referred to as the upper "Abo" dirty interval.

Paleoenvironmentally, this sequence comprises open marine subtidal sediments, coastal oolite shoals, low and high intertidal sediments, and low and high supratidal sediments. Each of these facies belts is characterized by the presence and/or absence of certain distinctive rock types and biota. Seemingly, this entire sequence is best interpreted as sedimentation occurring over a broad, low angle, distally steepened ramp.

Occurrences of dasyclad algae are restricted to thin, cyclic shelf carbonates of the upper "Abo" dirty interval and generally occur in those sediments thought to have been laid down within the low intertidal regime. The accurate placing of dasyclad-rich sediments within a paleoenvironmental sequence is made somewhat less difficult where relatively thin overlying and underlying sediments are characterized by distinctive attributes (e.g., rock type and biota).

References

Ahr WM (1973) The carbonate ramp: an alternative to the shelf model. Trans Gulf Coast Assoc Geol Soc 23rd Annu Conv, Houston, pp 221–225

Chuber S, Pusey WC (1969) Cyclic San Andres facies and their relationship to diagenesis, prosity, and permeability in the Reeves field, Yoakum County, Texas. In: Elam JG, Chuber S (eds) Cyclic sedimentation in the Permian Basin. West Tex Geol Soc Publ 69-56:135–150

Dixon J (1976) Patterned carbonate – a diagenetic feature. Bull Can Petrol Geol 24:450–456

Ginsburg RN (ed) (1975) Tidal deposits, a casebook of Recent examples and fossil counterparts. Springer, Berlin New York Heidelberg, 428 p

Grills R (1981) Depositional environments and porosity relationships in the Kingdom Abo field, Terry County, Texas (Abstr). PBS-SEPM Newsletter 26: No 6, February

James NP (1979) Shallowing upward sequences in carbonates. In: Walker RG (ed) Facies models. Geoscience Canada, Reprint Ser 1, No 10, pp 109–119

Jeary GL (1978) Leonardian strata in the northern Midland basin of West Texas. In: Energy Quest for the Southwest. West Tex Geol Soc Publ 78-69 SWS; 30–47

LeMay WJ (1960a) Oil accumulations along Abo reefing, southeastern New Mexico. In: Trans Southwest Fed Geol Soc 1:125–137

LeMay WJ (1960b) In southeast New Mexico Abo reef play yielding large oil fields. World Oil No 7:91–95

LeMay WJ (1972) Empire Abo field, southeast New Mexico. In: Stratigraphic oil and gas fields. Mem Am Assoc Petrol Geol 16:472–480

Mazzullo SJ (1982) Stratigraphy and depositional mosaics of lower Clear Fork and Wichita Groups (Permian), northern Midland basin, Texas. Bull Am Assoc Petrol Geol 66:210–227

McCaslin JC (1961) New Mexico's Abo touches Texas. Oil Gas J May 1 1961:184–185

Nottingham MW (1966) Abo reef buildup provides five stratigraphic trap zones. World Oil may 1966:107–110

Pitts JP (1977) Geologists predict Abo trend could ring the Midland basin. Dril Bit Nov 1977: 25–29

Read JF (1982) Carbonate platforms of passive (extensional) continental margins: types, characteristics, and evolution. In: Hales AL (ed) Geodyn Fin Symp, Tectonophys 1981:195–212

Runyan JW (1965) First New Mexican reef ditritus oil pools found down-dip from Abo trend. World Oil March 1965:99–106

Shinn EA (1968) Practical significance of birdseye structures in carbonate rocks. J Sediment Petrol 38:215–223

Silver BA, Todd RG (1969) Permian cyclic strata, northern Midland and Delaware basins, West Texas and southeastern New Mexico. Bull Am Assoc Petrol Geol 53:2223–2251

Todd RC (1976) Oolite-bar progradation, San Andres Formation, Midland basin, Texas. Bull Am Assoc Petrol Geol 60:907–925

27 Algal Marsh Deposits from the Upper Jurassic of Portugal

V.P. WRIGHT[1]

Introduction

Freshwater algal marsh deposits cover large areas of the low lying interiors and coastal zones of modern subtropical carbonate regions. In southern Florida these settings cover some 4000 km^2 (Gleason 1972) with Pleistocene equivalents in the same region covering over 10,000 km^2 (Halley and Rose 1977). Similar deposits also cover large areas of Andros Island in the Bahamas (Monty and Hardie 1976) and also occur along the coast of north east Yucatan (Brady 1978).

These algal marshes develop in areas where seasonal ponding of freshwater occurs to depths of a few centimetres to tens of centimetres. The waters are carbonate rich with the carbonate being derived from underlying limestones. Fine-grained calcite is precipitated on floating masses of algae (periphyton) and in benthic mats (Gleason and Spackman 1974, Monty and Hardie 1976). Deposits of carbonate mud develop after the decay and disintegration of these algal masses. These muds may be locally stromatolitic but are usually structureless and may contain gastropods, ostracodes, calcified algal tubules, and pollen. The muds show evidence of desiccation and contain mudcracks; microscopic rosettes and single crystals of calcite may also occur.

Despite several detailed papers on present-day algal marshes, records of ancient equivalents are rare. Halley and Rose (1977) have described similar limestones from the Pleistocene of Florida and the Lower Cretaceous of central Texas. Perkins (1977) has also described Pleistocene forms of Florida. The aim of this paper is to describe a sequence of freshwater algal marsh deposits from the Upper Jurassic of Portugal (the Vale Verde Beds), and to offer facies models for their occurrences.

Geologic Setting

Algal marsh deposits occur in the lower half of the Vale Verde Beds (earliest upper Oxfordian, Upper Jurassic) in the Mesozoic Lusitanian Basin of west central Portugal (Wilson 1979). The stratigraphic nomenclature of the Mesozoic in Portugal is disorganized, and few units are formally defined. Vale Verde Beds crop out over much of the basin (Fig. 1) and are termed Cabacos Beds in the southern half of the out-

1 Department of Geology, University of Bristol, Bristol BS8 1TR, Great Britain

Paleoalgology: Contemporary Research and Applications
(eds. D.F. Toomey and M.H. Nitecki)

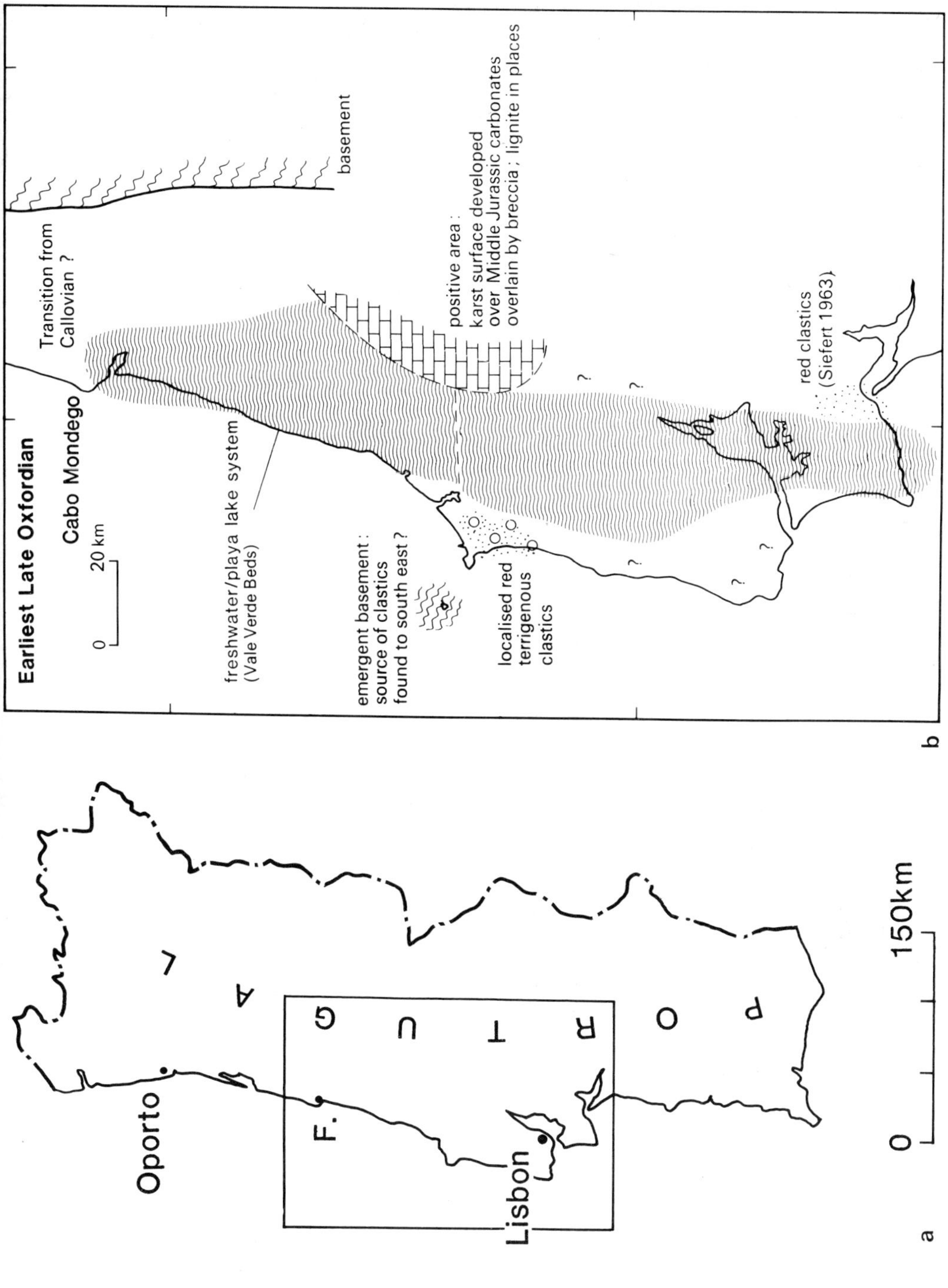

Fig. 1a,b. Index map showing: **a** Position of outcrop area in Portugal. **b** Outcrop area and facies types of the Vale Verde Beds and equivalents. *F* Figueira da Foz

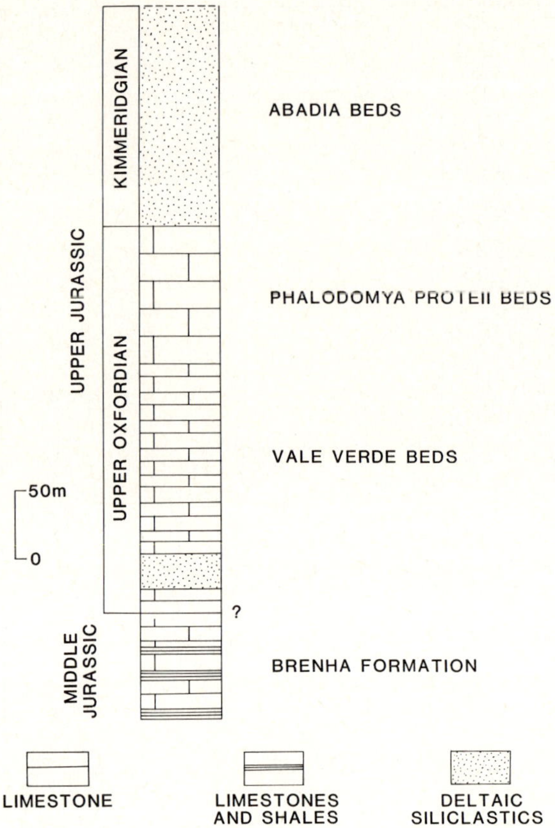

Fig. 2. Stratigraphic succession present at Cabo Mondego, Portugal

crop. The most complete sequence occurs along the coast at Cabo Mondego near Figueira da Foz (Fig. 1). Throughout the outcrop area the Vale Verde Beds (upper Oxfordian) overlie Middle Jurassic limestones and shales (Fig. 2) with the lower Oxfordian not being recorded. The contact is either a karstic surface or is apparently conformable.

The Vale Verde Beds are mainly a carbonate sequence, but siliclastics do occur and make up much of the unit in the west of the outcrop and in the south (Fig. 1). At Cabo Mondego the Vale Verde Beds are overlain by the *Phalodomya proteii* Beds (Fig. 2), which consist of skeletal wackestones interpreted as representing open bay deposits (Wilson 1979). The overlying Abadia Beds are Kimmeridgian in age and consist of restricted marine, fluvial, and deltaic cycles (Wilson 1979, P.M. Ellwood, in preparation).

The upper Oxfordian of the Lusitanian Basin represents a phase of fault and salt diapir movements, heralding a major phase of rifting in the Kimmeridgian, related to Atlantic opening. The Vale Verde Beds, earliest upper Oxfordian, occur over the entire basin with similar facies types, although there is some evidence locally of fault activity associated with alluvial fan and fluvial deposits. Later in the Oxfordian significant changes in basin configuration occurred with the development of slope car-

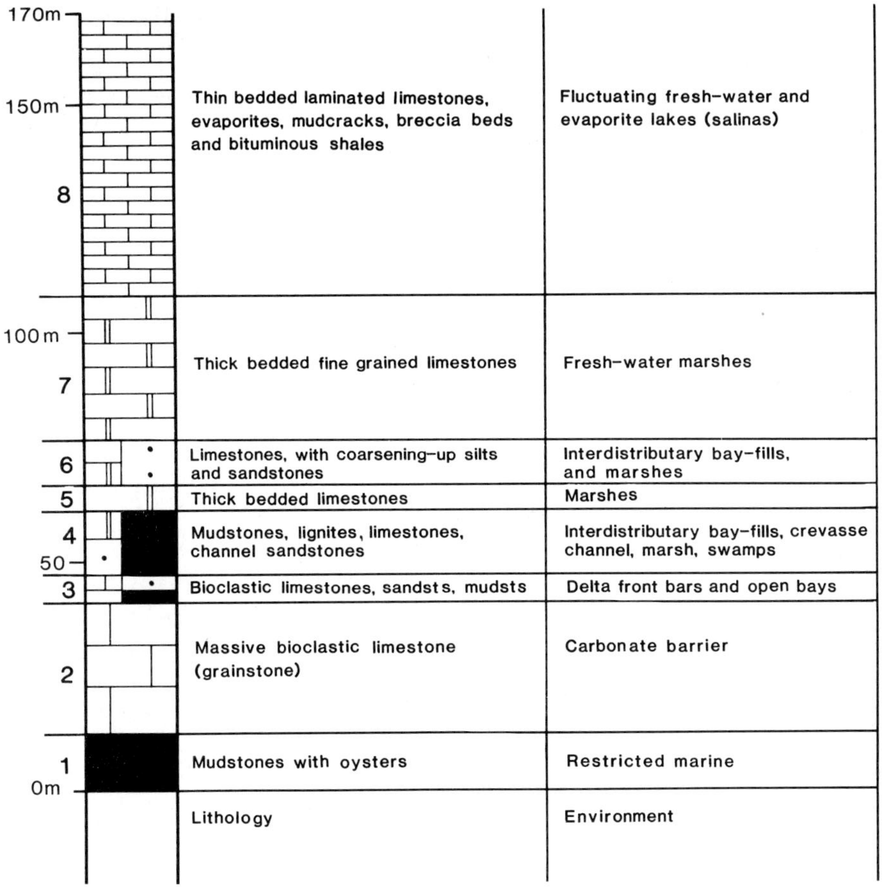

Fig. 3. Lithostratigraphic units in the Vale Verde Beds at Cabo Mondego, Portugal

bonates. By Kimmerdigian times the basin had a complex form in which coastal plain, slope and basinal sequences can be recognized (Wilson 1979).

Vale Verde Beds at Cabo Mondego

At Cabo Mondego a complete sequence from the Middle Jurassic (Callovian) to the Abadia Beds is exposed. The Callovian is represented by a sequence of limestones and shales (Brenha Formation) which grades, without any apparent lithological break, into the Vale Verde Beds.

The Vale Verde Beds here can be subdivided into eight units (Fig. 3) of which the top unit (8) is the most typical of the basin as a whole. A more detailed account of the whole sequence is in preparation, and this description is based on unpublished oil-company reports prepared by the Open University Portugal Research Group.

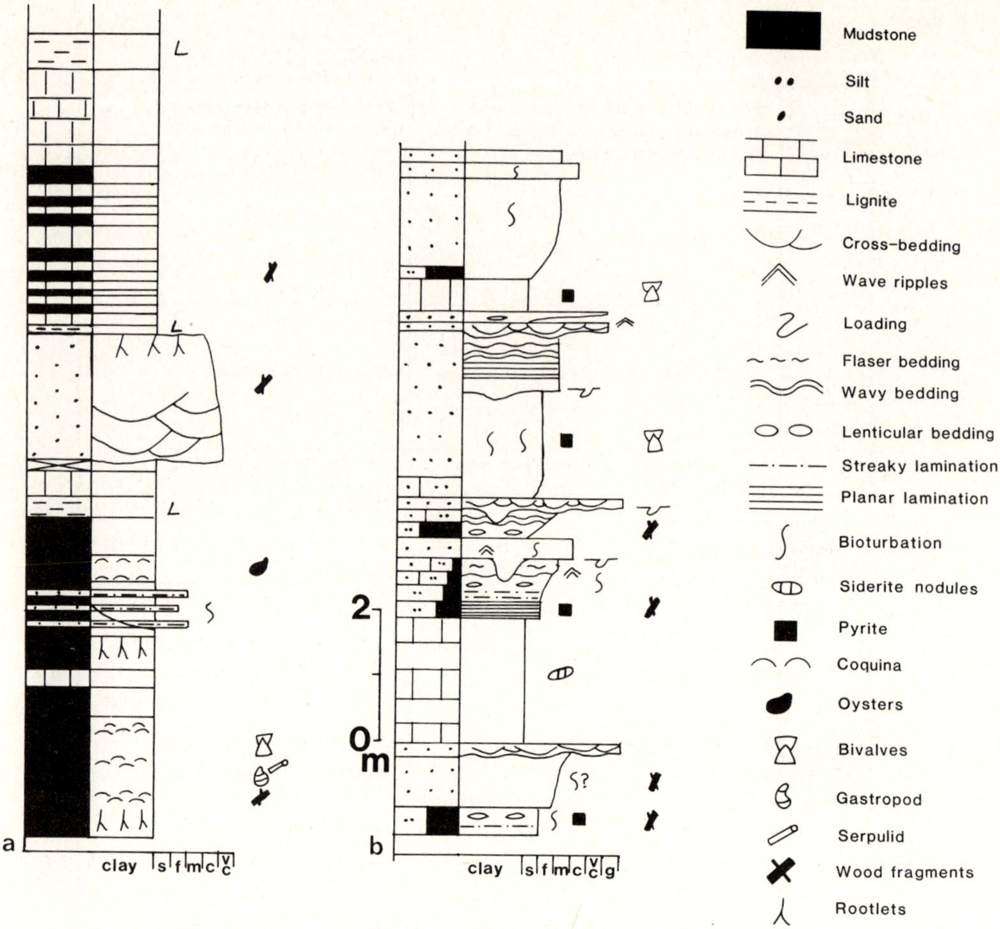

Fig. 4. a Schematic diagram of unit 4. **b** Schematic diagram of unit 6

Unit 1 consists of mudstones with a fauna of oysters, rarer brachipods, and echinoid fragments. It was probably deposited in a shallow marine reduced-salinity environment. Unit 2 consists of an initially coarsening-up carbonate sand-body that shows evidence of early lithification on its top surface. It is interpreted as a prograding carbonate barrier. Unit 3 is a complex sequence of marine limestones, lenses of sandstones with in situ oysters, and mudstones with in situ and abraded corals and oysters. This unit probably represents deposition in a series of lagoons with periodic small-scale delta progradations and subsequent salinity variations.

Limestones interpreted as algal marsh deposits occur in units 4 to 7. Unit 4 (Fig. 4a) contains fine-grained limestones within a sequence of mudstones with low diversity faunas of bivalves (including oysters), gastropods and serpulids. A fining-upward sandstone occurs with rootlets at the top of the unit overlain by a lignite horizon. The upper part consists of interbedded mudstones, limestones, and another lignite.

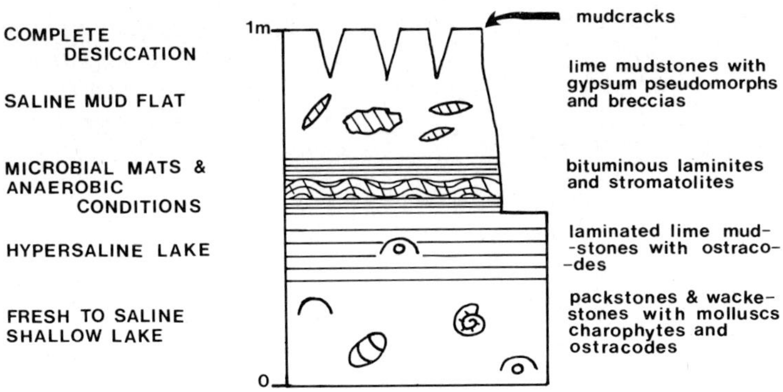

Fig. 5. Idealized cycle from unit 8

The low-diversity faunas of the lower half of the unit can be compared with the mesohaline communities described from other Portuguese Upper Jurassic marginal marine sequences by Fursich (1981). These communities are interpreted as having occupied brackish lagoons and delta-front areas. The mudstone lithologies also suggest a protected setting. The sandstone represents a channel deposit, probably a crevasse channel (see examples in Elliott 1978). The lower part was deposited in lowered salinity, interdistributary bays filled by crevasse channel deposition that resulted in the formation of a swamp (lignite). The upper half lacks any marine fauna, and the limestones contain *Unio* and *Planorbis* (Wilson 1979) and was deposited in a freshwater setting.

Unit 5 consists of a series of fine-grained limestones identical to those in units 4, 6, and 7. Lignite also occurs.

Unit 6 contains fine-grained limestones interbedded with coarsening silts and sandstones (Fig. 4b). The sandstones show evidence of wave reworking and have wavy and flaser bedding. They contain poorly preserved bivalves and are heavily bioturbated at some horizons. They are comparable to interdistributary bay-fill sequences (Elliott 1978).

Unit 7 consists of 30 m of thickley-bedded fine-grained limestones containing mudcracked surfaces at several horizons. The limestones are identical to those in the units below.

Unit 8, which contains the typical Vale Verde Beds lithologies, consists of a series of thin-bedded limestones with abundant charophytes, molluscs and ostracodes, stromatolites, evaporite-bearing limestones, evaporite breccias, and bituminous laminites (Wilson 1975, 1979). This sequence is interpreted as representing a series of coast marginal, shallow, perennial fresh to saline lakes. The lithologies are in part arranged in cycles as shown in Fig. 5, representing the desiccation of shallow freshwater lakes.

In summary the sequence consists of small-scale deltaic deposits developed behind a prograding-barrier complex and back-barrier lagoon. This is a setting similar to that along the present-day Texas coast where small deltas have built into lagoons developed on a strand plain behind a barrier complex (Kanes 1970, Donaldson et al. 1970). As the deltas prograded seaward, the delta plains became freshwater dominated (upper part of unit 4 and units 5 and 7). Unit 6 represents a return to a more active delta-plain environment.

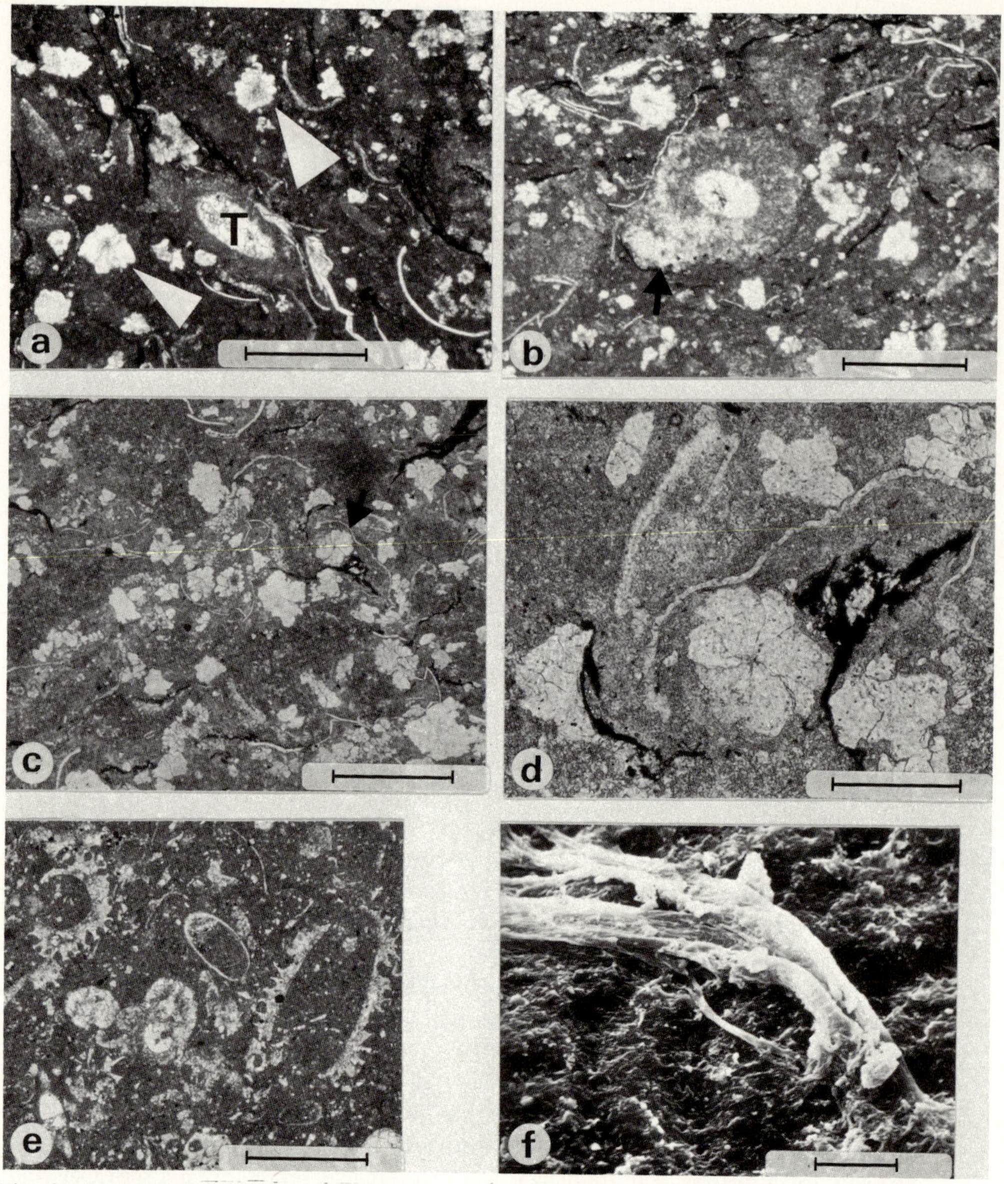

Fig. 6a–f. Various thin section photomicrograph from section at Cabo Mondego, Portugal. **a** Wackestone with ostracode valves (e.g., *lower right*) calcite rosettes (some are *arrowed*) and micrite-coated tube; *T* transverse section. Scale bar = 0.5 mm. **b** Wackestone with ostracode valves and rosettes (*white areas*). Micrite-coated tube in center shows partial recrystallization (*arrowed*). These tubular coatings often possess poorly preserved micritic tubes oriented perpendicular to the tube axis. The mottled appearance is due to micrite clots in the matrix that have a similar composition to the tube coatings. Scale bar = 0.5 mm. **c** Rosettes and more irregular crystal aggregates. Arrow indicates rosette shown in **d**. Scale bar = 0.5 mm. **d** Higher-power view of rosette arrowed in c. Note also ostracode valve and dark organic matter. Scale bar = 150 μm. **e** Fragments of charophyte stems in a bioclastic packstone-wackestone. Ostracode valves are also present. Scale bar = 0.5 mm. **f** Scanning electron micrograph of organic sheath material. Sample was fractured and lightly etched in 1% hydrochloric acid for 5 s. Scale bar is 10 μm long

Limestones in Units 4–7

All these limestones are dark blue-grey in colour, and are medium to very thickly bedded. They show rare laminations but are mostly mottled (bioturbated) or structureless, with mudcracked horizons consisting of large polygons, up to 30 cm in diameter. Contacts with the siliclastic units are sharp, but gradational contacts with some underlying lignites occur.

Under the microscope the limestones are predominantly lime mudstones or wackestones with some packstones. The matrix is a brown non-ferroan micrite that may contain mottles of grey micrite (Fig. 6a,d) up to 1 mm in diameter. These grey lumps contain poorly preserved narrow (10 μm or less) porostromate tubes, and may form the cortex of tubes (Fig. 6a,b) up to 300 μm in diameter. Besides these mottles the matrix is very organic-rich and in some beds contains light-brown crystals and rosettes of non-ferroan calcite (Fig. 6a–d). The crystals may occur as well-formed rhombs up to 30 μm in length, but more commonly as more irregular polyhedral crystals up to 100 μm in size. Some of these crystals are arranged into rosettes (Fig. 6d) but more commonly occur as irregular aggregates in which one or more of the component crystals are much larger than the others in the rosette (Fig. 6c). Some of the rosettes have mud-filled central pores. The matrix also cotain silt to fine sand-sized quartz.

The limestones contain a variety of bioclasts, with ostracodes (Fig. 6c) being the most abundant. Mollusc remains also occur, either as thin-shelled fragments (bivalves?), or as recognizable gastropod fragments. Poorly preserved charophyte remains occur including laminae packed with charophyte stems (Fig. 6e). These stems are only weakly calcified and resemble similar poorly preserved material from the carly Tertiary and late Cretaceous of France described by Freyet and Plazait (1982) and from modern lakes by Schneider et al. (1983). These stems are easily confused with dasyclad remains. Plant residue including spores is abundant in acid-digested samples.

Under the scanning electron microscope the micritic matrix is organic-rich with apparently well-preserved sheath-like algal material (Fig. 6f). Fresh samples were prepared for the SEM, and this organic material does not appear to be related to endoliths on near-surface samples. Acid digestion did not yield similar material, but such sheaths could easily have been disintegrated during digestion.

Interpretation

These limestones show many similarities to algal marsh deposits described by Wanless (1969), Gleason and Spackman (1974), Monty and Hardie (1976), and Halley and Rose (1977). In present-day freshwater algal marshes the partly calcified floating or benthic algal mats are rapidly transformed, on burial of only a few centimetres, into an amorphous, bioturbated calcitic mud. This mud contains rare bundles of porostromate tubes (Monty and Hardie 1976, p. 453; Brady 1978, p. 97), which can be fragmented and appear as denser clots in the mud. Calcified algal mats also coat higher plants and form tubular encrustations that can also accumulate in the muds (Monty and Hardie 1976, p. 453). Similar structures occur in the Vale Verde Beds (e.g., Fig. 6a,b). Rhombs, polyhedral crystals, and rosettes of calcite crystals also occur in the present-day algal marsh muds of Florida (Gleason and Spackman 1974, p. 163, 165; Monty and Hardie 1976, p. 451, 453). The rhombs described by these authors are smaller than those in the Vale Verde Beds, but similar rhombic calcite crystals are common in freshwater deposits. Kelts and Hsu (1978) describe similar rhombic crystals from Lake Zurich noting that, as in the Vale Verde Beds, irregular, polyhedral crystals are more common. Folk (1974) says that the maximum size for rhombicity in freshwater calcites appears to be 5–10 μm, above which the calcites become polyhedral. Gleason and Spackman (1974) describe their rosettes (dendrites) as being composed of single calcite crystals, which is not the case in these Jurassic forms. Irion and Muller (1968) describe rosettes strikingly like these described here, from present day freshwater tufas (their Fig. 10). Riding (1977) and Klappa (1979) also describe similar rosettes as being typical of the encrustation of algal filaments.

The steath-like organic material which is common at some horizons probably represents the remains of algal-sheaths in the original marsh muds. Although the fresh-appearance of this material is surprising, similar remains have been found in even older algal limestones (de Meijer 1969, Coron and Textoris 1974, Bathurst 1975).

Discussion

The non-marine limestones of the Vale Verde Beds show many features in common with Recent and Pleistocene freshwater algal marsh deposits. Monty and Hardie (1976) give detailed descriptions and facies associations for the Florida and Andros-types of algal marsh. The Florida examples occur in a transgressive sequence (Monty and Hardie 1976, p. 457; Wanless 1969), summarized in Fig. 7. On southwest Andros Island the algal marsh deposits occur in a regressive (prograding) sequence (Monty and Hardie 1976) (Fig. 7). This Jurassic example documents the occurrence in a prograding delta-plain sequence where the algal marshes develop at the final stages of the filling of interdistributary bays, including bays filled by crevasse channel sands (Fig. 7).

The lack of records of ancient algal marsh deposits is probably the result of such deposits not being recognized where they do occur and also of their low preservation potential. On transgressive coastlines algal marsh deposits are often removed during the transgression in all but the most protected areas (e.g., Biscayne Bay, Wanless 1969). Perkins (1977, p. 189) notes that in Florida, Pleistocene algal marsh limestones are best preserved in the upper parts of regressive cycles and not at the bases of transgressive units. In these cases early lithification by subsequent subaerial exposure may be important in preserving the marsh limestones. Along the northwest coast of Andros Island the coastal transgression has almost completely removed the underlying algal marsh deposits (Hardie and Ginsburg 1977).

These examples highlight the fact that algal marsh deposits are more likely to be preserved in regressive sequences. The Vale Verde Beds marsh deposits occur in a regressive sequence from marine limestones (units 1–3) to lacustrine (unit 8).

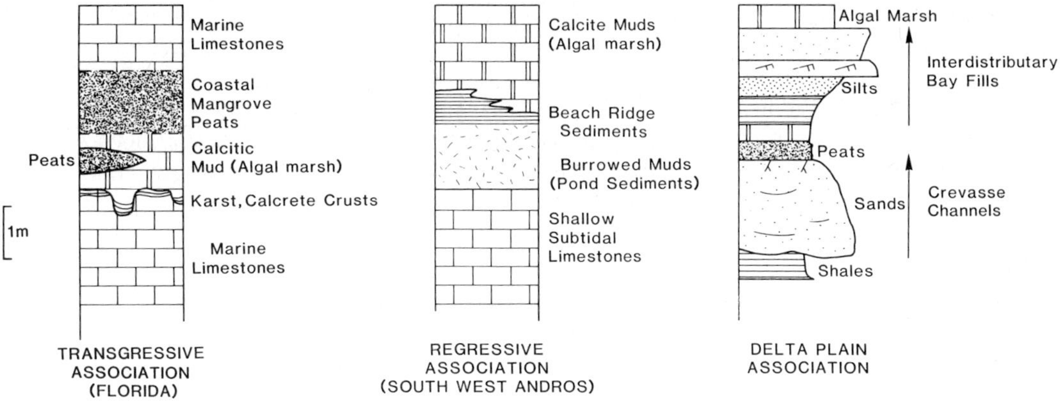

Fig. 7. Facies sequences for algal marsh deposits. The transgressive sequence is based on Florida (see Wanless 1969, Monty and Hardie 1976). This type of sequence has a low preservation potential and the peats and algal marsh deposits will be eroded except in protected settings. The sequence in Florida is developed over Pleistocene limestones.

The regressive sequence is based on southwest Andros. Algal marsh deposits may cap many shoaling-upward cycles in peritidal carbonates but are not recognized. This sequence along the southwest coast of Andros is also developed over Pleistocene basement (not shown).

The delta plain association has algal marsh deposits developed during the final stages of the filling of interdistributary bays. Such bays may be filled by a variety of deposits including crevasse channel sands and minor mouth bars. This sequence is based on the Vale Verde Beds of Portugal.

In all the above cases a major factor in the development of the algal marshes is the climate (see text)

The Pleistocene of Florida locally contains as many as seven freshwater limestones, as thick as 1.5 meters, but usually less that 1 meter thick (Halley and Rose 1977). The succession is complex and is interrupted by marine incursions and subaerial exposure caused by sea-level low-stands (Perkins 1977). The thickness of algal marsh limestones, especially in unit 7, must reflect deposition during long periods of very uniform climate with a balance between subsidence and ground water levels.

The association of freshwater algal marsh deposits and the evaporitic lake deposits of unit 8 requires comment. Algal marsh deposits develop in wet and rainy, tropical zones where short wet periods alternate with seasonal mild dry periods (Monty and Hardie 1976). The occurrence of fresh to saline lake deposits in unit 8 probably reflects the development of more prolonged periods of standing water (with charophytes and molluscs) coupled with periodic changes in the precipitation/runoff-evaporation balance leading to complete desiccation of the lakes. The overall change from unit 7 to 8 is more likely to reflect a subtle change in the precipitation-evaporation regime than a significant climatic change.

Summary

Non-marine limestones occur in the lower half of the Upper Jurassic Vale Verde Beds of Cabo Mondego, near Figueira da Foz, Portugal. They show many similarities to present-day and Pleistocene algal marsh deposits. The limestones consist predominantly of fine-grained calcite micrite interpreted as the products of calcification of floating and benthic algal mats. The limestones occur in association with siliclastic delta-plain sequences, and algal marsh formation was initiated by the filling of interdistributary bays.

The marshes developed under a climatic regime with a short wet season followed by a mild dry season. Under such conditions the algae thrived in the ponded marsh waters, which did not persist long enough for colonization by the higher aquatic plants.

Acknowledgments. I should like to thank Chris Wilson (Open University) for introducing me to the fascinating Jurassic sequences of Portugal. Dick Carlton provided invaluable help with the SEM work. The manuscript was ably typed by Beryl West. I gratefully acknowledge the award of a Royal Society travel grant which has enabled me to present this paper at the Third International Symposium of Fossil Algae.

References

Bathurst RGC (1975) Carbonate sediments and their diagenesis. Elsevier, Amsterdam, 658 p

Brady MJ (1978) Sedimentation and depositional history of coastal lagoons, north eastern Quintana Roo, Mexico. In: Ward WC, Weidie AE (eds) Geology and hydrogeology of northeastern Yucatan. New Orleans Geol Soc, New Orleans, pp 85–112

Coron CR, Textoris DA (1974) Non-calcareous algae in Silurian carbonate mud mound, Indiana. J Sediment Petrol 44:1248–1250

Donaldson AC, Martin RH, Kanes WH (1970) Holocene Guadeloupe delta of Texas Gulf Coast. In: Morgan JP, Shaver RH (eds) Deltaic sedimentation; Modern and ancient. Soc Econ Pal Mineral Spec Publ 12: Tulsa, Okl., pp 109–137

Elliott T (1978) Deltas. In: Reading HG (ed) Sedimentary environments and facies. Blackwell, Oxford, pp 97–142

Folk RL (1974) The natural history of crystalline calcium carbonate: effect of magnesian content and silinity. J Sediment Petrol 44:40–53

Freyet P, Plazait J-C (1982) Continental carbonate sedimentation and pedogenesis. Late cretaceous and early tertiary of southern France. Contrib Sedimentol, vol XII. Schweizerbart, Stuttgart, 214 p

Fursich FP (1981) Salinity – controlled benthic associations from the Upper Jurassic of Portugal. Lethaia 14:203–223

Gleason PJ (1972) The origin, sedimentation and stratigraphy of a calcitic mud located in the southern freshwater everglades. Ph D thesis, Pennsylvanian State Univ, 355 p

Gleason PJ, Spackman W (1974) Calcareous periphyton and water chemistry in the Everglades. In: Gleason PJ (ed) Environments of South Florida: present and past. Miami Geol Soc Mem 2, pp 146–181

Halley RB, Rose PR (1977) Significance of fresh water limestones in marine carbonate successions of Pleistocene and Cretaceous age. In: Bebout DG, Loucks RG (eds) Cretaceous carbonates of Texas and Mexico – applications to subsurface exploration. Rep of Invest 89. Bur Econ Geol, Austin, Tex, pp 206–215

Hardie LA, Ginsburg RN (1977) Layering: the origin and environmental significance of lamination and thin bedding. In: Hardie LA (ed) Sedimentation on the modern carbonate tidal flats of northwest Andros Island, Bahamas. John Hopkins Stud Geol 22. John Hopkins Univ Press, Baltimore, pp 50–123

Irion G, Muller G (1968) Mineralogy, petrology and chemical composition of some calcareous tufa from the Schwäbische Alb, Germany. In: Muller G, Friedman GM (eds) Recent developments in carbonate sedimentology in Central Europe. Springer, Berlin Heidelberg New York, pp 157–171

Kanes WH (1970) Facies and development of the Colorado River delta in Texas. In: Morgan JP, Shaver RH (eds) Deltaic sedimentation, modern and ancient. Soc Econ Palaeontol Mineral Spec Publ 15, Tulsa, Okla, pp 78–108

Kelts K, Hsu KJ (1978) Freshwater carbonate sedimentation. In: Lermann A (ed) Lakes – chemistry, geology and physics. Springer, Berlin Heidelberg New York, pp 295–353

Klappa CF (1979) Calcified filaments in Quaternary calcretes: organo – mineral interactions in the subaerial vadoes environment. J Sediment Petrol 49:955–968

Meijer JJ de (1969) Fossil non-calcareous algae from insoluble residues in algal limestones. Leidse Geol Med 44:235–263

Monty CLV, Hardie LA (1976) The geological significance of the freshwater blue green algal calcareous marsh. In: Walter MR (ed) Stromatolites. Elsevier, Amsterdam, pp 447–477

Perkins RD (1977) Depositional framework of Pleistocene rocks in South Florida. In: Enos P, Perkins RD (eds) Quaternary sedimentation in South Florida. Mem Geol Soc Am 147:131–191

Riding R (1977) Calcified Plectonema (blue green algae), a recent example of *Girvanella* from Aldabra Atoll. Palaeontology 20:33–46

Schneider J, Schroeder HG, Le Champion-Alsumardm Th (1983) Algal micro-reefs. Coated grains from fresh-water environments. In: Peryt TM (ed) Coated grains. Springer, Berlin Heidelberg New York, pp 284–298

Seifert H (1963) Beiträge zur Geologie der Serra da Arrabida in Portugal. Geol Jahrb Hannover 81:277–344

Wanless HR (1969) Sediments of Biscayne Bay-distribution and distribution and depositional history. Tech Tep 62-2. Inst Mar Atmos Sci, Univ Miami, 260 p

Wilson RCL (1969) A reconnaisance study of the Upper Jurassic of Portugal. Cienc Terra (UNL) Lisboa 5:53–84

Wilson RCL (1975) Some examples of shoaling deposits from the Upper Jurassic of Portugal. In: Ginsburg RM (ed) Tidal deposits: a casebook of recent examples and fossil counterparts. Springer, Berlin Heidelberg New York, pp 363–371

Diversity and Evolution

28 Diversity and Environments of Permian and Triassic Dasycladacean Algae

E. FLÜGEL[1]

Introduction

Paleoenvironments of ancient dasycladacean algae are generally inferred from the comparison with Recent dasycladaceans, which are a relatively inconspicuous element in marine algal floras (Fig. 1). The occurrence and distribution of Recent dasyclads depend on water temperature, substrate, salinity, water energy, and depth (Wray 1977, Valet 1979, see also Table 1). Most authors agree that the ecological requirements of fossil dasyclads were essentially the same as or at least very similar to those of their living descendants (e.g., Elliott 1968, 1978). Present-day distributions are explained by the dispersal of suitable shallow-marine habitats in the Mesozoic and early Cenozoic, followed by the mid-Cenozoic break-up of the Tethyan Sea and the separation of the Tethyan and the Caribbean algal provinces (Elliott 1981).

Diversity

Investigation of Permian and Triassic shelf carbonates from the Alpine-Mediterranean region, and the various environmental interpretations of dasycladacean limestones suggest that there are considerable variations and differences in the diversity and distributional patterns of dasyclads inhabiting inner and outer-shelf areas, as well as shelf-margin and reef environments.

Permian dasycladaceans, in contrast to early Tertiary, are especially useful for paleoecological interpretation because the Permian was a period of maximum development in the evolution of this algal group (Fig. 1). The Alpine-Mediterranean Triassic, on the other hand, offers excellent possibilities for the study of dasyclad communities following a significant world-wide absence in the Lower Triassic.

Permian and Triassic dasycladacean genera are regarded as long-lived because relatively few genera occur in both the Late Permian and the Middle and Upper Triassic (e.g., *Diplopora, Gyroporella, Macroporella,* and *Physoporella*), and because of distinct phylogenetic lines, seen within Permian and Triassic metaspondyle dasyclads

1 Institut für Paläontologie, Universität Erlangen-Nürnberg, Loewenichstraße 28, 8520 Erlangen, FRG

Paleoalgology: Contemporary Research and Applications
(eds. D.F. Toomey and M.H. Nitecki)

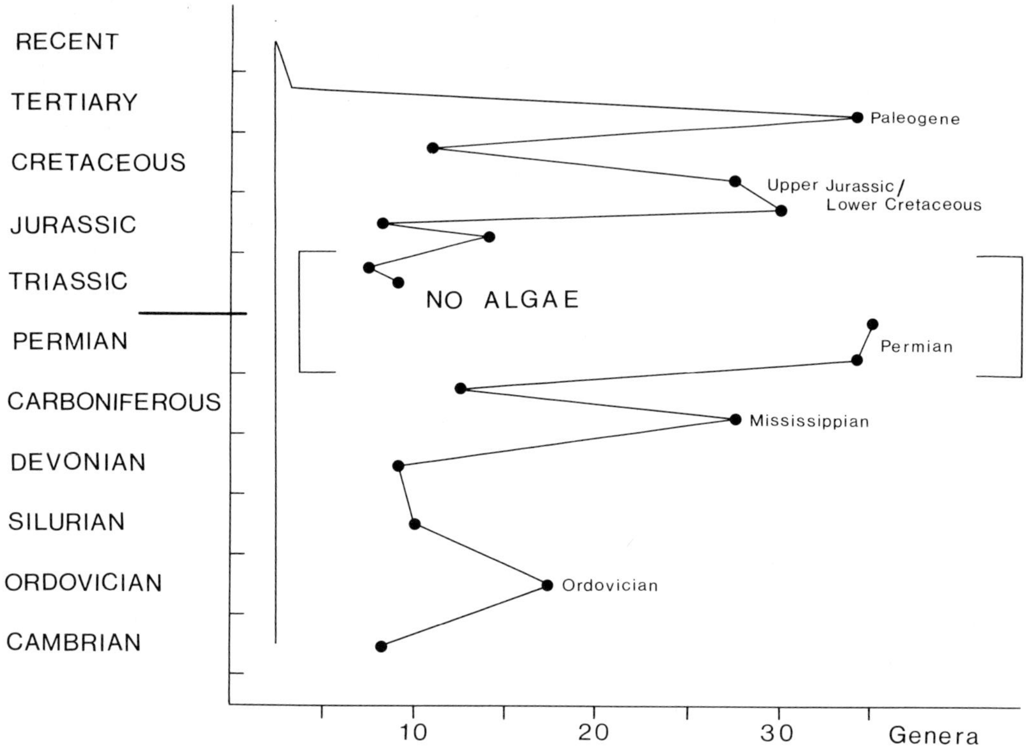

Fig. 1. Generic diversity of dasycladacean algae over time. Information is strongly biased because Late Paleozoic to Cenozoic dasycladaceans are much better known than Early Paleozoic dasyclads. Jurassic and Cretaceous genera (Bassoullet et al. 1978), Cenozoic genera (Deloffre and Génot 1982)

Table 1. Environmental factors affecting Recent dasyclads. (From Fott 1971, Valet 1979, Wray 1977, and others)

Water temperature	Tropical to subtropical; few species in warm waters
Substrate	Sand and mud; some species attached to calcareous shells and rocks
Salinity	Mostly normal marine; some species tolerate hypersaline to brackish waters (Acetabulariae)
Water energy	Lower energy, below intense wave agitation or in protected areas
Depth	Below low tide to about 30 m; commonly less than 5 m; few species down to 90 m

(Guvenĉ 1979). Biotopes of these genera, however, seem to have changed through time (Table 2), perhaps influenced by the regression-transgression pattern occurring at the Permian/Triassic boundary (Valentine 1973, Hüssner 1983). Other differences are the high generic diversity of Permian dasyclads, which gradually increased with time (Fig. 2), and the low constant generic diversity of Triassic dasycladaceans (Fig. 3).

Table 2. Permian and Triassic biotopes of *Diplopora*

Geologic time	Paleoenvironments	Occurrences
Upper Triassic	Coral sponge reefs; algal-foraminiferal sands near patch reefs	Upper Rhaetian Reef limestones, Northern Alps
	Near-reef lagoonal sands	Bedded Dachstein Limestone, Alpine-Mediterranean region
Middle Triassic	Platform carbonates with and without marginal reefs; lagoonal sands	Wetterstein Limestone, Northern Alps; Western Anatolia; Southern Spain
Late Permian	Inner shelf carbonates	Japan; Pamir Range

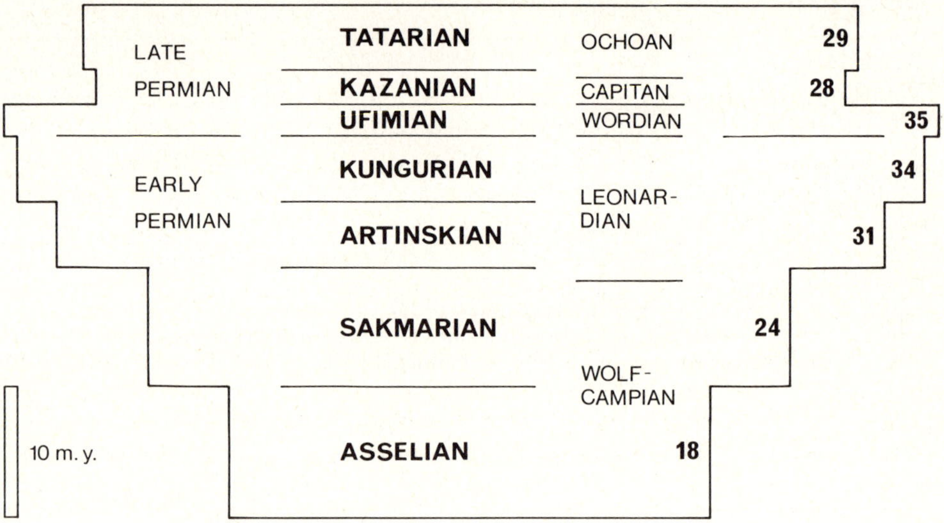

Fig. 2. Generic diversity of Permian dasycladacean algae

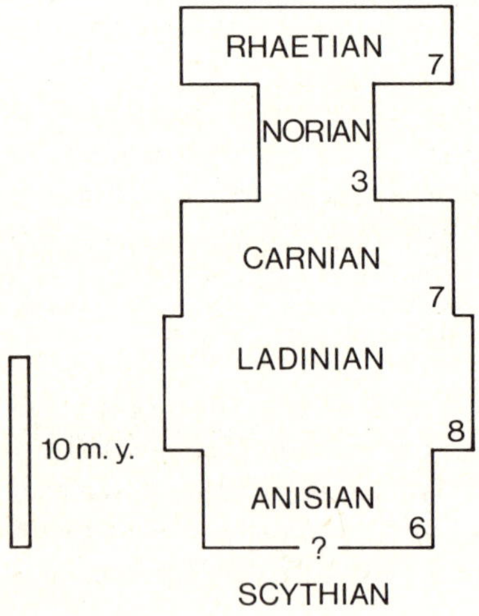

Fig. 3. Generic diversity of Triassic dasycladacean algae

Table 3. Permian and Triassic algal carbonates considered in this paper

Geologic age	Geologic formations	Geologic areas
Upper Triassic	Upper Rhaetian reef limestones	Salzburg area, Austria
	Norian Dachstein limestones	Northern Alps, Austria
	Carnian Pantokrator limestones	Hydra Island, Greece
Middle Triassic	Ladinian Wetterstein limestones	Northern Alps, Austria
	Ladinian lagoonal limestines	Betic Cordillera, Spain
	Anisian limestones	Southern Alps, Italy
Upper Permian	Changxing Reef limestone	Hubei, China
Middle Permian	Neoschwagerina limestone	Bled, Julian Alps, Slovenia
Lower Permian	Sakmarian Trogkofel limestone	Carnia, Southern Alps, Italy
	Sakmarian Shakhtau limestones	Bashkiria, Middle Asia
	Assilian/Sakmarian Duvansk reef limestones	Ural Mountains, USSR

I will use selected examples from the Lower and Middle Permian of the Southern Alps, the Urals, and Middle Asia, and from the Middle and Upper Triassic of the Alpine-Mediterranean region, to demonstrate major differences in diversity without comparable paleoenvironmental differences. This will be followed by comments on the possible causes of deviations from the Recent model. Sources are noted in Table 3.

Lower Permian examples show distinctly different species diversity patterns within shelf platforms and within different marginal buildups of relatively similar age (Fig. 4).

1. A common type, characterized by a high but fluctuating species diversity, is developed in extended shelf lagoons with protected as well as open-marine conditions. The percentage of dasycladaceans within the total algal flora (consisting of

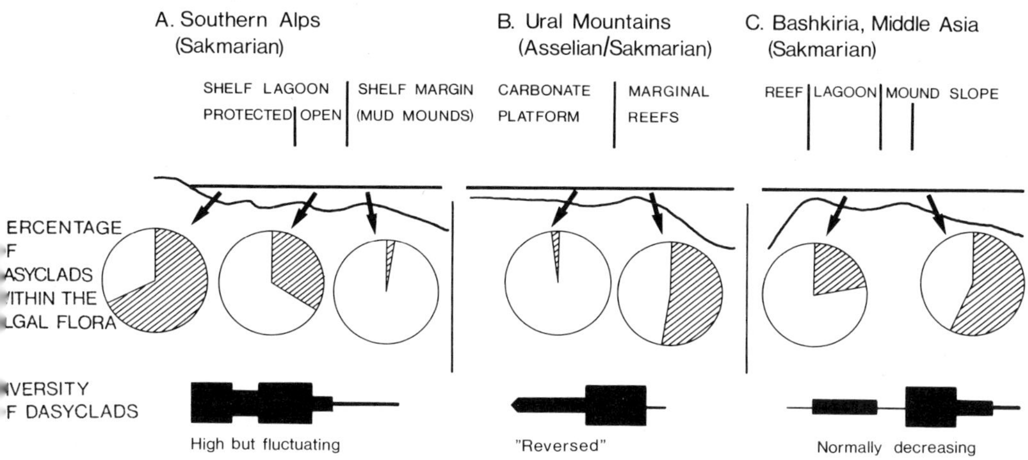

Fig. 4. Diversity patterns for dasycladacean algae in Lower Permian shelf platforms and marginal buildups. Southern Alps (Flügel and Flügel-Kahler 1980), Ural Mountains (Chuvashov 1974, 1983), Bashkiria (Kulik et al. 1978)

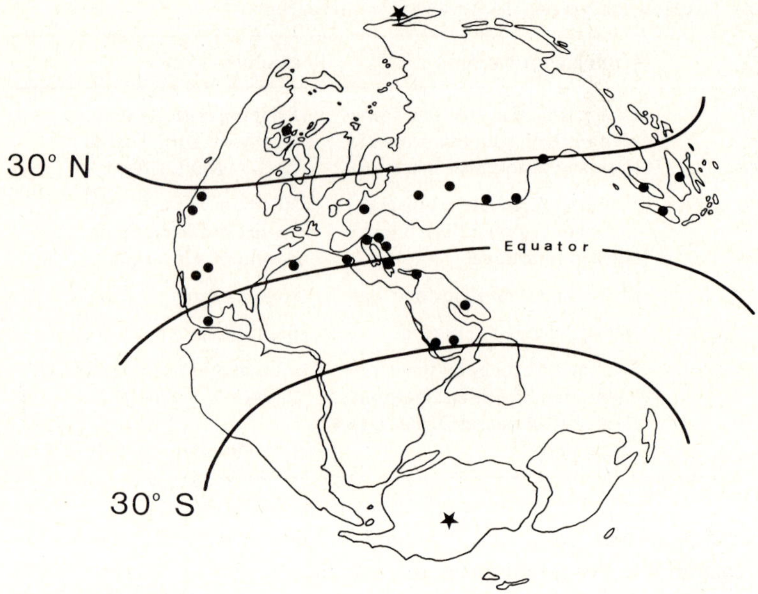

Fig. 5. Distributional pattern of Permian dasycladacean algae: Examples of Fig. 4 are within the same paleolatitudes

phylloid algae, solenoporaceans, and ancestral corallines) is high (up to about 70%) in bioclastic wackestones, packstones, and grainstones of the protected areas. It is relatively low in open-marine wackestones and grainstones, and significantly low in downslope mud mounds formed by phylloid algae of *Tubiphytes.* There is a fairly good correlation between dasycladacean diversity and textural rock types (Fig. 6), indicating a strong influence of the substrate on the composition of the dasycladacean floras. The Trogkofel limestones of the Southern Alps are good examples of this (E. Flügel and E. Flügel-Kahler 1980). Other examples are represented by Lower and Middle Permian algal associations, in the Velebit Mountains in Yugoslavia (numerous papers by Herak, Kochansky-Devid, and Milanović; see bibliography in Emberger 1976), and in the Middle Permian of Japan (Endo 1959, 1961) and Thailand (Endo 1969).

2. A "reversed" diversity – in comparison with type 1 – is represented by Asselian to Sakmarian carbonates of the Ural Mountains (Chuvashov 1974, 1983). High specific and generic diversities are found not in the shelf lagoons but within the marginal reefs (formed by *Palaeoaplysina,* and *Tubiphytes*), where the percentage of dasyclads is very high in contrast to the platform areas, which, with the exception of some ancestral corallines, are poor in algae.

3. Type 3 exhibits a constant low diversity in shallow lagoons bordered by reefs and mounds, and a gradually decreasing high to low diversity down the platform slope. Examples are the Sakmarian shallow-water carbonates in Bashkiria (Kulik et al. 1978), and some Middle Permian algal floras in southwestern China.

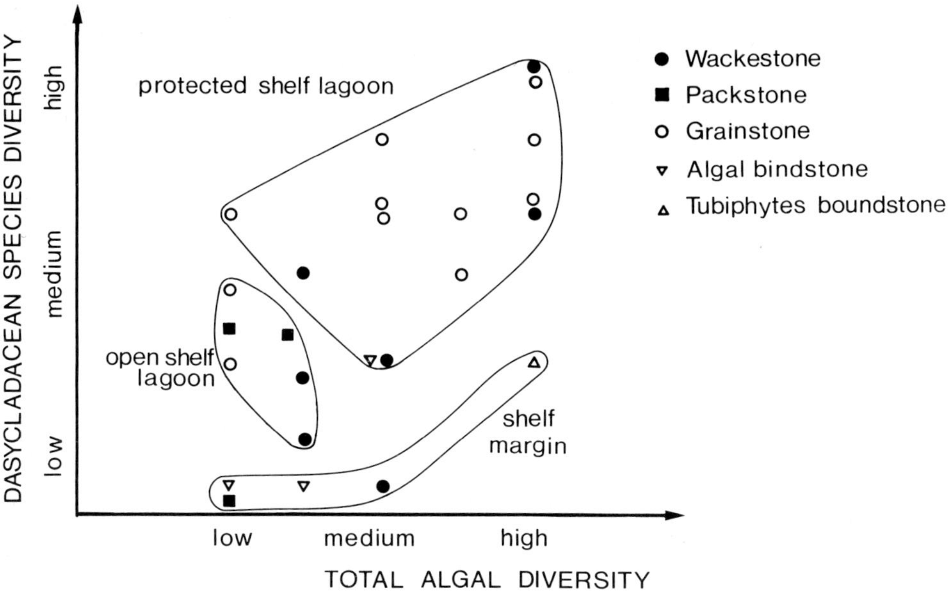

Fig. 6. Total algal diversity compared with the dasycladacean species diversity. Lower Permian Trogkofel platform carbonates and marginal buildups of the Southern Alps

Paleoenvironments

What are the factors which have influenced these different diversity patterns? Since we must assume rather similar paleolatitudes for the depositional sites of the algal carbonates (Fig. 5), differences in water temperature alone cannot have been responsible. Differences in water energy and in the substrate type are indicated by the correlation between carbonate texture types and the abundance, as well as diversity, of algae and dasycladaceans (Fig. 6).

Bioclastic wackestones and packstones generally have low to medium diversity for both the total algal diversity and the dasycladacean species diversity in type 1. Grainstones exhibit low to high total algal diversities, but more often medium to high dasycladean species diversities. Algal bindstones (formed by cyanophyceans) and *Tubiphytes/Archaeolithoporella* boundstones are generally poor in dasycladaceans but may be rich in other algae. This is also a common feature in other boundstones (e.g., Bled Reef, Slovenia).

In a very generalized way, this seems to demonstrate low to medium species diversities in low-energy carbonates, and medium to high diversities in high-energy carbonates.

Does this generalization make sense? These distributional and diversity patterns contradict many Recent examples where low-energy lagoons are the sites of prolific dasycladacean growth. The pattern visible in the Trogkofel example, however, also contradicts the depositional model (Wilson 1975), in which protected shelf lagoons are sites of predominantly micritic sedimentation.

A few of the mass dasyclad algal accumulations in grainstones may be allochthonous, and should not be considered any further. Other grainstone types, however, are characterized by particular dasycladacean genera (e.g., *Gyroporella* and *Macroporella*), which may have been especially adapted to higher-energy environments. All these species are cylindrical, rather small, and thin-walled. The same morphology is found in *Gyroporella* and *Macroporella* from identical microfacies in the Lower and Middle Permian of Japan. There is no agreement on the functional morphology of various branch types developed by dasycladacean algae during their evolution (Bassoulet et al. 1978). However, there are indications of an adaption of dasyclad morphology to water energy, different intensities of light, and sedimentation of mud. One of the best examples of this has been described by Zorn (1976, 1977) in Middle Triassic shelf carbonates from the Southern Alps, who used quantitative distributional data of Bechstädt and Brandner (1970).

Zorn found distinct differences in th branch morphology of dasyclads between those occurring in sediments strongly influenced by mud deposition and those living in clear-water habitats. The shape, size, and density of the branches seem to have been related to the requirements of life in biotopes with rapidly changing depositional conditions. This might also explain the "fluctuating diversity" in the protected inner-shelf environments of the Trogkofel platform (Figs. 4 and 6). Slight differences in the topography of this area could have been responsible for the low and high energy subenvironments inhabited by dasycladaceans of different diversities and consisting of different genera and species.

Zorn has stressed that some Triassic dasycladaceans might have flourished in hypersaline lagoons. This he supports by demonstrating the presence of salinity-tolerant species of extremely low diversity, mass occurrences, and the lack of stenohaline organisms. In the Triassic these criteria are absent until the late Ladinian (e.g., Esino Limestone of the Southern Alps and Wetterstein Limestone of the northern Alps), but what about the Permian? There we have monomict dasycladacean floras composed of only one species (e.g., Lower Permian *Anthracoporella* limestones; Middle and Upper Permian *Mizzia* limestones), but there is no evidence of hypersaline or brackish conditions. The occurrence of *Mizzia* in the carbonates of the northwestern Zechstein Sea of central Europe (Clark 1980) seems to indicate that euryhaline dasyclads were present during the Late Permian, but because of the strong adaptation of Recent dasycladaceans to changing salinities (Valet 1979), this unique example must be studied in more detail.

Conclusions

In conclusion, most deviations of the distributional patterns of Permian and of Triassic dasycladaceans from the Recent shallow-water model seem to have been caused by differences in the available substrates, rather than by adaptations to various salinities or to zones of different mean-water temperatures.

References

Bassoullet JP, Bernier P, Conrad MA, Deloffre R, Jaffrezo M (1978) Les algues Dasycladales du Jurassique et du Crétacé. Geobios Mem Spec 2:330

Bechstädt T, Brandner R (1970) Das Anis zwischen St. Vigil und dem Höhlensteintal (Pragser und Olanger Dolomiten, Südtirol). Festband 300-Jahr-Feier Univ Innsbruck, pp 9–103

Chuvashov BI (1974) Permskie izvestkovye vodorosli Urala. Akad Nauk SSSR, pp 3–76

Chuvashov BI (1983) Permian reefs of the Urals. Facies 8:191–212

Clark DN (1980) The sedimentology of the Zechstein 2 Carbonate Formation of Eastern Drenthe, The Netherlands. Contrib Sedimentol 9:131–165

Deloffre R, Génot P (1982) Determination generique des Dasycladales du Cenozoique a l'actuel. Symp Int Algues Fossiles. Bull Cent Rech Explor-Prod Elf-Equitaine 4(2):545–556

Elliott GF (1968) Permian to palaeocene calcareous algae (Dasycladaceae) of the Middle East. Bull Br Mus Nat Hist Suppl 4:111

Elliott GF (1978) Ecologic significance of post-Palaeozoic green calcareous algae. Geol Mag 115: 437–442

Elliott GF (1981) The Tethyan dispersal of some chlorophyte algae subsequent to the Palaeozoic. Palaeogeogr Palaeoclimatol Palaeoecol 32:341–358

Emberger J (1976) Les algues (Chlorophyceae, Prasinophyceae, Rhodophyceae) du Carbonifère et du Permien. Essai d'un inventaire bibliographique, géographique, stratigraphique. Bull Inst Geol Bassin Aquitaine Num Spec, 168 p

Endo R (1959) Stratigraphical and paleontological studies of the Later Paleozoic calcareous algae in Japan. XIV. Fossil algae from the Nyugawa valley in the Hida massif. Sci Rep Saitama Univ Ser B 3(2):177–207

Endo R (1961) Stratigraphical and paleontological studies of the Later Paleozoic algae in Japan. SVI. Fossil algae from the Ominagahama district. Sci Rep Saitama Univ Ser B Commem Vol R. Endo:77–118

Endo R (1969) Contributions to the geology and paleontology of Southeast Asia. Fossil Algae from Khao Phlong Phrab district in Thailand. Geol Paleontol Southeast Asia 7:35–85

Flügel E, Flügel-Kahler E (1980) Algen aus den Kalken der Trogkofel-Schichten der Karnischen Alpen. Die Trogkofel-Stufe im Unterperm der Karnischen Alpen, Carinthia II Sonderh 36: 113–182

Fott B (1971) Algenkunde, vol II. Fischer, Stuttgart, 581 p

Güvenć T (1979) Dasycladacées metaspondyles du Paléozoique supérieur et du Trias. Bull Cent Rech Explor-Prod Elf-Equitaine 3:625–637

Hüssner H (1983) Die Faunenwende Perm/Trias. Geol Rundsch 72:1–22

Kulik EL, Korolyuk IK, Rauser-Chernousova DM (1978) On stratigraphic and facial importance of Asselian and Sakmarian calcareous algae (on the example of the Shakthtau bioherm massif, Bashkiria). Voprosy mikropaleontologi. Akad Nauk SSSR 21:216–233

Valentine JW (1973) Evolutionary paleoecology of the marine biosphere. Prentice Hall, Englewood Cliffs, 511 p

Valet G (1979) Paleoecological approach to Dasycladales from the ecology of Recent forms. Bull Cent Rech Explor-Prod Elf-Aquitaine 3:855–857

Wilson JL (1975) Carbonate facies in geologic history. Springer, Berlin Heidelberg New York, 471 p

Wray JL (1977) Calcareous algae. Developments in palaeontology and stratigraphy, vol IV. Elsevier, Amsterdam, 185 p

Zorn H (1976) Über den Lebensraum fossiler Wirtelalgen in der Trias der Alpen. Naturwissenschaften 63:426–429

Zorn H (1977) Der Einfluß der Umwelt auf die Variation fossiler Arten. Nat Mus 107:1–5

29 Adaptations of Crustose Coralline Algae to Herbivory: Patterns in Space and Time

R.S. STENECK[1]

Introduction

Crustose coralline algae are among the most abundant organisms (plant or animal) to occupy hard substratum within the photic zone. In fact, the Holocene may be the "Age of Crustose Corallines," since corallines are more diverse and abundant than ever before. This is also a period of maximum intensity of herbivory. The radiation of crustose corallines and coralline grazing herbivores are temporally correlated. Ecological and paleontological research suggests that crustose corallines and herbivores have "diffusely" coevolved to attain their current preeminance.

Ecological research conducted in Maine, Washington, the Caribbean, and the Great Barrier Reef suggest there are several anatomical and morphological adaptations to herbivory. The adaptations are identified by finding convergently (= polyphyletically) evolved characters of similar function under different physical conditions (different oceans and climates) with different herbivore species but under the same intensity of herbivory. Specific adaptations include crust thickness, branching morphology, conceptacle shape, and protection of the meristem (region of growth). Today, geographic patterns of these characters correspond with geographic patterns in the distribution and abundance of certain herbivore groups, Paleontological studies indicate that the proportion of coralline species possessing certain adaptations progressively increased throughout the Phanerozoic corresponding to increased levels of herbivory over this period.

Nonarticulated coralline algae (crustose corallines) are an enigma. They are among the slowest growing algal forms in the world, they are readily overgrown by most other algae, and yet they may be the most abundant organism (plant or animal) to occupy hard bottoms within the photic zone. Although they live in a variety of habitats, they thrive in, and commonly dominate, environments of intensive herbivory (Paine and Vadas 1969, Van den Hoek 1969, Adey 1973, Vine 1974, Wanders 1977, Steneck 1978, Brock 1979, Paine 1980, Steneck 1982a,b). There is general consensus that herbivory is necessary to remove fouling epiphytes from corallines living in productive environments. However, not all herbivores have similar abilities, and coralline susceptibility to grazers is also variable. In this chapter I will focus on individual characters that allow corallines to exist under different intensities of herbivory. Specifi-

1 Department of Zoology and Oceanography Program, University of Maine at Orono, Darling Center Marine Laboratory, Walpole, ME 04573, U.S.A.

Paleoalgology: Contemporary Research and Applications
(eds. D.F. Toomey and M.H. Nitecki)

cally, I will report results of ecological studies involving vegetative (branching and crust thickness) and reproductive (conceptacle) morphologies, and an anatomical feature (meristem placement). I will than examine patterns in these coralline characters over time using the fossil record, in order to demonstrate that there is a correspondence between the evolutionary history of herbivores and the evolution of coralline adaptations to herbivory.

Overview

Herbivores. Recent studies on the ecology (Steneck 1982a,b, Steneck and Watling 1982) and evolution of herbivores (Steneck 1982b, 1983) suggest that there are 3 groups of functionally similar herbivores: (1) those that feed on algae but do not denude primary substratum (polychaetes, malacostracans), (2) those that denude primary substratum but cannot excavate calcium carbonate (some molluscs, urchins and fishes), and (3) those that excavate calcium carbonate (chitons, limpets, some urchins and parrotfishes). It is only the third herbivore group that has an ecological and evolutionary impact on coralline algae.

During the Paleozoic there were few if any calcium-carbonate-excavating herbivores (Fig. 1A). Throughout the Mesozoic and into the Cenozoic, various groups of carbonate-excavating herbivores evolved with successively "improved" mouthparts (such as limpets, echinoids and finally herbivorous teleost fishes). The evolution and persistence of each excavating herbivore group had the net effect of escalating grazing intensity (depth of grazing into the substratum) over time (Steneck 1983).

Calcareous Crusts. Corallines and their assumed parent taxon, solenopores, probably always required grazing to remain free of epiphytes (Steneck 1983). Correlated with the rise in intensity of grazing was a decline in solenopore abundance and diversity, and an explosive diversification in corallines (Fig. 1B). Along with the decline of solenopores was an apparent increase in the proportion of wounded plants (Steneck 1982,b 1983). The greater evidence of wounded individuals and the overall decline of solenopores has been ascribed to anatomical differences that make them more susceptible to excavating herbivores. Thus the success of corallines may be due to specific anatomical and morphological characters they possess that reduce their susceptibility to excavating herbivores.

The Approach. The following is a discussion of the adaptive value of individual anatomical, morphological, and reproductive characters of corallines. Each feature was studied relative to herbivore intensity (not to any specific herbivore) so that repeatable patterns among geographically and evolutionarily disjunct species assemblages could be identified and inferred as adaptive to a particular intensity of herbivory. This approach will be used to discuss trends in particular coralline characters in both space (ecological and biogeographical patterns) and time (geological patterns).

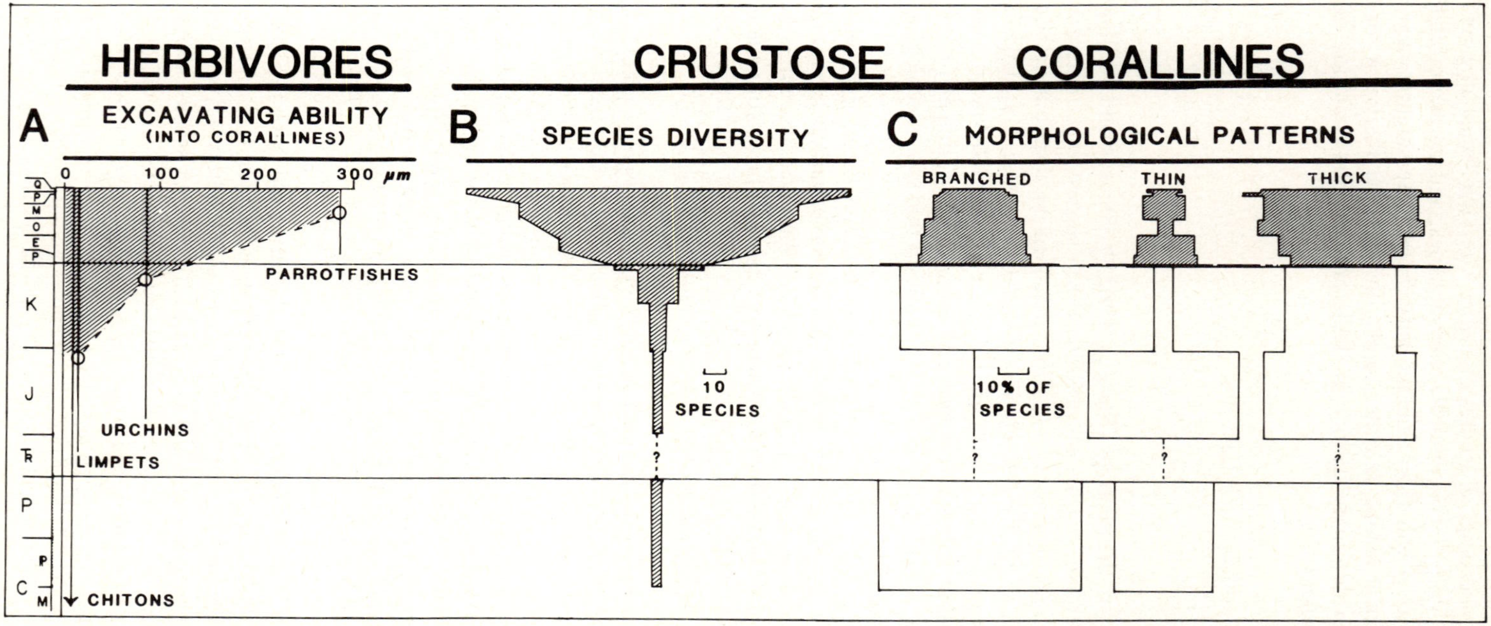

Fig. 1A–C. Evolutionary trends in herbivores and crustose coralline algae. (**A** and **B** from Steneck 1983a). The *circles, dotted line,* and *shading* in **A** indicate earliest record of herbivore abundance. *Vertical lines* indicate age of first appearance of herbivore groups. Excavating ability was determined from neontological studies (Steneck 1983a). Morphological trends (**C**) indicate the proportion of coralline species with branched, thin or thick crusts. A total of 527 fossil coralline species were used (literature cited in **Appendix 1** of Steneck 1982a). Only the Cenozoic portion of **C** was *shaded* to draw attention to morphological trends in corallines as they became abundant in shallow seas

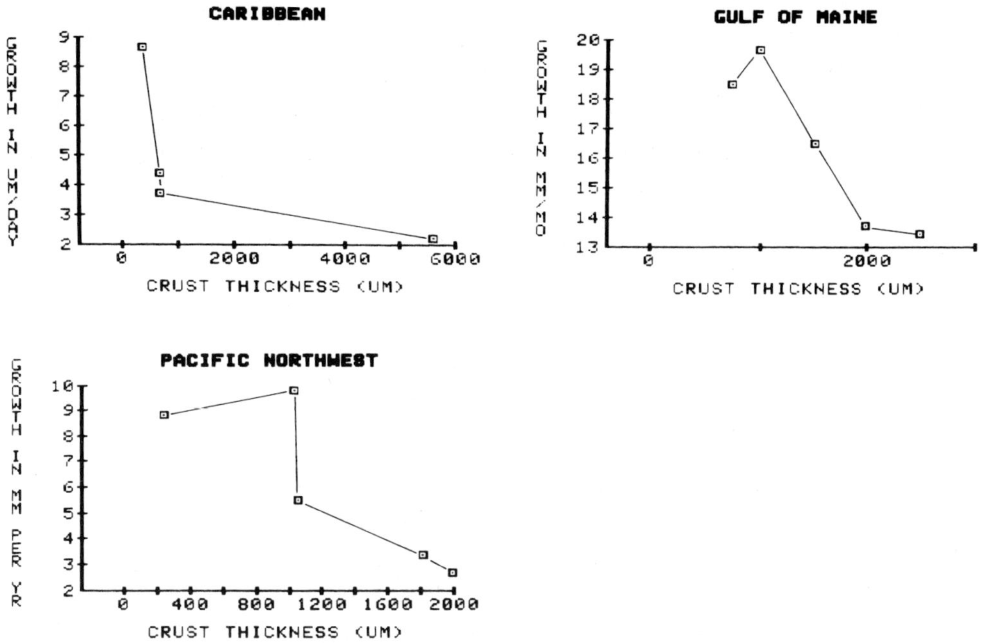

Fig. 2. Lateral growth rate in corallines as a function of crust thickness. Data are for (listed thinnest to thickest), *Phymatolithon laevigatum, P. rugulosum, Lithothamnion glaciale,L. lemoineae* and *Clathromorphum circumscriptum* from the Gulf of Maine; *Mesophyllum vancouveriense, Pseudolithophyllum whidbeyense, P. lichenare, Lithothamnion phymatodeum,* and *Lithophyllum impressum* from Pacific Northwest and; *Paragoniolithon solubile, Hydrolithon borgesenii, Neogoniolithon megacarpum* and *Porolithon pachydermum* from the Caribbean

Morphological and Anatomical Trends

Crust Thickness: Thick Crusts Dominate Zones of Intensive Herbivory. This feature varies greatly in corallines, ranging from a single cell layer that is about 10 μm thick (e.g., *Fosliella)* to more than a decimeter in thickness (e.g., *Clathromorphum,* Lebednik 1977). Many thick crusts are totally composed of living tissue, although only the top 200 μm is photosynthetic. Thus, the outer cell layers support the rest of the plant by translocating photosynthates (Steneck 1983). Why? Here I examine the advantages and disadvantages related to crust thickness.

For any given environment, thin crusts grow faster than thick ones (Fig. 2). Although this could affect substratum occupancy, thin crusts rarely dominate coralline communities in highly productive environments. In fact, thin crusts are abundant in disjunct environments such as some intertidal zones, under macroalgae canopies, or in subcryptic or deep water environments. Thus it seems, factors other than growth rate are important in determining the distribution and abundance of crust morphologies in nature.

Herbivory seems to be the best single predictor of crust thickness in nature (Fig. 3). Generally, the abundance of thick crusts increases with increasing herbivory regardless

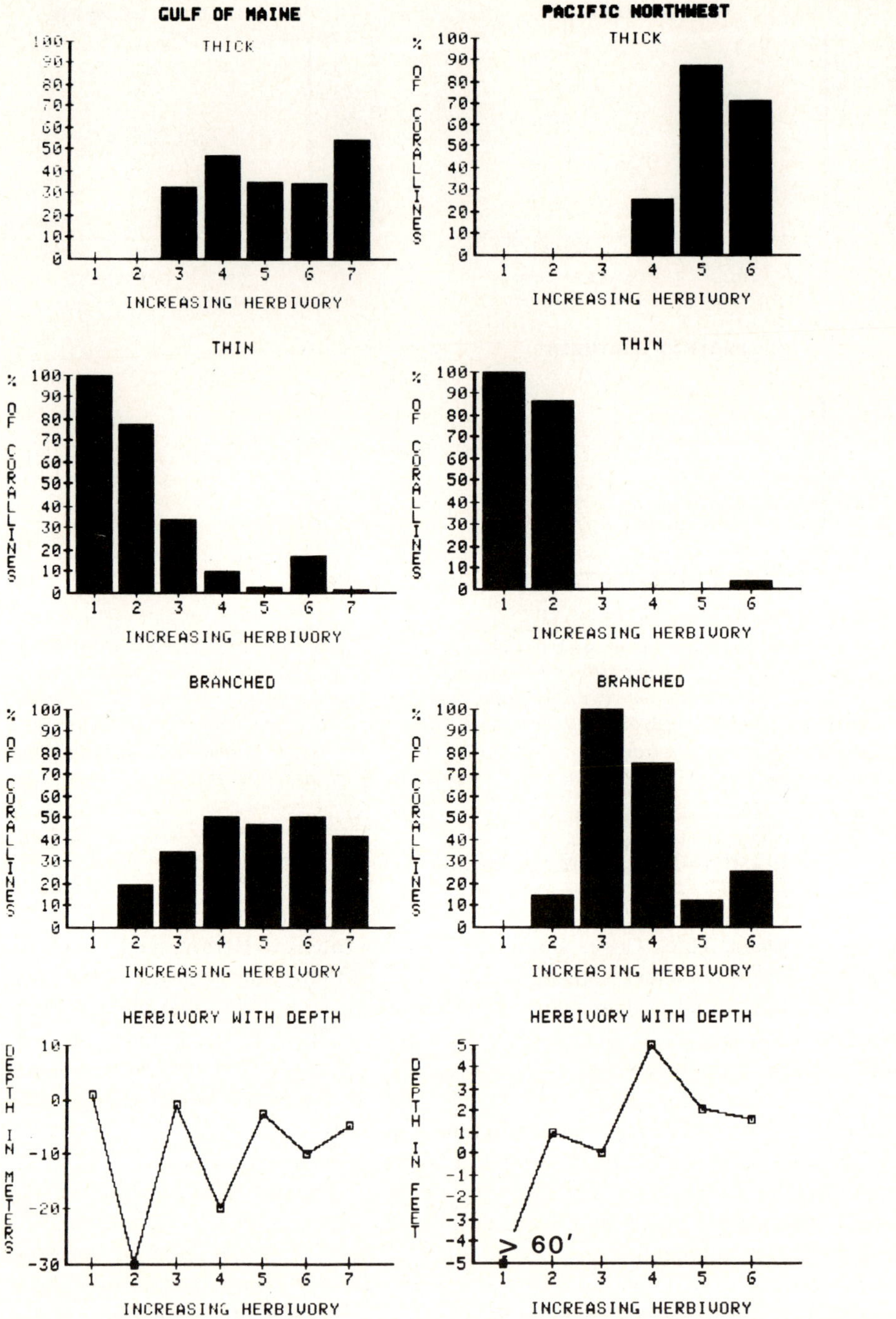

Fig. 3. A comparison of the distribution and abundance of different crust morphologies relative to levels of herbivory from four distinctly different biogeographic zones (see text). Herbivory at each location was ranked so that the higher numbers indicate more intensive herbivory. The same

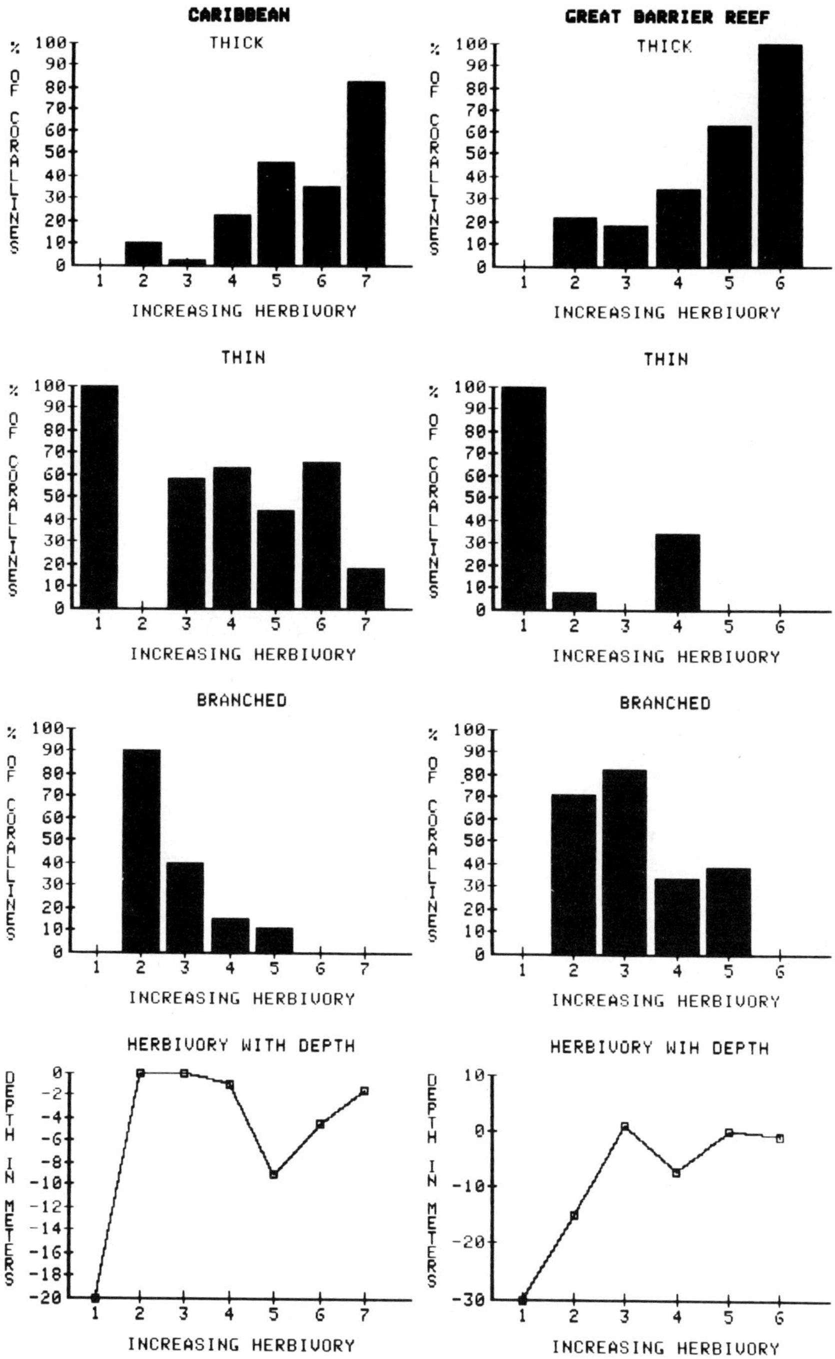

ranking between different biogeographic zones is not necessarily equal. Note that the relationship between depth and herbivory is weak

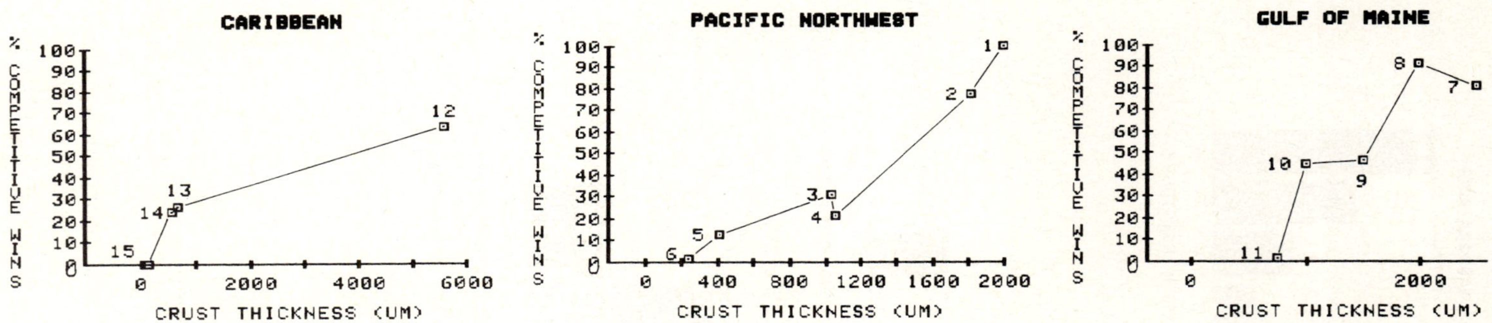

Fig. 4. Competitive ability relative to crust thickness in environments of intensive herbivory (numbers 6 and 7 in Fig. 3). Numbers indicate the crust species studied. They are: (1) *Pseudolithophyllum muricatum,* (2) *Lithophyllum impressum,* (3) *P. whidbeyense,* (4) *Lithothamnion phymatodeum,* (5) *Bossiella* sp. (crustose holdfast), and (6) *Corallina vancouveriensis* (crustose holdfast) from the Pacific Northwest; (7) *Clathromorphum circumscriptum,* (8) *Lithothamnion lemoineae,* (9) *L. glaciale,* (10) *Phymatolithon rugulosum,* and (11) *Phymatolithon laevigatum* from the Gulf of Maine; (12) *Porolithon pachydermum;* (13) *Neogoniolithon megacarpum,* (14) *Neogoniolithon imbricatum,* and (15) *Porolithon fragilis* from the Caribbean

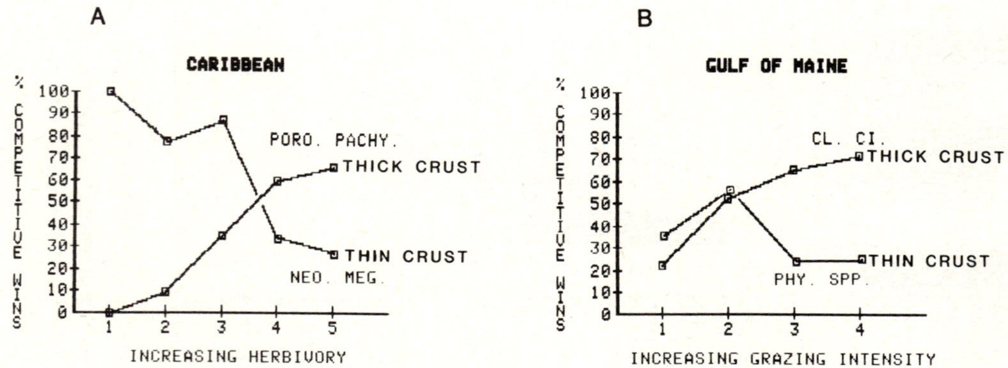

Fig. 5A,B. Competition mediated by herbivory. The competitive supremacy of thick crusts under conditions of intensive herbivory (Fig. 4) is reversed under lower levels of grazing (see text)

of water depth. To establish this relationship I examined four biogeographically-distinct floras (Gulf of Maine, Pacific Northwest, Caribbean, and Great Barrier Reef). At each location extensive quantitative abundance surveys and collections were conducted (data were collected at the species level; details in Steneck 1982b). Herbivore load was determined in the field by censusing the size frequency distribution of herbivores, by counting the observed number of bites per meter square (using time lapse movies), and by scoring each specimen for the abundance and type of herbivore graze marks left on the coralline thallus (technique in Steneck 1983). The regions were subdivided and ranked into 5 to 7 herbivore zones, ranging from lowest to highest levels of herbivory. Consistently, the abundance of thin crusts decreased and the abundance of thick crusts increased with increasing grazing intensity. There was no consistent relationship between herbivory and depth among sites (Fig. 3).

Recent studies demonstrated that some thick crusts withstand or even depend on intensive grazing for their existence (e.g., Steneck 1982a). However, surviving herbivory is only part of the story, since different corallines commonly grow as mosaics of species in competition for hard substratum. The predominant mechanism for competition is for one crust to overgrow another. When this happens, the overgrown portion of the crust dies. In time, one crust species can completely overgrow and kill another.

Survival and competitive ability of many corallines depend on the herbivory they experience. As grazing pressure varies in nature, so does the outcome of competing corallines. In particular, crust thickness is strongly correlated with competitive ability in environments of intensive grazing (Fig. 4), but under conditions of reduced grazing pressure, thinner crusts are competitively superior. This latter point was demonstrated in two places: the Caribbean and Gulf of Maine (Fig. 5). In the Caribbean (St. Croix) the two dominant crust species [*Porolithon pachydermum,* a thick crust, and *Neogoniolithon megacarpum,* a thinner crust (see Figs. 4 and 5A)] were sampled from five environments of different grazing pressures (as described above). The perimeter of overgrowth between the two species was measured for large collections (193 plants) at each station. The results (Fig. 5A) show a consistent decrease in competitive wins for *Neogoniolithon megacarpum* and a consistent increase in wins for *Porolithon pachydermum* along a gradient of increasing grazing pressure. This pattern was even observed on a single coral colony where herbivorous fish graze topographic peaks more than valleys. *P. pachydermum* competitively dominated the peaks but lost competitive battles in the valleys! A similar pattern was observed in Maine for *Clathromorphum circumscriptum* (a thick crust) and *Phymatolithon* spp. (thinner crusts) under conditions of variable limpet grazing pressure. Small cobbles are unstable and thus poor habitats for limpet survival. On such rolling cobbles, the thinner *Phymatolithon* spp. crusts competitively dominate. Whereas on cobbles cemented to fixed posts on which limpets were maintained, *Clathromorphum* dominated (Steneck 1982b, Steneck and Jackson, in preparation).

In a separate experiment (Fig. 5B) four sets of nine cobbles were subjected to a graded intensity of "artificial grazing" treatments (Steneck 1982b). The cobbles were placed in a raft so they had identical environmental conditions and no benthic grazers. The artificial grazing experiment involved four treatments. One group was left untouched, epiphytes were gently rubbed off another, and the remaining two sets were scrubbed with nylon and steel brushes respectively. The results (Fig. 5B) show a clear trend with increased grazing intensity of improved competitive ability for *Clathromorphum* (thick) but a poorer competitive ability for *Phymatolithon* (thin). This pattern, however, could be partly due to an anatomical characteristic of *Clathromorphum* (discussed below).

The consistent success of thick crusts under conditions of intensive herbivory is likely the result of these crusts maintaining a store of living tissue beneath the zone of photosynthesis (upper 200 μm). The "cost" for maintaining this nonphotosynthetic, tissue is evident in the slow growth rate of thick crusts (Fig. 2). When a deep grazing herbivore bites into a crust, it bares living, nonphotosynthetic, tissue that rapidly regenerates a meristem and the necessary photosynthetic apparatus. Thus, primary substrate is never exposed which would allow fouling microalgae to invade and impede (or inhibit) healing of herbivore induced wounds.

Branched Crusts: An Effective Defense Against Some Herbivores. Branched corallines (corallines with any sort of non-articulated protuberance) have several ecological advantages over other corallines. For a given area of substratum, branches increase the thallus surface area greatly, thereby increasing the photosynthetic and reproductive ability of the crusts. They also survive in silty environments. However, branched corallines are not universally abundant. This is partially due to the susceptibility of branches to injury from certain herbivores. Their susceptibility depends on their size and shape relative to the excavating ability of the herbivores.

Ecological studies have shown that stout branches in some nonarticulated corallines are an effective defense against deep grazing urchins (Steneck and Milliken 1981) and molluscs (Steneck and Paine, in preparation) but not against parrotfish (Steneck and Adey 1976, Steneck 1982b). Since parrotfish are confined to the tropics, the distribution and abundance of branched corallines are different there than in areas nearer the poles. For example, branched corallines are found in zones of most intensive herbivory in the Gulf of Maine and Pacific Northwest but not in the Caribbean or Great Barrier Reef (Figs. 3 and 6).

Stoutly branched crusts are effectively defended against excavating invertebrate herbivores because only the branch tips are grazed. This was documented in Maine where the predominant (and deepest grazing) herbivore is the sea urchin, *Strongylocentrotus droebachiensis.* Although urchin densities were more than three times greater on branched coralline species than they were on the unbranched crusts, only 15% of the branched thalli had graze marks. Thus, most of the crusts (85%) were unavailable to the herbivores (Steneck and Milliken 1981). Similar observations of graze mark patterns were made on corallines of the Pacific Northwest (Steneck and Paine, in preparation), although no species were in common between the two areas.

To test the hypothesis that branches enhance growth and survival by protecting most of the thallus from shallow excavating herbivores (chitons and limpets), an experiment was designed to give an unbranched coralline artificial branches (Steneck and Paine, in preparation). For this, plants of the unbranched coralline, *Pseudolithophyllum whidbeyense,* were transplanted into epoxy putty "islands". Around part of each thallus small (1 mm wide by 1–5 mm high) plastic "nubs" were implanted into the putty as artificial branches. The coralline grew radially from the transplants

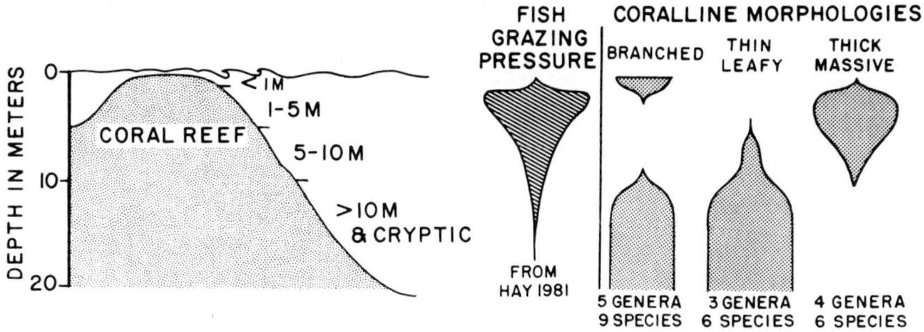

Fig. 6. The distribution of coralline morphologies along the Great Barrier Reef. Note that only thick crusts thrive under intensive fish grazing

over the rock-hard epoxy surface. After a year in the presence of molluscan herbivores, the growth rate of the coralline around and over the artificial branches was more than six times greater than it was in the absence of branches. In areas where the grazers were removed and kept off for a year, there was no difference in growth rates. Thus, branches enhance growth and survival of corallines frequently grazed by shallow excavating herbivores.

Branches are not, however, an effective defense against parrotfishes (Steneck and Adey 1976), which excavate corallines most deeply (Steneck 1983). Distribution and abundance patterns in corallines seem to reflect this (Fig. 6, Steneck and Adey 1976, Steneck 1982b,c). An important branched coralline in the Caribbean, *Lithophyllum congestum,* has a distribution restricted to areas of heavy wave action around mean low water (within 20 cm of MLW). The waves effectively keep the parrotfish away and branching forms persist. When branched crusts were transplanted to greater depths (where the plant can physiologically survive and grow), all of the branches were grazed off. Grazing frequency was so predictable on protuberances that a net "decrement" rate (or the rate at which calcium carbonate was removed) was calculated (Steneck and Adey 1976). A similar pattern was observed in the Great Barrier Reef (Fig. 6) where branched morphologies are commonly found above and below the zone of intensive parrotfish grazing (Steneck 1982b,c). Thus, because branching morphologies are susceptible to large grazing parrotfish, they are generally excluded from areas of intensive herbivory in the tropics (Figs. 3, 6 and 7). They tend to persist at, or slightly above, mean low water in zones of persistent turbulence, in shallow lagoons, at depth, or in cryptic environments where herbivory is low (Steneck 1982b,c).

Reproductive Structures: Sunken Conceptacles are Protected Against Invertebrate Herbivores. Most corallines produce spores or gametes within vegetative cavities called conceptacles. I will not discuss the exception [sporangial sori characteristic of *Archaeolithothamnium* (= *Sporolithon*)] in this paper. The ability of any coralline to disperse or recruit is partially dependent on the success of its disseminules to develop to maturity within intact conceptacles. There are two basic conceptacle morphologies: raised and sunken. The difference between these two morphologies is the location of the conceptacle roof at the time of reproductive maturity (just prior

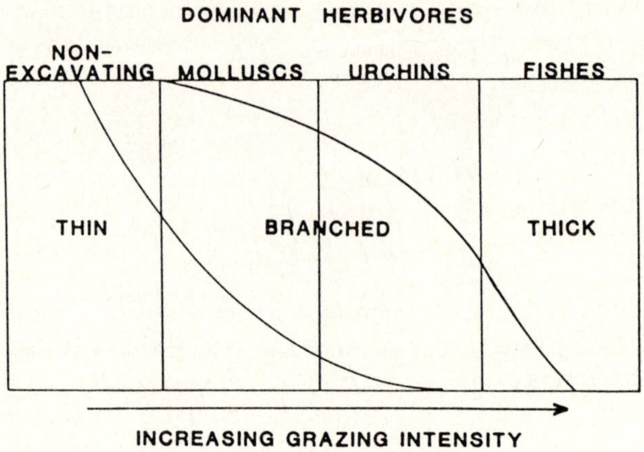

Fig. 7. A model summarizing patterns in coralline morphology with respect to grazing intensity

to release). When the roof is either flush with the thallus surface or sunken beneath it, the conceptacle is described as "sunken". When the roof and cavity create a mound above the thallus surface, the conceptacle is said to be "raised".

Of the three vegetative coralline morphologies (Fig. 3), thin and branched crusts tend to have raised conceptacles, whereas many of the thicker, unbranched crusts have sunken conceptacles. Among the thick crusts, there is a strongly negative correlation between species with raised conceptacles and the frequency and intensity of non-parrotfish herbivory on the crust surface. Most branched crusts, however, have raised conceptacles on the sides of the branches. Since branching morphologies are immune to herbivores in areas of non-parrotfish grazing (discussed above), the damaging effects of excavating herbivores on raised conceptacles on branches is minimal.

The pattern of conceptacle morphologies among unbranched corallines is more variable. Thin, herbivore-susceptible crusts generally have raised conceptacles. Although very small conceptacles could be sunken within most thin crusts, this has not been observed. In fact, it is common to find very large conceptacles on thin crusts (e.g., *Leptophytum laeve* is a thin crust with conceptacles exceeding 1000 μm in diameter!).

Thick crusts can have either raised or sunken conceptacles. However, most corallines with sunken conceptacles are found in zones of intensive invertebrate-grazing in non-tropical environments. In a study to test the effectiveness of raised conceptacles to damage from surface-scraping herbivores, a piece of spring steel was drawn over a crust surface with a force sufficient to remove only a few surface cells (similar to the forces exerted by urchins and deep grazing limpets). The same force when drawn over a cluster of raised conceptacles destroyed them. Thus the sunken conceptacle morphology seems to be relatively protected from invertebrate grazing pressure. The deep grazing parrotfish, however, would have little difficulty in grazing even the most deeply sunken conceptacles. This could explain the relative absence of sunken conceptacles in tropical waters.

Sunken Meristem: Deeply Sunken Meristem is Correlated With Grazing Molluscs. The meristem (region of growth) produces epithallial cells toward the upper surface and perithallial cells below. The epithallial cells protect the meristem from surfacial scrapes. In many cold-water habitats the epithallus builds up (called "multilayered"). This thick tissue can be deeply grazed without injury to the meristem. Research conducted in Maine (Adey 1973, Steneck 1977, Steneck 1982a) demonstrated that the especially thick epithallus of *Clathromorphum* protects this species from grazing by the limpet *Acmaea testudinalis.* This adaptation is so specific that when limpets are excluded, the plants die.

Coralline species with multilayered epithalli are common in non-tropical areas worldwide. Most of these species are similar to *Clathromorphum* ecologically in that they are thick and smooth and found in relatively shallow water (usually in the upper 10 m). They are also commonly associated with herbivorous molluscs, especially limpets. Examples include *Antarctiophyllum* in Antarctica, *Lithophyllum orbiculatum* and *L. incrustans* in Europe, *Lithophyllum* sp. in central California, *Pseudolithophyllum whidbeyense* and *Lithophyllum impressum* in the Pacific Northwest, and a *Lithothamnion* in South Africa. Close inspection of these corallines indicated that they were recently grazed by molluscs (usually limpets).

Ecological Trends in Corallines and Excavating Herbivores

The morphology of corallines seems to correspond to the intensity of herbivory in their environment. Figure 7 summarizes the qualitative relationship between the excavating ability of the dominant herbivore groups and the relative abundance of thin, branched, and thick corallines. Under low levels of grazing by excavating herbivores, thin crusts are the dominant growth form. At intermediate levels the abundance of branched crusts is maximal and thick crusts dominate under conditions of most intensive herbivory (Figs. 3 and 7). Correlated with the rise in abundance of massive crusts, is a higher proportion of crusts with sunken conceptacles and multilayered epithalli.

Evolutionary Trends in Corallines and Excavating Herbivores

Rise of Herbivore-Resistant Characters. Previous research established a temporal correlation between the evolution of (and changes within) excavating herbivores and the radiation of crustose corallines (Steneck 1983). Since this relationship is between two functional groups of organisms (i.e., calcareous algal crusts and excavating herbivores), rather than between any two species, this is an example of "diffuse coevolution" (sensu, Futuyma and Slatkin 1983). The concept of coevolution suggests that there has been an adaptive response in corallines to changes in herbivores. The ability of corallines to thrive and diversify under conditions of intensive herbivory lends support to the adaptive interpretation of certain unique coralline characters. How corallines acquired the adaptations is worth considering.

The fundamental characters, unique to corallines, that "paved the way" for their subsequent radiation included a thallus differentiated into several tissues (i.e., perithallus, hypothallus and epithallus), intercellular conduits (called "vegetative fusion cells" for translocating photosynthates), and conceptacles. Several adaptations to herbivory discussed above require these fundamental characters first. It was from this basic body plan that adaptive "fine tuning" was possible. For example, most branches require hypothallial tissue (called medullary hypothallus) for their formation, and conceptacles depend on vegetative fusion cells to nourish developing germ cells. Of course an epithallus must be present before a multilayered one can evolve. There is no evidence from the fossil record that these fundamental characters evolved gradually. They appeared suddenly with the evolution of the "ancestral coralline" *Archaeolithophyllum* in the Carboniferous (Wray 1964). This species could have been an "evolutionary dead end" since there is no record of corallines in the Triassic. If so, all of these characters evolved suddenly again in the Jurassic and have a continuous fossil record from that time on.

Correlated with escalating herbivory and the radiation of corallines (beginning in the Upper Cretaceous) is a decline in the proportion of the more herbivore-susceptible, thin and branched forms (Fig. 1C). This morphological trend was casually mentioned several times in the literature (e.g., Johnson 1962, 1963). In fact Johnson (1962) noted that most branched corallines became much less important in the tropics after the Eocene, which coincidentally is when parrotfish evolved! Thick and massive forms, however, continued to increase in abundance during this time. Thus, the ecological model (Fig. 7) for coralline morphologies relative to herbivore intensity also applies for evolutionary changes in herbivory. As herbivore intensity escalated (Fig. 1A), the relative abundance of coralline morphologies changed (Fig. 1C) in a way corresponding to Fig. 7.

The fossil record for conceptacle morphology is poor, since they, like all surface phenomena, preserve poorly. It is known, however, that the earliest coralline (or coralline-like) conceptacles in the Carboniferous were strongly raised structures (Wray 1964). Such structures would not withstand any degree of grazing by excavating herbivores. This corresponds well with the hypothesis that excavating herbivores were rare or absent in the Paleozoic (Steneck 1983). Several Mesozoic corallines appear to have had sunken conceptacles.

The multilayered epithallus is conspicuous by its absence from the fossil record. This character should be preserved, since it is commonly preserved when overgrown by another coralline or by itself. Today the genus *Lithophyllum* has many species with multilayered epithalli. Although epithalli of fossil *Lithophyllum* plants have been found (see Johnson and Adey 1965), I am not aware of *any* report of pre-Holocene corallines with this character. This point requires further study but there is evidence that the availability of suitable limpet habitat increased in the Quaternary (N. Sohl, pers. comm.) because of glaciation. If the multilayered epithallus is a specific adaptation to limpet grazing as suggested above, its abundance in several genera today could be due to a recent radiation in limpets during this same period.